Teleology, First Principles, and Scientific Method in Aristotle's Biology

OXFORD ARISTOTLE STUDIES

General Editors
Julia Annas and Lindsay Judson

PUBLISHED IN THE SERIES

Teleology, First Principles, and Scientific Method in Aristotle's Biology
Allan Gotthelf

Priority in Aristotle's Metaphysics
Michail Peramatzis

Doing and Being
An Interpretation of Aristotle's Metaphysics *Theta*
Jonathan Beere

Space, Time, Matter, and Form
Essays on Aristotle's Physics
David Bostock

Aristotle on Meaning and Essence
David Charles

Time for Aristotle
Ursula Coope

Aristotle on Teleology
Monte Ransome Johnson

On Location
Aristotle's Concepts of Place
Benjamin Morison

Order in Multiplicity
Homonymy in the Philosophy of Aristotle
Christopher Shields

Aristotle's Theory of Substance
The Categories *and* Metaphysics *Zeta*
Michael V. Wedin

Aristotle's *De Interpretatione*
Contradiction and Dialectic
C. W. A. Whitaker

Teleology, First Principles, and Scientific Method in Aristotle's Biology

Allan Gotthelf

OXFORD
UNIVERSITY PRESS

OXFORD
UNIVERSITY PRESS

Great Clarendon Street, Oxford OX2 6DP

Oxford University Press is a department of the University of Oxford.
It furthers the University's objective of excellence in research, scholarship, and education by publishing worldwide in

Oxford New York

Auckland Cape Town Dar es Salaam Hong Kong Karachi
Kuala Lumpur Madrid Melbourne Mexico City Nairobi
New Delhi Shanghai Taipei Toronto

With offices in

Argentina Austria Brazil Chile Czech Republic France Greece
Guatemala Hungary Italy Japan Poland Portugal Singapore
South Korea Switzerland Thailand Turkey Ukraine Vietnam

Oxford is a registered trade mark of Oxford University Press
in the UK and in certain other countries

Published in the United States
by Oxford University Press Inc., New York

First published 2012

British Library Cataloguing in Publication Data

Data available

Library of Congress Cataloging in Publication Data

Data available

Typeset by SPI Publisher Services, Pondichery, India
Printed in Great Britain
on acid-free paper by
MPG Books Group, Bodmin and King's Lynn

ISBN 978–0–19–928795–6

10 9 8 7 6 5 4 3 2 1

To the memory of Ayn Rand and David Balme

For Aristotle, *life* is not an inexplicable, supernatural mystery, but a fact of nature. And consciousness is a natural attribute of certain living entities, their natural power, their specific mode of action—*not* an unaccountable element in a mechanistic universe, to be explained away somehow in terms of inanimate matter, nor a mystic miracle incompatible with physical reality, to be attributed to some occult source in another dimension. For Aristotle, 'living' and 'knowing' are facts of reality; man's mind is neither unnatural nor supernatural, but *natural*—and *this* is the root of Aristotle's greatness, of the immeasurable distance that separates him from other thinkers.

Life—and its highest form, man's life—is the central fact in Aristotle's view of reality. The best way to describe it is to say that Aristotle's philosophy is '*biocentric*'.

This is the source of Aristotle's intense concern with the study of living entities, the source of the enormously 'pro-life' attitude that dominates his thinking.

Ayn Rand, 'Review of Randall's *Aristotle*' (1963)

But teleology nevertheless is the primary subject of [the *Parts of Animals*], and the other [biological] treatises likewise have primarily theoretical purposes. They are not just empirical studies in which some philosophical concepts are put to use. We go less far wrong if we regard them as studies of these concepts made through empirical data.

This suggests the possibility that we might gain light upon these concepts—such as substance, form, species, essence, *logos*—by examining their reference in biology, together with the arguments around them.

D. M. Balme, 'The Place of Biology in Aristotle's Philosophy' (1987a)

Preface

In 1961, midway through my undergraduate studies, I read Ayn Rand's *Atlas Shrugged*. I loved the novel. It glowed with respect for the mind, and for much else that I valued or came to value. Rand loved Aristotle, and that love permeated especially her later writings, from *Atlas Shrugged* on. She credited him with defining 'the *basic* principles of a rational view of existence and of man's consciousness', with rejecting Plato's otherworldly metaphysics, and with showing (in outline at least) how reason can come to understand *this* world based on sense-perception. Rand thus saw Aristotle as the father of science, and attributed the Renaissance and Enlightenment (and all of their products) to his influence:

> If we consider the fact that to this day everything that makes us civilized beings, every rational value that we possess—including the birth of science, the industrial revolution, the creation of the United States, even the structure of our language—is the result of Aristotle's influence, of the degree to which, explicitly or implicitly, men accepted his epistemological principles, we would have to say: never have so many owed so much to one man.[1]

I started reading Aristotle myself and found much of profound value philosophically, certainly more than I had found in the contemporary philosophy materials I studied for my regular courses. I had been fortunate to meet and get to know Rand, and I talked with her about Aristotle (and much else) on and off over the next fifteen or so years. Two ideas I would like to highlight here that attracted me to Aristotle from the beginning (in the first case, with the help of Rand's own discussion of the issues) were: (i) an understanding of causality and explanation in terms not of events and laws but of *things* with *natures* and *potentials*; and (ii) the idea that developed human knowledge takes the form of axiomatically structured bodies of understanding, which reflect a grasp of the essential natures and potentials of the relevant things (including their attributes), a grasp acquired through systematic sensory observation and a broadly inductive methodology.

And so began an adventure, which has not abated to this day and is reflected in the chapters of the volume before you. I set out to master what I could of Aristotle's philosophic and scientific thinking, of his understanding of human nature and psychology, and of his view of the life and actions proper to human beings as rational animals. I chose Columbia as the place to do my graduate work, in part so I could be in New York and study with Rand, and in part because John Herman Randall, Jr. was at Columbia; I knew that whatever the limitations of his *Aristotle* (1960), the book and the man had much to teach me.

[1] Rand 1961, 20 (2005, 17-18).

In my early work on Aristotle during that first year at Columbia, 1964-65, I was caught up in Aristotle's insistence that living organisms come to be, are structured, and function *for the sake of something.* Acorns and embryos are *goal-directed* in their development, as are mature organisms even in their vegetative functioning, even though there is no consciousness within to direct that development and functioning. How is this thesis to be understood?

I struggled with that issue for some years and eventually decided to make it my dissertation topic. What I took from Rand, and what I think in the end made all the difference to my 'irreducible potential for form' interpretation, was the insistence that we pose the problem in terms of *natures* and *potentials.* Readers of Part I of this volume—the chapters on 'Teleology, Irreducibility, and the *Generation of Animals* (*GA*)'—will find this insistence a pervasive theme, on which I draw in many different contexts. (See, first of all, chapter 1, sec. IV.)

Although I had worked out the heart of my interpretation of the teleology before I came across David Balme's work, reading his 1972 Clarendon Aristotle volume—*Aristotle's* De Partibus Animalium *I and* De Generatione Animalium *I (with excerpts from II. 1-3)*—was a revelation. Randall had introduced me to Aristotle's biological treatises and their underlying naturalism, but Balme taught me how to study them; and especially as I read on in his writings, he taught me their value—the value of the *whole* of the biological corpus—for a deeper understanding of Aristotle's philosophy. I met Balme in 1973 and we stayed in close touch as colleagues and then as friends until his death in early 1989; I was proud to edit his Festschrift in 1985. I tell some things about the man and his work in my introduction to that volume (Gotthelf 1985a), and discuss his work in some detail in some of the chapters below.

Thanks to Randall's influence, as I've mentioned, I had started reading in the late 1960s the main explanatory treatises in the biological corpus: *Generation of Animals*, for the teleology, and *Parts of Animals*, in part for the teleology, but in part to see to what extent my earlier-mentioned suspicion that Aristotelian explanatory science has a demonstrative structure was sound. In that first substantial read of *PA*, I thought I could detect glimmers: Aristotle seemed at times to move in his explanations *from* features which he clearly took to be more fundamental or essential to the animals in question *to* features of those animals that were less fundamental or essential. But in the limited time I had for that issue, I couldn't grasp the whole of Aristotle's project in *PA*. It wasn't until 1985 that I took on the problem directly, in a paper I developed for a conference G. E. R. Lloyd and I had organized at Cambridge. Chapter 7 of the present volume, which grew out of that conference paper, remains my statement of my view on this topic (especially when read in light of the elaborations provided in the chapter that follows it).

Although I am perhaps best known for my work on Aristotle's biology and natural philosophy, I have had strong interests throughout my career in Aristotle's metaphysics and ethics. The two chapters constituting Part III of this volume address metaphysical themes on which the biology sheds light and touch on aspects of my interpretation of the theory of substance in the central books of the *Metaphysics*, on which I hope to write more fully some day.

Personally, I view my interpretation of Aristotle's natural teleology, and my account of the broadly axiomatic structure of biological explanation, as my most important work. Third on that list are the various ways in which my work on particular philosophical topics may have contributed to the opening up, for other admirers of Aristotle, of the three main biological treatises for subsequent study. This includes especially the work I have done—much of it jointly with my long-time friend and colleague in so many Aristotelian endeavors, Jim Lennox—to help make the *History of Animals*, the largest and still least studied treatise in the Aristotelian corpus, more accessible to scholars. That work is presented in the three chapters of Part IV, 'Starting a Science: Theoretical Aims of the *History of Animals* (*HA*)'.

Part V, 'Aristotle as Theoretical Biologist', contains two chapters that are personal favorites of mine. 'Darwin on Aristotle' was motivated by a strong streak of hero-worship: I love the famous 1882 letter praising Aristotle that Charles Darwin wrote during the course of his first reading of Aristotle's biological work, and I was irritated beyond belief by an essay I read in the early 1980s that tried to debunk the letter. 'Debunk the debunker' was my motto, and I am confident I succeeded. I also had great fun doing some straight historical detective work, on details such as when, exactly, so-and-so died (see, for instance, chapter 15, n. 4).

The last chapter of Part V, which I consider also a 'coda' to the entire volume, for its synoptic character, is a lecture that I have delivered many times to general audiences. I was prompted to write it by an experience I had during my annual visits to Clare Hall, Cambridge in the 1980s, when I would describe my interest in Aristotle's scientific work to the other (often natural-science-oriented) visiting fellows. 'Oh, but he's not a *real* scientist', they would say, or 'Oh, but his work held back the course of science for 2,000 years'. My response, which I delivered for the first time at Clare Hall in 1987, is 'Aristotle as Scientist: A Proper Verdict (with emphasis on his biological works)'.

I have added to each chapter an opening unnumbered footnote in which I indicate the chapter's origin and/or its role in my thinking about Aristotle. With these introductory footnotes, with the grouping of the chapters into Parts, and with the extensive cross-references among chapters, I have not thought it necessary to add substantive introductions either to the five Parts or to the volume as a whole.

I have mentioned my interest in introducing scholars and students of Aristotle to the main biological treatises, not just as sources of philosophical insight, but as whole works in their own right. As indication that I think the chapters in this volume can do that, especially in Parts I, II, and IV, I have included in the titles of those Parts the name of the relevant treatise. Thus, in offering Part I as a body of study of Aristotelian teleology, I also offer it as a useful and reasonably systematic introduction to the *Generation of Animals*. Likewise, Part II, which provides a study of axiomatic, and more broadly explanatory, structure in *Parts of Animals*, can be a useful entrée into *PA*, as can Part IV into *HA*.

Although the original versions of some of the chapters in this volume go back to the 1970s and 1980s, several stem from a week-long 'intensive course' I gave in the summer of 1994 at Tokyo Metropolitan University, aimed at bringing professionals and advanced graduate students in northern Japan up to date on work on Aristotle's biology. Lasting some eighteen hours in total, the course was called 'Aristotle's Biological Enterprise and its Philosophical Significance'; the programs for each of the first four days are mirrored in the first four Parts of this volume. (The coda had been given at Keio University a week earlier, as introduction to the course.) The Tokyo course had been organized by Professors Shinro Kato and Shigeru Kanzaki, with assistance from Professors Kei Chiba and Toshio Kuwako, among others. A good deal of the material in the Tokyo course had been written up or outlined just a few months earlier for a seminar called 'Philosophical Issues in Aristotle's Biology' that David Charles, Jim Lennox, and I gave together at Oxford in Trinity Term 1994. I will on occasion refer in footnotes to those two very exciting events.

I have offered on the acknowledgments page my thanks to the publishers who are kindly permitting me to reprint earlier published work. I would like to express a special thank-you to two fellow scholars who graciously permitted me to reprint here all or part of papers I co-wrote with them: Mariska Leunissen, for chapter 5, and Pieter Beullens, for the appendix to chapter 12—all the more so because, in both cases, the original collaborative work was so extraordinarily pleasurable.

At a lunch in Oxford in the early to mid-1990s, soon after he had become philosophy acquisitions editor for Oxford University Press, and several years after *Philosophical Issues in Aristotle's Biology* had come out, Peter Momtchiloff asked me for a book of my own on the philosophical significance of Aristotle's biological work. I explained that my Tokyo lectures might be the basis for such a volume, and that, if I could pull it all together, I would love to publish with him. Peter's patience is legendary, and here we are some sixteen or so years later, with not quite a book but much more than a collection of papers. Thank you, Peter.

I would like also to thank Pat Corvini for meticulous pre-submission copy-editing, and the Oxford production crew for their post-submission work. Nigel Hope provided wide-ranging and very helpful copyediting and Michael Janes very valuable proofreading. My most special thank-you goes, however, to Elmandi du Toit, in charge of production, for her wonderfully sensitive and creative attention to my concerns throughout. I am also grateful to the Press for accepting my request to use for the book's cover the color called 'Cherokee red'; it is a tribute to my favorite architect, Frank Lloyd Wright, whose signature color it was.

The godfather of this volume, however, is my friend David Charles. In 2004, while in Pittsburgh for the conference that issued in Lennox and Bolton 2010, David suggested that I ought to think of publishing my Aristotle papers as a book, since they formed a unity in important ways. In the course of the nearly six years it has taken me to put this volume together, David has kindly provided me with written comments on all of the previously unpublished material. I also would like to thank the delegates of the Press for accepting the volume and Julia Annas, Lindsay Judson, and Peter Momtchiloff for admitting it into the Oxford Aristotle Studies series.

My time at the University of Pittsburgh these past seven years has been a rewarding one; I would like to thank especially the Anthem Foundation and the BB&T Charitable Foundation for their funding of the University's Fellowship for the Study of Objectivism, which has supported my position here. Through the years of final writing, assembly, editing, and proofreading, I was fortunate to have two outstanding research assistants at the University of Pittsburgh, first Greg Salmieri and then Keith Bemer, both of whom contributed greatly to the assembly of this volume. Greg is now an Aristotelian scholar and a philosopher in his own right, and this volume's name index (*s.v.* Salmieri, G.) will send you to some praise of his work. Keith is currently at work on a dissertation on Aristotle's *History of Animals*, which I expect will take us well beyond Part IV of this volume. Keith is also to be credited for the index locorum to this volume, and Peter Distelzweig for the subject index. Greg and Cassandra Brazié Love gave me valuable advice on this preface. I thank them all.

My debt to David Charles runs deeper than as catalyst for this book. I first met David in 1984, when he attended a seminar at Oxford I gave with John Ackrill. Our more than twenty-five years of electric and rewarding discussion since then, of the topics covered in this book, of much else in Aristotle, and of much else in philosophy, have been among the high points of my professional life.

I have already spoken of my joint endeavors with Jim Lennox. Our interactions and our warm friendship go back over thirty years, and his work has been a great influence on my own. Perhaps the best signs of this are the number of citations next to his name in this volume's name index—a far greater number than those for any other scholar (except David Balme)—and the fact that, of the thanks from me to other scholars for written comments on earlier drafts of my work, the greatest number are to him. I am also delighted to have this opportunity to express publicly to Jim—and to Rob Bolton, another friend and colleague from whom I have learned a good deal—my profound and humble thanks for conceiving and organizing the 2004 Festschrift conference at Pittsburgh for me, as well as the volume based on it, which, as I write this preface, has just appeared in print: *Being, Nature, and Life in Aristotle: Essays in Honor of Allan Gotthelf* (Lennox and Bolton 2010). Indeed, let me also take this opportunity to thank each of the scholars and friends who contributed to that volume, both for that contribution, and for their friendship, and for everything I have learned from them over the decades.

I end, where I started, with my formative debts to Ayn Rand and David Balme for helping me to see Aristotle's philosophic power and his love of the living. This book is dedicated to their memory.

December 2010

Contents

Part V. Aristotle as Theoretical Biologist

Acknowledgments

The previously published essays in this volume are reprinted with the kind permission of their publishers, from the following locations:

1. 'Aristotle's Conception of Final Causality', *The Review of Metaphysics* 30 (1976/77), 226-54 (with additional notes and a 'Postscript 1986', *Philosophical Issues in Aristotle's Biology*, ed. A. Gotthelf and J. G. Lennox, Cambridge: Cambridge University Press, 1987, 204-42).
2. 'The Place of the Good in Aristotle's Natural Teleology', in *Proceedings of the Boston Area Colloquium in Ancient Philosophy*, vol. 4 (1988), ed. J. J. Cleary and D. C. Shartin, Lanham, MD: University Press of America, 1989, 113-39.
3. 'Understanding Aristotle's Teleology', in *Final Causality in Nature and Human Affairs*, ed. R. F. Hassing, Washington, D.C.: Catholic University of America Press, 1997, 71-82—*significantly expanded.*
4. 'Teleology and Embryogenesis in Aristotle's *Generation of Animals* II.6'—*not previously published; scheduled, with permission of Oxford University Press, to appear also in the Festschrift for Heinrich von Staden, ed. A. Ferraces Rodríguez, K.-D. Fischer, and B. Holmes, forthcoming.*
5. '"What's Teleology Got to Do with It?": A Reinterpretation of Aristotle's *Generation of Animals* Book V' (co-authored with Mariska Leunissen), *Phronesis* 55 (2010), 325-56.
6. 'Teleology and Spontaneous Generation in Aristotle: A Discussion', in *Nature, Knowledge and Virtue*, ed. T. Penner and R. Kraut, *Apeiron: Special Issue* 22.4 (1989), Kelowna, B.C.: Academic Publishing and Printing, 181-93.
7. 'First Principles in Aristotle's *Parts of Animals*', in *Philosophical Issues in Aristotle's Biology*, ed. A. Gotthelf and J. G. Lennox, Cambridge: Cambridge University Press, 1987, 167-98.
8. 'The Elephant's Nose: Further Reflections on the Axiomatic Structure of Biological Explanation in Aristotle', in *Aristotelische Biologie: Intentionen, Methoden, Ergebnisse*, ed. W. Kullmann and S. Föllinger, Stuttgart: F. Steiner-Verlag, 1997, 85-96.
9. 'Division and Explanation in Aristotle's *Parts of Animals*', in *Beitrage zur antiken Philosophie: Festschrift für Wolfgang Kullmann*, ed. H.-C. Günther and A. Rengakos, Stuttgart: F. Steiner-Verlag, 1997, 215-30.
10. 'Notes towards a Study of Substance and Essence in Aristotle's *Parts of Animals* II–IV', in *Aristotle on Nature and Living Things: Philosophical and Historical Studies presented to David M. Balme on his Seventieth Birthday*, Pittsburgh: Mathesis Publications, 1985, 27-54.
11. 'A Biological Provenance', *Philosophical Studies* 94 (1999), 35-56; Dordrecht: Kluwer Academic Publishers (rights currently owned by Springer-Verlag).

12. 'Data-Organization, Classification, and Kinds: The Place of the *History of Animals* in Aristotle's Biological Enterprise'—*not previously published.*
    Appendix: A case for the ordering of the books comprising HA VII–IX and a question about the biological study of man that arises therefrom (excerpt from P. Beullens and A. Gotthelf, 'Theodore of Gaza's Translation of Aristotle's *De Animalibus*: Content, Influence, and Date', *Greek, Roman, and Byzantine Studies* 47 [2007] 474-77).
13. '*History of Animals* I.6 490b7-491a6: Aristotle's *megista genē*'—*not previously published; scheduled, with permission of Oxford University Press, to appear also in the Festschrift for Pierre Pellegrin, ed. C. Cerami, forthcoming 2012.*
14. '*Historiae* I: *Plantarum* et *Animalium*', in *Theophrastean Studies: On Natural Science, Physics and Metaphysics, Ethics, Religion, and Rhetoric*, ed. W. W. Fortenbaugh and R. W. Sharples, New Brunswick, NJ: Transaction Books, 1988, 100-35.
15. 'Darwin on Aristotle', *Journal of the History of Biology* 32 (1999). 3-30—*rights returned to author in 2000 by Springer-Verlag.*

CODA: 'Aristotle as Scientist: A Proper Verdict (with emphasis on his biological works)'—*not previously published.*

Abbreviations

Abbreviations of works by (or attributed to) Aristotle

An.	*De Anima*
APo.	*Posterior Analytics*
APr.	*Prior Analytics*
Cael.	*De Caelo*
Econ.	*Economica*
EE	*Eudemian Ethics*
GA	*Generation of Animals*
GC	*Generation and Corruption*
HA	*History of Animals**
IA	*De Incessu Animalium*
Juv.	*On Youth and Old Age*
Long.	*On Length and Shortness of Life*
MA	*De Motu Animalium*
Metaph.	*Metaphysics*
Meteor.	*Meteorology*
NE	*Nicomachean Ethics*
PA	*Parts of Animals*
Phys.	*Physics*
PN	*Parva Naturalia*
Pol.	*Politics*
Protrep.	*Protrepticus*
Resp.	*On Respiration*
Sens.	*On Sense and Sensible Objects*
Somn.	*On Sleep and Waking*

Abbreviations of modern works

Bonitz, *Ind. Arist.*	H. Bonitz. *Index Aristotelicus*, 2nd edn. (Graz 1955; repr. of orig. edn., 1870, vol. 5 of the Prussian Academy edition of *Aristotelis Opera*. Berlin)
Denniston	J. D. Denniston, *Greek Particles*, 2nd edn., ed. K. T. Dover (Oxford, 1954)
LSJ	H. J. Liddell and R. Scott, rev. H. S. Jones, *A Greek-English Lexicon*, 9th edn. (Oxford, 1940)
Smyth	H. W. Smyth, *Greek Grammar*, rev. G. M. Messing (Cambridge, Mass., 1956)

* For book numbering see below, p. 272 n. 33

PART I

Teleology, Irreducibility, and the *Generation of Animals* (*GA*)

1

Aristotle's Conception of Final Causality

I

Aristotle's teleology is a central component of his philosophy, and interpretations of it often heavily influence evaluations of the significance of his thought, both in science and in philosophy. Much has been written about this aspect of his philosophy, but surprisingly little sustained attention has been directed to what is clearly the fundamental question: what, precisely, does Aristotle mean when he asserts that something is, or comes to be, *for the sake of* something?[1]

II

If we are to answer this question with both historical accuracy and philosophical precision, how must we proceed?

The place to begin, of course, is with the text—ideally, with Aristotle's own statement of an answer. One would expect to find, somewhere in the vast Aristotelian corpus, a thorough analysis and explicit definition of this central notion. Surprisingly, it is not there to be found. Readers of the corpus will search in vain for a detailed analysis of what it is to be (or come to be) for the sake of something.

The longest continuous passages on final causality, *Physics* II.8 and (sections of) *Parts of Animals* I.1, while containing much that eventually proves helpful, do not address themselves directly to this issue. In each case, the purpose is to argue for the applicability to nature of a conception of final causality whose precise meaning and statement is largely taken for granted.

This paper is reprinted substantially as it appeared in 1976 (Gotthelf 1976/77). A 'Postscript 1986' and some notes were added for the 1987 reprinting (Gotthelf 1987b); those additional notes were placed, and remain, in square brackets. (For the present reprinting, a few small editorial changes have been made, and occasional references have been added, signaled by expressions such as 'see now', as in n. 1 below.)

[1] [Since this essay was first published much good work as been done on this issue. I comment on some of it in notes which follow, and in the appended Postscript 1986. For comments on post-1986 work, see now Gotthelf 1992 and chapter 3 below.]

In looking for such an analysis outside these chapters, one thinks first of the passages in which the final cause—'that for the sake of which'—is being introduced as one of the four (types of) causes.[2]

The main text is *Physics* II.3 194b32-195a3.[3] A careful reading of these lines, however, shows that here too Aristotle is not introducing or defining the concept of *being for the sake of* in terms of more familiar concepts. Rather, assuming a familiarity with the notion from its ordinary use, he is arguing for its status as a (type of) cause. More precisely, in fact, Aristotle is arguing for the inclusion of the *end* (*to telos*) in the list of causes. In a literal translation, the relevant portion reads:

In yet another way we call <something> a cause as <being> the end; this is that for the sake of which—e.g., of walking, health. For, why does one walk? We say, 'in order to be healthy,' and speaking so we think we have given the cause.[4]

[2] 'That for the sake of which' is a literal rendering of Aristotle's *to hou heneka*. *heneka* is an ordinary preposition in Greek, appearing mostly in the construction *A*-verb-*heneka*-*B*, with *A* the subject and *B* the genitive object (e.g., 'he walks for the sake of his health'). Aristotle formed abstract substantive expressions to designate the terms of the relation (*to hou heneka* for *B*, *to toutou heneka* for *A*) and the relation itself (e.g., *heneka tou*). I follow the traditional rendering of *aitia* and *aition* as 'cause'. I am not convinced by the arguments against this translation given by Gregory Vlastos in his 1969, 292-94. There is much to be said for the view that Aristotle is offering a conception of causality different from the Humean one that has so infiltrated the speech, and structured the intuitions, of many philosophers. One should not equate a particular *conception* of causality with the *concept* of causality as such. I cannot argue this here in any detail, though for some brief remarks see n. 51 below. [See also Gotthelf 1980, 368; Cooper 1987, 273-74; and Balme 1987c, 281.]

[3] *Metaph*. A.3 983a31-32 refers us to *Phys*. II.3; *Δ*.2 is an almost exact duplicate of part of II.3; and *APo*. II.11 merely repeats II.3's illustration.

[4] 194b24, 32-5. In canonical form the argument runs:

P_1 That for the sake of which is a cause.
P_2 The end is that for the sake of which.
———
The end is a cause.

P_1 is not actually stated, but is the implicit conclusion of a subsequent argument, embedded in the walking example which may be put:

P_3 An acceptable answer to a 'why' question states a cause.
P_4 Stating that for the sake of which is an acceptable answer to some 'why' questions.
———
That for the sake of which is a cause.

Neither P_3 nor P_4 is felt by Aristotle to need explanation or defense, and the wording of the example which generates P_4 makes clear that Aristotle is assuming that his readers are thoroughly familiar with the notion of being for the sake of, at least in its application to human action. Nowhere in the passages describing the four causes is a statement given of what it is to be (or come to be) for the sake of something. (In support of this reading, cf. the summary passage in the same chapter, at 195a24-26, and the similar arguments in the passages listed in n. 3.)

Failing to find an explicit account of the nature of final causality in the passages listing the four types of causes, one thinks next of looking through the corpus for passages which contain an explicit statement of conditions necessary and/or sufficient for something's being or (coming to be) for the sake of something. As it turns out, there are ten to fifteen passages which appear to offer this, but, once again, it can be shown that none of them gives what can be considered a definition—an account of what it is to be (or come to be) for the sake of something.[5]

III

Since the text does not provide us with an explicit answer to our main question, some indirect method of determining the answer is required. There are at least three such methods that might profitably be used.

(1) One might examine the passages just referred to in which, at least apparently, Aristotle gives conditions for something's being (or coming to be) for the sake of something, in the attempt to determine what particular conception(s) of final causality, if any, they presuppose.
(2) One might examine the various passages in which Aristotle is presenting arguments for his thesis that final causality is operative in nature, in the attempt to determine what particular conception of final causality is the subject of this thesis.[6]
(3) One might examine relevant texts throughout the corpus in the attempt to determine Aristotle's position on what is now known as the problem of (biological) reduction.

Of course, the manner in which one would proceed to do this last, and the point of such a procedure—i.e., the relevance to Aristotle's conception of final causality of his position on reduction, if he can even be said to have one—are by no means obvious, and eliciting them would require some sustained discussion. Since this

[5] Demonstrating this would require a longer discussion than is possible here. For now, I simply list the passages. There are eight or so distinct statements of conditions, two of which, notably, are repeated several times in different works. They are, in order of Bekker pagination: *APo.* II.11 95a6-9; *Phys.* II.2 194a28-32 (with II.8 199a8-9; b15-17; *PA* I.1 639b27-30; and 641b24-25); *Phys.* II.5 196b21-22; II.8 199a17-20; *An.* II.4 415b1-2 (with 20-21; *Phys.* II.2 194a35-36; *Metaph.* *Λ*.7 1072b2-3; and *EE* VIII.3 1249b15); *Metaph.* *α*.2 994b9-10; *Metaph.* *Δ*.17 1022a4, 6-8; and *Protrep.* Fr. 2 (Ross). [These passages are discussed in chapter 2 of my dissertation (Gotthelf 1975). Cf. also n. 13 below.]

[6] We should expect these arguments to have some such form as: 'Any phenomenon which is such-and-such is for the sake of something; some natural phenomena are such-and-such; therefore, some natural phenomena are for the sake of something.' The first premise, very possibly unstated in the text itself but easily formulable, might well be of some help in identifying what, for Aristotle, it is for something to be for the sake of something, if not itself already such an account.

method turns out to be the most fruitful—the others being more valuable for providing possible support to an interpretation already arrived at than for discovering and formulating a new one—let us turn directly to this discussion.

One might come to see the value of this method by means of the following considerations. First, in almost every passage in which Aristotle introduces, discusses, or argues for the existence of final causality, his attention is focused on the generation and development of a living organism.[7] In line with this, then, we ought to direct our attention to organic development and ask our main question of this type of case: what, precisely, does Aristotle mean when he asserts that the coming to be (or any stage in the coming to be) of a living organism is *for the sake of* the mature, functioning organism which results?[8]

Secondly, in considering this question we should remember that such an assertion is intended by Aristotle to convey an *explanation* of the occurrence and character of the stages in the development (and thus of the development itself), since that which the development is for the sake of is one of its 'causes'. Thus, a condition on any successful account of Aristotelian final causality is that it illuminate why Aristotle thinks reference to the end of a development might serve to explain the existence and character of that development.

This condition takes on a special importance when we remember, further, that in Aristotle's own study of organic development, *Generation of Animals*, he offers a detailed exposition of the 'mechanisms' involved in development, identifying sequences of changes in quality and place of the embryonic materials. In fact, it is often thought that *Generation of Animals* provides conclusive evidence that Aristotle himself believes that it is at least in principle possible to account for the development of a living organism fully in terms of features of the organism's constituent materials, features which make no reference to the development's outcome.[9]

[7] Cf. the passages listed in nn. 3 and 5. While the *Phys.* II.3 passage discussed above (see n. 4) and its counterparts (see n. 3) are restricted to examples of human action, the point of listing and describing the four types of cause is stated to be the contribution that knowing the types of causes will make to the investigations of the natural scientist (194b16-23); and in coming to understand that 'The nature is among the "for-the-sake-of-something causes"' (II.8 198b10-11, the chapter's opening lines), what the natural scientist learns is that the development of (the parts of) a living organism is for the sake of (their contribution to) its mature functioning. (Cf. Charlton 1970, 120-21, 122-23 and below n. 41.)

[8] That Aristotle's preoccupation is with organic development's being for the sake of something is no accident. It can be shown, once the correct interpretation of this use of 'for the sake of' is established, that all other uses of it are derived from and partially definable in terms of this one. See below n. 19.

[9] [Expressed most recently, in one form or another, in Nussbaum 1978 and (with reservations) Sorabji 1980. I have discussed Nussbaum 1978 in Gotthelf 1980 and 1981/82; for Nussbaum's latest views, see Nussbaum 1980, 1982, 1983, 1984. I discuss Sorabji 1980 in section II of 'Postscript 1986' (below, pp. 35-38, 39; cf. also the name index *s.v.* Sorabji, R.).]

If this is the case, we may properly wonder what the explanatory significance is of statements that refer the development's existence and character to its end, and thus in what precise sense the development is *for the sake of* the mature, functioning organism which issues from it. Alternatively, if Aristotle rejects, explicitly or implicitly, such a reduction, this too might well shed important light on what it is for a development—or for anything—to be *for the sake of something*.

Such, in brief, is the rationale for pursuing our main question through an examination of Aristotle's position on the problem of biological reduction.[10]

IV

Let us now give a more precise statement of the reducibility question, first in general, and then as it applies to Aristotle's own philosophical and scientific context. That will make us better able to discuss Aristotle's position on the question, and the implications of this position for our understanding of his conception of final causality.

The problem of biological reducibility is, for our purposes, the problem of determining the relationship between the laws of action of organic phenomena and the laws governing the living organism's material constituents. Specifically, can one account for a particular living process in terms of laws governing the material involved in the process, laws which make no mention of the end or goal of the process? In the case of organic development, can one give an account of the process of development solely in terms of laws of chemical interaction, laws which make no mention of the end to be realized, so that in principle one could give a chronological list of sets of chemical transformations of initial and added material such that the end result is a correct chemical description of the structure of a mature

[10] Some contemporary writers distinguish, under the general heading of biological reduction, two theses: (1) that teleological explanation can be reduced to non-teleological explanation; and (2) that biology can be reduced to chemistry (or physics, via chemistry). For example, Ernest Nagel, in a now classic discussion (1961, 398–446), argues that the first thesis is necessarily true, while the second is an empirical matter which is not currently known either to be true or to be false. Charles Taylor (1964, ch. 1), on the other hand, has argued convincingly that Nagel's case for the first thesis presupposes at key steps the truth of the second thesis. Taylor takes the position that the *first* thesis is an empirical matter, then argues that the claim that a system is teleological, that it 'inherently tends' in a certain direction, involves, essentially, the claim that such a reduction *cannot* be effected. 'It involves, for instance, the assumption that the basic level of explanation has been reached. For the claim that a system is [teleological] is a claim about the laws holding at the most basic level of explanation' (18). This view turns out to parallel Aristotle's own in important ways, and that is one reason the two reduction theses are not distinguished here. Furthermore, since all of Aristotle's distinctively biological 'laws' are teleological in character, thesis (2) for Aristotle reduces to thesis (1) at least with regard to the 'derivability' condition—see the next note).

living organism, without any of the laws of transformation making any reference to the end result? Such is the question stated generally.[11]

If one asks this question of Aristotle, one is immediately struck by the fact that Aristotle makes no use, and certainly no explanatory use, of the concept of *law* involved in the formulation of the problem. If we are to understand Aristotle's position on reducibility—and his teleology—we must understand *his* approach to explanation, his alternative to the concept of law, and also the equivalent in his system of what are for contemporary science the laws of physics and chemistry, the laws of the behavior of the material constituents of living organisms.

Now, Aristotle's central explanatory concepts are 'nature' (*phusis*) and 'potential' (*dunamis*). Compressing as much as possible, and simplifying somewhat, we may say that Aristotle's manner of explanation of natural changes and processes goes as follows.[12] Nature consists of individual entities, each of a specific kind, possessing various attributes, moving and changing in various ways. Some are living, some not; all are composed of the four simple bodies, the 'elements', which are themselves analyzable into combinations of elementary qualities and some sort of underlying matter. All these things which exist by nature—simple bodies, inanimate compounds, plants animals and their parts—move and/or change in ways characteristic of themselves if not impeded, which is to say, each has a *nature*. If an entity or material body is acting in such a characteristic way, it is a sufficient explanation of its action to identify that it is acting in accordance with its nature.[13]

[11] For a more rigorous and detailed analysis of what might constitute a reduction, see, e.g., Nagel 1961, 345-66, 433-35, summarized usefully in Ayala 1972, 1-6, or Schaffner 1974, 111-18. The problem as I have defined it here corresponds to the condition for a successful reduction which both Nagel and Schaffner call the condition of 'derivability'. Omitted in this essay is a consideration of the Aristotelian position on the second of the two formal conditions these writers lay down (which is called 'connectability' by Nagel and 'referential identity' by Schaffner). While a full-scale treatment of our subject would require discussion of this issue, such discussion is not possible in the present context, involving as it does the questions of the relationships of an Aristotelian entity (*ousia*) and its form, respectively, to the entity's constituent material, and of actions of (or identified in terms of) the entity to actions (i.e., motions and changes) of the entity's constituent materials. This involves many issues extraneous to our immediate concerns. Determining Aristotle's position on the 'derivability' issue will be sufficient to determine both his position on reduction and his conception of final causality. [On some of those 'extraneous issues' see Balme 1987d, Kosman 1987, Freeland 1987, the brief discussion of metaphysical issues in ch. 7 below, pp. 171-80, esp. 175-79, and the fuller study on which the latter discussion is based in ch. 10 below.]

[12] What follows is a fairly standard account, making use of central themes of several of Aristotle's works. The *loci classici* for 'nature' (*phusis*) and 'potential' (*dunamis*) are respectively, of course, *Ph.* II (with *Metaph. Δ*.4) and *Metaph. Θ* (with *Δ*.12). The exposition is so familiar that, with only one exception, references are cited only for quotation. The account will be slightly oversimplified in places, so as not to burden it with so many qualifications that the relevant points are lost; on this (and for the exception) see the next note.

[13] Cf., e.g., *Cael.* IV.3 310b15-19. Omitted here as irrelevant to our immediate purpose are the role of the efficient cause of locomotion to natural place and the role of the Unmoved

Not all motion and change, however, can be explained solely in terms of the nature of the moving or changing entity. Most motion and change, in fact, is due to *interaction*. Things act on other things, and things respond to the action on them of other things, in characteristic ways. In addition to having 'within themselves a source of motion-or-change and rest',[14] each natural thing has within itself 'a source of being moved-or-changed by another'[15] to which, in each case, corresponds a 'source' in that other of so changing the first. Thus, for Aristotle, in addition to having a nature, each natural thing has *potentials*—potentials to change certain other things in certain ways (which we may call 'active potentials') and potentials to be changed by certain other things in certain ways (which we may call 'passive potentials').[16]

Mover, for both of which see *Ph.* VIII. [There is a valuable discussion of these matters in Waterlow 1982.] Ignoring the latter, and the *Metaph.* *Λ* discussion of its role as the final cause of the rotation of the outermost sphere (especially 1072b2-3 and its sister passages—listed in n. 5 above), will be thought by some to be a serious mistake. Space limitations prohibit anything more than the following brief remarks. The passages which identify or refer to two ways in which *to hou heneka* is 'said' are intended to isolate the sense of 'that for the sake of which' which plays a technical role in Aristotle's philosophy from an ordinary use, approximating 'beneficiary', and as such are neither intended to nor do shed light on that technical sense. [For a somewhat different view see Kullmann 1985.] Secondly, the sense in which the action of (the *nous* of) the outermost sphere is for the sake of something is parallel to that in which human action, also directed by *nous*, is. If it can be shown that this sense is derivative from the sense in which organic development is for the sake of something, then an examination of the role of the Unmoved Mover as final cause cannot substitute for the examination undertaken herein—and in fact does not contribute anything to the attempt to answer our main question. On the priority of senses, see n. 19. [Charles Kahn has argued for an interpretation of Aristotelian teleology that gives a central role to the Prime Mover as final cause (Kahn 1985). In this interesting paper he maintains that the capacity of most living things to produce offspring one in form with themselves—which will be argued in this paper to be fundamental to Aristotelian teleology—is itself explained teleologically 'from general principles of cosmic perfection' (Kahn 1985, 195, 200). Among other passages, he appeals to *GA* II.1 731b18-732a1, *GC* II.10 336b25-34, and *An.* II.4 415a23-b7. There is not space to treat the issues Kahn raises at the length his discussion warrants (several are anticipated in Gotthelf 1975, 46-50, 260-61, 304-306 [esp. n. 161], 317-43); I can only make two points here. First, the thrust of these three passages is to assimilate generation to *self*-preservation (not the preservation of the species). (For discussion, see, e.g., Balme 1972 [1992], 96-97, Gotthelf 1975, 331-39, Lennox 1985a, and Balme, 1987c, 279-80.) Second, to relate the striving for self-preservation to the general striving for the better is not to make self-preservation an expression of some wider, cosmic goal the striving for which *explains* the striving for self-preservation. The striving for the better over which such passages generalize just *is*, in the end, a striving for self-preservation, and the arguments in each of these three passages *could* be stated without mention of 'the better' or of 'the divine' (cf. nn. 18 and 38 below, with references). But Kahn's wide-ranging paper deserves fuller attention than can be given here.]

[14] *Phys.* II.1 192b13-14.

[15] *Metaph.* *Δ*.12 1019a15-16; cf. *Θ*.1 1046a10-11.

[16] Any interaction will be, in fact, the actualization of a specific potential pair, an active and a passive, and there will be for any interaction a scientifically fundamental and ultimate way of describing it. We should note that the scientific account may not recognize a distinct potential,

These two concepts, 'nature' and 'potential', are *the* basic explanatory concepts for motion and change in Aristotle's scientific theories, and statements identifying the natures and potentials of things are the closest Aristotelian equivalents of 'laws' in modern science. Every process in nature, every motion or change, according to Aristotle, *is* action in accordance with a nature, *or* the actualization of a coordinate pair of active and passive potentials, *or* the sum of some combination of these. A natural motion or change is *explained* when it is shown to be the result of: action in accordance with one or more natures and/or actualizations of one or more irreducible potential-pairs. Explanation, to this extent, is for Aristotle subsumption (not under general laws but) under the natures and potentials of the acting and changing entities—and thus in large part consists simply of correctly identifying, in causally fundamental terms, *what* is actually happening.

V

As for the Aristotelian equivalent of modern physical and chemical laws, let us briefly consider the form that his account of the action of the material constituents of living organisms takes.

According to Aristotle, the material constituents of bodies, the 'so-called elements' (earth, water, air, fire), each has a nature and potentials. The nature of a simple body—its characteristic way of acting, if unimpeded—is locomotive, and its potentials—its characteristic ways of interacting with other such bodies—are primarily qualitative: each has a pair of primary qualities, and it is these that determine their interactions, including their mutual transformations.[17]

even though it might appear to be recognized in ordinary speech. For example, one could describe the collapsing of a row of dominoes as the actualization of the single (passive) potential of the row to be knocked over. This potential, however, consists of—is reducible to—the potentials of each of the dominoes (in virtue of the material of which it is made, and its shape) to be knocked over (and to knock the succeeding one over), and as such would not be, fundamentally or scientifically, a distinct potential. For a clear-cut conception of what for Aristotle would be the ultimate, irreducible potentials of the inanimate world, one would have to examine very closely the physical works, including *Meteor.* (esp. IV, now generally agreed to be authentic [cf. Furley 1983]), *GC* and *Cael.* It will always be an empirical matter which potentials are irreducible. (The potential of opium to cause sleep, for instance would *not* be—for Aristotle or for us—irreducible, and thus ascribing a 'dormitive virtue' to opium would for Aristotle *not* be explanatory.) [There is an interesting discussion of this point in Sorabji 1980, 171-72; I remark on it in my discussion of Sorabji in section II of 'Postscript 1986', p. 36 below). See also the quotation from Taylor 1964 in n. 10 above.]

[17] To simplify terminology we will designate the locomotive natures and qualitative potentials of the elements jointly as 'element-potentials'. For various reasons, the difference between a nature and a potential (though important in certain contexts) is not a great one, and Aristotle himself licenses this usage at *Metaph.* Θ.8 1049b6-11. (For the relationship between explanation in terms of natures and potentials—the type of Aristotelian explanation just expounded—and

To arrive at the Aristotelian form of the problem of reduction, then, one need only state the reducibility question in terms of element-potentials instead of laws of chemical interaction. With this substitution the question takes the following form: can one account for a particular living process completely in terms of the element-potentials involved in the process, making no mention of the overall end or goal of the process? In the case of development, can one give an account of the process of organic development solely in terms of element-potentials which make no reference to the overall outcome of the process, viz., the form of the mature, functioning organism? Or must at least one of the potentials involved in the account be irreducibly a potential for the development's end, for an organism of that form? If the latter is the case, one can think of the development as primarily the actualization of that irreducible potential, and any element-potentials additionally involved would be viewed as subsidiary to that one. Then, focusing on development, and with our discussion of the Aristotelian approach to explanation of natural motion and change in mind, we may put the reduction question as follows: *Is the development of a living organism the result of a sum of actualizations of element-potentials, or is it primarily the actualization of a single potential for an organism of that form, a potential the actualization of which involves the actualization of element-potentials, but is not reducible to them?*

VI

The answer to this question is the key to understanding the precise nature of Aristotle's conception of final causality.

It is the thesis of this essay:

(1) that Aristotle posed precisely this question, and that it was on this question more than on any other that he saw himself differing directly from his 'mechanistic' predecessors;
(2) that the evidence of the text—in the metaphysical and the physical, as well as in the biological, works—is that Aristotle's response to this question was an unwavering advocacy of the second alternative: that his view was that the development of a living organism is *not* the result of a sum of actualizations of element-potentials the identification of which includes no mention of the form of the mature organism, but *is* in fact the actualization primarily of a single potential for an organism of that form, an actualization which incorporates many element-potentials, but is not reducible to them;

explanation by causes, see n. 51 below. The connection between these two is, in large part, the deeper theme of this essay.)

(3) that this question and the question of whether a living organism develops *for the sake of* being a mature, functioning organism are *one and the same question*; and thus

(4) that the irreducibility to element-potentials of organic development is the core of the meaning of the assertion that the development is *for the sake of* the mature organism, and thus the core of Aristotle's conception of final causality.

In regard to this last, it is thus the thesis of this essay that Aristotle's concept of *coming to be for the sake of* may be defined as follows:

A stage in development, *A*, comes to be for the sake of the mature, functioning organism which results from the development, *B*, if and only if: (1) *A* is a necessary (or 'best possible') stage in a continuous change resulting in *B*, and (2) this change is (in part) the actualization of a potential for *B* which is not reducible to a sum of actualizations of element-potentials whose identification does not mention the form of *B*.[18]

[18] Two comments on this definition: First, the definition is formulated to apply only to those cases where the mature organism actually results. Since an organic development which does not succeed in reaching its goal is nevertheless for the sake of that end, a definition which applied to these cases as well would have to reformulate condition (1). For these cases, 'resulting in' would have to be replaced by something like 'tending to'. However, since 'tending to' for Aristotle can ultimately only be defined in terms of an irreducible potential for *B* (cf. Taylor's account, quoted in n. 10 above), there would be no need for condition (1) for these cases. In fact, condition (1) is superfluous, being a consequence of (2), and is included only to stress the contrast of this interpretation with those which do not make irreducibility a necessary condition of final causality for Aristotle. Second, the phrase 'best possible' is included to cover those (recognized by Aristotle to be) frequent cases where an organ or action is not actually *necessary* for the organism's life or well-being, but is one of a set of which one is necessary, and the most efficient of that set. This raises a number of considerations outside the scope of this essay, but the phrase needs to be included for completeness. See *PA* I.1 640a33-b3 and *IA* 2 704b10-17 for two representative texts.

[The considerations just mentioned include:

(1) the relationship between explanation by hypothetical necessity and explanation by the principle that 'nature does nothing in vain but always for each thing the best from among what is possible';
(2) the epistemological status of the latter principle; and
(3) the relationship between explanations that appeal to 'the honorable' and those that appeal to advantage.

On (1), and on hypothetical necessity generally, see n. 32 below, Cooper 1987, Balme 1987c, and chapter 7 below, pp. 174-75. On (2) see chapter 7 below. On (3) see Balme 1987c, 277 n. 5, and the discussion in Lennox 1985b, 149-55. They argue that appeals to 'the honorable' are a subset of appeals to advantage and do not represent a separate principle of explanation. A similar case can be made out, I believe, for appeals to 'living well' (*to eu zēn*). (Cf. also n. 13 above—and now ch. 2 below.) Since a naturalistic account can thus be given of the notion of the good with which Aristotle operates in his biology, it seems to me that the fundamental account of the final cause need not make use of that notion. While it is true that accounts more 'familiar to us' start there, it is not the case (contra, e.g., Kahn 1985, 197-98)

It is a corollary of this four-part main thesis that Aristotle's teleology—his doctrine that the development, structure, and functioning of a living organism is for the sake of something—is fundamentally *empirical* in character, and not an a priori doctrine brought *to* his investigation of nature.

The balance of this essay will consist of arguments in behalf of this thesis and its corollary.[19]

that Aristotle (in effect, arbitrarily) designates as the aim of a process that stage which he independently wishes to count as good. It is rather that the capacities actually at work *are* capacities for the mature state (*akmē*), where the mature state is identifiable in terms of the presence of maximal powers of self-maintenance without reference to independent normative criteria, *and* that for (what we would call) meta-ethical reasons the good is defined by reference to this same mature state. This raises the question of how one determines which capacities are actually 'at work'. This is a large issue, with important metaphysical dimensions, but surely it is clear that *Physics* II.2's poet spoke ridiculously (194a30-3) for more than one reason: if a living thing's aim were death one could not understand the elaborate process of continuous physical articulation and continuously expanding powers we call its development (the self-maintaining powers at each stage the same as those at earlier ones only more efficacious, etc.). The only question is whether the decline to death that regularly follows development to maturity is to be seen as part of the *dunamis* which is actualized in maturity, or is rather to be attributed to the natures and potentials of the materials of which the organism is made. I take it there are arguments (or premises from which one can construct arguments) in *An., PN, GC*, and *Metaph.* H and Θ in favor of the latter view (though as I say this is a large issue). (For a good discussion of 'self-maintenance' in Aristotle, see Nussbaum 1978, 76-78.) See also the new portion of n. 38 and section I of 'Postscript 1986', below—and, now, on all of these issues, chapter 2 in the present volume.]

[19] There are two other very controversial corollaries of the main thesis which can only be mentioned here:

(1) All uses of 'for the sake of' to be found in the Aristotelian corpus *other* than to denote the relationship between (a stage in) a development and its outcome are definable—and were understood by Aristotle to be definable—in terms of this primary use. This includes the notions of a *part* of an organism being for the sake of the whole, mature organism (i.e., the notion of a part having a *function*), of anything in the realm of human action—practical or productive—being for the sake of something, and of the action of heavenly bodies being for the sake of something (cf. n. 13 above). [For discussion of this dependence in the case of parts, see now section III of 'Postscript 1986', below.]

(2) The natural motions and changes of the simple bodies (earth, water, air, fire) when outside a living organism are *not* for the sake of anything. While these motions and changes are to (*eis*) a place or quality—very infrequently called a *telos*—they are not, and are never said to be, for the sake of anything (*heneka tou*). The latter concept is reserved for those processes of complex entities which are not reducible to element-potentials. Being the actualization of a single element-potential, each elemental motion or change *is*, of course, reducible to a set of element-potential actualizations—viz., the set consisting of that one element-potential actualization! Both *PA* II.1 646b5-9 and *Meteor.* IV. 12 389b29ff. concern elements as constituents of living organisms. Insofar as 'the movement of each body to its place *is* motion towards its own form' (*Cael.* IV. 3 310a33-4), that movement may be thought of as being *for the sake of* reaching that place—in as derivative a sense of 'for the sake of' as that passage's 'form' is derivative of the full-blooded organic forms of *Metaph.* ZHΘ. No such use of *heneka* is found in the corpus, however. [*Added 2010*: On the limited 'teleology' of the elements, according to

VII

A full case for this four-part thesis would contain at least three components: (1) an examination of Aristotle's own account, in *Generation of Animals*, of the nature and 'mechanisms' of the generative process—in order to put flesh on the so far very skeletal notion of an 'irreducible potential for form' and to provide support for the claim that, for Aristotle, organic development can be explained only by reference to such a potential and not solely in terms of element-potentials; (2) an examination, as suggested earlier, of the arguments Aristotle gives for his doctrine that final causality is operative in nature—in the attempt to elicit whatever information we can about the precise conception of final causality which is in fact the subject of this doctrine, against which information the present interpretation can be tested; and (3) an examination of the various apparent statements of conditions necessary and/or sufficient for something's being or coming to be for the sake of something, as also suggested earlier—for the same purpose.

In this essay we will have to limit ourselves to the first two tasks, in order to give them (minimally) sufficient attention. I will begin with a sketch of the theory of generation in *GA*, and a discussion of its implications for the reducibility question and the interpretation of Aristotelian final causality in terms of irreducibility. This will be followed by an examination of arguments to be found in each of the three chapters which are our main sources for Aristotle's arguments for natural teleology. After some brief remarks contrasting the interpretation offered here with the main alternative traditions of interpretation, I will conclude with an argument for the corollary of our main thesis, the claim that Aristotle's teleological doctrine is fundamentally empirical in character.

VIII

The purpose Aristotle sets for himself in *Generation of Animals* is to explain how reproduction—the generation and development of an offspring one in form with its parents—is effected. Across the first book he develops his thesis in a general way, in reaction to common opinion and to the views of his scientific predecessors.

His thesis, as it applies to the higher animals, is that the semen of the male acts on a material produced by the female, causing it to 'take shape' in the appropriate way.

Aristotle rejects the preformationist-pangenesis theories of some of his predecessors, to the effect that semen comes from the whole body, parts of the semen

Aristotle, see also, in the next chapter, the briefly noted contrast between Aristotle and Newton on the form of the principles stating the basic nature of matter (p. 65). Johnson 2005 misses the point(s) of the present paragraph, and the later one on Newton, in his criticism of my position on element teleology. See Henry 2007b (quoted below, pp. 83-84 n. 63), who explains the issues at stake.]

deriving from, resembling in miniature, and being directly responsible for parts of the body, generation being the enlargement and articulation of this composite mini-organism. The semen, Aristotle argues, does not contribute any *material* to the offspring. Rather, it carries, embodied in a 'motion' or set of motions, an active 'potential' to act on material supplied by the *female*, in just such a way as is required to effect reproduction to type.[20]

As the artist's tools, while in use, convey in their motions the specific potential to shape the material to the form he intends, so, Aristotle explains, the semen is the tool of (the father's) nature, conveying, in something like a motion, the potential to shape organic material to form, i.e. to a mature adult one in form with its parent.[21]

In Book II Aristotle offers a more systematic and detailed exposition and defense of his thesis. In the main theoretical passage of *GA* (II.1 734b19-735a4) he lays down the fundamental principle of explanation that must be used, then uses it to track down the identity of the 'potential' and 'motion' carried by the semen.

According to 'our first principle',[22] both the external agent of generation (the father) and its internal transmitter (the semen) must be adequate to the product. The former must be (or possess) in actuality that very (aspect of the) effect of which it is to be the cause, and the latter must possess the appropriate 'motion'. What it is for the 'motion' to be appropriate is not stated explicitly, but it seems clear that the purpose of the second part of the passage is to give some clarification of this:

[20] At 729b4-9 Aristotle states his view in the form of a question: '. . . or is it that the body of the semen takes no part, but the "potential" (*dunamis*) and the "motion" (*kinēsis*) in it does, for this is what is active, whereas that which is formed, which gets shape, is the [useful] remainder of the female residue?' Throughout this essay I have translated *kinēsis* in this way, for lack of any better consistent rendering. But the reader is to be warned that it is not an accurate translation. Not only does it refer to various types of change other than locomotion (change in quality, change in quantity), but in the biological works especially it often refers to an internal 'activity'—as if boiling water just removed from the fire were said to still contain a 'boiling motion'. This becomes especially significant in Book II, where Aristotle identifies the 'motion' in the semen with a special kind of heat; we are to think of this heat, as we are of the heat of the boiling water, not simply as a quality of the semen, but as an essentially active constituent of it. In II.2 Aristotle argues that semen is a kind of foam, with tiny bubbles of heat-bearing air. We have to understand *kinēsis* in such a way that it makes sense to call this heat a *kinēsis*. [For a later, and perhaps more precise, statement of the relationship between the *kinēseis* and the heat in question, see chapter 4 below, pp. 93 n. 9, 107-109].

[21] Cf. 730b8-23. Conception and development require that the material supplied by the female be appropriate, i.e., possess a passive potential for such an outcome. (For a representative text see II.3 737a24-25.) [As David Balme has pointed out, it is also part of the *GA* theory from the beginning that the female material as well carries 'movements' of its own, which play a role in the generative process. These are the very movements which convert her food into more of herself but which, because of her lesser heat, are unable to produce a new organism. See the important n. 14 in Balme 1987d, 293-94.]

[22] 734b20.

Just as we would not say that fire alone produces an axe or any other instrument, neither <should we say this of> a foot or a hand. Nor, likewise, of flesh, for even this has a certain function. Accordingly, hard, soft, tough, brittle, and all other such qualities belonging to the parts having soul—heat and cold may very well produce these; but never the organization (*logos*) by which one thing is flesh and another bone: only the 'motion' from the generating parent who is in complete actuality what that out of which <the offspring> comes to be is potentially <can produce this>—just as it is with the things which come to be according to an art: heat and cold make the iron hard and soft, but the *sword* <is made by> the 'motion' *of the instruments*, <this 'motion'> having a definition corresponding to the art, <which is> the source *and form* of the product.[23]

The semen's 'motion' is analogous to the motions of the artist's tools. As the motions of the tools have a definition corresponding to the art, i.e. to the form in the mind of the artist, so the semen's 'motion' must have a definition corresponding to the nature, i.e. to the form, of the parent.[24] That is to say, the semen's 'motion' *must be identified by reference to the form it is transmitting.*

To grasp the significance of this, let us recall that for Aristotle any specific 'motion' (and 'motion' as such) is to be defined in terms of its outcome: it 'gets its name *rather* from that-to-which than from that-from-which' it proceeds.[25] If the respective outcomes of two 'motions' are different in kind, then the two 'motions' themselves are different in kind. By stressing that the outcome of the semen's 'motion' is different in kind from the outcome of fire's heating—viz., an organization (*logos*) and not just a quality or a set of qualities—Aristotle is asserting that the semen's 'motion' itself is different *in kind* from the quality-generating 'motions' of fire. The semen's 'motion' is to be defined by its outcome: it is the fulfillment of the potential to generate an animal of a certain form or *logos*, qua potential; the form or *logos* is an inescapable part of its very definition.

If this is so, then the 'potential' which is manifested in the semen's motion also is to be identified by reference to the form being transmitted: it is, essentially, a potential for form, a potential *distinct from* and *not reducible to* any sum of qualitative and locomotive potentials. For, if it were reducible, so would the 'motion' be, and then heat and cold *could*, in the proper sequence of actions, produce 'the *logos* by which one thing is flesh and another bone'.[26]

[23] 734b28-735a3.

[24] That this is the point of the analogy is suggested most strongly by the last sentence of the full passage: 'For the art is source and form of the product, but in *another*; the 'motion' of the nature is in [the generated natural thing] *itself*, being *from* another nature, one having the form in actuality' (735a2-4).

[25] *Ph.* V.1 224b7-8. On the definition and individuation of motions see all of *Ph.* V.1-4. For the general definition of 'motion' see *Ph.* III.1-3. [Waterlow 1982 has some important discussion of these issues and of their relation to Aristotle's teleological commitments.]

[26] 734b33-4. The whole matter of the philosophical implications of Aristotle's theory of generation—his view of the manner in which form is 'transmitted' in generation, the nature and

IX

There is further evidence in *GA*, and in related passages elsewhere, that Aristotle viewed the 'potential' the semen carries as an irreducible potential for form, in the manner specified in the formulation of the main thesis of this essay.

In *GA* II.2-3 Aristotle takes up the question of the physical nature of the bearer of the 'potential' for form. In II.2 he offers a chemical analysis, as it were, of semen, an identification of its qualitative properties. Semen is 'a combination of *pneuma* and water, and the *pneuma* is hot air.'[27] In II.3 he asserts that it is the heat in the *pneuma* that is responsible for generation, and compares that heat with the more familiar heats of fire and the sun.

> Now, the 'potential' of all soul seems to be associated with a body different from and more divine than the so-called elements, and as the souls differ from each other in value and lack of value, so too this sort of nature differs. For within the semen of everything there is present that very thing which makes the semen fertile, the so-called 'hot'. This is not fire or that sort of potential, but the *pneuma* enclosed within the semen, that is, within the foamy part, and more precisely the nature *in the pneuma*, being analogous to the element of the stars. This is why fire generates no animal, and none is seen to be constituted in things subjected to fire, whether wet things or dry. But the heat of the sun and the heat of animals <do generate>—not only the heat conveyed through the semen, but also if there is some other residue of their nature, even this contains a life-source. Such things make plain that this heat in animals neither is fire nor has its origin from fire.[28]

This passage deserves much more attention and space than it can be given here. Its highlights are (1) its stress on the *difference* in character between the 'potential' for soul, and the heat which bears it, on the one hand, and the potentials of the elements and the elemental heat of fire, on the other; and (2) its stress on the similarity of generative heat to the heat from the sun, an analogy or comparison which has puzzled scholars.[29]

The doctrine of this passage nicely focuses the irreducibility question. It makes clear that if there is to be an irreducible 'potential' for form, it will be the 'potential' borne by the heat in the *pneuma*. We know this heat must be such as to act on the female's material in just such a way as to produce an organism of the relevant form. If this 'potential' is to be reducible, there will have to be a way of specifying the series of heatings, coolings, and movings around of material in the developing

character of the 'potential' and motion by which it is accomplished, etc.—deserves more extended treatment. [See Balme 1987c and 1987d, esp. 281-82 and 292-93 respectively, and now chapter 4 below. There is much work yet to be done.]

[27] 735b37-736a1.

[28] 736b27-737a7.

[29] On this and related matters concerning Aristotle's theory of generation, see Balme 1972 [1992], 158-65 *ad* 736a24-737a34 and above n. 26.

embryo which this heat effects, without referring to the form. On the face of it, the absence of any sort of quantitative chemistry, and the simplicity of the descriptions of qualitative 'potentials' and interactions that Aristotle countenances, make it difficult to believe that he thinks such a specification possible, and there is no evidence in his works that he does.[30]

On the contrary, there is at least one passage that suggests the reverse. In *An.* II.4, Aristotle considers whether fire could be the cause of nourishment:

Some think that it is the nature of fire which is the cause quite simply of nourishment and growth; for it appears that it alone of bodies is nourished and grows. For this reason one might suppose that in both plants and animals it is this which does the work. It is in a way a contributory cause, but not the cause simply; rather it is the soul which is this. For the growth of fire is unlimited while there is something to be burnt, but in all things which are naturally constituted there is a limit and a *logos* both for size and for growth; and these belong to soul, but not to fire, and to *logos* rather than to matter.[31]

[30] This also deserves fuller treatment, focusing on an analysis of Aristotle's conception of heat, its nature, potentials, and varieties. Balme 1972 [1992], 163-64 is helpful. Of interest is a little-noted passage in Theophrastus's *On Fire*, which clearly makes use of *GA* II.3 736b27-737a7, just quoted.

Now the heat which is natural to animate bodies, being infused into a comparatively large number of creatures and in a rather special way, becomes in a sense alive and capable of generating similar creatures. Even more so the heat from the sun. It is able to create animals and plants. It is mixed, to be sure, with air (occurring rather in it), but by reason of its gentleness (μαλακότητι) and fineness (λεπτότητι) has a certain appropriateness for generating life; it is not, like the heat of fire, hard (σκληρά) and caustic (περικαής). For that reason seeds subjected to fire do not generate, but those which are warmed by the sun to excess (ὑπερβολήν) generate and germinate. (Coutant 1971, §44, 28-31)

There are suggestions in this work of an attempt by Theophrastus to formulate some generalizations pertaining to the varying effects on various materials of varying degrees of (fineness of) heat, but they are not developed enough to indicate clearly a reductionist sympathy on Theophrastus' part, nor *a fortiori* on Aristotle's.

It is sometimes suggested that the analogy of embryonic development to the operation of the 'automatic marvels' (734b10, 741b9; Balme 1972 [1992], 157; Nussbaum 1976, 146-52; Balme 1987a, 18) makes clear that Aristotle thought of embryonic development as the result of a 'causal mechanism' involving only element-potentials. But it establishes no such thing. Briefly, the analogy is meant to illustrate how the male parent may be the efficient cause of the generation even though not in contact with the embryo (734b10), and that the development is a smooth, uninterrupted process (741b9), and is not meant to suggest anything about the character of the potentials being actualized. The common feature underlying the analogy, as Aristotle notes each time, is a complex but smooth sequence of actualizings of potentials immediately upon a single contact from outside. The analogy is in no way weakened by the fact that in the case of embryonic development some of these potentials will be irreducibly for form (e.g., the potential of the semen's heat to solidify the female-supplied material in just the shape necessary for an organism of that type, and irreducibly so, as we have just seen in our discussion of II.1 734b19-735a4. [On this, see now also chapter 3 below, p. 81.]

[31] 416a9-19.

There is no mention of *pneuma* here, but readers of the two passages from *GA* just quoted and discussed will find it hard to resist supplying it, especially in view of Aristotle's identification of the nutritive and reproductive faculties.

That this distinction between element-potentials and potentials for form (soul, *logos*, etc.) is at the root of the distinction between explanation in terms of material-efficient causes and explanation in terms of final-formal causes will be argued in the next three sections, as we examine the arguments Aristotle offers for the necessity of employing the final cause in the study of organic nature.[32]

X

The need for employing the concept of final causality—of being for the sake of something—in the understanding of nature is asserted and defended in three main places in the Aristotelian corpus: *Metaph.* A.7 (with 3), *Phys.* II.8, and *PA* I.1. In each case, the conception whose employment is defended is the one attributed to Aristotle by our 'irreducible potential' interpretation.

[32] A fuller examination of the evidence of the biological works would require discussion of the well-known set of passages in which Aristotle states that some part of an embryo or organism comes to be 'on the one hand of necessity, on the other for the sake of something'. I think that the key to understanding these passages is provided (as might be expected given the theoretical centrality of the passage) by *GA* II.1 734b19-735a4, as interpreted above in section VIII. My suggestion is that throughout Aristotle's biological works certain (qualitative) aspects of the embryo or organism are explained by element-potentials ('of necessity') and certain (organizational, functional) aspects by a potential for form ('for the sake of something')—and that the same aspect (e.g., the solidity of the bones mentioned in II.6 743a36-b18) is said to be both of necessity *and* for the sake of something only when that aspect contributes to the constitution of an aspect which is not there of material necessity (e.g., the *logos*—the bone itself, qua bone—as in 734b19-735a4). The 'biochemical' basis for the distinction in the type of causality involved in the generation of the different aspects of the embryo is to be found in the different aspects of the semen's 'motion', i.e. of its heat. The qualitative properties of the embryo are explained by the element-potentials that the *pneuma*'s heat has qua heat, the organizational properties by the 'purity' or 'fineness' (n. 30 above, esp. Balme loc. cit.) unique to this sort of heat (which 'fineness' is specifiable—I am arguing—only as being a potential to produce an organism of a certain sort). [See, on hypothetical necessity, Cooper 1987 and Balme 1987c. I take Cooper's thesis regarding the operation of material (or 'Democritean') necessity in organic nature to include, roughly, the following proposition: certain organic outcomes do not come to be of material necessity; those that do, do so *only* conditionally upon the coming to be of those that don't. This way of putting it shows that Richard Sorabji's criticism (1980, 173-74) of the view expressed in this note was misguided. The deer's shedding of horns (*PA* 663b12ff) is for the sake not (as Sorabji suggests) of greater lightness *per se* but of the contribution (*heneka opheleias*: 663b13) such lightness can make to the life of the deer, and *that* good when achieved is not achieved of necessity. As Cooper points out (1987, 265), Aristotle explicitly denies in the theoretical passages (and never asserts in the biology) that (e.g.) individual deer come to be and continue to exist of necessity.]

Metaphysics *A*

At A.3 983a33-b6, Aristotle explains in advance the purpose of his historical survey of the types of causes that his predecessors have employed: 'we shall either find another kind of cause or be more convinced of the correctness of those which we now maintain.' What is important to see is that the discussions that follow constitute arguments for the need to employ the four types of causes. If attempts by the best thinkers to account for certain natural phenomena without a given kind of cause are unsuccessful, while attempts making use of it are successful, that is good reason to maintain that this type of causality *is* operative in nature.

The case for the final cause is made in two parts. First, in chapter 3, we are introduced to the phenomena which, according to Aristotle, material factors alone—i.e., the natures and potentials of the elements—cannot account for, and which (therefore) gave rise to an awareness of the efficient cause. Then, in chapter 7, we are shown that the account in terms of the efficient cause was really a misguided attempt to formulate and employ the final cause.

The insufficiency of explanation wholly in terms of material causes became clear fairly soon, Aristotle states:

> For it is not likely either that fire or earth or any such element should be the reason why things manifest goodness and beauty both in their being and in their coming to be, or that those thinkers should have supposed it was; nor again could it be right to entrust so great a matter to spontaneity and chance. When one man said, then, that mind was present—as in animals so throughout all of nature—as the cause of order and of all arrangement, he seemed like a sober man in contrast with the random talk of his predecessors.[33]

These later thinkers, however, did not clearly distinguish the final from the efficient cause and thus never clearly saw the need for the former to account for 'goodness and beauty, . . . order and all arrangement.'

> For those who speak of mind or friendship class these causes as goods; they do not speak, however, as if anything that exists either existed or came into being for the sake of these, but as if movements started from these. . . . Therefore it turns out that in a sense they both say and do not say the good is a cause; for they do not call it a cause qua good [i.e., qua that for the sake of which] but only incidentally.[34]

The structure of this two-part argument is clear. The phenomenon to be explained is the order, the arrangement, the beauty and the goodness to be found in nature.[35] This cannot be explained solely in terms of element-potentials—neither individu-

[33] 984b11-18.

[34] 988b8-16.

[35] Though Aristotle does not here say so, it is clear from similar language in *PA* that he means both the regularity of the heavenly motions and the organization of living organisms: cf. I.5 644b22-645a36.

ally ('fire or earth or any such element') nor in combination ('spontaneity and chance'); it is unreasonable (un-'sober') to think it could, Aristotle asserts.[36] The only reasonable account, when clearly articulated, is one which acknowledges that the process by which instances of order and beauty come to be, is essentially or inherently a process *to* order and beauty. But, given Aristotle's definition of 'motion' as the actualization of a potential ('as such'), a process, which is *essentially* to some end, is precisely a process which is the actualization of a potential essentially—and thus *irreducibly* for that end. This, then, is what it is for something to come to be for the sake of something, according to the argument of *Metaph.* Λ.

XI

Physics *II.8*

This notorious chapter is the only one in the entire corpus devoted completely and in detail to a defense of natural teleology. It opens with a clear statement of its purpose, sketches the alternative position, then offers a series of arguments and remarks in opposition to this alternative and in support of its own thesis.

The alternative to teleology sketched in the opening section of the chapter is precisely the view that the development of a living organism is the necessary outcome of a sum of actualizations of element-potentials.

> For everyone brings things back to this cause [*sc.* necessity: cf. 198b11] saying that because the hot is by nature such as to be thus, and similarly the cold and everything of that sort, therefore these things [*sc.* natural things: cf. 198b12] of necessity come to be and are. . . .
>
> The problem thus arises: why should we suppose that nature acts for the sake of something and because it is better? Why should not everything be like the rain? Zeus does not send the rain in order to make the corn grow: it comes to be of necessity. The stuff which has been drawn up is bound to cool, and having cooled, to turn to water and come down. It is merely concurrent that this having happened, the corn grows. . . . What, then, is to stop parts in nature too from being like this—the front teeth of necessity growing sharp and suitable for biting, and the back teeth broad and serviceable for chewing the food, not coming to be for the sake of this, but by coincidence? And similarly with the other parts in which the 'for the sake of something' seems to be present.[37]

[36] While Aristotle does not argue on behalf of this charge here, he would surely do so along the same lines as he does in rejecting fire as the cause of nutrition in *An.* II.4 (quoted and discussed in section IX above). See also n. 38 and my analysis of the first argument in *Phys.* II.8, in section XI below.

[37] 198b12-29.

Thus, when Aristotle announces in the lines immediately following these that this view, though plausible, *cannot* be true, it is the possibility of an account of development solely in terms of element-potentials that he is rejecting.[38]

While Aristotle here characterizes his own proposal only as the view that development is 'for the sake of something and because it is better', its nature seems clear. For one may ask how Aristotle could conceive an alternative *account* of development—given his general view of explanation sketched above—other than as one which holds the development to involve the actualization of an irreducible potential for form. If every process *is* the actualization of one or more potentials,[39] and organic development is *not* the actualization only of a sum of element-potentials, then it must be the actualization of a potential irreducibly for a mature organism of the relevant form. These two accounts are the only kinds of explanation available to Aristotle.[40]

Thus, the very way Aristotle poses the alternative he is rejecting in the name of final causality suggests his view of its nature and lends support to the 'irreducible potential' interpretation.

[38] 198b34. That this is so becomes certain if we consider how Aristotle could possibly characterize the alternative view as maintaining both that development is due to *necessity* and that it is due to *chance* (in the wider sense which encompasses chance and spontaneity). For a good discussion of this see Guthrie 1965, 163-64 and 414-19. [This language is nicely explained in Cooper 1987, 251. However, Cooper has maintained (1987, 250-53) that the line of argument outlined in this passage is independent of Aristotle's thesis that material natures are insufficient to explain the formation of precisely structured living outcomes. According to this line, Cooper claims, even if material natures *were* sufficient in that way, they still would not explain why these outcomes are *good* for the organism which possesses them; on the materialists' view, that they are beneficial would be a *coincidence*. But it is not clear that this is right. His point, I take it, is that even if material factors explained the regular production of, say, just this type of eye, and of just that type of eye-covering, they would not explain why the type of eye-covering that is in fact produced is just the kind needed for the successful functioning of just the type of eye produced. But, if the material natures were such as *regularly* to produce, by themselves, just such eyes *and* just such eye-coverings (etc.), then they would be such as to produce just such well-covered eyes (and just such well-organized living things); no *further* explanation would be needed. It is not as if they *could* have produced bad eye-coverings, or none, in this circumstance. Of material necessity, when there are these eyes, there are also those eye-coverings. Once one sees that the material natures are such as to produce such wholes, the only further question one could be asking would be why material things have the natures they do, and as Cooper rightly argues in his discussion of material necessity, this was not a question Aristotle thought one could ask. For this reason, I am still inclined to read this first argument in *Phys.* II.8 as presupposing the thesis that material natures are insufficient by themselves to generate living organisms or their parts. (On the role of the concept of 'the good' in Aristotle's teleology, see also the bracketed addition to n. 18 above, and section I of 'Postscript 1986' below—and now chapter 2 below.) James Lennox (1982) has argued that the contrast between sexual and *spontaneous* generation is central to understanding the contrast between teleological and non-teleological processes in Aristotle. I do not think that is right, and have discussed his paper in chapter 6 below.]

[39] In the broadened sense that includes natures, as above n. 17.

[40] Cf. Charlton 1970, 114-17.

So does the first argument Aristotle offers in support of his teleological account of organic development.[41] The argument may be summarized as follows:

> Organic development is either for the sake of something or by chance; it is not by chance (since chance outcomes are irregular, organic outcomes regular); therefore organic development is for the sake of something.

More explicitly than in the opening statement of the problem, the available modes of explanation are said to be just two: explanation by element-potentials ('chance'[42]) and explanation by final cause ('that for the sake of which'). The case against the former is given in the parenthetical subargument: while the processes leading to a given chance outcome neither always nor for the most part issue in outcomes of the same type, the processes leading to the existence of a mature organism of some type always or for the most part do issue in an organism of that type. Thus the development of a living organism cannot be the result of a series of actualizations of element-potentials. Rather, it is a process for the sake of its outcome, which must mean, if our account of Aristotelian explanation is correct, that it involves the actualization of a potential irreducibly for an organism of that type.

This argument of Aristotle's has often been criticized. It is charged that he equivocates, switching senses of 'chance' from, roughly, 'mechanical sequence' to 'unusual sequence'. His argument is alleged actually to proceed: either for the sake of something or by chance (by 'mechanical' sequence); because usual not by chance (by unusual sequence); therefore, for the sake of something.[43]

I submit, however, that this is not an equivocation, because Aristotle's description of chance outcomes as 'unusual' is not ultimately a matter of language: what appears to be an appeal to ordinary language is in fact a factual claim.

W. D. Ross writes of this premise that, according to Aristotle,

> it is of the nature of chance events to be the exception, while *everything* or almost everything in nature has the appearance of adaptation to purpose. That end-like results should be *constantly* produced in the absence of final causation would be too extraordinary a coincidence to be credible.[44]

[41] 198b34-199a7. That the subject of this argument is organic development, and not natural things as such, is argued successfully by Charlton 1970, 120-21. [David Furley has argued against Charlton on this matter in his 1985. In editorial comments on his penultimate draft, I proposed a simpler reading of 198b36, which if right defeats his argument. Furley summarized my argument in his n. 3, then, oddly, argued that my thesis must be wrong because it has the same problem Charlton's does, the very problem which my reading enables Charlton to avoid. See now the excellent discussion of this passage in Irwin 1988, 522 n. 18.]

[42] Interpreted as above, n. 38.

[43] The harshest such criticism is that of Cherniss 1935, 250-52; cf. Charlton 1970, 123.

[44] Ross 1936, 43.

Ross's last sentence might well have been intended as a summary of the argument for final causality in *Metaph. A*.3 and 7, and even a restatement of 984b11-15, where, as we recall, Aristotle writes that

> it is not likely that fire or earth or any such element should be the reason why things manifest goodness and beauty, both in their being and in their coming to be, or that those thinkers [*sc.* the early Presocratics] should have supposed it was; nor again could it be right to entrust so great a matter to spontaneity or chance.

The argument in *Phys.* II.8 is neither an equivocation nor an appeal to ordinary language. It is a restatement of the argument in *Metaph.* A.3, and has the same force. Given the simplicity of Aristotle's chemistry, he can only believe that the outcome of organic development is too complex, too orderly, possessing too much of limit, *logos*, and form, to be the result merely of the unlimited, relatively indefinite natural action and interaction of the elements. The development must be *for the sake of* its outcome—i.e., essentially and irreducibly a development *to* order—i.e., to form. The development, in short, must be the actualization of an irreducible potential for form.

XII

Parts of Animals *I.1*

This chapter is a lengthy and structurally complex one, which develops several themes pertaining to method in biological science.[45] One of its dominant themes is the importance of final causality for the study of the parts of animals. Though the subject of study is the structure and functioning of the parts of a mature, functioning animal, the argument for the necessity of employing final causality derives from the character of the coming to be of these parts. An examination of Aristotle's argument here should therefore be of value in determining his view of what it is to come to be for the sake of something.

The relevant conclusion is stated near the end of the chapter:

> There are then these two causes, the *for-the-sake-of-which* and the of-necessity. . . . That there are two modes of causation, then, and that we must at least attempt to state both, is now clear, as it is that all those who do not do this say virtually nothing 'about nature,' for the nature of a thing is more an origin than its matter. . . . The reason why our predecessors did not arrive at this method of procedure is that the what-it-is-to-be and the defining of the being did not exist.[46]

[45] Balme's masterly analysis (1972 [1992] *ad loc.*) of the structure of this chapter is invaluable here. [See now also Lennox 2001a *ad loc.*]

[46] 642a1-2, 13-27.

The position of those who spoke only of matter and necessity was described in some detail in an earlier passage.[47] These were the thinkers who searched 'for the material origin and that sort of cause', explaining everything from the origin of the universe to the formation of nostrils by reference to 'the underlying matter's having some definite nature by necessity, e.g., that of fire, hot, that of earth, cold, the one light, the other heavy.' Clearly, they attempted to explain everything in terms of element-potentials.

But, just as Empedocles did not realize that complex structures do not come about at random, but as the result of the actualization of a specific potential for a form that includes that structure,[48] so he and the others did not realize in general that:

> if man and the animals and their parts exist by nature, then we must have to say of each part—of flesh, bone, blood, and all uniform parts, and similarly of the non-uniform ones such as face, hand, foot—in virtue of what each of them is such as it is, and according to what sort of potential. For the 'out of which' is not enough . . . we have to speak also of . . . what sort of thing it is in respect of its form; for the nature in respect of form is of more fundamental importance than the material nature.[49]

Thus explanation solely in terms of element-potentials will not account for aspects of the form; this requires a second 'mode of causation'—an appeal to that for the sake of which these parts have come to be—which requires an awareness of the phenomenon of form and of the proper 'sort of potential.'[50]

The doctrine of *PA* I.1 in regard to final causality, then, is essentially the same as that of each of the other texts we have investigated—*GA* II, *Metaph. A*, and *Phys.* II. Processes for the sake of something are distinguished from those that are not by the presence in the one case, and the absence in the other, of a potential for form. Since reference to this potential is necessary, and mention only of element-potentials insufficient, this potential is evidently one which is *not* reducible to element-potentials. Thus, for a living organism of a certain form to come to be for the sake of something is precisely for it to result from a sum of actualizations of potentials, one of which—and the most explanatorily important of which—is an

[47] 640b4-28.

[48] Earlier at 640a19-27, quoted in the next note.

[49] 640b18-28. Cf. 640a19-27: 'So Empedocles was not right when he said that animals have many of the characteristics they do because "it just happened this way" in the process of coming to be—e.g., even the spine, he says, is such as it is because "it happened to get twisted and thus break". He was unaware, first, that the productive seed must have a specific (*toiautēn*) potential and, secondly, that the producing agent is prior, not only in *logos* but also in time—for a human begets a human, and therefore it is on account of the parent being of such a sort that the child's coming to be is such as it is.'

[50] Thus, 'the reason why our predecessors did not arrive at this method of procedure is that the what-it-is-to-be (*to ti ēn einai*) and the defining of the being (*to horisasthai tēn ousian*) did not exist' (642a24-27).

irreducible potential for an organism of that form, a potential (transmitted via the male parent's semen) to act on the material supplied by the female parent in just such a manner as to produce an organism one in form with the parents.[51]

XIII

Such is the case for the main thesis of this essay. The interpretation of Aristotle's conception of final causality presented here may be designated the 'irreducible potential' interpretation.

Most interpretations that have been offered, heretofore, fall into two main traditions. The first, which may be called the 'immaterial agency' interpretation, maintains that Aristotle understands natural teleology fundamentally on an analogy to human purposive action, in a way that implies that the developing embryo is (or embodies) some sort of conscious or quasi-conscious agent, directing the flow of materials, guiding its development to maturity.[52]

[51] The argument of these sections (and of this essay as a whole) suggests an account of Aristotle's four causes in terms of the concepts of potentiality and actuality. Every process is the actualization of a pair of potentials, active and passive. (A nature may be thought of as an internalized potential-pair.) The bearer of the active potential (the 'whence the source of motion') is the efficient cause. When the outcome of a process is so organized (has a *logos* such) that it has a distinctive manner of acting (an *ergon*) not reducible to the qualitative changes of its material constituents, then the active and passive potentials must be irreducibly potentials for form. This form, which must be specified in the identification of these potentials and their actualization (the 'what it is to be' both of the outcome and of its development), is the formal cause. And since the process, the development, is irreducibly *for* that form, that form ('that for the sake of which') is the final cause. Such a conception of causality presupposes, of course, the reality of natures and potentials. Aristotle's case against Hume here may in part be found in *Metaph. Θ*.3, but a fuller and stronger case can be constructed from other texts. [There is an excellent discussion of Aristotle's concept of 'the nature of a thing', and its role in his natural philosophy, by Sarah Waterlow Broadie in Waterlow 1982, esp. chs. 1-2. Professor Broadie does not discuss the implications of her analysis for our understanding of Aristotle's conception of a cause, but it would be well worth someone's trying to draw out those implications. Her analysis, for one thing, provides an explanation of why accounts of *aitia* in Aristotle simply in terms of necessary and/or sufficient conditions (as Sorabji 1980, ch. 2 argues) must fail, and why it does not follow that we must fall back on epistemologically oriented analyses (such as Sorabji offers) which explicate *aitia* in terms of explanation. (Cf. n. 2 above.) I hope to explore this issue elsewhere.]

[52] Advocates include Zeller (1897, I. 459-61) and Collingwood (1945, 83-85). I include in this category interpretations which speak of an 'unconscious desire' on the part of the embryo (e.g., Rist 1965), since 'unconscious desire', as psychologists use the term, is possible only to beings capable of conscious desire. Thus, as meant in its ascriptions to embryos, this characterization is actually not metaphorical and illuminating, but self-contradictory. [Robinson 1983 defends a form of the 'immaterial agency' view in the course of a critique of the contrary view in Nussbaum 1978. His argument rests, I think, on a confusion between soul as efficient cause and soul as final cause, and more generally between the *dunamis* which is efficient cause and its object (what, as we say, the *dunamis* is *for*) which is final cause (as per n. 51 above.]

Such a view has little, if any, textual plausibility; it rests heavily on Aristotle's (in fact, quite infrequent) use of the metaphors of striving and desiring, and a somewhat casual reading of the comparisons of a nature to an artisan in *Physics* II.8 and the biological works. While this sort of interpretation has the merit of taking Aristotle's teleology seriously and as a real alternative to the mechanism of his predecessors, its insensitive pressing of the nature/art analogy results in an Aristotelian mortal sin. Ultimately, the 'immaterial agency' interpretation makes nature a subcategory of art—i.e., of consciously directed, intentional action—thereby reversing the actual Aristotelian priority of nature to art, and thus rendering them both unintelligible.

The second, and more recent, interpretation may be designated the 'explanatory condition' interpretation. It maintains that the 'final cause' is in no sense actually a *cause*, though it plays a role in a certain type of explanation; and that its role in the explanation of organic development is to identify that for which the stages of development are necessary and (thus) that in terms of which they must be identified if they (and the development they constitute) are to be rendered 'intelligible'. The rooting of an acorn, for instance, is *for the sake of* the oak tree which results, according to this interpretation, if and only if the rooting is (1) a necessary condition of the existence of the oak tree and (2) 'intelligible'—fully 'understood'—only if identified by reference to the oak tree (e.g., as the rooting of a seed-for-an-oak-tree).[53]

Such an account works from a textual base that renders it initially plausible, but in viewing the specification of the end as necessary only to facilitate an (undefined) demand for 'understanding' or 'intelligibility', and not as demanded by the irreducible character of the potential actually possessed by the relevant 'seed', this interpretation appears to imply that the existence of final causality is compatible with the reducibility of the development to element-potentials. As a consequence, this interpretation seems to commit an Aristotelian mortal sin of its own: it appears to permit chance events to be for the sake of something. For instance, on this interpretation, it seems, Empedocles' human-headed bull[54] came to be for the sake of something: the coming together of the head and body were necessary for the existence of the bull, and these organic parts seem fully 'intelligible' only if reference to their ability to function in the calf is included in (or entailed by) their account.[55]

[53] Randall 1960 is often read this way (and not without warrant)—e.g., by Toulmin and Goodfield, who give an admirably clear statement of this interpretation in 1966, 83-85. Cf. Wieland 1970, sec. 16, 'Zum Teleologieproblem' (English translation in Wieland 1975).

[54] One of *ta bougenē androprōira* (*Phys.* II.8 198b32; cf. DK 31 B61).

[55] My hesitation in attributing this consequence to the interpretation is due to the vagueness of its notion of 'intelligibility'. If a process were taken to be 'intelligible' only in terms of its end if and only if the potential being actualized was irreducibly for that end, then the objection would not hold. Even in that case, however, the 'irreducible potential' interpretation would constitute a significant advance, as I go on to indicate. I am indebted to Richard Schuldenfrei for bringing this possibility to my attention.

Aristotelian teleology, in fact, is neither vitalist and mystical, nor 'as if' and mechanical. The notion of an irreducible potential for form supplies the proper content to the awareness that, for Aristotle, organic development is actually direct*ive*, without implying (as the 'immaterial agency' interpretation does) that it is direct*ed*; and it identifies the ontological basis of the awareness that the existence and stages of a development can be understood only in terms of its end—by establishing that the *identity* of the development is its being *irreducibly* a development to that end, irreducibly the actualization of a potential for form.

XIV

Aristotle's teleology—his thesis that the development, structure, and functioning of a living organism are for the sake of something—is a central tenet of his thought. It is a corollary of the 'irreducible potential' interpretation of his conception of final causality that this thesis is *factual* or *empirical* in character: it is a conclusion drawn from observation of nature and not a premise brought to it.

Philosophers of science today are in increasing agreement that the question of reduction is an empirical one; they insist that one cannot legislate the precise form of the laws in which our understanding of nature is expressed. Aristotle's attitude is similar: he does not attempt to legislate a priori the particular form which a successful account of the natures and potentials of living organisms must take. His arguments for his teleological doctrine make this clear. What he insists is that the facts as we have observed them, and the identifications of the natures and potentials of things which these observations have led us to, entail the irreducibility thesis which is at the core of the concept of final causality asserted to obtain in nature. Though the simplicity and the non-mathematical character of Aristotle's chemistry (and physics) eliminates for him any real possibility of a successful reduction to element-potentials of the complexities of the organic world, this makes his thesis no less empirical—for his view of the inanimate world is equally subject to revision. There is nothing in the fundamentals of Aristotle's philosophy, and nothing in his philosophical or scientific method, which would prohibit the adoption of a reducibility thesis, should the scientific evidence be judged to warrant it.[56]

[56] [In an important discussion of teleological explanation in Aristotle (Waterlow 1982, chs. 1-2), which I was delighted to find has much in common with my own, Sarah Waterlow Broadie takes me to task for the exaggeration involved in my 'empiric Aristotle' (91). She argues that the commitment to irreducibility of the sort I have argued here (and on which she has much to say that it highly illuminating) is bound up in the very notions of substance and the nature of a thing. Since these notions are surely among 'the fundamentals of [Aristotle's] philosophy', if teleology goes so go some fundamentals. (Johnson 2005, 183 n. 39 makes a similar point, although oddly he ignores what I said in the present note, first published in 1987, or in the last section of chapter 3 below, first published in 1997, and referred to at the end of this note.) It is not clear to me,

But Aristotle did not believe that the evidence warranted it. On the contrary, we have seen that he thought that a reduction of biology to chemistry could not be accomplished, and thus that organic development involves the actualization of an irreducible potential for form—which is to say, that it is a process *for the sake of* its end, irreducibility of potential for form being the core of Aristotle's conception of being *for the sake of something*.[57]

Postscript 1986

I

Though I would write this essay somewhat differently if I were writing it today, I still agree with virtually[*] all of its main claims. There are three that are crucial.

however, that reducibility of the sort Aristotle denies would render the notion of nature inapplicable at the elemental level or, more importantly, that there would not be other grounds on which the substantiality of living things or even their teleological character could be grounded. There are weaker sorts of teleology that we could imagine Aristotle legitimately retreating to, upon being faced with, and coming to accept, a more quantitative and structural chemistry. Such teleologies could certainly give no trouble to his concept of the nature of a thing, and would probably sustain his theory of substance. (The thesis 'T1' that Bradie and Miller [1984] describe as 'the core of Aristotle's teleology' [143] is an example of such a weaker teleology. Their mistake, as I see it, is in claiming that this is the thesis which Aristotle actually argues for, and not the stronger 'T2'. They seem to assume [cf. 142 bottom] that Aristotle wanted to leave open the possibility of there being a deeper level of matter to which potentials for form *were* reducible, but there is surely no evidence for this. What is puzzling is how, if that were Aristotle's position, it would still be classified as 'Irreducible Compatibilism', as Bradie and Miller correctly argue earlier it is, rather than (e.g.) the unwanted 'Supererogatory Compatibilism'. Though I must take issue with them here, I am gratified by their endorsement of many of the claims of this essay. [See now ch. 3 below, for further discussion of their interpretation.])]

The issues raised by Professor Broadie require further investigation. Though I stand by my original statement, I happily concede that in her very rich study she has shown the question more complicated and subtle than I had made it out to be.]

[57] [Some of the ideas developed in the Postscript and additional notes, as well as the main themes of this chapter, were presented to several meetings of a graduate course I gave with Professor J. L. Ackrill at Oxford University in Trinity Term 1984, on 'Substance and teleology: some issues in Aristotle's biological works'. I am grateful to all those who participated in discussions on that occasion, both inside and outside of class, and especially to Professor Ackrill and to David Charles, both of whom forced me to sharpen some formulations—and some thoughts. In addition, I have benefited over the years, since the original publication of this essay, from discussions of these ideas with more people than I can name, but I have benefited most, I think, from exchanges with David Balme, James Lennox, and (more recently especially with) David Charles. Warm thanks are due also to Richard Sorabji and Sarah Waterlow Broadie for discussing some of my views in print, especially to Sorabji, who was the first and with whom I had in addition some valuable correspondence.]

[*] Cf. n. 56 above. [This sentence, written in 1986, still holds true today, in 2010.]

(i) *Ontological irreducibility.* The development, structure, and functioning of living organisms cannot be wholly explained by—*because it is not wholly due to*—the simple natures and potentials of the elements which constitute these organisms. No sum of actualizations of what I have called 'element-potentials' is sufficient by itself for the production of those complex living structures and functionings for which Aristotle offers teleological explanation.

(By 'simple' natures and potentials I mean those whose essential specification does not make reference to the form or nature of the living organism as a whole. I do not mean to deny, and am inclined to affirm, that the elements have (irreducible) capacities to be worked up into such wholes, and that such capacities may be seen as part of their natures as the elements they are, on which cf. *Metaph.* *Z*.16 and *Meteor.* IV.12, and Balme 1987c, 284-85 and Kosman 1987, 389.)

(ii) *The dependence of teleological explanation on ontological irreducibility.* If some sum of actualizations of element-potentials *were* by itself sufficient for the production of some outcome, that outcome would *not* be subject to teleological explanation for Aristotle.

(iii) *The way ontological irreducibility generates and legitimates teleological explanation.* This is a two-part claim, involving (a) a thesis about the nature of explanation generally, and (b) a thesis about the form of a teleological explanation:

(a) Explanation, for Aristotle, works fundamentally by identifying the natures and potentials whose actualization actually constitutes, or is producing, the phenomenon being explained.

(b) A teleological explanation explains the presence of some phenomenon by showing it to be necessary for the presence of some complex end which a potential being actualized is irreducibly *for.*

(To be 'teleological', we will agree, an explanation must have some such form as '*A* is present/occurs because *A* is necessary or best for some end *B*'. Typically an 'end' is defined as a good outcome. While it will in fact always be true that an end is something good, this is not, on claim (iiib) here, part of Aristotle's concept or analysis of something's being an end. Rather to be an end is to be what a potential being actualized is actually and irreducibly for. This is a factual or objective matter and does not depend on imported normative notions. [These claims require, and will get, further explanation below, where I also show how (iiib) incorporates the 'best possible' clause.] As above, the potential is 'irreducibly' for some complex end when the production of that end is not due wholly to the actualization of element-potentials.)

Claim (i) is agreed to in one form or another by Waterlow (1982), Lennox (1982), Robinson (1983), Bradie and Miller (1984), Charlton (1985), Balme (1987c) and Cooper (1987), among others. It was disputed by Nussbaum (1978) and Sorabji (1980), who thus also rejected claim (ii). I discuss Nussbaum's views in Gotthelf 1980 and 1981/82, and examine Sorabji's in section II of this Postscript. The challenge for interpretations which jointly deny (i) and (ii) is to show how teleological explanation could have any force as explanation if a reduction to element-potentials were in principle available. For, given Aristotle's realism, there will not be some merely *pragmatic* function that certain language can perform which other language for the very same aspect(s) of a phenomenon does not; and it seems to me that any authentic explanatory function which the teleological level would be thought to have could be performed instead by the corresponding account at the material level, if a reduction were available (cf., e.g., the new part of n. 38 above). These are, of course, only assertions on my part, and are best taken, as I say, as (friendly) challenges to other interpretations. (For the positions of Balme and Cooper [fellow contributors to the Teleology section of Gotthelf and Lennox 1987] in particular, on claim (ii), see Balme 1987c and Cooper 1987.)

My approach to the issue claim (iii) addresses has been to ask why the fact that something, *A*, is necessary for some end *B*, should explain the presence of *A*, and how, thus, *B* would be *responsible* for *A*. I start from claim (iiia)—that explanation, for Aristotle, is essentially by natures (*phuseis*) and potentials (*dunameis*). This seems clear to me both from the *Physics* and *Metaphysics* passages cited in n. 12 above, and from the pervasive use of these concepts in key explanatory passages throughout the scientific treatises, including, especially, *GA* and *PA* I, focused on in this chapter. This being so, we have to ask, for any phenomenon to be explained, which nature(s) and/or potential(s) are being actualized in the course of the phenomenon's generation or functioning. Now, the identity of a nature or potential is given in part by its object or end (i.e., by what it is irreducibly *for*), so we have to ask what that object or end is in each case. If, in accordance with claim (i), the production of a complex organic outcome is *not* the sum of actualizations of element-potentials, then the nature or potential being actualized must be a potential for that outcome (or for some larger end that includes that outcome), in other words: a potential for form. But *that* makes the explanation teleological, because it puts into the explanans an irreducible reference to an outcome for which the explanandum is antecedently necessary.

The use of teleological explanation in cases of complex organic outcomes is thus legitimated by the fact that those outcomes are the result of actualizations of potentials that are in fact for (and thus whose descriptions make inescapable reference to) complex living forms as such. On this picture, the fact that something, *A*, is necessary for some end, *B*, will *explain* the presence of that *A* because (i) a potential for *B* is being realized, and (ii) if a potential for *B* is being realized, anything necessary for *B* is also being realized. (That is part of what it *is* for a potential for *B* to be being realized.) If this account of how teleological explanation is legitimated is to be successful, however, two conditions will have to be satisfied. First, it will have to be true, in some objective way, to say that *B* is somehow *responsible* for *A*. Second, the fact that the potential for *B* is being realized will have to be a basic fact, so far as explanation is concerned. Both conditions, I would like to argue, are in fact satisfied, on the picture I have sketched.

It might be said that the first condition is automatically satisfied if there is an explanation since an *aitia* just is that which explains, for Aristotle, but I have rejected this view in n. 51 above. Though I do not, as I have acknowledged in that note, have a good account of what an *aitia* is for Aristotle generally, it seems clear to me that when an (irreducible) potential for some end is being realized, that end is responsible for the process leading up to it as that which the process is (irreducibly) *for*. That is to say, the end is a sort of intentional object of the process, much as the object of a desire (or perhaps intention) is the object of the action aimed at satisfying that desire. The end has a real status as aim, and was the end of the process even if the process failed to reach that end. Objects that perform such real functions clearly deserve the label '*aitia*', according to Aristotle; in fact, it could be said to be a condition on any successful account of Aristotle's conception of 'being responsible for' that it make clear why it should be appropriate to view such objects (of potentials; of desires or intentions) as actually *responsible* for their effects. (This is not a demand to collapse the final into the efficient cause. It is the potential [or its bearer *qua* bearer of that potential] which is the efficient cause, while the final cause is the object of the potential, what, as we say, it is 'for', and this distinction must be maintained [cf. the new part of n. 52 above].) Note 51 above, with its bracketed addition, contains, I think, some pieces of such a general account of an Aristotelian cause, but I do not yet know how to put them together.

So much for the first condition my account of Aristotelian teleological explanation must satisfy if it is truly to count as explanation for Aristotle. As for the second condition, if *A*'s being necessary for *B* explains *A*'s presence because an irreducible potential for *B* (which brings *A* along, as it were) is being realized, I will need to show that the presence of an irreducible potential for form is basic so far as explanation is concerned. For, otherwise, whatever explains the presence of such a potential will be more basic, and will be in fact what grounds teleological explanation. (For a view that there is something more basic, see, in particular, John Cooper's interpretation, in Cooper 1987.) In arguing for irreducibility (claim (i)), I argued that the realization of *B* cannot be seen as the necessary consequence of the actualization of element-potentials alone, but I need also to show that my claim to the presence of an 'irreducible potential for form' is the most fundamental way to put the ontological consequences of irreducibility. That it is, is in part established, I think, by the argument for viewing natures and potentials as the fundamental explanatory concepts in Aristotle (as summarized two paragraphs above), but in part involves a crucial metaphysical claim—an ascription to Aristotle of a sort of ontological 'individualism'. The claim is, roughly, that the basic fact about the living world is the existence of individual living organisms with the capacity to produce other individuals like themselves. According to this view, other facts at this level of generality which might be thought to ground teleological explanation (e.g., the permanence of species, or the pervasive seeking of the good), are to be seen as *consequences* of the existence of self-sustaining, reproducing individuals. Some such view is defended by David Balme in Balme 1987c and 1987d, and at least suggested by the arguments in Lennox 1985a and 1987b. If such a view can be made out, as I firmly think it can, then the fundamental analysis of teleological explanation will in fact be one in terms of 'irreducible potentials'.

Now, Charles Kahn (1985, 197–98 and n. 16) suggests that such a view gives only part of the picture, because one will need to 'build into the analysis of actualization the required normative component', and will need as part of one's analysis of teleological explanation an independent notion of the good. I have explained in the new portion of n. 18 above why I do not think this is right. I would like here to say just a little more on this issue. (My fullest discussion is now in chapter 2 below, though it does not entirely substitute for what follows.) Supporters of what we might call 'the normative analysis' of Aristotelian teleology suggest that the basic pattern of a teleological explanation is something like:

A is there because it is good that *A* be there.

This is generally said to divide into two alternatives:

(a) *A* is there because it is necessary for the presence of some good, *B*;
(b) *A* is there because it is good (better, best) for some good, *B* (where *A* is not necessary for *B*).

The good, according to this view, thus enters independently into the analysis at two places. I have argued above (n. 18) that the goodness of the end is *not* an independent constituent of the analysis, nor what centrally establishes that end as the end. Rather the process really is towards a fully developed state, specifiable independently of its goodness, and establishable as an end without reference to its goodness. That all such ends are good is a separate fact about the world, itself a consequence of a 'metaethical' analysis of the good that analyzes it in terms of ends (and not ends in terms of it).

This view can be made more plausible by performing a thought experiment, in which one rejects the standard Aristotelian analysis of the good, and asks if teleological explanation will still be applicable. Let us imagine that the biological world is just as it is now, except that successful life was in no way a good thing. Let us suppose (*per impossibile* perhaps) that the existence of complexity was intrinsically evil, and the destruction of living bodies therefore intrinsically good (and *not* because of the will of some non-material being). In such a world, living organisms would seem to develop and function for the sake of outcomes that were in no way good, and yet, we may ask, if the production of these outcomes could not be explained by element-potentials alone, would not Aristotle still wish to offer teleological explanations for such developings and functionings? The 'evil' lung, for instance, forms because irreducible potentials for 'evil' viviparous land animals are being actualized, and such animals require lungs to function.

If the thought-experiment sits hard, it is because, as good Aristotelian ethical theorists, we cannot sustain this imaginary notion of the good. But *that* is because of something about the nature of the good, not something about the nature of ends as such. The 'normative analysis' of ends, then, which is perhaps 'first to us', is not 'first in nature'. 'First in nature' is an analysis in terms of potentials for complex living outcomes specifiable without reference to independent normative notions.

What, then, about pattern (b) above, where one explains *A*'s presence not because it is necessary for *B* but because it is the best of several things, one of which, but no particular one of which, is necessary for *B*? There are two points to make here. First, the operative notion of 'best for' in all cases here is something like 'greatest contribution to the life of', so that we do not have what I have been calling an independent notion of the good. (On this point, see above n. 18, with references.) Secondly, as others have pointed out (e.g., Sorabji 1980, 158-59), this pattern of explanation rests on the proposition, roughly, that 'nature always acts for the best'. I take this principle to state a basic fact about the sort of *ends* that exist in nature (and thus the sort of outcomes that the irreducible potentials are irreducible potentials *for*), viz., that in cases where there are options in the organization of living things, the ends aimed at are the 'best-possible organizations'. Both development and functioning are aimed at the realization and maintenance of an organism of maximum efficiency and capacity within what is possible to the basic nature of the organism in question. (That there is such a 'basic nature' against which such efficiency can be measured I argue in chapter 7 below; see also chapter 8.) In that way, pattern (b) has the same logical form as pattern (a). Explanations of the form '*A* is best for outcome *B*' are to be understood, at a deeper level, as being of the form '*A* is necessary for outcome *best-organized-B*', where 'best' has the fully naturalistic analysis I have indicated. That there are nonetheless *two* patterns of explanation derives, theoretically, from the fact that the second pattern requires an additional premise about nature, over and above the irreducibility premise needed for the first pattern, and practically, from the fact that the scientific work involved in producing explanations according to the different patterns is in some ways noticeably different. (For further support for my approach here, see chapter 2 below.)

Supporters of the 'normative analysis' of the form of Aristotelian teleological explanation point to Andrew Woodfield's analysis of teleology (Woodfield 1976) as one that captures important aspects of Aristotle's own analysis. For the reasons just given, I think rather that, so far as the basic form of analysis is concerned, Charles Taylor's account (cf. above, n. 10) is closer than Woodfield's. (On certain aspects, Larry Wright's [1976] account is even closer: cf. section III of this Postscript, below.)

II

In *Necessity, Cause, and Blame* (1980), Richard Sorabji argues that, according to Aristotle, a full material-level account of teleological processes is in principle available. One of the larger themes of his book is the relation between cause and necessity, and in that connection he explores the claim that the use of the final cause is compatible with the existence of necessitating causes. He thinks it is, and devotes one chapter to showing that, for the most part, Aristotle thought that individual outcomes in nature have necessitating (material) causes, and a second to showing that a proper understanding of how teleological explanations work reveals their compatibility with such causes.

In chapter 9 of *Necessity, Cause, and Blame*, Sorabji sets out very nicely the problem that has engaged scholars: in several theoretical passages Aristotle appears to deny that individual natural outcomes have necessitating causes, while Aristotle insists time and again in the biology that outcomes are both for the sake of something *and* of necessity. Sorabji rightly objects to the view presented in Balme 1972 [1992] that material necessity is a disguised hypothetical necessity, presaging some of the objections developed more fully by John Cooper (1987, 260-62), and concludes that we in fact have conflicting claims here. We can only, he says, attempt to offer an explanation for the conflict, and the one he proposes is that Aristotle has confused weaker claims, primarily that necessitating causes do not explain, with the stronger claim that they do not exist (150-52, 162, 173-74).

This explanation does not seem possible, however. For, in one of the theoretical passages Sorabji cites in which Aristotle denies that individual natural outcomes occur of unhypothetical necessity (*PA* 639b21-640a8), Aristotle appears quite clear on the difference: one reason he gives for there being no explanation of natural outcomes in terms of simple necessity alone is precisely that such necessity does not exist in nature. This is not an unconscious slide but a deliberate move, and it is from non-existence to non-explanation, not the other way.

Himself convinced that individual outcomes in nature have necessitating causes according to Aristotle, Sorabji attempts in his chapter 10 to show that this is fully compatible with their being for the sake of something. (Sorabji typically speaks of organs and processes having a 'purpose', a term I want to avoid for its [to me unavoidable] implication of design or intention.) Making use of the important *PA* 640a33ff., he finds three distinct patterns of teleological explanation, the first two of *parts* of animals (as necessary, given certain basic features of the organism, or as better, given such features), the third of *processes*, including organic development. In each case he seeks to show that the availability of a teleological explanation for an animal feature (part or process) does not preclude there being necessitating causes of that feature.

The first mode of teleological explanation, Sorabji argues, works according to the following generic pattern: Given *x*, it is only to be expected that *y*. He refers to

Aristotle's explanation of the camel's extra stomachs (*PA* 674a28-b15), writing that:

> The camel needs to eat tough and thorny food, and therefore it has an extra series of stomachs for digesting.... The explanation seems to work by taking the existence of the camel, together with certain characteristics, such as the thorny diet as given. Once these things are given, extra stomachs are *only to be expected*. (156, italics in original)

But there is nothing in such an explanation that precludes the existence of necessitating causes (162-63), and even when these causes are fully known 'in many contexts, and for many purposes, we shall always want teleological explanations' (165).

My own view, mentioned in n. 19 above, is that this pattern (and the second, of which Sorabji gives a similar account, adding as given that 'nature does nothing in vain' [158]), works only when, and only because, the generation of that part was itself teleological. 'Coming to be for the sake of', on this view, is logically prior to 'being for the sake of'. My arguments for this perhaps paradoxical-sounding thesis are now presented below, in section III of this Postscript. If successful, they show that Sorabji's first two patterns are not independent of the third. And since this third pattern works, in the case of organic development, only because such development is not the result of material necessity alone (as I have argued above), the first pattern is *not* compatible with the existence of necessitating causes.

Sorabji himself thinks, however, that the third pattern *is* compatible with the existence of necessitating causes. He accepts, with his typical generous acknowledgement, the account of this pattern presented in the present chapter. As he puts it, to say that the development of an organism, for instance, is for the sake of the mature organism that issues from it is to *identify* which of the various possible *dunameis* (of the whole organism, of the materials) is the one at work (171-72). But having said this, he soon qualifies it, in my view fatally. It is not that the material *dunameis* are not operative and necessitating, he says; it is that they are not considered explanatory (174). But that way of putting it is incompatible with his own formulation of my account of how these explanations work. If the outcome is of material necessity, then the material capacities, in sum, are sufficient, when actualized, to produce such an outcome, and in this case *have been actualized*. But this is just to say that *they* are the *dunameis* at work, and that the *dunamis* of the whole organism *is not*. My account of the *dunamis* of the whole allows for *dunameis* of the materials to be at work—anything hypothetically necessitated by the actualization of the *dunamis* for the form of the whole will be so necessitated only *given* the basic natures and potentials of the elements; but it does not allow that they be *sufficient* for the outcome. If they *are* sufficient, then there is simply no place for a *dunamis* of the whole.

Sorabji could modify my view to say that explanation here consists in selecting not which *dunameis* are *at work* but which are *explanatory*. But the result would largely be trivial. In fact, this exposes what I find to be another weakness in Sorabji's overall account of the relation of teleology and necessity. He says much about how outcomes might be the result of material necessity yet not be explained in that way but only teleologically. But he does little to show what features would make the one explanatory and the other not. Surely if an outcome is known to be of material necessity, then given the antecedents, it is 'only to be expected' that the outcome will occur. Why should that be sufficient for explanation in the case of parts, but not here? Sorabji refers several times to the relativity of explanation to the purposes of the questioner, but to many this will seem an un-Aristotelian thesis, and he mentions no passages which require it. (Cf. 11, 29-31, 56, 58-59, 158, listed in his index, s.v. 'explanation: . . . relative to the question asked'; also 165-66.)

What, then, of the reservations Sorabji expressed (172-74) to my own interpretation of these matters? Following David Hamlyn, Sorabji objects first to calling the relevant *dunamis* by itself the explanation, since 'if anything explains, it is rather the whole *complex* of facts into which the [relevant] *dunamis* fits' (172). But I never denied this or implied the opposite. As Sorabji observes there, Aristotle himself is happy to call the relevant *dunamis* the efficient cause; the mistake would be in equating *cause* with *explanation*, which Sorabji does here probably because of the arguments to that effect which he has given earlier (40-42 and chapter 2 generally). The full explanation (*logos* or *apodeixis*, not *aitia*) would certainly involve more than just the efficient cause. As I have argued elsewhere, Aristotle views the various causes here as factors in one explanation, not as distinct explanations (cf. Gotthelf 1980, 373-74).

Sorabji then (173) objects to my thesis just mentioned that teleological explanation of parts is derivative from teleological explanation of development, and offers a counterargument. I deal with this in section III below.

Finally, he objects to my treating the absence of a full material-level account as central to Aristotelian teleology, on grounds of his earlier argument that necessitating material causes *are* present even if not explanatory. Even this latter claim must be tempered, he observes, since 'Indeed, Aristotle uses a tattoo of words implying that the necessitating causes which he cites for biological arrangements really do explain them' (174, with nn. 62, 63). In the bracketed addition to n. 32 above, I responded to Sorabji's earlier claim that the pervasive language of 'both for the sake of . . . and of necessity' implies the existence of material-level causes sufficient to necessitate biological arrangements, referring to discussions in Cooper 1987 and Balme 1987c. Sorabji himself provides, several chapters back (1980, 51, n. 24), the basis on which the explanatory language he points to here can be understood:

> The denials [of explanatory value to material causes] are themselves sometimes nuanced: e.g., at *GA* v 8, 789b20, the word 'through' (*dia*) seems to allow physiological causes explanatory value in producing biological parts in those cases where the parts serve no

purpose; and even when the causes produce useful parts, they can be called explanations of the material kind (*hôs hulê aitia*, 789b8). Cf. *Phys.* II.9, 200a6, a9-10: things do not occur 'through' (*dia*) material causes, 'except in the sense of occurring through matter' (*plên hôs di' hulên*) . . .

To be explanatory as matter is to be explanatory as that which, given its elemental natures and potentials, is necessary *if* some end not itself necessary is to be achieved—i.e., if some potential for form, not reducible to element-potentials, is to be actualized. This is a hypothetical, not a simple, necessity, and does nothing to imply that material factors explain (or even necessitate) *by themselves.*

III

Hence we should if possible say that because this is what it is to be a man, therefore he has these things; for he cannot be without these parts. Failing that, we should get as near as possible to it: we should either say altogether that it cannot be otherwise, or that it is at least good thus. And these follow. And because he is such a thing, his coming-to-be necessarily happens so and is such. And that is why this part comes to be first, and then this. And this is the way we should speak of everything that is composed naturally. (*PA* I.1 640a33-b4, tr. Balme)

In this well-known passage Aristotle describes the various patterns of explanation of animals' parts which a truly educated man should expect to find in a properly developed natural philosophy (cf. 639a1-15). The basic distinction is between (three types of) explanation of the being of parts—their being present in or possessed by certain animal kinds—and explanation of their coming to be, i.e., of the process by which they come to be present in or possessed by these animal kinds. Such explanations are teleological, establishing, respectively, that the part *is* or *came to be* for the sake of something.

Elsewhere, Aristotle speaks as if he recognizes that distinction. He concludes his first argument in support of the use of teleological explanation in *Physics* II.8, for instance, with the words: 'The "for the sake of something", then, is present in things which by nature come to be and are' (199a7-8). In *GA* V.1 he excludes non-universal *pathē* from such explanation by writing that:

Whatever features do not belong to the nature <of an animal generally> nor are peculiar (*idia*) to any particular <animal> kind—of features of this sort none either are or come to be for the sake of anything. (778a30-2)

Of those things which do either belong to the nature of an animal generally or are peculiar to some animal kind, Aristotle says a few lines later that:

Each one, then, is for the sake of something, and yet comes to be both on account of this cause and on account of the others [*sc.* material and moving causes: b9] . . . (b10-12; for more on the opening methodological passage of *GA* V.1, see now chapter 5 below)

And the division of labor between *PA* and *GA* pretty much mirrors this distinction, *PA* explaining the presence of parts teleologically, *GA* explaining their formation.**

In n. 19 above I suggested that *being for the sake of* was thought by Aristotle to be definable in terms of *coming to be for the sake of*, and I have repeated this claim in section II of this Postscript. I want here to give my reasons. The idea has struck some people as peculiar, on grounds that Richard Sorabji well expresses, referring to my suggestion:

> This seems to me to *reverse* Aristotle's order of priorities. In the passage with which this chapter starts (*PA* 1 1, 640a10-b4), reference to a *dunamis* occurs only in explaining processes, and the passage in question makes the explanation of processes *subordinate* to the explanation of why man has the parts he has. (Sorabji 1980, 173)

Now, in the passage quoted above Aristotle says explicitly that it is 'because he is such a thing [that] his coming to be necessarily happens so and is such', so in one sense Sorabji is surely right. But that is only part of the story, and a part, I shall argue, which already *presupposes* that coming to be is for the sake of something.

The thesis I wish to defend is that a part, *A*, *is* for the sake of doing something, *B*, if, and only if, *A* in fact does *B*, *and* has *come to be* for the sake of doing *B*. Since coming to be for the sake of is explicated on my account without reference to being for the sake of (above, p. 12), one can say, as I have, that the latter is defined in terms of the former.

Perhaps it should be stressed that this account is meant to hold only in those cases where giving what *A* is for the sake of is meant to give an *aitia*, and thus an

** This is not to say that *PA* is unconcerned with generation. *PA* is concerned with why animals have the parts they do, and animals have the parts they do *partly* because some such parts are needed (or are best) for the functioning of the whole and *partly* because the available materials permit those parts: *tēi men heneka tou . . . tēi de ex anangkēs*. *PA* will thus have things to say about the 'necessary natures' of the available materials, and the constraints these natures place on what parts could be formed to fulfill a needed function, but it will not focus on the generative process as such, and its timing and stages, and how these are necessitated by the nature of the result to be produced: those are the focus of *GA*, and *GA*'s initial focus is not so much the final cause of generation but its efficient cause: what antecedent movements (as it were) produce offspring one in form with their parents and how they do it. (Cf. *GA* I.1 715a11-18, II.1 733b23-32, etc.) Since the movements Aristotle identifies are those borne by the heat transmitted from the father, and involve, in addition to low-level material *dunameis*, a *dunamis* for form, teleological explanation will be inevitable, the end being the object of the *dunamis* that is the efficient cause, what, as we say, that *dunamis* is 'for'. And teleological explanation does appear in *GA*, both for generation as such and its general features (II.1 731b18-732a11, 6 742a16-b17, 743a36-b3), and for particular formations, timings, and sequences (743b3-18, 744a35-b9, etc.). But it is introduced selectively and is not the primary focus of the work. Still, insofar as they are concerned with for-the-sake-of relationships, *PA* is primarily concerned with *being* for the sake of, *GA* with *coming to be* for the sake of. (For more on *PA* as just described, see Lennox 1997b; for more on *GA*, chapter 4 below.)

explanation. When Aristotle says early in *Phys.* II.5 that 'among things which are neither necessary nor for the most part there are some to which it can belong to be for the sake of something' (196b19-21, after Charlton), and when he says of the tripod that fell open accidentally so as to be ready for sitting on that 'it was for the sake of sitting, but it did not fall for the sake of sitting' (196b17-18), one might wish to interpret him to be using a purely *descriptive* sense of 'is for the sake of' which merely labels some doing or capacity of the thing as useful in some way, without implying that that is why it is there. I think myself that these passages are to be read differently (cf. Gotthelf 1975, 62-72, 285, n. 119), but in any case I want to make it clear that it is only explanatory uses of 'for the sake of' that are at issue here.

In fact, it is the chance cases that first bring out the plausibility of this thesis. One needs to ask whether, in cases where a part did something which contributed to life or well-being but had not come to be for the sake of doing so, Aristotle would be willing to say that the part *is for the sake of* that contribution, and thereby *explains* its presence in the organism. It seems not. To see this we might consider Empedocles' human-headed bull (n. 54 above). If, for instance, the neck muscles of the torso of the animal were strong enough (and the only muscles strong enough) to support the head that on the theory it accidentally joined up with, would Aristotle say that those muscles *are* in the animal *for the sake of* supporting the head? If one puts oneself in Aristotle's place, one is likely to say 'No'.

Aristotle himself seems to say not. In the well-known first argument of *Phys.* II.8 (above sec. XI), he generalizes from the case of teeth which arise by chance so as to be serviceable in some way, to all the parts in which 'the "for the sake of something" seems to be present' (198b27-29) but on this view would not. Now Aristotle's focus is on coming to be, and this sentence might well be speaking only of coming to be for the sake of. But it is perhaps more natural to read it as speaking of parts which appear to *be* for the sake of something, and in that case Aristotle (and his opponents, as he characterizes them) would seem to be assuming that, on the view under consideration, these parts cannot be said to *be* for the sake of something, precisely because they have not come to be for the sake of that but rather by chance.

Aristotle's use here (198b28) of *huparchein* ('to be present') suggests this, as does the way Aristotle states the conclusion of his refutation of the opponents' view, at 199a7-8, the one place here where he explicitly mentions *being* for the sake of something:

> The 'for the sake of something', then, is present in things which by nature come to be and are.

That is to say, the argument presented here establishes, and is apparently *needed* to establish, that things in nature *are* for the sake of something as well as come to be for the sake of something. Without the part's having *come* to be for the sake of something it simply isn't a candidate for *being* for the sake of anything.

But, *PA* II–IV is full of arguments that conclude to some part's being for the sake of some function. Do these arguments refer to their parts having come to be for the sake of anything? On the face of it, no. Aristotle either takes as common knowledge, or argues to, the contribution a part makes to life or well-being, and from that concludes directly that the part is for the sake of making that contribution. He speaks, as noted, of the materially necessary factors in the part's coming to be but not of its having come to be for the sake of anything.

However, *PA* II–IV is a study of animals, not of explanation. The theory of explanation which is at work in those books is itself developed, among other places, in *PA* I.1, and in that chapter, I would like to argue, Aristotle makes the dependence we have been discussing clear.

Much of I.1 is devoted to the final cause, in one way or another, and there is much argument in its behalf. It is striking that all of this argument, and most of this discussion, is devoted to the causes of *coming to be* and the notion of *coming to be* for the sake of. Richard Sorabji once suggested in correspondence that this can be explained by the preoccupation of Aristotle's opponents with coming to be and his concern to respond to them. But I think there is a deeper explanation, captured in the priority thesis I have been arguing for here. For, in the passage quoted at the head of this section, in which Aristotle presents the proper method for explaining parts, observe the 'hence' (*dio*). That this is the way to explain parts follows from something—from what?

As Balme notes, '"Hence" resumes the argument from a26' (1972 [1992], 87 *ad* 640a33; on a27-33 as parenthetical, cf. 86 *ad loc.*). The argument begins at a15 in defense of an answer Aristotle gives to the question (a10-12) whether one should describe first an animal's coming to be, explaining the character of its being by reference to that, or the reverse. Aristotle's answer is, of course, the latter.

> For in house-building too it is more the case that *these* things take place because the form of the house is *such*, than that the house is such because it comes to be in this way. For coming-to-be is for the sake of being, not being for the sake of coming-to-be. Hence Empedocles was wrong in saying that many attributes belong to animals because it happened so in their coming-to-be, for instance that their backbone is such because it happened to get broken by bending. He failed to recognize, first, that the seed previously constituted must already possess this sort of capability (*toiautēn dunamin*), and secondly that its producer was prior not only in definition but in time; for it is the man that generates a man, and therefore it is because *that* man is such that *this* man's coming-to-be happens so. (a15-26, tr. Balme)
>
> *Hence* we should if possible say that because this is what it is to be a man, therefore he has these things; . . . (a33ff., my emphasis)

The appropriateness of teleological explanation of the presence of parts by reference to the animal's form or essence clearly derives from the appropriateness of teleological explanation of their coming to be. If parts came to be of material necessity (and thus animals were as they were because of the way they came to be—the force of the rejected alternative, 'being for the sake of coming to be'), teleological explanation of the presence of these parts by reference to their contribution to the being of the animals of which they are a part would not be warranted. Given that they *come to be* for the sake of such a contribution, their *being* can be *explained* by reference to that contribution.

Two questions, perhaps, arise at this point. First, what difference is there, then, between teleological explanation of the being or presence of a part and teleological explanation of its coming to be? And second, if the relationship is as I say, why does Aristotle speak so insistently in this context of the priority of being over coming to be?

The answer to the first question has been covered in my note above (p. 39) on the difference of focus between *PA* and *GA*. A teleological explanation of the being of a part will make reference only to its current contribution to the being of the animal of which it is a part, and not to the stages by which that part is formed. As noted there, such an explanation may make reference to constraints on the character of the part that derive from the available matter, but you will not attempt to explain why, for instance, 'this part comes to be first, and then this' (b2–3). Teleological explanation of coming to be will do that, and as noted, there are numerous such explanations in *GA*.

As for the second question, it is important to note that Aristotle's remarks about the priority of being over coming to be all have to do with the proper explanation of the *coming to be* of animals and their parts. The explanation of the parts' *being* is not involved at all. (*That* is what comes in the 'hence' passage.) To explain coming to be teleologically is to explain it by reference to (the needs of) the formed organism, and thus by reference to the organism's being. Since 'coming to be *is* for the sake of being, not being for the sake of coming to be', such teleological explanation is proper, and grasping the animal's being will be prior to engaging in such explanation.

Now, of course, one can show that a certain sort of animal must have a certain part, if it is to exist at all, or to exist well, without having shown that the coming to be of the animal is for the sake of something. And particular teleological explanations of coming to be will, as Aristotle says in the passage with which we began, build on such analyses. But it is important to see that such an analysis will not be a *teleological* explanation (or any other sort of *explanation*) of the part's presence, *if* the part has not come to be for the sake of the contribution the analysis shows it to make.

We can perhaps best see this by returning to Sorabji's account of these matters, quoted just above, and in particular to his rendering of the working of Aristotle's explanation of the additional stomachs in the camel (quoted in section II above, p. 36). 'The explanation seems to work by taking the existence of the camel, together with certain characteristics, such as the thorny diet, as given. Once these things are given, extra stomachs are *only to be expected*' (Sorabji 1980: 156; emphasis in original). The last sentence by itself is true enough. But why should it count as an *explanation*? If stalactites always produced stalagmites, and stalactites if they existed at all always existed, then given the existence of stalactites, it would be only be expected that there would be stalagmites. But what does the explaining is not the fact of universal correlation (compare Sorabji's barometer at 1980, 155), but our understanding of the *process* by which the one produces the other. Similarly with Aristotle's animals, it is the *process* by which the part comes to be that explains its presence; but *for him that process is irreducibly for an end*. The identity of the process is given by the capacity of which it is the actualization, and that capacity is an irreducible capacity for the production of an entity of the relevant kind, thus necessitating the relevant parts. Given that the process *is* irreducibly for the end, one is entitled to see the analysis of what parts are needed by that end as an *explanation* of the presence of those parts, and of course as a *teleological* explanation. Being for the sake of is thus to be defined in terms of coming to be for the sake of.

(The argument at 640a10–26 seems partly the basis for, and partly summarized in, the brief, difficult discussion of method at 645b14–28 in chapter 5. It is at an important stage in that later discussion that Aristotle makes what may be the most explicit statement of the thesis I am here ascribing to him. In Balme's translation:

> Consequently the body too is in a way for the sake of the soul, and the parts are for the sake of the functions in relation to which each has naturally grown (*pros ha pephuken hekaston*). (b19–20)

However, I cannot make sense of the argument of which this is an intermediate conclusion, and thus am not entirely sure how to take this statement.)

(Spontaneously generated organisms might appear to provide a counterexample to the thesis I have been arguing for in this section: their coming to be would seem to be of material necessity and not for the sake of anything, yet the presence of at least some of their parts seems to get teleological explanation in *PA* IV. I discuss this matter in chapter 6 below.)

There is an interesting parallel between this analysis of 'being for the sake of' and a popular (and generally very plausible) contemporary analysis of the notion of the function of something. (Aristotle's notion of what something is for the sake of is pretty much equivalent to our notion of the function of something; 'is for the sake of' does more theoretical work in the biology than *ergon* does, though the latter is certainly present.) According to Wright's definition, 'The function of *X* is *Z iff*: (i) *Z* is a consequence (result) of *X*'s being there, and (ii) *X* is there because it does (results in) *Z*.' (Wright 1976, 81; cf. his 1973, 161.) Clause (ii) captures the distinction, important for the notion of a thing's function, between the function and anything the thing might do by accident. As we have seen, a similar distinction is important to Aristotle as well, and for him too, the function of *X* is *Z* only if '*X* is there because it does (results in) *Z*'. (Cf. chapter 7 below, p. 168 n. 42.)

While Wright elucidates the 'there because' in that condition in terms of natural selection, Aristotle, with his non-evolutionary perspective, can only speak of what 'the nature does' in embryonic development. He elucidates the 'there because' in the definition of function, as I have been arguing, in terms of another (and thus prior) teleological notion, that of 'coming to be for the sake of': the function of *X* is *Z* for Aristotle *iff* (i) *Z* is a consequence (result) of *X*'s being there, and (ii) *X* came to be for the sake of doing (making possible, etc.) *Z*. Identifying the contribution to self-maintenance an organ makes thus identifies its function (and explains its presence) *only* on the condition that the organ came to be for the sake of making that contribution.

Since only individuals can come to be, the thesis defended in this section makes the individual prior to the species for purposes of explanation.[***] This appears to be in harmony with David Balme's analysis of the general relation of individual to species, in his important 1987d (cf. above, sec. I of this Postscript, pp. 32-33), and perhaps provides additional support for that analysis.

[***] I owe this characterization of my position to David Furley (in comments he provided on a portion of my dissertation, in the mid-1970s).

2

The Place of the Good in Aristotle's Natural Teleology

In chapter 1 (and elsewhere) I have offered an interpretation of Aristotle's conception of final causality in terms of his conception of an 'irreducible potential for form'. I have argued that final causality is operative in nature, and teleological explanation thus appropriate, only when there is being actualized a potential for a complex organic outcome which is not ontologically reducible to a sum of actualizations of potentials of the organism's elemental constituents. At no place in this analysis do I refer to the goodness of that complex organic outcome.[1]

Some recent writers have suggested that the absence of any reference to goodness in the analysis of Aristotelian ends is a mistake. Thus, one of the most influential recent discussions of teleological explanation, John Cooper's 1982 Owen Festschrift paper, 'Aristotle on Natural Teleology', begins as follows:

> Aristotle believed that many (not, of course, all) natural events and facts need to be explained by reference to natural goals. He understands by a goal (*hou heneka*), whether natural or not, something good (from some point of view) that something else causes or makes possible, where this other thing exists or happens (at least in part) because of that good.

In a note at this point, after citing some passages, Cooper states:

> That the concept of a goal is the concept of something good is a view Aristotle inherited from Plato's *Phaedo* (cf., e.g., 97c6-d3, e1-4, 98a6-b3, 99a7-c7); unless one bears the connection between goal and good clearly in mind one will fail to understand much that Aristotle says about natural teleology, and many applications he makes of it . . . Andrew Woodfield (1976, pp. 205-206) correctly notes that according to Aristotle all teleological

This chapter is based on the public lecture I delivered as part of my visit to the Boston Area Colloquium in Ancient Philosophy in 1988 (see n. 8 below). It was published in the fourth volume of their proceedings, in 1989. The chapter develops a dimension of my interpretation of Aristotle's teleology (and his theory of the good) which, though present from the beginning, was first expressed in print in chapter 1's 'Postscript 1986', sec. I.

[1] By the 'goodness' of S here and throughout I mean merely S's being good (in whatever way, and for whatever subject, if any, it is). I do not mean to restrict the term to moral virtue.

explanations are claims that something happened *because it is good*, and makes this theme central to his own unifying account of teleological description.[2]

In a paper published in the 1985 Balme Festschrift, Charles Kahn makes a similar claim, and he too cites Woodfield:

The following schema is borrowed from Andrew Woodfield, whose analysis seems to me the most insightful. (The formulation here is limited to the simpler case of natural teleology, as distinct from the intentional finality of art and choice). *A is (or occurs) for the sake of B* may be analyzed as:

(i) B is good (for the relevant subject)
(ii) A contributes to, or is necessary for, B
(iii) Therefore, A occurs (the relevant subject has or does A).[3]

As these commentators have noted, Aristotle himself frequently conjoins the terms *telos* ('end') and *hou heneka* ('that for the sake of which') with *agathon* ('good').[4] And there are several passages where he appears to suggest that reference to the good is fundamental to our understanding of the end or final cause. For instance, in *Metaph. A*.7 he characterizes his predecessors' limited grasp of the final cause as follows:

That for the sake of which actions and changes and movements take place, they assert to be a cause in a way, but not in this way, i.e. not in the way in which it is its *nature* to be a cause. For those who speak of reason or friendship class these causes as goods; they do not speak, however, as if anything that exists either existed or came into being for the sake of these, but as if movements started from these. . . . Therefore it turns out that in a sense they both say and do not say the good is a cause; for they do not call it a cause *qua* good but only incidentally.[5]

The way in which it is the nature of that which is a *hou heneka* to be a cause is as a *good* which is aimed at, and not as an efficient cause which happens to be good. On one reading of this passage, at least, a final cause is essentially something good.

The frequently cited *Physics* II.2 passage in which Aristotle explains why death is not the goal of life even reads as if an antecedent conception of goodness serves as a criterion for determining which state is the end or goal of a process:

Now, the nature is the end and that for the sake of which. For whenever to a continuous change there is a definite end, *this* is last and that for the sake of which. (And so even the poet could write, laughingly, 'He has reached the end for the sake of which he was born'—for not everything last can be an end, but only the best.)[6]

[2] Cooper 1982, 197 (1987, 245).
[3] Kahn, 1985, 197.
[4] Or *beltion* ('better') or *eu* ('well'), etc. Cooper lists several of these passages in 1982, 197 n. 1 (1987, 245 n. 4).
[5] 988b6-15; tr. Ross.
[6] 194a28-33.

Finally, as part of the argument in *De Somno* which establishes what the final cause of sleep is, Aristotle even writes:

The waking state is an end, for perceiving and thinking are an end for all those who have them—for these are best and the end is something best.[7]

Now, I of course agree that for Aristotle a natural goal or end is always something good and is a (final) cause in virtue of that about it which makes it good, but I do not believe that the fundamental account of what it is for something to be an end for Aristotle must—or indeed should—refer to the goodness of that end. The account of what it is to be an end, and the criterion for determining what stage of some end-directed process is its end, can be given—ultimately *must* be given—without reference to the goodness of the end in question. The concept of an end, on my view, is rather to be understood by reference to Aristotle's conception of an actuality (and of the potentiality for certain actualities). Because of his analysis of the good, the very same states (or activities) which are ends will also be goods, but they will not be ends *in virtue of* being goods.[8]

In his paper just cited, however, Charles Kahn suggests that such a view is unsatisfactory because one will need in any case to 'build into the analysis of actualization the required normative component':

The schema adapted from Woodfield has the advantage of making explicit the normative or evaluative status of the *telos*, which Aristotle will usually describe as 'the good', 'the best', or 'the divine' (*to beltion, to ariston, to kalon, to theion*, and so on). It is true that Aristotle also speaks of actuality (*energeia, entelecheia*) as the goal in teleological contexts, and we can give a partial explication of his notion of *telos* in terms of the degrees of actuality, as the more or most complete realization of a thing's potentialities.[9] But this strategy can give an adequate account of Aristotle's conception only if we build into the analysis of actualization the required normative component. For potentialities are typically two-way possibilities, and a living organism is always potentially a dead body. As Aristotle remarks, it would be ridiculous to describe this final stage in the life cycle as 'that for the sake of which' the thing was born.[10]

[7] 455b22-5.

[8] At this point in my Boston Colloquium lecture I presented an overview of my interpretation of Aristotle's natural teleology, drawing heavily on section I of the Postscript 1986 of what is now chapter 1 above. Much of Dory Scaltsas's commentary (Scaltsas 1989) concerned this now omitted portion, and Prof. Scaltsas kindly agreed to adjust his references to Gotthelf 1987b (the relevant pages of which are 230-34 [chapter 1 above, pp. 29-34]). Later in the present chapter I introduce those aspects of his discussion which are most directly relevant to my rejection of the normative analysis of Aristotelian ends, and I comment on aspects of that discussion both in n. 41 below and in an appendix.

[9] Here Kahn refers in a footnote to the 1976/77 version of what is now chapter 1 above.

[10] *Phys*. II.2 194a30-3. The quoted passage is from Kahn 1985, 197-98.

The truth of Kahn's conclusion aside, however, his argument is not right. In *Metaph.* Θ.4 Aristotle states that it is only rational potentialities that are 'two-way possibilities' ('for contraries', he says); non-rational potentialities are each of *one* state (*mia henos*: 1046b4–6). And in *Metaph.* H.5 he specifically denies that 'a living organism is always potentially a dead body':

> It's a particular aporia too why the wine is neither matter of the vinegar nor potentially vinegar (after all, vinegar does come-to-be 'out of' it), nor the living being potentially a dead body. No: rather [these] corruptions are per accidens, it's the *matter* of the animal that is itself by way of *its* corruption a potentiality or matter for a dead body, likewise the *water* [not the wine, that's matter] for vinegar; for they come-to-be 'out of' these [vinegar from wine, corpse from animal] in the fashion that night [comes-to-be] 'out of' day.[11]

The potential being actualized in the development of a living thing is for the mature state (*akmē*), the full development of the organism's soul or form. If a normative analysis of ends is required, that can only be because a normative analysis of a potential for form is required, and thus a normative analysis of form itself: on this view, that state will count as the form (and thus as the object of the *dunamis* that grounds teleological explanation) which is by independent standards the best stage.[12] We need to show, by contrast, that one can both identify that state (viz. the full development of the organism's soul or form), *and* establish its status as object of the potential being actualized in organic development, without use of any conception of the good.

I want first to establish the plausibility of the enterprise of seeking a non-normative analysis of ends via a non-normative analysis of objects of higher-level *dunameis* in Aristotle by exploring some problems in the normative analysis itself.

These problems are all variations on one theme, viz., that ends cannot be analyzed in terms of goodness because goodness must be analyzed in terms of ends and, ultimately, in terms of a notion of actuality which is not itself analyzed in terms of goodness.

I would like to show, specifically, that for Aristotle the goodness of something, at least in biological contexts, is regularly its capacity to contribute to the continued life (*to zēn*) of the organism which has or performs or undergoes that something, where the notion of what it is to live does not itself rest on a prior notion of the good. This needs to be done in stages. First I argue that there is a standard use in the

[11] 1044b34–1045a3; tr. Furth 1985. Cf. *An.* I.4 408b18–24: 'The intellect seems to be born in us as a kind of substance and not to be destroyed. For it would be destroyed if at all by the feebleness of old age, while as things are what happens is similar to what happens in the case of the sense-organs. For if an old man acquired an eye of a certain kind, he would see as well as even a young man. *Hence old age is not due to the soul's being affected in a certain way, but to this happening to that which the soul is in, as is the case in drunkenness and disease*' (tr. Hamlyn, emphasis added).

[12] Some such view of form is presented in Preus 1979. By 'independent' I mean here and elsewhere 'not itself defined or analyzed in terms of the notion of an end'.

biology of good/better/best in which they have this sense. Then I consider three concepts, which are sometimes used by Aristotle to designate biological ends: (i) living *well* (*eu zēn*); (ii) being valuable (*timion*); and (iii) being divine (*theion*). These concepts are often thought to extend beyond the notion of living by incorporating a prior and distinct notion of the good. I attempt to show that these concepts are in fact constructed either out of this notion of living or out of the notion of actuality under which Aristotle in any case subsumes living, which notions are not defined in terms of a prior notion of the good. I suggest that Aristotle's notion of goodness is rather itself defined in terms of the notions of actuality and end (and that Aristotle's theory here is not circular).

I then turn to my claim that a non-normative account can be given of the object of an irreducible *dunamis* for form, and thus of the final cause in Aristotle, and I outline such an account. I close with some remarks about the evidence earlier cited for dependence of end on good, and about what in general is the place of references to the good in Aristotle's discussions of natural teleology.

Prima facie case against the normative analysis of ends: *Nicomachean Ethics* I

The first problem with the claim that Aristotle would define an end in terms of goodness is that he surely seems to do the reverse in *NE* I. The first sentence of *NE* makes it clear that the concept of the good finds its conceptual roots in the fact that humans, at least, strive for ends. Among the things that are first to us about goodness, from which we may build our analysis of its nature, is the fact that it is a concept applicable to objects *qua* ends of action:

> Every art and every inquiry, and similarly every action and choice, is thought to aim (*ephiesthai*) at some good; and for this reason the good has rightly been declared to be that at which all things aim (*hou pant' ephietai*).[13]

The logic of Aristotle's argument across *NE* I is notoriously difficult to unravel, but one very plausible reading has it that the argument's aim is to determine the nature of the human good by way of determining the nature of the natural human end. Thus, the notorious conditional at the start of chapter 2 begins and ends: 'If, then, there is some end (*telos*) of the things we do, . . . clearly this must be the good and the chief good (*tagathon kai to ariston*).' And one criterion for the 'chief good' in chapter 7 is that it be a 'complete end' (a *telos* which is *teleion*: 1097a28). Of course, this notion of completeness is itself explained in terms of the difficult notions of the object-of-choice and the object-of-pursuit (*to haireton, to diōkton*), and one might imagine that these are to be explained in terms of the apparent good. But the

[13] 1094a1-3; tr. Ross/Urmson.

discussion of the parallel notion of the object-of-wish in III.4 suggests that such an approach would be wrong: in addition to the apparent object-of-wish there is a *natural* object-of-wish (cf. 1113a21), as there are *natural* goods (cf. *EE* 1248b26ff.[14]). These natural objects-of-wish and natural goods will appear good to the good man (*spoudaios*), but their natural goodness is not *constituted* by their so appearing: as Aristotle says, the *spoudaios* or *phronimos* is the measure (for those who need a measure)[15] *because* he sees the truth.[16]

Finally, the famous function argument makes clear the dependence of goodness on natural ends: 'in general, for all things that have a function, the good and the "well" is thought to reside in the function', and the function is always a certain kind of life. Of course, the good for S is to perform S's function *well*, and we will need to see if this adverb imports a distinct standard not constructible from the notion of the function itself. But, *prima facie* one might think it does not, since the notion of function already carries a reference to an immediate end, and thus would seem to carry an internal standard of success or failure. To perform a function is to achieve the end. To perform it *well* would be to achieve that end fully; to perform it less than well, to fall short of the end in some way. For instance, the function of *NE* I.7's lyre-player is not merely to make some plucking motions, but to produce music that itself serves some natural function, e.g. enjoyment or honoring a god. A lyre-player plays well if the music he produces is fully satisfying to man or god; less than well if he falls short of that end. The esthetic criterion that the good lyre-player meets is already contained in the function of the lyre-player as such.[17]

'The good' and 'the better' in the biology

The biological works will support this thesis of the dependence of good on end, so let us turn to them. We will see that the goodness of a state or process of an organism lies in its contribution to the life—to the preservation and realization of

[14] This passage, together with 1228b18ff., suggests the identity of the class of things good *phusei* with the more frequently mentioned class of things good *haplōs* (for references, cf. Cooper 1975, 127 n. 37).

[15] 1095b9ff.

[16] 1113a29-33; cf. 1138b18-34, 1139a26-31, 1140b4-6, 20-1, etc., and generally Cooper 1975, ch. 2.

[17] Something like this must be the meaning of Aristotle's remark at 1098a8-10: 'a so-and-so and a good so-and-so have a function which is the same in kind', though this is an unorthodox use of 'same in kind'. A few lines later Aristotle (or some scholiast) explains 'well' in terms of 'excellence'—'any action is well performed when it is performed in accordance with the appropriate excellence' (a14-15)—but in II.6 excellence is itself explained in terms of 'well': 'every excellence both completes (*apotelei*) the thing of which it is the excellence, making it [be] itself well, and makes (*apodidōsin*) the function of that thing be done well' (1106a15-17; I have modified the first part of the Ross/Urmson translation). See also *Juv.* 469a28-31, quoted below, pp. 52-53.

the being—of that organism, where the concept of life is not itself defined in terms of a prior concept of good. The goodness of the life itself will derive rather from the life's being an (approximation to full) actuality.

We may start from an endorsement of at least part of this thesis by David Balme, in his 1987 'Teleology and Necessity'.[18] Comparing Aristotle with Plato on 'the analysis of causes into two kinds, those acting for the sake of an end and those acting of necessity', Balme writes:

> Aristotle too expresses the distinction between the causes as between necessity and 'the better' or 'the good,' although he makes it clearer than Plato does that 'good' is not an extrinsic value-judgment but means the useful or advantageous from the animal's viewpoint.[19]

And in a note he writes 'The useful:' and cites a series of passages in *PA* and *Resp.*[20] Each of these passages explains the presence of a part by reference to its usefulness (*chrēsimon, chrēsimōtaton*) to the organism, which usefulness consists of its *permitting* or *facilitating* some activity which is necessary for its existence as the sort of thing it is. For example:

> In birds the so-called 'beak' is the mouth; for they have this instead of lips and teeth. This differs in them according to the uses (*chrēseis*) and protective functions (*boiētheiai*) <it has>. For in the case of the so-called crook-taloned birds all have a crooked beak because (*dia*) they are flesh-eaters and do not eat vegetable food; for such <a beak> is useful for (*chrēsimon pros*) mastery <of their prey> and is more able to exert force (*biastikōteron*) . . . , while in the case of the other <birds>, the beak is for each useful in relation to its mode of life (*bios*), e.g. in woodpeckers it is strong and hard . . .[21]

The crooked beak permits mastery of prey and facilitates the exertion of force. Now, the term 'good' (or 'better' or 'best') is not used in any of the passages that Balme cites, but I take it his point is that these passages could equally well have used 'good' (or 'better'), as others do, and that this suggests that (nuances apart) 'useful' and 'good' have here essentially the same meaning.

Balme is certainly right that being useful is a way of being good for Aristotle, and that cases of usefulness are subsumed under such generalizations as that certain things happen because it is better (*dia beltion, heneka beltiou*). Aristotle sometimes connects these terms, and occurrences of one seem replaceable with the other. For instance, in *PA* IV.12, speaking of the web-footedness of some birds, Aristotle says:

[18] Balme 1987c.

[19] Ibid. 277.

[20] '*PA* II. 654a19; III. 662a33, b3, 7; IV. 677a16, 678a4-16, 683b37, 684a3, 685a28, 687b29, 691b1; *Resp.* 476a12' (Balme 1987c, 277 n. 5). The rest of the note, on 'the valuable' (*timion*) will be discussed below.

[21] *PA* 662a34-b7.

So these <feet> result *of necessity* on account of these causes; but they are 'on account of the better' in that <these birds> have such feet for the sake of their mode of life (*bios*), for they live in the water, and while their wings are useless (*achreioi*), these feet are useful (*chrēsimoi*) for swimming. (694b5-9)

And in *IA* 8 Aristotle cites his familiar generalization that 'A nature makes nothing without a point but always for the best for each thing from among the things that are possible to it, preserving the particular being and essence of each', and then uses it to explain the footlessness of snakes: given their particular being and essence, feet would be useless (*anōphelē* here; 708a9-20). This same generalization explains why only man possesses hands, given his intelligence, in the well-known passage on hands in *PA* IV.10, since only man can make of them the multiple uses they permit (*echrēsato, chrēsimon*: 687a15-23). Finally, as examples of occurrences of 'better' which could easily be replaced by 'useful', here are three from *PA:*

The differences that there are among <different specimens of> each of these <uniform parts> are for the sake of the better.... The thicker and hotter blood is more conducive to sensation and thought. The analogue to blood has difference in the same way, and so bees and other such animals are of a more intelligent nature than many of the blooded <animals>...[22]

One must admit that it is the same with the other [uniform] parts and with the non-uniform: difference is present—some for (*pros*) the functions and in general the being of each of the animals, others for 'the better or worse', e.g. the eyes of some are hard, of others fluid, and some have eyelids, others not; in both <the difference> is for greater accuracy of vision.[23]

The kidneys are present in those [animals] that have them not of necessity but... where the [residue] is somewhat more abundant, to help the bladder perform its function better.[24]

The various passages I have quoted are interesting for another, related reason. They make it clear that, where something's presence is said to be 'better' than its absence, its contribution is to enhance the performance of a function that would be performed without it (if not as well). The bladder would excrete the waste fluid anyway, but the kidneys enable it to do it more thoroughly or quickly. The bird would see anyway, but a more fluid eye gives its vision a greater accuracy. All blooded animals have strength, but thicker and hotter blood makes its owners strong*er*. All animals have some discriminative power, but the right consistency of nutritive fluid gives bees *more*. The crooked beak enables crook-taloned birds to exert *more* force. And so on. At one point in *Juv*., having noted that 'we see that in

[22] 647b29-30, 648a2-7.

[23] 648a13-19. On the occurrence of *cheiron* here, see Balme 1987d, 299 n. 45.

[24] 670b23-7. I will return to the omitted *tou eu kai kalōs heneken* in a moment.

all things nature produces what is best' (here *to kalliston*), Aristotle uses as the measure of 'best' that result which 'most enables each of the parts fully to perform its proper function' (*malista apotelei tōn moriōn hekateron to hautou ergon*: 469a28-31). In all these cases a part is not 'better' because it has added some capability which meets some standard of goodness 'extrinsic' (as Balme would put it) to survival needs; rather its betterness consists precisely in its facilitating the fulfillment of those needs. Even where the feature whose presence is better adds a function, as perhaps the webbed feet make possible swimming (and not just make it go better), its betterness is its use in enabling the animal to fulfill the needs of its distinctive mode of life.

'The well'

But, some animals do not only live, but live well—and Aristotle says of some parts that they are not merely for the sake of living (*to zēn*), but are for the sake of living *well* (*to eu zēn*). Does that not introduce a standard of the good independent of the life of the organism?

I don't think so. Though a fair number of passages in the biology speak of 'the well' (*to eu*), very few speak of '*living* well' (*to eu zēn*). In fact, I know of only one: a passage in *PA* II.10, which runs as follows:

> Now, plants are by nature stationary and thus do not have many sorts of non-uniform parts: to few activities (*praxeis*) correspond the exercise (*chrēsis*) of few organs. . . . Those which have, in addition to life, sensation have a polymorphous nature, and more in some than others, and the most varied in those whose nature has a share not only of living but also of living well. Of this sort is the *genos* of men, for it alone of the animals known to us has a share of the divine, or it most of all. . . .[25]

One might take this passage as implying that having a share of the divine is a *necessary* condition of having a share of living well, but the surrounding context suggests that this might only be a sufficient condition.[26] In any case, the import of this passage is not clear,[27] *and* animals which have such shares are not here said to have their extremely varied structure *for the sake of* living well. The passages often

[25] 655b37-656a8.

[26] Man is brought in here more as a transition to explaining the organization of the rest of the treatise around man (656a8ff.) than as exemplifying something fundamental to the claim he exemplifies.

[27] What is required over and above sensation for 'living well'? If having a share of the divine is a necessary condition, then probably reason is (though *bees* are once said to have something of divinity in them, too: cf. *GA* 761a5), and maybe having a share of living well means having a share of happiness, as per *NE* 1095a18-19. If having some divinity is only a sufficient condition, we are even more in the dark as to what the significance of 'living well' is here. The exploration of 'the well' as such may help us with this. See below, n. 38.

thought of as claiming that some animals have certain of their parts for the sake not of living but of living well—for example, towards the end of *An.* and at the beginning of *Sens.*[28]—actually declare the parts to be present for the sake of 'the well' (*to eu*). We ought, then, to turn to that expression, to ask if it imports a distinct standard of the good over and above the requirements of animal life as such, and to see what light it might shed on this one occurrence of *to eu zēn*.

The passage on the kidney, which I quoted a moment ago, makes clear that at least sometimes 'the well' functions much as does 'the better'. For, the remark that the kidneys were for the sake of the better was merely concretizing the initial claim, omitted from my quotation, that the kidneys are present (not of necessity but) for the sake of the well (*tou eu kai kalōs heneken*: 670b23-4). The same can be said for *kalōs*, which in fact stands in for 'the better' in the general characterization of explanation by the better in *PA* I.1:

> Hence we should if possible say that because this is what it is to be a man, therefore he has these things; for he cannot be without these parts. Failing that, we should get as near as possible to it: we should either say altogether that it cannot be otherwise, or that it is at least *kalōs* thus . . . (640a33-b1)

Aristotle's practice with *kalōs* bears this out: it is often used, as is *eulogōs* and *orthōs*, to characterize a nature's 'selection' of that one from among several possible means of fulfilling a function which does the job particularly well. For instance, he explains in *PA* II.10 how the nature has acted reasonably (*eulogōs*: 656b32) and well (or finely, *kalōs*: 656b27, 657a11) in placing the sense organs as they are; had the organ of smell, for instance, been divided into two, as that of hearing is, it would not have been able to perform its function (*ouk an epoiei to hautēs ergon*: 657a5-6).

In *GA* I.4 Aristotle explains that testicles are not necessary for generation (or they would have been present in all cases), so that it must be *better* that they are there than not there, and proposes that the delay in producing sexual desire which their winding presence results in prevents that desire from interfering with other functions. (Many animals have not got much else to do besides feed and reproduce when the mood strikes them, but the higher animals at least need a balance (*dei sōphronestera einai*: 717a27) in their functioning, and testicles in the case of reproduction, and intestines in the case of nutrition, permit that balance. The higher animals would perhaps survive and reproduce without testicles, but they would not realize all of their natural functions. The term *eu* is not used here, but this passage is a model for those in which it is.

Thus, in *An.* III.12, and again in 13, Aristotle notes that while touch and taste are universal to animals because necessary in the detection and securing of nutrition, the other senses are not, but are there rather for the sake of 'the well' (434b24,

[28] *An.* 434b22-29, 435b17-25 (cf. 420b19-22); *Sens.* 436b13-437a17.

435b21). But his argument is peculiar: in both places, he infers to the lack of necessity of the other senses from the fact that some animals do not have them, but immediately goes on to argue that in those that do their presence is necessary.[29] The corresponding passage in *Sens.* makes clear Aristotle's point, however: the distance receptors (smelling, hearing, seeing) are present for the sake of survival (*sōtērias heneken huparchousin*):

> But in those that happen to be intelligent they are also (*men* . . . *de*) for the sake of the well, for they bring to our attention many differences, from which wisdom, both theoretical and practical, arises.[30]

Other organs said to be present for the sake of 'the well' are in the same situation: lips and tongue are necessary for taste and thus for life, but in certain animals they also make possible speech and thus communication.[31]

Now one might think that things are for the sake of 'the well' when they contribute to rational activity, the value of which for Aristotle is *not* restricted to its causal contribution to survival. But we must not fail to notice that the 'speech' and 'communication' that the tongue makes possible is not only human speech (*logos*) but vocal articulation and communication generally (e.g. among birds: 660a27-b1). Survival, for Aristotle, is, after all, preservation of one's *being*,[32] and while some organs make possible the preservation of those aspects of the being of an animal that it shares with all animals, others make possible the preservation of its distinctive being over and above that. The notion of being 'for the sake of the well' may be meant to capture that fact. Notions of 'better' and 'well' enter when there are options in the survival of the animal. 'Better' selects the more efficient options from among real possibilities for an animal's particular being,[33] but 'the well' may pick out options that exist for animals as such, even if they are not real options (but are necessities) for the being of the animal kind in question.[34]

Even if this hypothesis is right, why should the animal be doing *well* in having features that go beyond what is necessary to animal life as such? It does seem as if some standard is involved over and above the being of the organism, if not in dictating that the function which the organ serves is part of the end, at least in viewing the organism as better off for having that organ as part of its being. The significance of this can be best seen if we take a look at our next category of end-designating expression: the 'honorable' (as *timion* is usually translated).

[29] Cf. 434b22-29, 435b4-25. The argument that their presence is after all necessary leads Richard Sorabji (1980, 157-58) to dismiss them as true examples of 'luxury organs'.

[30] 436b18-437a3.

[31] 659b27-660a13, 660a14-b1.

[32] *An.* 416b14 and surroundings.

[33] In the passage from *IA* 8 quoted above, note the *idia ousia* at 708a11-12.

[34] Cf. Sorabji 1980, 157-58.

The 'honorable' (or 'valuable')

Here we may start from the remainder of David Balme's note on 'the useful' which I quoted earlier. After listing passages for 'the useful', Balme writes:

> Sometimes he gives precedence to 'the valuable' (*timion*: e.g. *PA* 2, 658a22; 3, 672b20; 4, 687a15). But the part's value derives from its usefulness, not vice versa: *IA* 706b14; *PA* 3, 672b15.[35]

Balme's translation of *to timion* as 'the valuable', rather than as the more common 'the honorable' or 'the noble', is very helpful. It prevents us from automatically reading into the use of the term nuances that should be argued for, and which arguably are not present. For, I think Balme is right that, for instance, the upper parts are more valuable than the lower (672b20), and the heart than the rest (665b20, 744b12), and intelligence than hands (687a15), *because* in each case the former is more central to the life and being of the animal than the latter, while the latter are regularly there because they contribute to the functioning or realization of the former. And, as Aristotle himself points out in the *IA* passage Balme refers to, locations are to be designated as more or less valuable because of the organs that are placed there. As James Lennox has pointed out in an important discussion of this passage,[36] Aristotle alludes to Plato's view in the *Timaeus* (45a), that the origins of animal motion are placed in the front because that is a more 'honorable' place and 'more honorable' parts should be in 'more honorable' places, and corrects it:

> It is in fact well (*kalōs*) to speak of these matters the other way round, and say that it is because the origins are in these [places] that these [places] are more honorable than the opposing [ones].[37]

There is a small set of occurrences of 'the valuable' that do not fit easily with this interpretation, however. In one, in *GA* II.3, Aristotle ranks soul-faculties as more or less valuable, apparently indifferent to whether these souls are within an animal or across different kinds of animal:

> Now, the potential of all soul seems to be associated with a body different from and more divine than the so-called elements, and as souls differ from each other in value and lack of value, so too this sort of nature differs. (736b29-33)

[35] Balme 1987c, 277 n. 5.

[36] Lennox 1985b, 149-54.

[37] 706b14-16. The parts (*moria*) mentioned at the end of the sentence are the parts of the body designated by upper/lower, front/back, etc., i.e., the places (cf. b11 and what precedes it). It is difficult for this interpretation that Aristotle says that the Platonic way of looking at it is *eulogōs*, and that his way is *kalōs*, which would seem to suggest that the two approaches are on a par. *kai* does sometimes mean 'in fact', and I count on that, and an assumed Aristotelian tone of voice, in endorsing Balme's and Lennox's reading of the passage; on *kai* cf. Denniston, 317.

If Aristotle is speaking only of the relative value of the soul-faculties within any given kind of animal, one could make a case for the view that the more valuable faculty is more valuable to the being of the animal in that it is more uniquely characteristic of it, but even this ranking within ends is different from the ranking of relative means to relative ends we have been discussing. Insofar as Aristotle is speaking across kinds, this passage would be of a piece with the rest of this set of passages, in each of which he clearly is ranking animal kinds. There is first the well-known exhortation to biological study in *PA* I.5, in which Aristotle attempts to justify the study of objects that are 'less valuable' than those he and his students had previously been studying, viz., the stars (645b22-25). Then, in *GA* II.1 (732a17), in explaining why some animals emit semen, Aristotle refers to these as 'the more valuable animals': their designation as such appears to be connected with their ability to move about, their greater heat, and perhaps their capacity for knowledge, referred to in the previous chapter and said to be something of great value (731a30-b8). And, in speaking of spontaneously generated animals in *GA* III.11, Aristotle offers to explain 'the difference in greater or lesser value of the thing being formed' (762a24-25) in terms of the material which is enclosed by the 'frothy bubble'; so the ranking operates even on the lower levels of animals.

Can we derive this ranking of animals as more or less valuable from the notion of life, as we have attempted to do for the notion of 'the well'? Not directly, since we are comparing types of lives and not relative contributions to a given type of life. But perhaps we can say this much. This notion of one form of life as more valuable than another seems much like the common conception today of one form of life as 'higher' than another. In speaking of one form as higher than another, one might be putting an independent premium on one mode of functioning over another, perhaps on anthropophilic grounds, but there is another perspective one might have in mind. On this perspective, one life-form is 'higher' than another if it has all the capacities the other has, and then some; the 'then some' may be another capacity, or it may be a more powerful version of the former capacity.[38] This could well be Aristotle's meaning in speaking of some animal kinds as more valuable than others, and if so it would not involve the importation of any criterion of value other than contribution to life—although 'is more valuable' would not mean 'makes a greater contribution to life'.

[38] On this picture, those forms which rise above the basic minimum for survival as animals could be said not just to live but to live *well*. Why the line is drawn where it is would, however, still remain unclear: Why, for instance, on this picture, did Aristotle not view plants as living and all animals as living well? I don't have a clear answer for this. I would like to think the issue is not an important one, and varies with the context, much as does the criterion for which animals are 'complete' (*teleia*) and which are not (cf. chapter 7 below, pp. 168-69 n. 45); but I don't find any evidence of that.

'The divine'

However, this does not seem fully to capture the premium Aristotle clearly puts on knowledge or understanding, and on the capacity for it. This capacity surely gets its value in part from the *divinity* it provides man, and as Aristotle says in *GA* II.1, 'that which is valuable and divine is always, in virtue of its own nature, the cause of the better in such things as admit of being better or worse' (731b25-7). So, we had better take a look both at this notion of divinity and at its role as an end. Does the notion derive in some way from the notion of life, or does it provide an independent standard for what will count as an end for Aristotle?

The *GA* II.1 passage is explaining the unendingness of sexual generation as part of its attempt to explain the existence of distinct sexes. In understanding this passage and the similar ones in *An.* II.4 and *GC* II.10,[39] it is crucial to remember that reproduction is, for Aristotle, an extension of self-preservation: the animal is unable to preserve itself eternally, so it does the next best thing towards its preservation—it leaves behind something essentially like itself. The aim is *self*-preservation, not preservation of the species. (The latter is a consequence of the former.) And insofar as eternality confers divinity, the organism is aiming at a form of divinity, as Aristotle himself says in the *An.* passage:

> It is the most natural function in living things, such as are complete and not mutilated or do not have spontaneous generation, to produce another thing like themselves—an animal to produce an animal, a plant a plant— in order that they may partake of the everlasting and divine in so far as they can.[40]

So, in aiming at divinity, animals are not aiming at something other than self-preservation; divinity here *just is* eternal self-preservation, eternal self-continuation.

And perhaps reason is, of faculties, most divine, because its exercise most embodies that crucial feature of divinity, eternal self-continuation. To see that, let us remind ourselves what it is after all that defines divinity for Aristotle. To be divine is to be godlike, so let us ask what features render Aristotle's god what he is; and since his life is best of all, what features render his life best. Is his life best because it involves (is constituted by) the exercise of reason—or is the exercise of reason best because it embodies certain features more definitive of bestness?

In fact, of course, it is the latter. God's life is best, according to *Metaph.* *Λ*, because it is *eternally* actual:

> The first mover, then, of necessity exists; *and in so far as it is necessary, it is good*, and in this sense a first principle. . . . On such a principle, then, depend the heavens and the world of

[39] 731b18-732a1; 415a23-b7; 336b25-34. [40] 415a26-b1.

nature. And its life is such as the best which we enjoy, and enjoy for but a short time (*for it is ever in this state*, which we cannot be), since its actuality is also pleasure.[41]

Apparently, goodness for Aristotle is logically connected with actuality, with being fully in actuality what one is. The cosmos is such as to require a being which is always fully actual, and that being's activity, whatever it turns out to be, will be the best possible, precisely as being something which can be and is eternally actual. What can that being's nature be, if it is always to be actual? Well, it must be *thought*, since only thought can be always actual, though only under one condition, viz., that it is not attached to any body. For a body would entail deterioration, and thus the need for sleep, and, eventually, death. Other beings can only approximate to the goodness of this being, but each does so not by attempting to mimic it—say, by trying to be an unmoved mover, or an object of love—but by realizing its own nature as much as possible. Other beings imitate the prime mover insofar as it is (or, as some philosophers put it, 'under the description') something 'fully actual', something 'fully itself'. Living organisms pursue actuality *by* pursuing life: 'in the case of living beings, to be is to live' (*An.* II.4, 415b13). Thus, in sum, references to divinity, eternality, value, and wellness in the description of the end of a living thing do not add distinct features to the notion and aim of self-maintenance, of *life*. They rather bring out and underscore aspects of the nature of that notion.[42]

[41] 1072b10-16; emphasis added. The omitted words, which are quoted in Dory Scaltsas's comments (Scaltsas 1989, 147), distinguish three senses of 'necessary'; the relevant sense is clearly 'that which cannot be otherwise but is *absolutely* necessary'. Dory suggests that the relevant one is 'that without which the good (*to eu*) is impossible', but that cannot be right. It would make the prime mover's existence contingent on the existence of other good things (or refer to the epistemological necessity within Aristotle's system that without the prime mover one could not explain the goodness of the world—but the parallel passage in *Metaph.* *Δ*.5 makes clear, if it needs making clear, that that cannot be meant). The necessity of the prime mover's existence has just been *inferred* (*ara*: 1072b10) from the preceding lines, and in those lines Aristotle has argued that the prime mover 'can in no way be otherwise than it is' (1072b8; compare *oudamōs* with *haplōs* at b10).

1072b13-30 is difficult but does not count against the interpretation of the good here offered. On grounds that are not entirely clear, Aristotle assumes that the prime mover's activity will be like our own best activity, then infers from its ability (and necessity) to be engaged in this activity always that its activity will be better than ours, in fact best of all. This is surely compatible with (even if it does not entail) the dependence of goodness on actuality.

[42] Cf. *Pol.* II.2 1261b9-10: 'Surely the good for each thing is what preserves each thing.' The scale of value and wellness might derive from the connection between goodness and actuality: those beings or faculties are more valuable whose life is most fully actual most of the time. Something like this might be the force of the *Cael.* passages Charles Kahn discusses (Kahn 1985), if one gives them a more 'minimalist' and naturalistic reading than Kahn is willing to.

The anonymous reader of the earlier version of this chapter asks what content there is to the surely Aristotelian proposition that the full actuality of a living thing is its good, if goodness is defined in terms of actuality, as I claim it is. It is perhaps clearer in the final version of this chapter, than it was in the version the reader saw, that I do not mean goodness to be exhausted by actuality: as I suggest in the text just above, the assertion that some state of a living organism is good for it is an assertion not

Ends and potentials: a non-normative analysis

This too-brief excursion into *Metaph. Λ* supports the main line of argument of this chapter: if goodness is essentially actuality, the full realization of one's actual ends, it cannot serve to define, even in part, what those ends are or what it is to be an end. What then does?

I return to my own analysis of Aristotelian teleology, referred to at the beginning of this chapter, and presented in chapter 1. To say that the development of a living thing is aimed at the full realization of its powers (e.g., above, p. 47) is, for Aristotle, not merely to say that it is moving through stages, and that the stage 'being at the height of its powers' is the best state for it.[43] Rather, that stage is the best state for it precisely because in that state it *is* at the height of its powers, and because its nature *is* to aim at that state. The living processes that constitute that development *really are* aimed at that state; they really are—as my analysis has it—(stages in) the actualization of a potential for being in that state. There just *are* real *dunameis* for mature states in the world, and when these *dunameis* are actualized, developments are

only that it contributes to the full actuality of that living organism, but *also* that that full actuality is something *aimed at* by that organism. What that comes to, on my view, is that the (full set of) soul-capacities whose exercise constitutes the organism's actuality are irreducible capacities for form in the sense described in chapter 1 above. Thus, to the reader's question, 'Is the [teleological] explanation complete when we say "X occurs because X contributes to the full actuality", or is the latter's goodness the reason why it is an end for the sake of which X occurs?', I answer: neither. For the teleological explanation to be complete, one must add to the schema 'X occurs because X contributes to the full actuality' the clause 'and the full actuality is something aimed at'—i.e., on my view, 'and is the object of an irreducible *dunamis*.' (For the way irreducible *dunameis* generate teleological explanations, see chapter 1 above, pp. 30-31.)

The reader asks too whether I think the analysis of goodness in terms of actuality and aims is 'Aristotle's deliberate position or . . . simply all that A. *needs* in order to apply his teleology in the ways he does in the *Physics* and biological works.' It is Aristotle's deliberate position, I think, but to establish that I would have to provide a fuller discussion of the relevant passages in *Metaph.* and in *NE* I and X than I have. I take the argument of this chapter to be a *prima facie* case for the stronger thesis.

The reader asks, finally, whether the eternal motion caused by the prime mover is not an eternally continuous actuality, and thus whether the unique goodness of the prime mover derives rather from its thinking than from its actuality. I would like to think that the outermost sphere's eternal motion *is* an eternally continuous actuality and thus that Aristotle has no real need for the prime mover in his system (as he may have thought when writing parts of *Cael.*), but I don't think that in writing *Λ* and the central books of the *Metaph.* Aristotle thought so. More precisely, I think he viewed the sphere's eternal motion as he viewed all motion: as an 'incomplete actuality'. This is part of what I have in mind in saying below that a full discussion of these issues requires a close look at the discussion of actuality in *Metaph. Θ*.6, because it is there that Aristotle distinguishes full actualities from incomplete ones, classing thinking under the former and motion under the latter. Cf. 1048b18-36 and Kosman 1969 and 1984 (though see now Burnyeat 2008 for a cautionary note as to the passage's proper home).

[43] I use 'state' as an all-purpose term and not as designating a *hexis* rather than an *energeia*; in fact the 'state' in question will always be an *energeia*.

aimed at those states.[44] The only questions that remain are how one tells in any case that such a *dunamis* exists, and how one tells what the ends of such *dunameis* really are.

To those questions I would like to propose the following outline of a reply. *Dunamis* is an explanatory notion in Aristotle. One establishes the existence of a *dunamis* by showing that reference to it is necessary if one is successfully to explain certain facts (here, certain facts about the living world). That one must appeal to a *dunamis* to explain certain facts does not *constitute* the existence of that *dunamis*; that would be too anachronistic (and unappealing) a notion for Aristotle. But that is how we know that a *dunamis* exists. Now everything in the living world is made of the (so-called) elements, and these have certain *dunameis* in their own right. So explanation must start from there, in the sense that in trying to determine what *dunameis* there are in the world, one must ask if there is anything one knows exists that cannot be explained by those elemental *dunameis*. If there is—as my 'ontological irreducibility' thesis claims—that then establishes the existence of a *dunamis*—*for just those features of organisms which cannot be explained by the organism's constituent materials alone*. Since it is precisely the capacities for successful living which are not explainable in terms of element-potentials, it is these for which there must be potentials over and above the element-potentials. And why are these potentials potentials for the mature state, and not for youth, old age, or death? Surely because all the others are explainable from the existence of a capacity for the mature state, but the latter is not explainable from them. If the potential were for death, we would have no explanation of growth and development (and death can in any case be explained from element-potentials). If it were for some state of lesser efficacy relative to survival than the mature state, we would have no explanation of the development from that state to the state we now call maturity (if the state in question came before the mature state), nor of the development to maturity that occurred before that state was reached (if that state came after maturity). But movement to maturity does occur, and needs explanation. Only a potential for development *to maturity* would explain that. Such a potential must then be present, and must be being actualized.

The place of the good in Aristotle's natural teleology

There is, of course, much more to be said about these issues. There is more to be said about the concept of form, for instance; and about its connection with irreducibility and actuality; and about the *Metaph.* Θ.6 conception of an activity that is an end in itself; and about Aristotle's theory of thought. But all these issues are for another time. I would like to close with some remarks on my initial

[44] Cf. chapter 1 above, Postscript 1986, sec. I.

question: what *is* the place of the good in Aristotle's natural teleology? If it is not to provide an inescapable criterion for something's being an end, what is it?

Well, perhaps it is to provide an *escapable* criterion. That is to say, one function of the frequent use of the language of the good in introducing the notions of *end and that for the sake of which*, might well be to serve as a pointer towards these more fundamental notions of end and actuality. What is first to us will not be what is first in nature, but *we* must start with what is first to us. So Aristotle regularly picks out the end of a process or part for us by speaking of the good it contributes to or realizes. Doing so enables him to pick out ends correctly and simply, but this is not because the state or activity's being good is part of what it is for it to be an end; rather this is because it is more easily recognizable than the ontological complexities which underlie it.[45]

To put the same point another way, we are familiar with normative language from our own behavior, and we can understand the common structures underlying both human action and natural teleology best if we start from them. But (assuming I am right that natural teleology is definitionally and naturally prior to the teleology of human action) does this mean that once we understand these common structures, we can dispense with normative language?

I think not, for two reasons. First, it remains a valuable heuristic. Given certain ontological premises, the simplest way to find out what end a part or process has is to ask what good it serves, what benefit it provides. Given that the process or part *is* for the full realization of the organism's life-capacities, we can ask what contribution it makes towards, what benefit it provides for, what use it has in the achievement of, and how its presence is better for, that realization. But, there is a

[45] Most of the passages cited or referred to at the beginning of this chapter, as evidence for the normative analysis, occur as part of dialectical or other contexts in which Aristotle is moving *to* rather than *from* his teleological theory. Only the *Somn.* passage assumes the theory, and in that case there is no reason, given the argument of this chapter, not to take its use of good to determine end as heuristic. John Cooper's insistence that 'unless one bears the connection between goal and good clearly in mind one will fail to understand much that Aristotle says about natural teleology, and many applications he makes of it' (Cooper 1982, 197 n. 1; Cooper 1987, 245 n. 4) is apparently based, in part at least, on the following considerations: (i) his analysis of the first argument in *Phys.* II.8; (ii) the occurrence of explanations by reference to 'the better', and the variety of types of such explanation; and (iii) the occurrence of explanations by reference to 'the honorable'. I discuss (i) in chapter 1 above, p. 22 n. 38, and (ii) and (iii) in this chapter, above. (On (ii) see also chapter 1 above, pp. 33–34.) Employment of Woodfield's analysis to understand Aristotle is complicated by Woodfield's peculiar stipulation, in discussing teleological descriptions of intentional actions, that 'S believes that G is a good' in his analysis means no more than 'S wants to do G' (1976, 204, 207). This creates problems for the claim that Woodfield offers a unified analysis that could be of help in understanding Aristotle. On these problems, and on Woodfield 1976 generally, see Wright 1978 and Nissen 1984. The close connection between desire (in the human case) and *dunamis* (in the biological) provides the key to understanding the way in which Aristotle's own analysis is in fact unified (cf. *Phys.* 199a18-20 and chapter 1 above, p. 13 n. 19 and p. 32).

second, deeper point that flows out of this. Normative language arises in the context of human action, but it picks out features of the structure of that action that are equally present in cases of natural teleology. We do not yet know this when we learn and first use that language, but the fact remains. As Aristotle puts it rather cryptically in *Phys.* II.8, in arguing from the teleology of art to the teleology of nature, 'the relation of the earlier to the later is the same' in both (199a18-20). So long as there *are* aims in nature, there is a good in nature, and language picking out good things as good is not out of place. And so such language is there. But goodness is derivative from endness and actuality, and that should not be forgotten either.[46]

Appendix: response to Scaltsas

In this Appendix, I would like to respond to Dory Scaltsas's stimulating comments (Scaltsas 1989). Scaltsas suggests that teleological relations cannot be analyzed in terms of the potential–actual relation, but I do not find his arguments convincing, and I think he misunderstands certain aspects of my position. Scaltsas assumes throughout that I take 'the notion of *being for* something as the fundamental teleological notion' (Scaltsas 1989, 141), and then offers two arguments. First, he equates the 'being for' relation with the means–end relation, and argues that since 'the money in my pocket is the means for buying a ticket, but is not potentially a ticket' (ibid.), the teleological relation 'is incompatible with', and thus cannot be analyzed in terms of, the potential–actual relation. Then he argues that 'there is no distinction between the way that elemental potentialities *are for* their end' (ibid.), so, if one is teleological the other is too.

However, I have always insisted that the fundamental teleological notion is that of *coming to be for the sake of*, and that *being for the sake of* is to be defined in terms of it.[47] I have argued that:

> A stage in development, *A*, comes to be for the sake of the mature, functioning organism which results from the development, *B*, if and only if: (1) *A* is a necessary (or 'best possible') stage in a continuous change resulting in *B*, and (2) this change is (in part) the actualization of a potential for *B* which is not reducible to a sum of actualizations of element-potentials whose identification does not mention the form of *B*. (chapter 1 above, p. 12)

On the other hand, I have maintained (patterning the formulation on the language of Larry Wright's definition of *function*) that for Aristotle:

[46] I would like to thank Dory Scaltsas and everyone else who participated for the lively discussion that capped our Colloquium session, as well as David Balme, William Charlton, Mary Louise Gill, Aryeh Kosman, James Lennox, and an anonymous reader, for helpful written comments on that version of this chapter. For a response to Scaltsas's comments, see the Appendix which follows.

[47] Chapter 1 above, pp. 6, 13 n. 19, 36, 37, 38-44.

A part *X* of an organism is for the sake of its functional contribution *Z* to the life of that organism if (roughly) (1) *Z* is a consequence (result) of *X*'s being in the organism, and (2) *X* came to be for the sake of doing (making possible, etc.) *Z*.[48]

Coming to be for the sake of is thus prior to *being for the sake of.* The means–end relation, in my view, is thus *not* the fundamental teleological relation, and teleology is not to be connected with potentiality via it. It is thus irrelevant to the truth of my interpretation whether money is or is not potentially a ticket. My primary claim is rather that in all cases in which some *stage of development* (e.g., the formation of the heart, or of the kidneys, or of the horns) is for the sake of the mature, functioning organism, there is being actualized a potentiality for a complex outcome of which the thing formed (e.g., the heart, the kidneys, the horns) is a necessary or 'best possible' part, and that the part's *coming to be for the sake of* that complex outcome is (in part) its being a stage in that actualization. Nothing Scaltsas has said rules that out.

As for money and tickets, surely the question is not whether the money is potentially the ticket but whether it is potentially the buying of tickets. But, whatever the answer to that question is, the relation of money to tickets is too far distant from that of organ to organism to be taken as a paradigm for Aristotle's analysis of *being for the sake of.*

Scaltsas's own analysis of *being for the sake of* (Scaltsas 1989, 144), it will be noticed, is very close to mine, capturing the essence of my first clause with the words 'a functional constituent', and of my second with 'is produced . . . as . . .'. He appears to differ on the analysis of the second clause, i.e., on the analysis of *coming to be for the sake of.* By denying that irreducibility is relevant to the notion of *coming to be for the sake of,* Scaltsas appears to suggest that the teleology of organic development itself is compatible with the availability of a full account of the causation of that development in terms of element-potentials. If that is what he thinks, then he will have to meet the '(friendly) challenge' I made in the excerpt from the paper reprinted as chapter 1 above, which I originally included in this chapter, viz., 'to show how teleological explanation could have any force *as explanation* if a reduction to element-potentials were in principle available.'[49] Scaltsas's reply comes, in effect, in his statement that: 'What chewing is is not determined by what teeth are, even if chewing depends on teeth. Rather, what teeth are is dictated by what chewing is' (145). Now, it is true that for Aristotle, as Scaltsas reminds us, a part is to be defined in terms of its function, and cannot exist as that part if it cannot perform that function (144 n. 8), but the question is why Aristotle held that. In the fullest discussion of that issue and related ones concerning parts, in *PA* I.1, Aristotle derives the definitional thesis from a teleological claim

48 See chapter 1 above, pp. 43-44, for my extrapolation of this formulation from Wright's definition of *function.* My overall argument that coming to be for the sake of is prior to being for the sake of is developed across pp. 38-44.

49 Above, pp. 30-31; cf. my criticism of Richard Sorabji's attempt, pp. 35-38.

which itself is made to rest on the inadequacy of explanation in terms of element-potentials! Functional relations thus do not ground teleological relations, they presuppose them.[50] And it is interesting to note that in criticizing those who thought one could reduce the process to element-potentials, and thus dispense with teleology, Aristotle more than once refers to a distinctive potentiality in the seed of which his predecessors were unaware (640a22-23, 641b36-642a1).

As for Scaltsas's second argument, it is true that elemental natures are specified in terms of outcomes rather than (e.g., as for Newton) in terms of mathematical relations of earlier to later stages. But it remains the case that for Aristotle there is a directiveness *over and above that.* It is for this latter directiveness—directiveness over and above that of the element-potentials—that Aristotle reserves his teleological language: that much is a pre-interpretive fact (chapter 1 above, p. 13 n. 19). What distinguishes potentials for form from elemental potentials on my view is just that actualizations of the latter but not actualizations of the former are irreducible to elemental potentials (ibid.)!

Finally, I am not sure what to say about Scaltsas's use of one of Hilary Putnam's arguments against traditional (Aristotelian!) realism to ascribe dualism to Aristotle. *Aristotle* would surely not accept Putnam's argument, at least not in its present form, since he would not accept a possible worlds analysis of causal relations (or of anything else, I suspect), and certainly would not accept the premise that a world with different natural laws than the present one is a possible one (compare Kripke 1980, 164). Surely, Aristotle would vigorously deny that the notion of similarity presupposed in the statement of causal relations involves intention in any way, and it is not at all clear that he would be wrong. But even if Putnam is right, Aristotle isn't for that reason a dualist, he is merely *wrong about causal relations.* As I see it, we must distinguish an ascription of a commitment to Aristotle which follows from premises he would clearly accept (whether or not he himself has drawn these implications) from an ascription of a commitment which follows only from positions of ours. Scaltsas's ascription is of the latter type—the less interesting of the types, in my view (the more so as I don't myself find the Putnam argument convincing).

Scaltsas wisely asks what my wider concern is in exploring Aristotle's conception of the good and its relation to teleology. It is not, as he suspects it is, to deny dualism (at least not to deny dualism in the weaker sense of a commitment to the existence of properties over and above physical ones, if one wants to call that 'dualism'): with my irreducibility thesis I long ago insisted that the biological realm is ontologically autonomous for Aristotle. The aim is rather, as the remarks in the previous paragraph suggest, to insist, with regard both to the good and to teleology, on a straightforward (metaphysical) realist interpretation of Aristotle. On teleology,

[50] The relevant passage is 640a9-642a21, especially 640a14-26, 640b29-641a17.

one version of the view I am criticizing suggests that a teleological process chugs along, as it were, moved in fact by material-efficient causes, with each state equally a result of such causes. *We*, however, select out one state as 'good' and insist on explaining the process by reference to that state, as necessary or most efficient if that state is to be reached. On another version of this view, the good is, in effect, an unanalyzable intrinsic property of some stage in the process, and early stages happen (equally non-analyzably) *because* they are needed or most efficient for that stage. But I don't think that for Aristotle teleology is an explanatory structure imposed by us on the world, nor that either the 'because' relation involved in teleological explanation or the conception of the good is for him unanalyzable.

As I said above, it is for Aristotle a (metaphysically real) fact about the world that *dunameis* for form exist and are actualized, and thus a (metaphysically real) fact that some non-intentional processes aim at ends; and it is a fact that these ends are real states of the organism, specifiable apart from the fact of their goodness. I am thus attributing to Aristotle a realism about teleology and a naturalism about the good. (There are two restrictions on the latter. The naturalism must be construed quite broadly, in terms of a general [but still good-independent] notion of actuality; and the good must be understood not merely to be actuality: it is actuality *which is aimed at* [in the way that the analysis of 'the chief good' in *NE* is to specify]. Both end and actuality must be included in the analysis of the good. But with these restrictions in mind, we can still speak of an ethical naturalism.)

It is this naturalism and this straightforward realism of Aristotle's that it is my concern to preserve in our understanding of his thought on the issues of this section of the present book, as well as the (appropriately sophisticated) empiricism that goes along with them (on which see both the first chapter, above, pp. 28-29, and the Coda below). Copernicus's revolution Aristotle would have been willing to accept; Kant's (and its modern variants), he would have fought to the philosophical death.

3

Understanding Aristotle's Teleology

Final causality is the mode of causation that characterizes human action, and it extends throughout the living world. I believe, too, that it is the goal-directed character of life—and the biological basis of that goal-directed character—which provides the foundation for the very concept of value.[1] Understanding final causality is thus essential to understanding man's nature as the distinctive biological entity he is, and to understanding a great part of the natural world of which he is a part, and how his good is to be found in that world.

Aristotle shared this view of teleology's importance. His teleology—his thesis that final causality is operative both in human action and in the development, structure, and functioning of living organisms generally—is central to his philosophical thought, and an understanding of Aristotle's teleology has been a central concern of my own research over the years. In various chapters of the present volume, and elsewhere, I have offered an interpretation of Aristotle's teleology: its logical structure, its ontological basis, its empirical character, its relation to his theory of the good, its consequences for the axiomatic structure of biological science, and its implications for his metaphysical theory of substance and essence.[2] There has been, in addition, an enormous literature on these topics since 1976.[3] In this chapter, although I will have occasion to summarize my own line of interpretation and to comment briefly on other lines, I want rather to focus on three other matters. First, I would like to step back from the various controversies regarding Aristotle's teleology to identify, and

This chapter was presented as a Matchette lecture at Catholic University of America in 1992, and published in Hassing 1997. The typology of interpretations was developed in 1994, for presentation first to my Oxford seminar with David Charles and James Lennox and then to the Tokyo course (above, p. x). It also formed part of my contribution to an APA Pacific Division Symposium in 1995 on 'Aristotle's Teleology: The State of the Question', with Susan Suavé Meyer and Mohan Matthen, organized by David Depew. (See also n. 26 below.) The expanded discussion of Meyer's line of interpretation, which I add late in the chapter, is based in part on my discussion of her line in that 1995 presentation, and in part on valuable written comments from, and discussion with, David Charles.

[1] Binswanger 1992; Rand 1964, ch. 1. See now Gotthelf and Salmieri forthcoming, chs. 3 and 4.

[2] See Gotthelf 1975 and, in the book before you, chapters 1, 2, 6, 7, 8, and 11.

[3] See, e.g., Gotthelf 1992 and the bibliography in Leunissen 2010b.

group, the range of issues involved—the range of issues on which one has to have a position if one can claim an understanding of Aristotle's theory, quite apart from what interpretation one ends up with. I find three such groups of questions, which I call 'Analysis', 'Basis', and 'Extent''. Second, focusing on the questions I group under 'Basis', which I consider to be the most fundamental of the three sets of questions, I will provide a typology of interpretations I have found in the recent literature starting *c.*1970. Third, I would like to comment at some length on one particular line of interpretation other than my own which raises in a very interesting way the question of the relevance of Aristotle's natural teleology today: Is Aristotle's conception of final causality, as I interpret it—a conception developed in the context of a very different physics, chemistry, and biology—still applicable in today's scientific context?[4]

Three groups of questions

First, then, let us turn to the range of issues on which all full interpretations must decide. I divide them into three, which, as mentioned, I call: Analysis, Basis, Extent. According to my own interpretation, these three sets of issues are logically connected, even interwoven, so the division is not a radical one; nevertheless, I think the issues still need distinguishing, since whatever their connections they must each be worked through separately, and in the recent literature they are occasionally discussed in isolation from each other.

Under 'Analysis', I include questions about the basic categories of teleological causation Aristotle appears to find, and how they relate to each other. For instance, Aristotle finds teleological causation at work both in human action and in nature, in living organisms. Is the pattern of causation the same in both cases? Teleological causation in human action involves desires and beliefs. In the famous example in *Physics* II.3 with which Aristotle introduces the fourth way of being a cause, viz., as being an *end* or *aim*, a health-sustaining digestion is the aim that causes the walk after dinner (194b33-35). The aim, as discussions in various other treatises make clear, is embodied in a desire (and in the conception of the good it represents), and gets translated into a walk only through the belief (craft-knowledge in the case of the doctor) that walks after dinner facilitate a health-sustaining digestion. But the development of a living organism also occurs for the sake of some end or aim the living organism has. How is that aim embodied? Surely not in any desire or conception of the good, or any belief about means, since the embryo has no desires or beliefs, and Aristotle does not recognize any sort of cosmic *design*.[5] So how are these two cases of teleological causation to be analyzed, and how are they related?

[4] As indicated in the opening unnumbered note to this chapter, I will also expand my comments on another line of interpretation with which I disagree. For the main theses and supporting arguments of my own line, see the two preceding chapters in the present volume.

[5] In the sense in which design implies a designer: see n. 12 below.

Furthermore, in the case of natural organisms, Aristotle not only speaks of processes of formation as being for the sake of the whole organism that results, but also of parts of developed organisms (or the features of these parts) as being for the sake of the whole organism (by way of being for the sake of some activity that contributes to the life or well-being of the whole organism). These are not obviously the same patterns. Let's call the former mode of teleological causation 'coming to be for the sake of' and the latter 'being for the sake of'. How are they each to be analyzed and what is their relationship to each other? In my own work, for example, I have argued that for Aristotle, paradoxical as it might sound, 'being for the sake of' is defined in terms of 'coming to be for the sake of': a part is for the sake of its contribution to the whole only if it has come to be for the sake of that whole.[6] Others have suggested that it is the other way round.[7] Whatever the answer, the questions of how these different ascriptions of teleological causation are to be analyzed and of what their relationship is, both fall under my heading, 'Analysis'.

Some scholars have suggested that what helps unite the various patterns is the fact that in each case the end is something good. While accepting of course that all ends are indeed good (or apparently good), I have argued in the previous chapter against the view that a proper analysis of what it is to be an end need refer to that fact. In any case, I locate the question of the place of the good in Aristotle's analysis of teleological causation under the heading 'Analysis'.

David Charles in a 1991 paper has argued that *Phys.* II.8 provides no basis for precise answers to any of these questions (except possibly the last), and I agree.[8] I said as much in my dissertation and in chapter 1 above. I have suggested that the route to answering the questions that fall under my heading, 'Analysis', is rather to turn first to the questions that fall under the heading 'Basis'. For only after we understand the overall ontological basis of Aristotle's ascriptions of teleological causation to nature will we be in a position to provide a proper analysis and comparison of each of the categories.

Under the heading 'Basis', I group all questions that have directly to do with the question: what fact or facts ground or license Aristotle's use of teleological explanation? On my own view, as I have already intimated, there is an *ontological* basis for that use—something different *in the world* between those cases to which teleological explanation is applicable and those cases to which it is not, and something different about the causation of the phenomena in question. But since some people deny this, at least for Aristotle's natural teleology, and insist that the difference is only epistemological or pragmatic, that is, that it depends somehow on our own mode of understanding, the question needs to be put more generally. I will review a variety of answers that have been put forward to this question shortly, but let me

[6] See chapter 1 above, Postscript 1986, sec. III.
[7] See, for example, Sorabji 1980, 173.
[8] Charles 1991.

first indicate what I have in mind for the third category of issue, which I call 'Extent'.

It is uncontroversial that Aristotle applies final causality to human action, and to the development, structure, and functioning of living organisms generally, and it is uncontroversial that in most cases the end in question is the good (however ultimately analyzed) of the agent or organism whose development, structure, or functioning is being explained. But in recent writings, some scholars have argued that Aristotelian teleology extends beyond those cases, downwards to the inanimate, and upwards to the prime mover. Thus, Charles Kahn has revived the older view that the prime mover is a direct teleological cause not just of the motion of the outermost sphere but of all natural motion in the cosmos, both living and inanimate.[9] And David Furley has argued that in *Phys.* II.8 Aristotle supports the view that rainfall is for the sake of something.[10] David Sedley has endorsed Furley's analysis but has gone beyond it in claiming that the rainfall is for the sake of man. Joining this passage (198b16-199a8) with the notorious one in the first book of the *Politics*, in which Aristotle states that plants are for the sake of animals and animals for the sake of man (1256b15-22), as well as with other passages, Sedley has argued that Aristotelian teleology is *anthropocentric*, with man the ultimate beneficiary of all sublunar teleological action. God, or the prime mover, is the ultimate aim or aspiration of teleological action, but man is its ultimate beneficiary.[11] These are, needless to say, deeply important questions for our overall understanding of Aristotle's thought.

As with the issues under 'Analysis', I do not myself think in the end one can decide how to read the key texts that bear on questions of 'Extent' without appeal to one's views on the questions under 'Basis'. So I will focus in this chapter on that issue. I have found it helpful in considering the recent literature on the ontological basis of teleological explanation to group the various positions that have been defended according to the typology which follows.[12]

[9] Kahn 1985.

[10] Furley 1985.

[11] Sedley 1991 with Sedley 2000, and for a fuller, more rounded statement, Sedley 2010.

[12] The typology includes studies from about 1970. A full survey, and grouping, of interpretations, reaching back to the late ancient world, would be a book in itself. However, I'd like to say just a word on an earlier interpretation that Charles Kahn (in an essay which falls within our temporal range) reports on: the view that Aristotle's teleology rests on design (in the sense in which that requires a designer). Although this view was held by several scholars in nineteenth-century Germany, it has no basis in the text and is in conflict with Aristotle's conception of the prime mover. (See Kahn 1985, 185.) At the beginning of his commentary on Aristotle's *Phys.* II.8, Thomas Aquinas, whose work on Aristotle I greatly respect, seems at first glance to be attributing such a view to him, but arguably he is simply asserting his own view, without distinguishing it, as he more typically does, from Aristotle's. See Thomas Aquinas 1963, 115-16. I should note as well that not everyone who writes on Aristotle's teleology takes a view on its *basis*, as I have explained that term.

Five lines of interpretation

1. *Strong irreducibility*

Living organisms and their parts do not come to be by material necessity alone. The use of teleological explanation is sanctioned by the absence—and indeed in-principle unobtainability—of such a material-level account, since that absence alone entails that the form of the living organism has a real causal role (as aim) in the organism's generation.

This view, which is mine, is also defended by Charlton, Waterlow [Broadie], Cohen, and several others, and, as one of Aristotle's two lines of argument for natural teleology, by Cooper.[13] In my view, the absence of a full material-level account requires the presence of an irreducible potential for form, and this irreducible potential provides a *primitive directiveness upon* an end; this primitive directiveness upon an end is the ontological basis of Aristotle's natural teleology. Some others (for example, Charlton, Cooper) treat something other than a generative potential for form as ontologically basic for natural teleology.[14]

2. *The regulative/pragmatic view*

Living organisms and their parts *do* come to be by simple material necessity alone; material-efficient causes are the only actual *causes* involved. Teleological explanation is sanctioned rather by the fact that material-level accounts fail to satisfy some transcendental, pragmatic, or otherwise mind-based need we (sometimes) have for a certain type of explanation, a need which teleological explanation *does* satisfy.

[13] Charlton 1970 and 1985; Waterlow [Broadie] 1982; Cohen 1989; Cooper 1982 and 1987. A summary statement of my own view may be found in chapter 1 above, Postscript 1986, sec. I.

[14] e.g., Cooper 1987, 249: '[the] fundamental postulate . . . that the world *permanently has* whatever species it contains; that is, . . . that it so governs itself as to preserve in existence the species of well-adapted living things that it actually contains'; Charlton 1985, 146-50: it is a 'primitive' (149) fact, in the case of living things that they act in certain ways '*because* [these actions] are necessary' (147) for the existence and/or good of those things. As Andrea Falcon has pointed out to me, the difference between my view and Cooper's could be put this way: my own view takes as primitive a certain *biological* fact, while Cooper's takes as primitive a certain *cosmological* fact. I should mention, though, that on my view the eternality of the species is *not* a primitive fact for Aristotle: it is a *consequence* of a prior fact, namely that 'it is the most natural function in living things, such as are perfect and not mutilated or do not have spontaneous generation, to produce another thing like themselves—an animal to produce an animal, a plant a plant—in order that they may partake of the everlasting and divine insofar as they can' (*An.* II.4 415a26-b1; cf. *GA* II.1 731b24-732a1). And this prior fact is a biological one—a fact about the soul of a living thing, which is the locus of the 'irreducible potential for form' at work in the living thing's goal-directed action. Though it's clear that Charlton intends the line of his that I have quoted about the primitiveness of a certain type of explanation to reflect a certain primitive fact *about the world*, it is not clear what he means that fact to be (other than 'a fact which is such that it mandates this form of explanation').

Wieland defends a Kantian version of this theme.[15] Nussbaum speaks of the superior simplicity and generality of teleological explanation over material-level explanation.[16] There is an ontological element to Nussbaum's line, although it is not developed (it gets more developed in her later papers[17]), and her interpretation there remains significantly pragmatic or epistemological. Sorabji speaks of 'the relativity of explanation to the questions asked'.[18]

3. Limited irreducibility

Living organisms and their parts do indeed come to be of material necessity, at least if they (and their actions and processes) are described materially (including geometrical properties).[19] But certain feature(s) of these organisms and their parts are not explained by that fact, namely their *goodness* (or: the organism's status as an *organism*). The irreducibility of the outcome's goodness (or organismic status) is what sanctions the use of teleological explanations, which do explain those features of organisms and their parts. That material-efficient causes necessitate but do not explain (or fully explain) teleological outcomes is a consequence of the fact that for Aristotle explanation most properly proceeds from essence, and essence is given by the organism's goal and thus good, and not by its material constitution.

This view is defended by Charles.[20] Cooper finds a *per impossibile* version of this argument in the first argument in *Phys.* II.8, supplementing the basic line of argument found in II.9 (see line of interpretation 1 above): even if, *per impossibile*, organisms and their parts came to be of material necessity, that would still not explain the fact of the good arrangement of the parts in these organisms.

[15] Wieland 1970, §16 (Engl. tr. Wieland 1975). He speaks there of teleology as a 'concept' of reflection'. For others who have read Aristotle this way, see Johnson 2005, 182-83. Kant does not, to my knowledge, attempt to compare his own view of biological teleology closely with Aristotle's. Kant's own account of the basis of teleological explanation in inherent forms of judgment is of course found in his *Critique of Judgment* (Kant 2000, in which the title is rendered, nicely, as *Critique of the Power of Judgment*). On Kant's own account of biological teleology especially, cf. also McLaughlin 1990, Steigerwald 2006, and, for a wider perspective, Guyer 2005.

[16] Nussbaum 1978, 70.

[17] Nussbaum 1980, 1982, 1983, 1984.

[18] Sorabji 1980, 29-31 and 158.

[19] This view is sometimes expressed more cautiously, by saying (i) that there is nothing in Aristotle account that rules out the possibility of these things coming to be of material necessity and (ii) that Aristotle's discussion does sometimes push towards an account of these things' coming to be in terms of material necessity. But the caution cannot be so great that the view slips out from under this line of interpretation.

[20] Charles 1988. See the previous note. As I indicate in the 'Assessments' section, Charles's version of this approach has developed in fruitful ways. Thus now, he has indicated, he would say that the best explanation of living kinds (and of anything else) proceeds from essence. In living kinds their essence is given by their goal and good. Thus the best explanation of living kinds is given by their goal and good (and not by their being necessitated by material-efficient causes).

4. *Weak irreducibility*

While Aristotle did in fact believe that living organisms and their parts do not come to be by material necessity alone, and while he did believe that the presence of an irreducible potential for form is what sanctions the use of teleological explanation, his defense of the existence of such an irreducible potential does not depend on the unavailability of a full account in terms of material necessity. On this view, there will be an irreducible potential for form so long as there is a 'program' in the seed which controls and regulates organic development ensuring that it results in an organism of that form, a 'program' which is defined in terms of that form. Whether the operation of such a 'program' is reducible to current physico-chemical laws is irrelevant to the question of its irreducibility in the weaker sense which is relevant here. That weaker sense is roughly that biological explanation must make reference to such a program.

This I understand to be the view presented by Bradie and Miller[21] and (to some extent) in Balme's last writing on the subject.[22] Bradie and Miller compare it to the contemporary analysis of teleological explanation offered by Ernest Mayr.[23] I discuss their interpretation below, in the final section of this chapter.

5. *Intrinsic causes/anti-eliminativism*

Whether or not Aristotle thought living organisms and their parts come to be by material necessity (in any of the ways advocated or denied in the preceding lines of interpretation) is *irrelevant* to his endorsement of teleological explanation. This endorsement depends rather on a denial that the coming to be of living organisms and their parts is *accidental*. (One could deny accidentalness either because one thought there was a primitive directiveness upon the relevant forms *or* because one thought there was a material mechanism which guaranteed that such forms would be a regular outcome. This line of interpretation maintains that it is the latter—the presence of a 'reliable mechanism' for generating organic outcomes—that is the basis of Aristotle's denial of accidentalness.[24]) That denial derives from Aristotle's insistence, against his accidentalist rivals, that the coming to be of an organism and its parts has an *intrinsic (efficient) cause* of the appropriate sort.[25] This line of interpretation claims that Aristotle regards his rivals as eliminativists (as I will

[21] Bradie and Miller 1984.

[22] Balme 1987c.

[23] Mayr 1988, ch. 1; see also the references to Mayr in Bradie and Miller 1984, 143-44.

[24] Susan Meyer, a proponent of such a view, writes: 'My proposal, like Cooper's, supposes that the permanence of the species plays a central role in Aristotle's teleology. But unlike Cooper's, my proposal takes this permanence to be explicable; *the reliability of the mechanism* that perpetuates the species explains the species' permanence' (Meyer 1992, 811-12 n. 21; my emphasis).

[25] I follow Meyer here in translating *aition kath' hauto*, often rendered '*per se* cause', as 'intrinsic cause'.

explain shortly), and thus makes Aristotle an *anti-eliminativist*, not an *anti-reductionist*, as interpretations 1, 3, and 4 hold.

There are two versions of this view:

(5a) The appropriate sort of intrinsic efficient cause is something 'program'-like which insures a regular outcome, much as in the contemporary view defended by Mayr (see interpretation 4 above). There, material supplied by the female is of such a character that it responds to the material supplied by the male in much the same way that the structure of moveable parts inside the 'marvelous automata' (or: 'self-moving marvels')—children's toys that Aristotle refers to several times[26]—responds to the release of the tension holding it fixed: a simple initial motion produces a 'pre-programmed' sequence of motions generating a complex, structured result (in the automata perhaps a dance, in biological generation the sequence of development leading to an organism).[27] The formal character of the outcome, the regularity of its production, and the epistemic need to define the initial material structure in terms of its ability to produce such a result, allow us to speak of that result as a final cause. This view appears to be defended by Mohan Matthen[28] (and is probably what the Bradie–Miller view turns into when one asks for a definition of 'program' and an explanation of the sort of irreducibility involved).

(5b) The appropriate sort of intrinsic efficient cause is one which not only guarantees that a living organism of a certain form regularly results, but one in which that intrinsic efficient cause is present *because the result it produces is good*. This view is defended by Irwin and Meyer.[29] Irwin claims that Aristotle's critique of the Empedoclean 'selectionist' position shows that he, Aristotle, held that the intrinsic efficient cause present in the seed is present *because* its presence in the parent contributed to the good result which is the parent's survival and reproduction; Meyer claims that the Empedoclean discussion at least shows one possible way in which Aristotle *could* have supported the claim that the intrinsic efficient cause is present because it leads to some good result. The appeal to selective success in past generations as the basis of teleological explanation in present generations is a central part of the contemporary 'etiological' analysis of teleological explanation, developed first by Harry Binswanger and Larry Wright.[30]

[26] 'Marvelous automata': *ta automata tōn thaumatōn* (*GA* II.1 734b9-12), *ta automata thaumata* (*GA* II.5 741b7-9, in the dative), *ta automata* (*MA* 7 701b2-3). The flavor of the term *automata* here is perhaps 'self-moving' (in the sense of moving-on-its-own), so these are 'the self-moving marvels', 'the marvels that move on their own'. For the history of the use of the term in natural contexts prior to Aristotle see Johnson 2005, 104-5 with nn. 25-28. See further, below, pp. 75-76.

[27] I have a different view of the import of the two *GA* II passages; see below, pp. 75-76.

[28] Matthen 1989, though I am not sure I clearly understand his position as presented there.

[29] Irwin 1988, ch. 5; Meyer 1992.

[30] Binswanger 1990 and 1992 (first presented in his 1973 Columbia University dissertation); Wright 1976. It is interesting, in view of Meyer's explicit claim that Aristotle was an anti-

Assessments[31]

I have commented on versions of line 1 other than my own, and on versions of line 2, elsewhere.[32] Lines 3 and 5 each invite sustained discussion. In the present context I will comment only briefly on line 3, giving much more attention to line 5b in Meyer's version. My primary reason for focusing more on Meyer is that her view has been quite influential in persuading people to turn away from the discussions of reductionism that were to the fore in the 1970s and 1980s, and neither Charles nor I (nor others who have defended any sort of irreducibility interpretation) have responded in print to Meyer's critique of irreducibility interpretations. By contrast, the fundamental differences that Charles and I have with each other's views are largely on the table, most especially in Charles 1988 and in several of the chapters in the first Part of this volume, including: chapter 4 on the extent to which Aristotle entertains in *GA* the possibility of a 'full material level story'; and chapter 2, in which I give a lesser role to the good in my analysis of ends and account of the ontological basis of Aristotelian natural teleology than Charles does; and of course in chapter 1, which contains the basic statement of my interpretation. But I do start here with some concerns I have about the line of interpretation Charles embraces, building on what I said in the original version of this chapter (Gotthelf 1997b).

In regard to that line of interpretation, which I have labeled as line 3, there is a *prima facie* problem for the view that Aristotle holds there is a full material-efficient causal account but gives primacy to the teleological account over the material-efficient one on grounds that teleological factors go into the essence of animals and their parts. After all, according to a common reading of the *Analytics* and *An.* I.1, we determine what goes into the essence by *first* seeing what is explanatorily primary. But, I submit, if there is a full material-efficient account, then *that* account is explanatory prior to any account that might supervene on it. Charles 2000, Part II *passim*, esp. ch. 10, presents a subtle and interesting view of the relation between essence and explanation. Charles's thesis is that there are extra-explanatory determinants of essence (as well as explanatory ones). Most generally, an essence, and the definition which expresses it, aims to capture *what* something is. And though the

eliminativist rather than an anti-reductionist, that one section of Binswanger 1992 is titled 'Reduction vs. Elimination'. Though I will not focus on this point here, note that, on the face of it, the tense of 'leads' in the previous sentence in the text above is wrong: it seems it should be 'led'. It is not obvious how Irwin's or Meyer's Aristotle would justify the needed switch in tenses. On how Binswanger and Wright address this in their own accounts of biological teleology, see the references below, in chapter 15, p. 367 n. 35.

[31] The section is new to this reprinting. I move material on lines 3 and 5 from an extended footnote into the text and add substantial further discussion of line 5b.

[32] For references, see the present book's name index, s.v. Cooper; Charlton; Wieland; Sorabji; Nussbaum.

reasons why something is as it is are intimately connected with what that thing is, they do not exhaust it, observes Charles. (This is particularly clear, he says, where there are two distinct causal stories for the same phenomenon; at most one of the causes will be definitionally prior and be a candidate for capturing the essence of the phenomenon.) As he writes, speaking of the background to constraints on (essential) definition to be found (e.g.) in the *Topics*, 'All competent thinkers will have a theory of the world as containing kinds, standing in natural genera, with essences which are prior and unities (in the ways suggested)'.[33] However, he does not in that discussion consider how or even whether such a view supports his 1988 approach to teleological explanation, nor does he address there the concerns expressed in the present paragraph (which first appeared in print in 1997).[34]

We now turn to line of interpretation 5b, in Susan Meyer's version. I take it that there are three questions we need to consider:

(i) Does Aristotle see his predecessor opponents of teleology as eliminativists rather than reductionists?

[33] Charles 2000, 254.

[34] On the extra-explanatory determinants of essence, see also, now. Charles 2010b. Charles there observes that, when considering the teleological and efficient causal accounts of the presence of a house, Aristotle argues that the teleological cause is definitionally prior to the efficient cause. Again, part of Charles's case for distinguishing definitional priority from causal priority is just those cases where one has two causal stories for the same phenomenon, for one cannot then read off the essence from the starting point of the causal line, as there is not just one causal line. Although the argument here is restricted to artifacts, there are suggestions in Charles 2010c for applying it to natural outcomes as well. I received Charles 2010c too late to take proper account of it here. (I thank Charles for the occasion of its publication.) Notwithstanding this argument, I am not yet convinced that any of the determinants of essence Charles lists are, in the end, extra-explanatory (or, more precisely, extra-causal); I would ask that we consider what the source of the relevant definitional priority is. But that aside, I don't see how Charles's view is to work. Let us grant that successful animal generation and self-maintenance, when they occur, are necessitated materially. While I certainly agree that, on that premise, animals would still remain organisms and well-organized, I don't see how they could be *essentially* organisms or *essentially* well-organized. For, on that premise, the animals' organic or well-organized status would *not* be causally primary to them: it would be consequential upon their being materially necessitated, and causal priority is a necessary condition of 'basic essence' according to Charles. (See condition [I] at Charles 2010b, 294 and join with that what I say about the way, on such a picture, (e.g.) the production of good results could be explained in terms of material necessity in the bracketed portion of n. 38 of chapter 1 above, p. 22.)

On another matter, Francis Fallon has suggested to me that certain versions of this line of interpretation are in danger of collapsing into a version of line 2, the regulative/pragmatic line of interpretation, for which it is a fact about us rather than about the world that demands the use of teleological explanation in (e.g.) biological cases. (Mariska Leunissen suggests as much regarding Charles's view in her 2010b, 23 n. 35, with n. 34.) One way to avoid this collapse, as Charles has suggested, would perhaps be to insist that Aristotle determines what counts as the nature or essence of a living thing by reference not to facts about us but to facts about the world. In any case, the issue Fallon raises deserves more attention.

(ii) Even if he sees them as eliminativists, does that mean that he himself did not take his natural teleology to be based on an irreducibility thesis?

(iii) Does Aristotle hold (as Meyer maintains he does) that the intrinsic efficient cause of (e.g.) the coming to be of an animal and its parts is a material 'mechanism' that regularly (or 'reliably') produces such outcomes?

I am going to answer all three questions in the negative.

(i) An eliminativist with regard to some type of thing denies the existence of that type of thing. Aristotle's materialist predecessors do not deny the *existence* of animals or their parts. They do, from Aristotle's standpoint, deny their status as *substances*, and deny that they have *natures*.[35] Meyer's claim that Aristotle viewed his materialist predecessors as eliminativists must, then, be delimited in this way.[36] Let us say that Aristotle viewed them as *substance*-eliminativists about animals and their parts. Does it follow that he did *not* view them as reductionists? I don't see that it does.

We need first to distinguish between two questions we might ask:

(a) What view did Aristotle take his materialist predecessors to have of their own position?

(b) What view did Aristotle think they were committed to, once certain premises he accepts (but they may not) are added to their claims?

Meyer focuses on question (b). She argues that, since Aristotle himself held that only *intrinsic* causes provide explanation, and he took his predecessors to deny that animals and their parts have intrinsic causes, he concluded that on their view animals and their parts would have no explanation, and could not be considered to be substances or to have a nature. Fair enough. But, let us focus on question (a), and ask if Aristotle actually took *his predecessors themselves* to hold that necessitating factors which were not intrinsic causes were not the sort of causes that could provide explanation. It is doubtful that he did, and several passages in *Phys.* II.8 suggest the opposite. Meyer herself refers to Aristotle's characterization of his predecessors at the beginning of *Phys.* II.8; let us have a look at his words:

> We must first state why <a thing's> nature is among the causes (*aitiai*) for the sake of something, and then, in regard to the necessary, how it is situated in that which is natural. For everyone brings <things> back to this cause (*aitia*), <saying> that since the hot is by nature such as to be *thus*, and similarly the cold and everything else of that sort, *these* of necessity are and come to be. For if they mention any other cause (*aitia*), as one does love and strife and another mind, they just touch on it, and then goodbye.[37]

[35] Meyer (1992, 813-14) cites *Phys.* II.1 193a21-26 and 193b32-33.

[36] Cf. Meyer 1992, 825.

[37] *Phys.* II.8 198b10-16 (tr. in part from Charlton). For those who think there is a distinction in the senses, either in Greek philosophy generally, or in Aristotle in particular, between the senses of *aition* and *aitia*, it is perhaps worth noting that Aristotle all three times here uses *aitia*

In response to this, Meyer writes:

> [I]t must be conceded that there is an intuitive and familiar conception of explanation according to which the objector's claim is true. On this conception of explanation, to identify the conditions sufficient for a phenomenon is to explain it. But this conception of explanation is not Aristotle's, and the question at issue is not how *we* might characterize the views of the thinkers whom Aristotle identifies as the opponents to his teleological thesis. The question is how Aristotle conceives their objection to natural teleology. In Aristotle's view, a genuine explanation cites an intrinsic cause. And he goes to some trouble to point out, in the passages the standard incompatibility argument tends to cite, that the materialists' proposal fails to satisfy his conditions for intrinsic causation.[38]

But that is not right. What is at issue is neither how *we* might characterize their views *nor* what Aristotle takes them to be committed to once certain premises of his own are added to their position, but how Aristotle thinks *they would characterize their own views.*[39] *They* believe that they are offering materialist *explanations* of the production of animals and their parts, by reference to the comings-together of independent chains of necessity. So—returning to our two questions—even if, in answer to question (b), we maintain that Aristotle took their position to amount to a kind of eliminativism, there is no reason to accept, in answer to question (a), that he took them to view themselves as eliminativists—and good textual reason to think he did not do so. Indeed, they appear to be presented as offering a reductionist account of the phenomena under review.

(ii) Does it follow, from the fact that Aristotle himself saw his predecessors' views as entailing eliminativism (regardless of how they saw themselves), that Aristotle took his own commitment to natural teleology to rest on his rejection of their eliminativism and not on an irreducibility thesis? I don't see that. I think we must distinguish the defense Aristotle offers in any particular context against particular opponents from (what he took to be) the ontological basis of his own thesis. Meyer herself speaks of the 'dialectical context' of *Phys.* II.8.[40] Aristotle's focus there is on the implications of his predecessors' thesis that animals and their parts are accidental by-products of materially necessary chains. Their problem, in his eyes, is that accidental by-products do not occur *regularly* ('always or for the most part', 198b35-36). So they must come to be for the sake of something. The form of Aristotle's argument here in the first part of *Phys.* II.8 is *disjunctive syllogism*, and arguments of the form 'p or q; not p; therefore q' do not tell us *in virtue of what* q is true; that is not their purpose. Aristotle is not pro-teleology *because* he is anti-eliminativist (in the relevant, limited sense). Indeed, I want to suggest

rather than *aition*. Either way, it is hard to believe Aristotle doesn't take his predecessors to think they are offering *explanations* of the production of animals and their parts.

[38] Meyer 1992, 823.

[39] My thanks to David Charles for the suggestion that Aristotle's opponents themselves might not have thought that causes need to be intrinsic to be explanatory.

[40] Meyer 1992, 824.

that he is anti-eliminativist (in that sense)—and pro-teleology—*because he is himself anti-reductionist.*[41]

I have often quoted, as the leitmotif of my interpretation of the ontological basis of Aristotelian natural teleology, the following passage from *PA* I.1:

So Empedocles was not right when he said that animals have many of the characteristics they do because 'it just happened this way' in the process of coming to be—e.g., even the spine, he says, is such as it is because 'it happened to get twisted and thus break'. He was unaware, first, that the productive seed must have a specific potential (*toiautēn dunamin*) and, secondly, that the producing agent is prior, not only in *logos* but also in time—for a human begets a human, and therefore it is on account of the parent being of such a sort that the child's coming to be is such as it is.[42]

While the first argument in *Phys.* II.8 does not, as we have seen, identify *in virtue of what* the generation is teleological, this passage, taken in context, does: the direction of the generation is towards (e.g.) a *human being* because the seed the parent has supplied to the conceptus 'has a specific potential' to develop in that direction. This potential is a potential for an organism of a certain form (whence my 'potential for form'). Spotlighted here in *PA* I.1, this is the *dunamis* that, as we have already seen (in chapter 1) and will see again (in the next chapter), does so much explanatory work in the fundamental *GA* account of the efficient cause of animal generation.

The alternative accounts Aristotle describes, here of the vertebral character of the spine, and a bit later of the hollowed-out character of the stomach and the nostrils, are accounts in terms of material necessity, identifying processes which, their authors appear to think, occur regularly when animals are in fact produced. And, here, as at the opening of *Phys.* II.8, these accounts are presented as the carrying out of a program to give explanations systematically in terms of the natures and potentials of their elemental constituents.[43] But (for reasons we'll come back to) Aristotle thinks such a program can't be carried out; the causal accounts are just not available. He thinks we need to postulate, over and above these element-potentials, a specific

[41] Regarding the relationship between irreducibility and anti-substance-eliminativism, I will in fact argue, in chapter 11 below, that the general theory of substance in the central books of the *Metaphysics* rests precisely on such an irreducibility thesis, and on the teleological consequences Aristotle draws from that thesis. See esp. pp. 249-51.

[42] 640a19-29. Cf., e.g., above, chapter 1, n. 49.

[43] 640b4-15: 'Now those who in ancient times were the first to philosophize about nature were thinking about the material origin and that sort of cause—what and what kind of thing is matter, how does the universe come to be out of it, and with what cause of movement (such as strife or love or mind or spontaneity), assuming that the underlying matter has a certain kind of nature by necessity—fire a hot nature and earth a cold one, the former light and the latter heavy. For this is how they generate the universe. And they speak similarly of the coming to be of animals and plants: they say, for example, that when water flowed in the body a hollow stomach came to be, together with all the receptacles of food and residue, and when breath made its way through, the nostrils were forced open' (tr. Balme).

potential for form, not reducible to those element potentials. Aristotle postulates potentials primitively directed upon form—potentials which as so directed underwrite *teleological* explanation—on grounds precisely of an irreducibility thesis.[44]

Meyer is thus wrong to reject the views of those (such as Charles and myself) who see an irreducibility thesis at the basis of Aristotle's natural teleology. She is also wrong, I submit, to criticize those (such as me but not Charles) who take that irreducibility thesis to rest in part on a denial of what she calls 'the thesis of necessity'.[45] Let's have a closer look.

Meyer certainly agrees that Aristotle's explanatory account of the generation of animals and their parts must include reference to the form of the offspring; it could not be reduced to an account specified in terms of (what I have called) element-potentials.[46] But she thinks this is so, not because Aristotle rejects the thesis of necessity in this context, but because no such purely material account would be giving an intrinsic efficient cause of the generation. An intrinsic efficient cause of the generation specified in terms of material necessity *would* be available *only if* one identified, among the materially necessary interactions Aristotle recognizes as operating in that generation, a 'reliable mechanism' for producing the relevant animal and its parts, a 'mechanism' identified in terms of that animal's form.[47]

Now, as I indicated in the original published version of this discussion,[48] it is certainly true that *for us* (with contemporary biochemistry, etc.), the non-accidentalness and consequential teleological character of the production of animals and their parts is compatible with their having come to be by material necessity. This is in effect what is claimed in the contemporary 'etiological' accounts of biological teleology such as those of Binswanger and Wright I have already mentioned.[49] But—and I want to stress this again—it simply does not follow from any of that that *Aristotle himself* thought that animals and their parts could be produced by material necessity. To maintain that view of him, one would need evidence that Aristotle thought that 'reliable mechanisms' of the sort Meyer speaks of were within reach—which takes us to the third of our three original questions.[50]

[44] I comment below, p. 85 n. 66, on those rare cases, such as the formation of the omentum (*PA* IV.3), where a part is formed of material necessity, and only then used by the formal nature. (See Lennox 2001a, 290-92.)

[45] 'the thesis that natural phenomena result of necessity from the activities of the material elements: earth, air, fire, water' (792).

[46] e.g.: 'Aristotle's 'defense of natural teleology commits him to the rejection of the reductionist thesis generally attributed to his opponents' (Meyer 1992, 821).

[47] This at least appears to be Meyer's position. I don't think she is clear on *how* the form of the offspring enters into the specification of the 'reliable mechanism' that underwrites the intrinsic efficient causality here. Cf. Meyer 1992, 820-22.

[48] Gotthelf 1997b, 79 n. 27.

[49] Above, p. 74.

[50] Above, p. 77.

(iii) Let us ask, then, whether Aristotle actually thinks that the intrinsic efficient cause of (e.g.) the coming to be of an animal and its parts is a material 'mechanism' that regularly (or 'reliably') produces such outcomes. Interestingly, Meyer gives us absolutely no reason for thinking that is so. She does not discuss *GA* II.6, the one long stretch in that treatise where one might expect Aristotle to mention such a materially based mechanism, if he thought there were such. My next chapter is devoted to a close look at *GA* II.6 to see what sort of account Aristotle in fact gives of the formation of the various parts in the embryo. That examination will make it clear that there is no textual basis for Meyer's proposed 'reliable mechanisms'. It will also help us to understand what exactly is the contribution of material necessity to animal generation and how that contribution interacts, *in a single unified account* of embryonic development, with the formal-efficient causation provided by the new offspring's formal nature, its now inherent potential for form.

For many scholars, however, the availability to Aristotle of a reliable mechanism for animal generation will be thought to have been settled with his use, twice already in *GA* II, chs. 1 and 5, of an analogy I've already mentioned, to the 'marvelous automata'.[51] So I would like to turn briefly to this analogy, and to the general question of the availability to Aristotle of 'material mechanisms' underlying animal generation.

I have long thought[52] that the standard reading of the automata passages in *GA*—a reading that takes the initial motion to act only once at the beginning of the development, causing some motion which then separately causes another motion, etc., as in the case of a sequence of dominos that fall one after the other—is incorrect. As more recent scholarship has confirmed, the texts suggest rather that the initial source of motion, the active potentiality that resides first in the father's seminal material and then primarily in the fetus's heart, is continuously active, all the way through the development. This seems crucial to the point of the analogy, which is to show how the father (as source of the full 'potential for form') can be the ultimate intrinsic efficient cause of the production of the offspring while not in contact with it at any point during its development.[53] Here, I will only refer readers to that scholarship,[54] and turn to the frequent use in discussions of Aristotle's views on animal generation of the language of 'mechanisms'.

In a very interesting recent paper, 'Teleology Without Tears: Aristotle and the Role of Mechanistic Conceptions of Animals',[55] Sylvia Berryman explains

[51] Above, nn. 26, 27. [52] e.g., chapter 1 above, n. 30.

[53] The two *GA* II passages use the analogy in somewhat different ways. See the reference in the previous note, and the studies mentioned in the following note.

[54] Henry 2005, 36-40; De Groot 2008, esp. 58-62; and, indirectly, Berryman 2003, 357-60, and 2009, 72-75.

[55] Berryman 2007.

convincingly that the term 'mechanism' should not be used in situations where a materially necessary sequence or set of sequences results in accidental outcomes, but only in cases where regular outcomes are (for the most part) guaranteed by an internal structure analogous in its operation to a machine (even if there is no account of how such a structure first came to be). Such an internal mechanism, she observes, would count as an intrinsic efficient cause of the sort that might license talk of, and explanation in terms of, goals. Citing Henry 2005 and her own 2003 paper,[56] Berryman seems to agree that Aristotle's comparison of aspects of the early generative process in animals to the 'automatic marvels' is *not* meant as an analogy of that sort. Such analogies only come post-Aristotle, with the tradition of writings on the art of mechanics leading up to the time of Hero of Alexandria in the first century AD (a tradition that begins with ps.-Arist. *Mechanica*), and the concurrent development of complicated, somewhat 'self-moving', machines, such as are referred to by Hero.[57]

Now, Berryman herself sees Aristotle as attempting to navigate between the 'Scylla' of his predecessors with their non-teleological accidental outcomes and the 'Charybdis' of 'those who insist that teleological analyses include components—powers, natures, capacities—needed to fill efficient-causal gaps in the material account, and whose function cannot be analysed in terms of the material properties in which they are realized'.[58] This latter option packages together 'immaterial agency' understandings of goal-directedness (see above, chapter 1, pp. 26-27), such as Plato's in the *Timaeus*, and my own irreducibility interpretation; and she sees Aristotle as in 'tension' between the accidentalist option and this packaged option, struggling, as it were, for some middle ground, and perhaps longing implicitly for a more 'mechanical' understanding of the underlying material processes, of the sort that only began to come after his death.

I don't myself find any such struggle (or longing) in Aristotle. Note, for instance, that Aristotle sees the difference between his own view and that of Plato's in the *Timaeus* as so fundamental, that he doesn't even, in *Metaph.* A, count Plato as having reached an awareness of the final cause.[59]

Indeed, I see Aristotle as having charted, fully successfully in his view, a middle course between the anti-teleology of his materialist predecessors and the Platonic 'immaterial agency' view, a course to a *natural* teleology, in which there is *scientific reason* to postulate a distinct causal power 'whose function' (to borrow Berryman's words), while making use of 'material properties in which they are realized', cannot be 'analysed in terms of' them.[60]

[56] See n. 54 above.
[57] Berryman 2007, 365-66; cf. 2003, 361-65.
[58] Berryman 2007, 352.
[59] See Lennox 1985d and the excellent discussion in Johnson 2005, 116-27.
[60] Berryman 2007, 352; on 'scientific', see the Coda to this volume.

I intimate several times in this volume, including in the last section of this chapter, that it would have taken something like modern biochemistry and evolutionary theory to dislodge Aristotle from the primitive directiveness inherent in his 'strong irreducibility' position into a weaker view, something like the recent etiological approach to biological teleology developed by Binswanger and Wright. But Berryman has raised for me the question of whether exposure to a later ancient attempt at a mechanistic account of animal generation would have begun the dislodgement, had Aristotle been exposed to it. In the end, I doubt that it would. Later mechanisms are or appear 'self-moving', but they don't produce, or develop into, anything different and more complicated than they are, as embryos do. That process would require, for Aristotle, significant qualitative change, and it's hard to imagine how he might come to think an internal *mechanism* could produce or control that. And the models Aristotle has for qualitative interaction—models of the blending and cooking up of earth–water–air–fire mixtures, and of the 'overpowering' by one contrary of another in the case of the elementary powers, the 'hot', 'cold', 'moist', and 'dry'—just don't lend themselves to the building up of precise, delimited structures.[61] It's rather the beginnings of a change to a structural chemistry that I *think* would be crucial here. But either way, these speculations give us no reason to see Aristotle himself in any sort of struggle about the ontology of natural teleology, nor should he have been. In short, I suspect Professor Berryman's 'tears' are unnecessarily anachronistic, and I am sure Aristotle would have found them misplaced. However, her discussion does imply that at least one key assumption of Meyer's influential paper, regarding the availability of 'reliable mechanisms', is false (and indeed anachronistic).

Two recent books have endorsed Meyer's account of Aristotle's defense of his natural teleology: Monte Johnson's *Aristotle on Teleology* (Johnson 2005) and Mariska Leunissen's *Explanation and Teleology in Aristotle's Science of Nature* (Leunissen 2010b). It will be interesting to look briefly at the cases they make.

Johnson twice cites approvingly Meyer's claim that Aristotle's teleology does not rest on an irreducibility thesis, but gives no argument himself for the thesis that the issue at stake between Aristotle and his opponents is eliminativism.[62] He does offer separate reasons of his own in favor of banishing the reducibility question, and with it interpretations such as mine, but gives no reason for endorsing Meyer's alternative.[63]

[61] We will look at one possible exception to this (viz., the determination of shape from bone length) in the next chapter, pp. 113-15, though it will turn out not to be an exception after all. Spontaneous generation is examined in chapter 6. Some remarks in von Staden 1997 might suggest that my claim here is overstated. I will discuss those remarks in a note to be added to the version of chapter 4 below that will appear in Ferraces Rodríguez, Fischer, and Holmes forthcoming.

[62] Johnson 2005, 98, 187 n. 46.

[63] Johnson dismisses my own view after a very brief discussion. First, he criticizes my irreducibility approach on grounds that elemental interaction is itself teleological, according to Aristotle, so that there is no non-teleological level in relation to which biological teleology may

Leunissen provides a somewhat more sensitive examination of this aspect of Aristotle's teleology. She begins with an assertion of her agreement with Meyer that:

> we should dismiss the question of whether or not teleology and necessity are mutually compatible or reducible to each other as not being the main point for Aristotle, and [focus], conversely, on the issue of intrinsic versus incidental causation....

Then she writes:

> Following this approach, I believe that the reason why Aristotle contrasts his own teleological worldview with that of his materialist predecessors is that the latter deny that high-order natural phenomena, such as animals and plants, have a privileged ontological status.... In short, for Aristotle an animal is the primary example of a natural substance, whereas for the materialist an animal is merely a coincidental conglomeration of elements. The heart of the debate does not pertain (or at least not directly) to the question of reductionism, but rather to that of eliminativism: the mistake of the materialists is not just that they reduce plants and animals to their material elements, but more that, as a consequence of this reductionism, they deny that plants and animals are substances at all.[64]

As did Meyer, Leunissen focuses on Aristotle's grounds for rejecting the accidentalist implications of his opponents' denial of natural teleology. But notice what she

or may not be reducible. I noted the obvious response to this line of criticism many years ago, in chapter 1 above, n. 19. Devin Henry offers a similar reply in his review of Johnson's book: 'Johnson's treatment of Gotthelf's position is also somewhat unfair. For there is an important ontological dimension to Gotthelf's thesis that is untouched by the question of the teleological explicability of the elements (or their *dunameis*). Gotthelf raises the issue about the ontological relation between the developmental capacities of organisms and the basic capacities of the elements that compose them. The key question here is whether or not the nature that controls growth and development can be reduced to a set of elemental *dunameis*. If not, Gotthelf argues, then it must be a distinct potential which is *irreducibly for* an organism of some specific form. This nature would impose its own *primitive directedness* on the process of development that could not be accounted for solely in terms of a sum of actualizations of element-potentials. What is irreducible to element potentials, here, is not final causation *per se* but the directedness of the process onto the adult form' (Henry 2007b, 196). As Henry remarks earlier on the same page, any 'teleology' ascribed to the elements would be a weaker form of teleology than the sort of teleology my position ascribes to biological processes, and the reducibility of the latter to the former remains a real question. Later in his book, Johnson briefly suggests that Aristotle's endorsement of spontaneous generation defeats my interpretation of his teleology, but does not comment on my lengthy discussion of this issue in the 1989 paper he cites (now chapter 6 below). That said, Johnson's book contains much valuable historical material, and (as I have already mentioned) an excellent account of why Aristotle refuses to recognize Plato as a discoverer of teleological explanation; and he has some good criticism of those who extend Aristotelian teleology in inappropriate ways (e.g., Sedley, n. 11 above), and I can recommend the book for those features, among others.

64 Leunissen 2010b, 24.

takes to be Aristotle's view of the *source* of their eliminativism, and at what stage of that reasoning she has Aristotle take issue with them. As we have just seen her say:

[T]he mistake of the materialists is not just that they reduce plants and animals to their material elements, but that, as a consequence of this reductionism, they deny that plants and animals are substances at all.

But what of her endorsement only one paragraph back of Meyer's view that we should '*dismiss* the question of whether or not teleology and necessity are mutually compatible or reducible to each other'?[65] How can the question of reductionism be dismissed if the 'mistake of the materialists' is, as we've already seen, and as Leunissen correctly realizes here, a *consequence* of their reductionism? It is their reductionism that is their cardinal sin, and is the root of the eliminativism that Aristotle ascribes to them.[66]

[65] My emphasis.

[66] Leunissen has told me that one of the reasons she was attracted to Meyer's view that Aristotle's focus is on anti-elimination and not anti-reducibility is its apparently superior ability to accommodate two phenomena Aristotle recognizes: spontaneous generation and cases of what she has called 'secondary teleology', cases in which materially necessary by-products of earlier teleological processes are subsequently made use of by the formal nature. I have shown in chapter 6 below how I would accommodate spontaneous generation; let us look here at secondary teleology. Primary teleology, as I have been arguing, has as its basis the irreducibility of the formal *dunamis* to the *dunameis* of the constitutent elements. To identify the ontological basis of secondary teleology, I start with the following three theses, drawing on what Leunissen and I have written together on this matter in chapter 5 below, p. 132 and n. 40: (i) the formal *dunamis* that 'makes use of' of material byproducts of an earlier primary teleology is the very same *dunamis* that was in operation in the earlier case; (ii) that *dunamis*'s 'making use of' those materials is not itself of material necessity (e.g., the formal nature's diversion of the earthy material in embryonic ruminants to make the horns does not itself occur of material necessity, but only because it serves the end of defense); and (i) and (ii) are true notwithstanding that (iii) those materials originate from materially (and not conditionally) necessary processes and they open up additional options for the formal nature (which is in part why the teleology is *secondary*).

Because the teleological character of secondary teleology is grounded in the same irreducible potential for form that the primary teleology is, the ontological basis of the former is the same as that of the latter. This is true, I want to maintain, even in those (quite rare) cases, such as the production of the omentum (PA IV.3), where processes of material necessity are sufficient actually to produce the part 'ready to function' and the formal nature has no need to act further upon those materials (677b30-32). But even in those cases, the formal nature is at work 'making use' of that materially necessary result. For, did the omental structure not serve some good end, it would not have been allowed to persist; its formation would have been interrupted and its materials diverted. Letting happen is, after all, a kind of making use, and is also due to the *dunamis* for form.

So, in secondary teleology too we have in operation the causal agency of an (irreducible) formal nature—a potential for form—whose outcomes are not due to 'reliable mechanisms'; and it is this, and not any concern with erroneous views of his opponents regarding elimination, that ontologically grounds the teleology for Aristotle. I thank Leunissen for discussion of the issues of this note.

Leunissen's understanding of the logic of Aristotle's position, even if it is not fully worked out, is to be preferred to Meyer's, and it is perhaps no accident that, when she seeks to identify the primary pattern of intrinsic causation involved in Aristotelian teleological processes, she speaks specifically of 'realizing an internal, preexisting potential for form'.[67]

In sum, then—to end this part of the assessment of interpretation 5b—whatever the merits of Meyer's analysis of Aristotle's reasons for rejecting his materialist predecessors' non-teleological accounts of animal generation, I see no reason to accept that analysis as an account of *the ontological basis of* Aristotle's natural teleology.

Aristotle and contemporary thought

Let us, finally, look in more detail at interpretative line 4, and in particular at the version of this line developed in the 1984 paper by Michael Bradie and Fred D. Miller, Jr. referred to above (n. 21). This line of interpretation is of special interest, since it raises important questions regarding the relationship of Aristotle's natural teleology to contemporary biological teleology.

In a review of the literature to 1983, Bradie and Miller conclude (i) that advocates of line 1 are right to claim that, for Aristotle, material-efficient causality could not by itself produce living outcomes, and (ii) that I was right to stress the role of a potential for form in grounding teleological explanation. They agree with me too that Aristotle thought the potential for form was not reducible to potentials of the material elements. But they distinguish two irreducibility theses:

(i) The potential for form is irreducible to the potentials of earth, air, fire, and water.

(ii) The potential for form is irreducible to any potential on any level of explanation.[68]

They say that I have argued successfully that Aristotle held (i) but that my claim that (i) was the core of Aristotle's teleology requires Aristotle to also hold (ii). However, they claim, there is no evidence that Aristotle ever held (ii), and in fact it would be anachronistic to attribute (ii) to Aristotle, since that would require him to rule out in advance theories he could not have imagined, such as contemporary DNA theory. That leads them to distinguish two claims about movement involved in Aristotle's teleology:

[67] 2010b, 25. See also 4 n. 10 where she notes that her 'characterization of 'primary teleology' 'builds on the understanding of teleological causation in the case of animal generation provided by Gotthelf (1987b [= chapter 1 above])'. There is much I have learned from her work as well, and (notwithstanding certain important differences, on some of which see Gotthelf 2010), I heartily recommend her book.

[68] Bradie and Miller 1984, 147.

(T1) In a life process, the goal is produced by means of a potential existing from the outset, through movement conforming to a formal principle ('containing a certain *logos*').

(T2) The source of the movement through which life processes occur is vital heat, which cannot be identified with the potentials of the simple sublunar elements.

About these claims they write:

> Claim T2 has been thoroughly discredited by modern biology, but T1, which is the core of Aristotle's teleology, has been vindicated by modern biology. For the point of T1 is that life processes are self-regulating in virtue of inherent forms or structures. The type of movement required on Aristotle's account for a potential for form is the type of movement exemplified by the DNA molecule. The genetic 'program' contained in the molecule's structure directs and limits the organism's growth in the manner set forth in Aristotle's biological writings.[69]

In outline, my response encompasses four points.

(1) I find their claim that it is anachronistic to attribute (ii) to Aristotle puzzling, for there is a single, simple premise, available to Aristotle, that gets one from (i) to (ii). It is that the level of earth, air, fire, and water is the *basic* level.[70] To deny Aristotle that premise, and to attribute to him the view that there might be a more basic level from which development could be accounted for wholly by material-efficient causes, is what strikes me as anachronistic, on two counts: First, there is no tentativeness whatsoever in Aristotle's theory of the elements (nor is the sort of general denial of certainty that is common today any part of Aristotelian epistemology). Second, as both Waterlow [Broadie] and Cohen point out, all of Aristotle's instincts regarding matter theory are in the direction of the view that the lower you go the less structure you get, and the less possibility of explaining complex, highly structured, highly functional outcomes.

(2) Bradie and Miller do indeed make clear, with their T1, that there is a non-trivial generic similarity between Aristotle's account of the basis of teleological explanation and the contemporary genetic program account. But Aristotle could not have projected such an account and thus did not himself countenance such a generic version, much less endorse it. But *we* can see it, and we can call any view that endorses T1 *Aristotelian* (but not 'Aristotle's').

(3) Actually, point 2 requires qualification: as suggested by the parenthesis attached to the description of line of interpretation 5 above (p. 74), the Bradie and Miller view would seem, if pressed in the ways indicated, to turn into a version

[69] Ibid. 148.

[70] Here, since we are talking about the simple *bodies*, I ignore the fact that for Aristotle these bodies, these 'so-called elements', are themselves analyzable as combinations of the qualities or powers hot, cold, moist, dry.

of line 5, and in particular into a version of 5a. But for reasons both Irwin and Meyer on the one hand, and Binswanger on the other, have suggested, 5b is to be preferred over 5a, both in the contemporary context and as the generic 'Aristotelian' view. The point may be put simply. If there is a regular direction upon an end, guaranteed by an intrinsic efficient cause of a material sort, this seems only to *mimic* teleological causation; what we need is that the intrinsic cause's activity that leads to the end occurs *because* it leads to that end. In my interpretation, as I elaborate in my Postscript 1986, sec. I, the addition to Aristotle's ontology of potentials for form entails that when some potential for form is being realized, everything needed for that form is also being realized. These latter needs, then, occur *because* the form is coming to be. But if an actualization of such a potential is only a redescription of some set of actualizations of material-level potentials, I don't see that anything happens or comes to be *because* it is needed for an end. Under these circumstances, teleological explanation would seem to be applicable only if the 'program' itself is present solely because of the success of its outcome in previous generations, and this is interpretation 5b.

(4) Taking the previous three points into account, then, the best we can say on behalf of the Bradie and Miller thesis is that, in the face of contemporary physics and biochemistry, Aristotle would very likely *retreat* to a version of 5b. But it remains the case (given Bradie and MIller's declared agreements with me together with point 1 just above) that, on Aristotle's view, a 'potential for form' involves a *primitive* directiveness upon an end, not a directiveness that is in any way derivative from any material level 'mechanism' or 'structure' or 'intrinsic efficient cause'. The existence of such a *primitive directiveness*, made possible by a strongly irreducible potential for form, is the core of Aristotle's teleology, and is what differentiates that teleology from the various modern theories to which some interpreters have tried to assimilate it and with which it indeed has important generic similarities.[71]

From a contemporary standpoint, these generic similarities are not to be dismissed, and the contemporary physicist and biological theorist Max Delbrück was perhaps only slightly exaggerating in insisting that Aristotle ought to be awarded a posthumous Nobel Prize 'for the discovery of the principle implied in DNA'.[72] And comparing contemporary biological teleology with Aristotle's own may help illuminate for us the generic features of our favored theory. But if we are to be historically accurate, and to understand what it was that was missed by the early modern critics of Aristotelian teleology, for whom the only alternatives were mechanism or design, we need to understand what it is that is *distinctive*

[71] For a presentation of his current view, which takes into account some of the discussion of the present chapter, see Miller 1995, 336–46.

[72] Delbrück 1971.

about Aristotle's natural teleology.[73] And if I am right in thinking that much else in Aristotle's thought, from his ethical theory to the theory of substance in the central books of the *Metaphysics*, depends centrally on his natural teleology, understanding the teleology aright will be crucial for understanding much else in his philosophy, and thus for learning the great amount there still is for us to learn from Aristotle today.

[73] For further discussion of this, see the comparison of Darwin's and Aristotle's approaches to biological teleology in chapter 15 below, pp. 366-68; cf. also the coda to the present volume, pp. 390-91, and chapter 1 above, Postscript 1986, sec. III.

4

Teleology and Embryogenesis in Aristotle's *Generation of Animals* II.6

Two formulations of the aim of *GA*

Because Aristotle's *Generation of Animals* is to a large extent 'one long argument' (to borrow Charles Darwin's characterization of his *Origin of Species*[1]), let us place *GA* II.6 in the context of the whole of *GA*, identifying what this treatise's overall aim is and how teleology is embedded in that aim, and thus in this treatise.

We are told of *GA*'s aim in at least two places. The first is in *PA* I.1, in a passage we have already quoted in the previous two chapters; the second is in the opening paragraph of *GA*.

In *PA* I.1, having argued that coming to be is for the sake of being, and not the reverse, Aristotle writes of what follows from that proposition for the study both of animal parts and of their generation. With good reason, scholars standardly take these statements as governing the forms treatises like our *PA* II–IV and *GA* are to take. First *PA* II–IV:

I knew at the time I submitted my dissertation that my treatment of Aristotle's teleology would not be complete without a careful examination of the bearing of *GA* II.6 on the topic. Work towards that end began finally in 1994, with presentations to that year's Oxford and Tokyo seminars (above, p. x). While preparing a version some ten years later for presentation at the University of Texas at Austin, I received helpful written comments from Devin Henry. Although the thesis and much of the argument of the chapter as it now stands are significantly different from that of those earlier versions (from which it will be differentiated as the chapter progresses), they are in the same broad spirit, and I remain thankful both to Devin and to those who participated in discussion on those various occasions. More recently, comments from Mariska Leunissen on specific passages in II.6, as noted, were very helpful. Extensive discussion of later drafts of the chapter with David Charles, and then of the penultimate version with Andrea Falcon, significantly improved the final version. Special thank-yous to them both. With some additional notes, this chapter will also appear, in due course, in the Festschrift volume for Heinrich von Staden (Ferraces Rodríguez, Fischer, and Holmes forthcoming).

[1] Darwin 1859, 459. I developed this theme in a presentation under that name to the Fifth Pittsburgh/London Workshop on Aristotle's *Generation of Animals*: Understanding the Methodology at Work in *GA*, organized by Devin Henry and myself at the University of Western Ontario, 27 May 2010. Andrea Falcon and I are currently carrying that project further, as we work together on a co-authored paper, provisionally called '"One Long Argument?"—The Unity of Aristotle's *Generation of Animals*'.

(T1) Hence we should if possible say that because this is what it is to be a man, therefore he has these things; for he cannot be without these parts. Failing that, we should get as near as is possible to it: we should either say altogether that it cannot be otherwise, or that it is at least good thus. And these follow.[2]

Then *GA*:

(T2) And because he is such a thing, his coming-to-be necessarily happens so and is such. And that is why this part comes to be first, and then this.[3]

From the nature of what is to be produced, certain things follow (of conditional necessity) regarding how that product must come into being. *And that is why* the parts come into being in the order in which they do. I will connect the first of these two lines with the opening stretch of *GA* in a moment, but it is striking to notice, in view of the second line, how much of *GA* II.6 (and indeed II.1-6) is indeed, and explicitly, concerned with the order in which the parts come to be.[4] Perhaps this shouldn't be a surprise, since one can't say much about *how* or *why* an animal's *genesis* occurs as it does without knowing in what order the parts come to be. Still, I am struck by the extent of Aristotle's preoccupation with this question, and will return to it later.[5]

The opening stretch of *GA* (715a1-18) announces that we have already studied the final, formal, and material causes of each of the parts of animals except the generative ones. That leaves a *PA*-like, three-cause study of the generative parts—and an overall study of the efficient cause. This latter study and the study of the generation of each animal, Aristotle says, somewhat enigmatically, 'are, in a way (*pōs*), *the same thing*'.[6] A primary aim of our treatise, *On the Generation of Animals*,

[2] 640a33-b1 (tr. Balme). For the ways in which this passage models the explanations in *PA* II–IV, see above, pp. 12 n. 18, 35, 38 and below, 124 n. 16, 154, 175, 191, and chapter 10, 219-20 nn. 4 and 6, *et passim* ch. 10.

[3] 640b1-3 (tr. Balme). Bekker's Greek reads: *ἐπεὶ δ' ἔστι τοιοῦτον, τὴν γένεσιν ὡδὶ καὶ τοιαύτην συμβαίνειν ἀναγκαῖον. διὸ γίγνεται πρῶτον τῶν μορίων τόδε, εἶτα τόδε.*

The next sentence is presumably retrospective, covering the instructions for *PA* and *GA*: 'And this is the way we should speak of everything that is composed (*sunistamenōn*) naturally.'

[4] II.1 733b23-734b4, 735a12-26; II.5-6 741b15-37, 742a16-b17, b32-743a21, 743b18-36 and ff.

[5] This preoccupation of Aristotle's in *GA* II (and especially in II.6) is especially brought home in Mariska Leunissen's 'Order and Method in Aristotle's *Generation of Animals* II', her as yet unpublished presentation to the Fifth Pittsburgh/London GA Workshop (above, n. 1). She gives a full account of the different sorts of priority at work in Aristotle's account in II.6, and of how those different senses shape the construction of the chapter. For my own briefer discussion, see below, pp. 101-13.

[6] *GA* I.1 715a14-15. I have purposely not supplied an object for 'the efficient cause'—whether of the animals or of their parts (or both)—since the Greek reads only: *περὶ αἰτίας δὲ τῆς κινούσης τίς ἀρχή. τὸ δὲ περὶ ταύτης σκοπεῖν καὶ τὸ περὶ τῆς γενέσεως τῆς ἑκάστου τρόπον τινὰ ταὐτόν ἐστιν*. But I am going to assume it is both, since 715a9-11 says as much for the other three causes.

then, is a study of 'the source of the moving cause' of animals and their parts, that is: a study of *what agent(s), and of what actions of theirs, bring animals and their parts into existence?*

If we now bring to bear on this conclusion the two *GA*-aim-stating sentences we isolated in *PA* I.1 as T2, what can we add? Starting from its second sentence, we may add the words: *And in what order?*—a sensible addition, which keeps us within the ambit of an efficient-causal account. The first sentence, however, reads: 'And because he is such a thing, his coming to be necessarily happens so and is such.' This is, as already mentioned, an account in terms of conditional necessity, and this brings our attention to the (immediate) *final cause* of generation: the production, or existence, of an animal of a certain nature. A full account of the generation of animals will, I propose, *not only* include *but also* intimately relate the final and the efficient causes of generation. The need for the final-cause dimension of the account will explain why Aristotle says that an account of the efficient cause by itself is an account of generation only *pōs*, 'in a way'. With the teleological dimension in place, the account of generation is, in Aristotle's eyes, complete.

Towards the theory in *GA* II.6

That there is an intimate connection, at least in the case of generation, between the final and the efficient cause will not be of surprise to readers of the taxonomy of interpretations of final causality offered in the previous chapter. For, if we connect the operation of a final cause with the operation, as efficient cause, of a *dunamis* directed upon such an end, then that connection of causes is just what we should expect. And if we are to get illumination on the nature of a final cause from a study of *GA* (as explained in chapter 1 above, pp. 6–7, 14ff., 31–32) the route to this will be to follow closely *GA*'s account of the efficient cause of generation, looking in particular for any move that is made, via conditional necessity, from the nature (or essence) of the animal produced to the nature of the process required to produce it.

Yet it is striking that explicit mentions of final causation occur rarely in *GA*, once the review of the generative parts is completed by the end of I.16. A final cause of generation itself is offered at the beginning of II.1, followed by an account of the final cause of the separation of the male and female principles. Final causes are not mentioned again until II.4, at 738b1, when a final cause is given for the diversion of female nutritive residue to the uterus to serve as generative material, and then at 739b29 for the formation of membranes around the outer surface of the *katamēnia* in the conception process. However, when Aristotle turns to his account of embryological development in II.6, he draws at 742a16–b17 on a set of general theses about various types of priority involved in final causation to guide him in the empirical identification of the sequence in which parts are formed in the embryo, and to help explain why the sequence occurs in the order it does. The aim here is the efficient-causal account, but the efficient *archē* within the embryo (viz., the heart or its analogue), Aristotle claims, must stand in various relationships to the end, the *telos*, of the generative process—although quite what those relationships are is difficult to determine from this very difficult passage. Still this passage, so outlined, appears to give us *precisely* what the *PA* I.1 passage said should be our guide in determining the character and order of the process by which the animals come into being. So this II.6 passage will have to be our focus. But let us build up to it step by step, starting from *GA* I.1, with our eye on what Aristotle's account of the efficient cause of generation has to tell us about the nature of final causation, as operative in animal generation.

First, some context. In chapter 1 above, I made two general claims about the *GA* theory: (i) that Aristotle does not attempt to offer (the outlines of) a full material-level account of the generative process, but rather appeals at key places to a *dunamis* that is *primitively* directed upon the form the offspring is to embody; and (ii) that the material bearer or basis of that *dunamis*, sometimes identified as a distinctive 'vital' heat (*psuchikē thermotēs, zōtikē thermotēs*[7]) and sometimes as 'movements' (*kinēseis*[8]),[9] has no material-level description of its vital potentials, being different in this respect from 'ordinary' heat found in ordinary air and fire. In his valuable 1988 paper, 'Aristotle on Hypothetical Necessity and Irreducibility', David Charles questioned both of these claims, and that invites us to take a close look at the passages in question to see if there is some way to decide the matter. Though there is a reasonably plausible case to be made for a reading like Charles's, in the end the weight of the evidence in *GA* II.6, I shall argue, is firmly in favor of my two claims.

[7] 732a18-19. 739a11, 762a20; 739b23, cf. 737a2-5. See too Peck 1961, liv note *c* and Freudenthal 1995, 19-34 *et passim*.

[8] 729b5-6 and ff., and T3, T4, T7, and T10 below; cf. Peck 1961, liv and Freudenthal 1995, 27-29.

[9] In a very few passages, and always in passing, Aristotle indicates what the relationship between these two *archai* might be. Here are two such passages (emphases mine):

(i) '[T]he nutritive soul-capacity, just as, in the animals and plants themselves later on, it produces growth from the food, using as tools heat and cold (for the soul-capacity's movement is *in these*...)...' II.4 740b29-32 (part of T7 below).

(ii) 'The heat is present in the spermatic residue *having its* movement and actuality in an amount and character...' II.6 743a26-28 (part of T10 below).

These passages give rise, however, to several interrelated questions which I don't know how to answer: What is it for heat either to have movements *simpliciter*, or to have them *in* it? Is every case of (vital) heat*ing* the action of 'movements' that are either of or in the agent's heat? Is every generative moving of another thing a case of heating it? The best discussion of the relationships among the Aristotelian notions involved in these questions is Freudenthal 1995, ch. I, sec 1 and ch. III, sec. 2 (though he says relatively little about the generative 'movements'); however, there is more that still needs to be said, if we are to fully understand Aristotle's conceptions of heat in general, of 'vital' heat in particular, and of generative 'movements'—and thereby his precise view of the efficient causation involved in generation.

A focus on these claims makes clear the need, right at the outset, to make a distinction we have not yet made in regard to efficient causes. Scholars, myself included, often tend to view efficient causes as *material-efficient causes*, cases where the elemental bodies or powers act or interact in a certain way of (what we often call) material necessity. But Aristotle at least as often speaks of *form* and *soul* as source of change, and there are what we might call *formal-efficient causes*, cases where the agent is a formal nature, a *dunamis* for form, and so forth. It is precisely an open question whether in every case in which the agent is said to be a formal nature, there is a material-efficient causal process—a process operating wholly of material necessity—that underlies it. So, in following Aristotle's account of the efficient cause of generation through *GA* I and II in hopes of gaining light on his account of final causation, we will need always to be clear whether we think Aristotle is speaking of material-efficient or of formal-efficient causation (or somehow of both).

Let us return, then, to *GA* I.1. At 715a19 Aristotle begins his argument from the three-part *phainomenon* ('appearance') that some animals are produced from the (i) *coupling* of (ii) *male* and (iii) *female*, i.e., that generation sometimes occurs inside one animal ('female') as a result of some sort of action upon it by another animal ('male') during a process in which relevant parts of each come into contact ('copulation'). This often, though not always, involves a distinct active agent, a 'seed' (*sperma*), supplied at least in part by the male. We will thus understand such generations (and, it turns out, the spontaneous ones as well) when we understand more precisely the respective contributions to this process of the female and the male, including the extent to which each contributes 'seed'. The female evidently contributes some sort of material, and argument from observable data confirms this; this material counts as 'seed' to some extent (which will be important in the discussion of family resemblance in Book IV), but it is evidently incomplete, since females do not generate by themselves. By contrast, although in most kinds of animals the male contributes 'seed', sometimes it does not; this is a crucial observation, since it supports the general claim, extending to all animal kinds that reproduce (a claim which Aristotle argues for on other grounds as well), that the male's power (*dunamis*; cf. 729b4-6) to act on the female material is not exercised by its contributing any additional *material*. Aristotle consolidates his results and states his hypothesis in a key statement late in Book I at 730b8-23:

(T3) One may also grasp from these cases how the male contributes to generation. For not every male emits seed, and in those that do emit it, the seed is no part of the fetus that is being produced, just as nothing comes away from the carpenter to the matter of the timber, nor is there any part of carpentry in the thing produced, but the shape and the form is produced *in* (*egginetai*) the matter *from* the carpenter *through* the movement (*kinēsis*): his soul (in which is the form) and his knowledge move his hands or some other part in a certain sort of movement—different when that which is produced from them is different, the same when it is the same—the hands move the tools, and the tools move the matter. Similarly the male's nature, in those that emit seed, uses the seed as a tool containing movement in actuality (*energeiāi*), just as in the things produced by art the tools are in movement; for the movement of the art is in a way in them.[10]

Further argument in Book II chapter 1 supports and substantially elaborates this hypothesis, moving from the complexity and specificity of the generative product to the nature of its immediate efficient cause. The key passage is 734b17-735a4:

(T4) It is clear, then, that there is something which makes < the parts >, but not by being a certain 'this', present in a completed state at the beginning [sc. but rather by being a *dunamis* and *kinēsis*, 734b8, 11, cf. 729b5-6].

Now, we must grasp how in the world each part is produced from that beginning, by taking first of all as our starting point that everything that is produced by nature or by art is produced *by* a thing that is <whatever it is> in actuality, *out of* what is potentially of that sort. Now the seed and the movement and source it has are such that when the movement stops each part is produced having soul. For it is not face nor flesh unless it has soul: after their death it will be equivocal to say that the one is a face and the other flesh, as it would be if they were made of stone or wood. The uniform parts and the instrumental parts are produced simultaneously. And just as we would not say that fire alone makes an axe or any other tool, neither should we [say this] of foot or hand. Nor, likewise, of flesh, for it too has a function. Now, hard and soft and supple and brittle, and whatever other such conditions (*pathē*) that belong to the parts containing soul—heat and cold would make those, but would not go so far as to make the definition (*logos*) in virtue of which the one is now flesh and the other bone: that is due to the movement (*kinēsis*) from the generator who is in *full actuality* (*entelecheiāi*) what the thing out of which the product comes is *potentially*, just as is the case with the things produced according to an art; for the hot and the cold make the iron hard or soft, but what makes it a *sword* is the movement of the tools which contains a definition (*logos*) belonging to the art. For the art is source and form of the thing produced, but in another; the movement of the nature is in the thing itself, being from another nature which has the form in actuality (*energeiāi*).[11]

[10] In this and subsequent notes, the Greek text provided is from H. J. Drossaart Lulofs's OCT edition (1965): (730b8) *λάβοι δ' ἄν τις ἐκ τού των καὶ τὸ ἄρρεν πῶς συμβάλλεται πρὸς τὴν γένεσιν· οὐδὲ* (b10) *γὰρ τὸ ἄρρεν ἅπαν προΐεται σπέρμα, ὅσα τε προΐεται τῶν ἀρρένων, οὐθὲν μόριον τοῦτ' ἔστι τοῦ γιγνομένου κυήματος, ὥσπερ οὐδ' ἀπὸ τοῦ τέκτονος πρὸς τὴν τῶν ξύλων ὕλην οὔτ' ἀπέρχεται οὐθέν, οὔτε μόριον οὐθέν ἐστιν ἐν τῷ γιγνομένῳ τῆς τεκτονικῆς, ἀλλ' ἡ μορφὴ καὶ τὸ εἶδος ἀπ' ἐκείνου ἐγγίγνεται* (b15) *διὰ τῆς κινήσεως ἐν τῇ ὕλῃ, καὶ ἡ μὲν ψυχὴ ἐν ᾗ τὸ εἶδος καὶ ἡ ἐπιστήμη κινοῦσι τὰς χεῖρας ἤ τι μόριον ἕτερον ποιάν τινα κίνησιν, ἑτέραν μὲν ἀφ' ὧν τὸ γιγνόμενον ἕτερον, τὴν αὐτὴν δὲ ἀφ' ὧν τὸ αὐτό, αἱ δὲ χεῖρες τὰ ὄργανα, τὰ δ' ὄργανα τὴν ὕλην. ὁμοίως δὲ καὶ ἡ φύσις ἡ ἐν τῷ ἄρρενι* (b20) *τῶν σπέρμα προϊεμένων χρῆται τῷ σπέρματι ὡς ὀργάνῳ καὶ ἔχοντι κίνησιν ἐνεργείᾳ, ὥσπερ ἐν τοῖς κατὰ τέχνην γιγνομένοις τὰ ὄργανα κινεῖται· ἐν ἐκείνοις γάρ πως ἡ κίνησις τῆς τέχνης.*

[11] (734b17) *ὅτι μὲν οὖν ἔστι τι ὃ ποιεῖ, οὐχ οὕτως δὲ ὡς τόδε τι οὐδ' ἐνυπάρχον ὡς τετελεσμένον τὸ πρῶτον, δῆλον. πῶς δέ ποτε ἕκαστον γίγνεται* (b20) *ἐντεῦθεν δεῖ λαβεῖν, ἀρχὴν ποιησαμένους πρῶτον μὲν ὅτι ὅσα φύσει γίγνεται ἢ τέχνῃ ὑπ' ἐνεργείᾳ ὄντος γίγνεται ἐκ τοῦ δυνάμει τοιούτου. τὸ μὲν οὖν σπέρμα τοιοῦτον, καὶ ἔχει κίνησιν καὶ ἀρχὴν τοιαύτην ὥστε παυομένης τῆς κινήσεως γίγνεσθαι ἕκαστον τῶν μορίων καὶ ἔμψυχον. οὐ γάρ ἐστι πρόσωπον*

This distinction between the power of heat and cold to produce the *pathē* of the parts and the power of the 'movement' supplied by the father to produce the parts themselves will be crucial, and we will return to it shortly. At this point in the text it raises the question for Aristotle of what the material nature of the semen (the male seed, *gonē*) must be for it to be able to carry the generative *dunamis*. In a long and careful discussion in II.2, which makes use of the chemical theory presented in *Meteorology* IV, Aristotle infers from the unusual reactions of semen to heat and cold that it must contain '*pneuma*', which in its ordinary usage refers to breath or warm wind, but which had apparently begun to take on a role in the medical tradition as an important constituent of living bodies, given the evident centrality to life of breath and its warmth.[12] At 735b37-736a16 Aristotle concludes:

(T5) The seed, then, is a combination of *pneuma* and water, and the *pneuma* is hot air; that is why it is moist in nature, because it is composed of water. . . . [W]hile it is inevitable that one seed is earthier than another and it is so most of all in those which have much earthy stuff in virtue of the bulk of their bodies, it is thick and white because *pneuma* is mixed in. . . . The cause of the whiteness of the seed is that the semen is foam, and foam is something white, especially that which has the smallest constituents, each bubble too small to be seen . . .[13]

Since air is already warm and moist, *pneuma* must carry an additional heat. The nature of this heat is the subject of a crucial passage in the next chapter, at 736b29-737a7 and 18-24:

(T6) Now, every soul-potential seems to be associated with a body different from and more divine than the so-called elements; and as souls differ from each other in value and lack of value, so too this sort of nature differs. For within the seed of all things there is present that which makes the seed fertile, what is called 'the hot'. But this is not fire or that sort of

(b25) *μὴ ἔχον ψυχὴν οὐδὲ σάρξ, ἀλλὰ φθαρέντα ὁμωνύμως λεχθήσεται τὸ μὲν εἶναι πρόσωπον τὸ δὲ σάρξ, ὥσπερ κἂν εἰ ἐγίγνετο λίθινα ἢ ξύλινα. ἅμα δὲ τὰ ὁμοιομερῆ γίγνεται καὶ τὰ ὀργανικά· καὶ ὥσπερ οὐδ' ἂν πέλεκυν οὐδ' ἄλλο ὄργανον φήσαιμεν ἂν ποιῆσαι τὸ πῦρ μόνον οὕτως οὐδὲ πόδα* (b30) *οὐδὲ χεῖρα. τὸν αὐτὸν δὲ τρόπον οὐδὲ σάρκα· καὶ γὰρ ταύτης ἔργον τί ἐστιν. σκληρὰ μὲν οὖν καὶ μαλακὰ καὶ γλίσχρα καὶ κραῦρα καὶ ὅσα ἄλλα τοιαῦτα πάθη ὑπάρχει τοῖς ἐμψύχοις μορίοις θερμότης καὶ ψυχρότης ποιήσειεν ἄν, τὸν δὲ λόγον ᾧ ἤδη τὸ μὲν σὰρξ τὸ δ' ὀστοῦν οὐκέτι, ἀλλ' ἡ κίνησις ἡ* (b35) *ἀπὸ τοῦ γεννήσαντος τοῦ ἐντελεχείᾳ ὄντος ὅ ἐστι δυνάμει ἐξ οὗ γίγνεται, ὥσπερ καὶ ἐπὶ τῶν γιγνομένων κατὰ τέχνην· σκληρὸν μὲν γὰρ καὶ μαλακὸν τὸν σίδηρον ποιεῖ τὸ θερμὸν* (735a1) *καὶ τὸ ψυχρόν, ἀλλὰ ξίφος ἡ κίνησις ἡ τῶν ὀργάνων ἔχουσα λόγον [τὸν] τῆς τέχνης. ἡ γὰρ τέχνη ἀρχὴ καὶ εἶδος τοῦ γιγνομένου, ἀλλ' ἐν ἑτέρῳ· ἡ δὲ τῆς φύσεως κίνησις ἐν αὐτῷ ἀφ' ἑτέρας οὖσα φύσεως τῆς ἐχούσης τὸ εἶδος ἐνεργείᾳ.*

[12] Balme 1992, 161. Cf. Van der Eijk 2000-2001, s.v. (in each volume) breath (*pneuma*); Van der Eijk 2005, s.v. *pneuma*.

[13] (735b37) *ἔστι μὲν οὖν τὸ σπέρμα κοινὸν πνεύματος καὶ ὕδατος, τὸ δὲ πνεῦμά ἐστι θερμὸς ἀήρ· διὸ ὑγρὸν τὴν φύσιν ὅτι ἐξ ὕδατος. . . .* (736a5) *μᾶλλον μὲν γὰρ ἕτερον ἑτέρου σπέρμα γεωδέστερον ἀναγκαῖον εἶναι, καὶ μάλιστα τοιοῦτον ὅσοις πολὺ γεῶδες ὑπάρχει κατὰ τὸν ὄγκον τοῦ σώματος, παχὺ δὲ καὶ λευκὸν διὰ τὸ μεμῖχθαι πνεῦμα. . . .* (a14) *αἴτιον δὲ τῆς λευκότητος τοῦ σπέρματος ὅτι ἐστὶν ἡ γονὴ ἀφρός, ὁ δ' ἀφρὸς λευκόν, καὶ μάλιστα τὸ ἐξ ὀλιγίστων συγκείμενον μορίων καὶ οὕτω μικρῶν ὥσπερ ἑκάστης ἀοράτου τῆς πομφόλυγος οὔσης . . .*

potentiality but rather the *pneuma* enclosed within the seed, that is, within the foamy part, and more precisely the nature in the *pneuma*, being analogous to the element of the stars. That is why fire generates no animal, and none is seen to be constituted in things subjected to fire, whether wet things or dry; but the heat of the sun and that of animals do—not only that conveyed through the seed but also if there should be any other residue of their nature, even this too will contain a life-source (*zōtikē archē*). Such things make it evident, then, that the heat in the animals is neither fire nor has its origin from fire.

. . .

Since the seed is a residue and is moving with the same movement as that with which the body grows when the final nutriment is being apportioned out, when it goes into the uterus it constitutes and moves the female's residue with the same movement as that with which it itself is actually moving. For that too is residue and contains all the parts potentially, though none actually.[14]

It is prima facie odd that, having distinguished at II.1 734b19ff. (T4 above) the causal power of the 'movement' supplied by the father from the causal power of heat and cold, Aristotle here presents that movement as somehow resident in the semen's distinctive *heat*.[15] Let us for the moment, however, continue with our narrative and note that Aristotle has now completed the presentation of the broad lines of his theory of the efficient cause of generation: the father, as ultimate efficient cause, acts on a material supplied by the mother, the (useful part of the) *katamēnia*, but does so through his semen which carries, embedded somehow in a distinctive heat, a set of movements which initiate a process leading in normal circumstances to the production of an offspring one in form with the father. What makes that production possible is that the semen and *katamēnia* are both residues of nutritive material; as a result the set of movements supplied by the father are the

[14] (736b29) *πάσης μὲν οὖν ψυχῆς δύναμις ἑτέρου σώματος ἔοικε κεκοινωνηκέναι καὶ θειοτέρου τῶν καλουμένων στοιχείων· ὡς δὲ διαφέρουσι τιμιότητι αἱ ψυχαὶ καὶ ἀτιμίᾳ ἀλλήλων οὕτω καὶ ἡ τοιαύτη διαφέρει φύσις. πάντων μὲν γὰρ ἐν τῷ σπέρματι ἐνυπάρχει ὅπερ ποιεῖ γόνιμα εἶναι τὰ σπέρματα, τὸ καλούμενον* (b35) *θερμόν. τοῦτο δ᾽ οὐ πῦρ οὐδὲ τοιαύτη δύναμίς ἐστιν ἀλλὰ τὸ ἐμπεριλαμβανόμενον ἐν τῷ σπέρματι καὶ ἐν τῷ ἀφρώδει πνεῦμα καὶ ἡ ἐν τῷ πνεύματι φύσις, ἀνάλογον οὖσα τῷ* (737a1) *τῶν ἄστρων στοιχείῳ. διὸ πῦρ μὲν οὐθὲν γεννᾷ ζῷον, οὐδὲ φαίνεται συνιστάμενον ἐν πυρουμένοις οὔτ᾽ ἐν ὑγροῖς οὔτ᾽ ἐν ξηροῖς οὐθέν· ἡ δὲ τοῦ ἡλίου θερμότης καὶ ἡ τῶν ζῴων οὐ μόνον ἡ διὰ τοῦ σπέρματος, ἀλλὰ κἄν τι περίττωμα τύχῃ τῆς φύσεως* (a5) *ὂν ἕτερον, ὅμως ἔχει καὶ τοῦτο ζωτικὴν ἀρχήν. ὅτι μὲν οὖν ἡ ἐν τοῖς ζῴοις θερμότης οὔτε πῦρ οὔτε ἀπὸ πυρὸς ἔχει τὴν ἀρχὴν ἐκ τῶν τοιούτων ἐστὶ φανερόν . . .*

(737a18) *τοῦ δὲ σπέρματος ὄντος περιττώματος καὶ κινουμένου κίνησιν τὴν αὐτὴν καθ᾽ ἥνπερ τὸ σῶμα αὐξάνεται* (a20) *μεριζομένης τῆς ἐσχάτης τροφῆς, ὅταν ἔλθῃ εἰς τὴν ὑστέραν συνίστησι καὶ κινεῖ τὸ περίττωμα τὸ τοῦ θήλεος τὴν αὐτὴν κίνησιν ἥνπερ αὐτὸ τυγχάνει κινούμενον κἀκεῖνο. καὶ γὰρ ἐκεῖνο περίττωμα, καὶ πάντα τὰ μόρια ἔχει δυνάμει, ἐνεργείᾳ δ᾽ οὐθέν.*

[15] I have suggested in chapter 1 above (p. 19 n. 32) that the distinction between these two kinds of heat (or between these two aspects of the vital heat) reflects the fact that vital heat is indeed *heat* and so contains both material and formal powers or potentials. Compare the illuminating discussion in Freudenthal 1995, 31-34 and cf. n. 4 above and the text to which it is attached. I return to this prima facie puzzle below, p. 103.

very same ones that within himself carry the potentiality to maintain his own form by converting his food into more of himself in the nutritive and growth process. That last point is a stroke of genius on Aristotle's part, explaining how the offspring can—and will in the paradigm case—be like the father; and it provides, incidentally, one part of the basis for the claim which he makes, several times in *GA* as well as in *De Anima*, that the nutritive soul and the generative soul are one and the same potentiality.[16]

In II.4 Aristotle analyzes in more detail the formation of the male and female contributions, as residues of the nutritive process, discusses the initial constitution or 'setting' (*kuēsis, sunistēsis*) of the embryo and its early formation, and turns to the question of its development and differentiation (*diakrisis*, 740b13). Another crucial passage follows, with marked similarities to T4 above (734b19ff.), which is not surprising given their similar aims; the present passage, however, builds in the results of the theory developed since that earlier passage. The reference is 740b12-741a2, a passage that gets rather convoluted towards the end, but whose sense remains clear. I translate literally:

(T7) The differentiation (*diakrisis*) of the parts is produced, not as some think, because it is natural for like to be carried to like . . . , but because the residue of the female is potentially such as the animal is by nature, and the parts are in it potentially and not at all actually; it is due to this cause that each of them is produced, and because when the active and the to-be-acted-upon come into contact, in the manner in which the one is active and the other to-be-acted-upon (I mean by 'manner' the how and where and when), straightaway the one acts and the other is acted upon. [Cf. *Metaph.* Θ.5 1047b35-1048a7.] The female, then, provides the matter, the male the source of the movement. And just as things which are produced by the agency of an art are produced by means of tools—it's truer in fact to say 'by the *movement* of the tools' (this <movement> being the actuality of the art, and the art being the shape of that which comes to be in another)—in this way the nutritive soul-capacity, just as, in the animals and plants themselves later on, it produces growth from the food using as tools heat and cold (for the soul-capacity's movement is *in* these, and each thing is produced according to a certain definition [*logōi tini*]), in this way also at the very beginning the thing produced by a nature is constituted. For it is the same matter by which it grows and from which it is first constituted, so that the productive capacity <for growth> is also the same as the initial one (but greater than it). If then this is the nutritive soul, it is also that which generates, and this is the nature of each <organism> and present in all plants and animals . . .[17]

[16] e.g., *GA* II. 735a17-18, 740b35-37 [most of which is quoted in T8 below], and *An.* II.4 416a19. The other part of the basis for identifying the nutritive and generative faculties is their having the same goal, which is the maintenance in being (in whatever way possible) of the organism. Cf. chapters 1 and 2 above, pp. 32-33 and 58.

[17] I take the passage's long and convoluted antepenultimate sentence (b25-34) to be drawing, within an overarching parallel between artistic and natural (i.e. biological) production, a parallel between the production of growth in an already formed animal or plant and the initial production—i.e., constituting—of that animal or plant. Aristotle's main claim here is the

How shall we translate *logōi tini*—*by* or *according to* a certain definition? In line with 735a1-2 (in T4 above), *hē kinēsis hē tōn organōn echousa logon ton tēs technēs*, perhaps the meaning is 'each thing is produced by a movement whose definition (and thus nature) specifies the shape and form of the product' (cf. chapter 1 above, p. 16).

The latter part of this passage, where the parallels with T4 are found, is important for both of our questions: it bears on the question of the nature of the heat which carries the 'movement' responsible for generation and development, and it bears on the question of whether Aristotle intends to offer a full material account of the generation of animal parts to complement his teleological account, or whether the teleological account, though it makes use of material necessities, is predicated on the unavailability of a full material-level causal account. We will return to both questions shortly.

Since female birds can produce unfertilized eggs that have some life in them ('wind-eggs'), and since a few animal kinds have females and no males, Aristotle

overarching parallel—that just as an art produces its products via tools, that is, via the movement of its tools, so too the nature of the organism works via *its* tools, viz., heat and cold, or more precisely, now, via the movement it imparts to heat and cold. Aristotle wants to make this latter claim in regard to the formation of the offspring from its very conception, but seems to think the claim is clearer in the case of the development of an organism already given birth to; so he draws a second parallel, on the natural side of the overarching art-nature parallel, between, in effect, the forming embryo and the growing child. Just as the relevant aspect of the child's nature, its nutritive soul-capacity, imparts movements to its body's heat and cold, which movements maintain and develop its body and faculties, thus using its heat and cold as 'tools', so too this same nutritive capacity, in its initial form, uses heat and cold as tools, to form the embryo. The sentence reads as if Aristotle has switched focus away from the art–nature parallel with which he began, but because the nature side of that parallel is a nature's use of tools in the formation of the embryo, and that use, he thinks, is for the listener clearer in the case of a child's growth, he reaches the former by paralleling it to the latter. The last sentence of the passage then offers some justification for the growth–embryogenesis parallel. Here is the Greek text of the entire passage: (740b12) *ἡ δὲ διάκρισις γίγνεται τῶν μορίων οὐχ ὥς τινες ὑπολαμβάνουσι διὰ τὸ πεφυκέναι φέρεσθαι τὸ ὅμοιον πρὸς τὸ ὅμοιον (πρὸς* (b15) *γὰρ πολλαῖς ἄλλαις αἷς ὁ λόγος οὗτος ἔχει δυσχερείαις συμβαίνει χωρὶς ἕκαστον γίγνεσθαι τῶν μορίων τῶν ὁμοιομερῶν, οἷον ὀστᾶ καθ' αὑτὰ καὶ νεῦρα καὶ τὰς σάρκας καθ' αὑτάς, εἴ τις ἀποδέξαιτο ταύτην τὴν αἰτίαν)· ἀλλ' ὅτι τὸ περίττωμα τὸ τοῦ θήλεος δυνάμει τοιοῦτόν ἐστιν οἷον φύσει τὸ* (b20) *ζῷον καὶ ἔνεστι δυνάμει τὰ μόρια ἐνεργείᾳ δ' οὐθέν, διὰ ταύτην τὴν αἰτίαν γίγνεται ἕκαστον αὐτῶν, καὶ ὅτι τὸ ποιητικὸν καὶ τὸ παθητικὸν ὅταν θίγωσιν, ὃν τρόπον ἐστὶ τὸ μὲν ποιητικὸν τὸ δὲ παθητικόν (τὸν δὲ τρόπον λέγω τὸ ὧς καὶ οὗ καὶ ὅτε), εὐθὺς τὸ μὲν ποιεῖ τὸ δὲ πάσχει. ὕλην μὲν οὖν* (b25) *παρέχει τὸ θῆλυ, τὴν δ' ἀρχὴν τῆς κινήσεως τὸ ἄρρεν. ὥσπερ δὲ τὰ ὑπὸ τῆς τέχνης γιγνόμενα γίγνεται διὰ τῶν ὀργάνων—ἔστι δ' ἀληθέστερον εἰπεῖν διὰ τῆς κινήσεως αὐτῶν· αὕτη δ' ἐστὶν ἡ ἐνέργεια τῆς τέχνης, ἡ δὲ τέχνη μορφὴ τῶν γιγνομένων ἐν ἄλλῳ—οὕτως ἡ τῆς θρεπτικῆς ψυχῆς δύναμις, ὥσπερ* (b30) *καὶ ἐν αὐτοῖς τοῖς ζῴοις καὶ τοῖς φυτοῖς ὕστερον ἐκ τῆς τροφῆς ποιεῖ τὴν αὔξησιν, χρωμένη οἷον ὀργάνοις θερμότητι καὶ ψυχρότητι (ἐν γὰρ τούτοις ἡ κίνησις ἐκείνης, καὶ λόγῳ τινὶ ἕκαστον γίγνεται), οὕτω καὶ ἐξ ἀρχῆς συνίστησι τὸ φύσει γιγνόμενον. ἡ γὰρ αὐτή ἐστιν ὕλη ᾗ αὐξάνεται καὶ ἐξ ἧς* (b35) *συνίσταται τὸ πρῶτον, ὥστε καὶ ἡ ποιοῦσα δύναμις ταὐτὸ τῷ ἐξ ἀρχῆς· μείζων δὲ αὕτη ἐστίν. εἰ οὖν αὕτη ἐστὶν ἡ θρεπτικὴ ψυχή, αὕτη ἐστὶ καὶ ἡ γεννῶσα· καὶ τοῦτ' ἔστιν ἡ* (741a1) *φύσις ἡ ἑκάστου ἐνυπάρχουσα καὶ ἐν φυτοῖς καὶ ἐν ζῴοις πᾶσιν . . .*

rounds out his presentation with a closer examination of the female contribution to the formal side of generation. He turns finally in chapter 6 to a detailed account of the formation of individual animal parts. This is the *diakrisis* ('articulating out', 'differentiation') that was described generally in the first part of T7, indicated, at 741b7, with the verb *diorizetai* (perhaps 'delineates' or even, in its physical meaning, 'defines'). And it is here, in the detailed accounts of the formation of individual parts, that I am thinking we might look for the reach that material-efficient causation has (and doesn't have) in Aristotle's explanatory account of the generation of animals and their parts.

One view, then, would be roughly this. It would hold that II.6 is meant to point towards the basic pattern of a full material-level account, in which sufficient conditions are given for the production of a sufficient number and range of animal parts to make clear that such an account is always in principle possible. Material-efficient causes would *necessitate* what are clearly treated as teleological outcomes, but would *explain* those outcomes only insofar as they are identified as formal and there because these outcomes are good, etc.[18]

To anticipate, the alternative view we will look at holds that II.6 aims to provide not the 'bottom-up' material-level side of embryogenesis (as I am calling the development of the embryo and its parts), but rather a single, unified account of the entire embryogenesis—an account in which the embryo's formal nature (its potential for form) is playing the central efficient-causal role, by 'making use of' material-efficient agents such as heat and cold and the necessary natures of the materials from which the embryo is being constructed. According to this view, these material-level contributions, though necessary, are by no means sufficient by themselves for embryogenesis. On this view the respective contributions of the formal nature and of the material agency of heat and cold correspond more or less to a natural reading of what we have already been given in general terms in T4 above, and their interplay to what we have already been given in T7 (though we get useful amplification of them both in a passage from II.6 itself we will call below 'T10'). I will defend this second view.

Advocates of the first view would maintain that T4 and T7 are consistent with their thesis that Aristotle sees the relevant material agency as *sufficient* for embryogenesis. We shall see. I provide in a footnote the relevant parts of the two passages, for ease of reference.[19]

[18] I mean such a general formulation to capture the views both of Charles 1988 and Meyer 1992 (on which more shortly). See the previous chapter for a fuller and more precise characterization of their views, and of some important differences between them.

[19] (T4-excerpt) Now, hard and soft and supple and brittle, and whatever other such conditions (*pathē*) that belong to the parts containing soul—heat and cold would make those, but would not go so far as to make the definition (*logos*) in virtue of which the one is now flesh and the other bone: that is due to the movement (*kinēsis*) from the generator who is in *full actuality* (*entelecheiāi*) what the thing out of which the product comes is *potentially*, just as is the case with

The embryogenetic account in II.6

GA II.6 is a lengthy chapter. My tendency for a long time had been to focus in initially at the point in the chapter at which Aristotle has established that the first part to be produced is the heart, which then serves as the 'source' for the rest, because it is in this stretch of text that he appears to be identifying the parts he discusses as being formed of material necessity. Indeed, the discussion opens at 743a1-5 with an indication that the focus will be on the formation, by material agency, of the uniform parts:

(T8) From the heart the blood-vessels extend <throughout the body>, as when artists sketch out preliminary figures on the walls; for the parts are situated around the blood-vessels, because they are generated out of them. The generation [out] of the uniform parts occurs by the agency of cooling and heat, for some things are constituted and solidified by the cold and some by the hot.[20]

It ends at 745b21-22, however, with the remark:

(T9) We have stated, then, how *each* of the parts is constituted, and what the cause of their generation is.[21]

I underscore 'each': not each of the uniform parts, but each of the *parts*. And yet, in the discussion that takes place between T8 and T9, there is no mention of heterogeneously structured non-uniform parts, such as hand or foot. That fact also helps firm up the OCT editor's bracketing of *ek* ('out') at a4 in T8. Though found in all the main manuscripts, *ek* is not read in the earlier ninth-century Arabic translation nor in William of Moerbeke's thirteenth-century Latin translation (itself thought to rest on MSS earlier than any we have). And, the *ek* is just not plausible. Aristotle's standard examples when he is distinguishing between uniform and

the things produced according to an art; for the hot and the cold make the iron hard or soft, but what makes it a *sword* is the movement of the tools which contains a definition (*logos*) belonging to the art. (734b31-735a4)

(T7-excerpt) And just as things which are produced by the agency of an art are produced by means of tools—it's truer in fact to say 'by the *movement* of the tools' (this <movement> being the actuality of the art, and the art being the shape of that which comes to be in another)—in this way the nutritive soul-capacity, just as, in the animals and plants themselves later on, it produces growth from the food using as tools heat and cold (for the soul-capacity's movement is *in* these, and each thing is produced according to a certain definition [*logōi tini*]), in this way also at the very beginning the thing produced by a nature is constituted.... If then this is the nutritive soul, it is also that which generates, and this is the nature of each <organism> and present in all plants and animals. (740b25-741a2)

20 ἐκ δὲ τῆς καρδίας αἱ φλέβες διατεταμέναι, καθάπερ οἱ τοὺς κανάβους γράφοντες ἐν τοῖς τοίχοις· τὰ γὰρ μέρη περὶ ταύτας ἐστίν, ἅτε γιγνόμενα ἐκ τούτων. ἡ δὲ γένεσίς ἐστιν [ἐκ] τῶν ὁμοιομερῶν ὑπὸ ψύξεως καὶ θερμότητος· συνίσταται γὰρ καὶ πήγνυται τὰ μὲν ψυχρῷ τὰ δὲ θερμῷ.

21 πῶς μὲν οὖν ἕκαστον συνίσταται τῶν μορίων εἴρηται, καὶ τί τῆς γενέσεως αἴτιον.

non-uniform parts, as T4 above exemplifies, are, for the uniform parts, flesh and bone, and for the non-uniform parts, heterogeneously structured ones such as hand and foot. But, in the lines immediately following T8, his attention is directed to the formation of flesh (and other uniform parts that harden out of it, such as nail and hoof) and of bone (and sinew), the former resulting from a cooling process, the latter from a kind of internal 'baking'—precisely what T8, without the *ek*, is announcing he will attend to. Whether or not, in the remainder of II.6, Aristotle ever turns to heterogeneously structured non-uniform parts like hand or foot, there is certainly no project of systematically explaining their formation *out of* such uniform parts as flesh and bone. The non-uniform parts may be said to be formed *along with* the uniform ones, i.e., *as* the latter are forming, but not *out of* them.[22] So the *ek* must go.[23]

One reason why the formation of non-uniform parts is not discussed between T8 and T9 *may* be that Aristotle does not think they are formed of necessity by the agency of heat and cold. Indeed, at one time I hypothesized that Aristotle's view was that uniform parts were formed wholly by material necessity but non-uniform parts were not; they, *à la* T4, were due to the formal nature (via specific movements, directed to form, that it initiates in the material). This view was appealing, since it appeared to deny that the embryo as a whole is formed of material necessity; instead, only its simpler (i.e., uniform) parts are.

But there are two problems with that picture, as I have come to realize. First, note that T4 says, of the divide between what is caused by the material agency of heat and cold alone and what is caused by the formal nature, that flesh falls on the latter side, along with hand and foot ('for even [flesh] has a function', 734b30-31). *And*, in the lines that immediately follow in T4, lines which I have singled out above, what is ascribed to the material agency of heat and cold is the production, not of the whole of each uniform part, but only of their *pathē*—their *conditions*, such as their being 'hard and soft and supple and brittle'; and, to be precise, that passage does not reflect the uniform/non-uniform distinction, since what is produced by

[22] Cf., e.g., II.1 734b27-28 (in T4 above): they come to be *hama*, simultaneously. At best, one could say that the stretch between T8 and T9 addresses non-uniform parts that are composed of one uniform part (cf. *PA* II.1 on sense-organs and viscera, and Furth 1987, 34-36). But these are not discussed under that heading (in which case one could imagine Aristotle explicitly beginning a turn to the treatment of the non-uniform parts), and only the eye is discussed at any length, at 743b32-744b12; even there the main focus is on why the eye, an important part formed out of the best of the nourishment and thus, one would expect, first, is in fact formed last. The other sense organs are mentioned once in that discussion (743b34-744a6) and then briefly just after the eye, in a passage asserting generally that more and less valuable parts are formed, respectively, from higher and lower qualities of nutriment ('the bodies of the other sense organs', 744b23).

[23] It is most likely a case of dittography—see the *ek* in the previous line—encouraged by an interpretative confusion. We mustn't, however, forget that earlier *ek*: we will need to explain which parts are generated *out of the blood-vessels*. On the face of it, these are the uniform parts mentioned in the next line, the one we have been discussing; but see p. 106 below.

heat and cold are the *pathē* of each of 'the parts . . . having soul' (b31-33), i.e., the *pathē* of *both* the uniform *and* the non-uniform parts.

That is the first problem for thinking that our II.6 passage might be attributing the uniform parts (though not the non-uniform) to material agency alone: it is the *pathē* (of both the uniform and the non-uniform parts), and not the uniform parts per se, that are caused by the agency of heat and cold alone. The second problem has to do with what actually happens in the stretch of text from T8 to T9. For not only is there no attempt by Aristotle in that stretch to show that non-uniform parts are formed of material necessity from uniform parts (or in any other way), but even the crucial part of the causation in the explanation of uniform parts is not said to be due to material necessity, but to 'the nature' and 'the source of movement'. Some aspects of the formation of the 'eye body' are ascribed to (material) necessity (e.g., at 744a13) but the eye itself is formed when 'the source of movement takes control (*kratei*)' (744a33-34), which is late for an upper part (744a33-35); the eyelid, needed only when there is an eye, is also formed late, since 'a nature makes nothing superfluous or without a point' (744a36-38). Likewise, the use of higher-quality nutriment for more valuable parts, just mentioned, is said to be the work of a nature internal to the forming animals.[24] And notice the agency of the nature (*hē phusis*) at 745a32, in the discussion of teeth. There is really very little that is ascribed to material necessity alone. For, even in Aristotle's own context, quite apart from any consideration of the views of his predecessors, accounts of formations wholly in terms of material necessity are in danger of treating those formations as accidental (743a21-23, to be discussed momentarily as part of T10). This danger is avoided, we will see, not by an appeal, such as Susan Sauvé Meyer makes,[25] to a 'reliable material mechanism'—there does not seem to be any such thing to appeal to, and in any case we hear of none. (We'll return to this shortly.) The danger is avoided by appeal—in the very same causal account that incorporates such material necessity as is at work—*to the agency of the developing animal's formal nature*. In addition to the mentions of such agency I have already just cited, there is the important explicit general remark, at 743a17-b5, that I will label 'T10'.

To set up this passage, before I quote it, we need to return to a puzzle I noted earlier in comparing T4 and T6, but did not try to resolve.[26] T4 distinguishes the agency of the 'movement' supplied by the formal nature from the material agency of heat and cold (734b31-36). T6 speaks of a soul- (or form-) specific generative heat which is somehow associated with those formative movements (736b29-737a1, 18-22). The question of whether heat and cold are separate material agents, or rather express the agency of the formal nature, arises again in subsequent

[24] *οὕτως ἐν τοῖς γιγνομένοις αὐτοῖς ἡ φύσις ἐκ μὲν τῆς καθαρωτάτης ὕλης σάρκας καὶ τῶν ἄλλων αἰσθητηρίων τὰ σώματα συνίστησιν.*

[25] See the previous chapter, pp. 74-75, and below, p. 113 n. 44.

[26] Other than to make some back references (in n. 15), which were not meant to be decisive.

passages. In T7 the formal nature is said to 'use heat and cold' as tools, because its formative movement and definition is resident in them (740b25-35). In T8, as we noted when we introduced that passage, heat and cold appear to be spoken of as material agents, having certain causal powers as such, e.g., to solidify (743a1-5); however, their being said in the same breath also to 'constitute' (a5) leaves open the possibility that here too they may be 'tools' of the formal nature; the appearance of 'constituted' again in T9 (745b21) perhaps suggests as much. The puzzle, then, is twofold: (a) to what extent, and when, are references to the agency of hot and cold references to a material-efficient agency that is to be distinguished from any agency (or explanatory role) of the formal nature; and (b) in those cases where heat and cold are said to be 'tools' of the formal nature, being 'used' by it for its own more global ends, how is the agency of heat and cold here to be understood, and how does its material dimension stand in relation to its role as tool of the formal nature?

Our next passage, 743a6-b5, which I am calling 'T10', is a very long one, but it should help us with those questions. It follows immediately upon T8 (which, again is 743a1-5), and is the first part of the stretch we have been discussing, from T8 to T9. To set the context, I repeat the immediately preceding lines, which we have called 'T8', and then identify the beginning of this new material:

(T8) From the heart the blood-vessels extend <throughout the body>, as when artists sketch out preliminary figures on the walls; for the parts are situated around the blood-vessels, because they are generated out of them. The generation [out] of the uniform parts occurs by the agency of cooling and heat, for some things are constituted and solidified by the cold and some by the hot.

(T10) We have spoken previously elsewhere concerning the differences between these, and have stated what sorts of things are dissolvable by fluid and by fire and what sorts of things are not dissolvable by fluid and cannot be melted by fire. Now, as the nourishment oozes through the blood-vessels and the passages in each of them—just as water does when it stands in unbaked earthenware—flesh, or its counterpart, is formed (*gignontai*), being constituted by the cold (which is why it is dissolvable by fire). As that nourishment wells up, the excessively earthy stuff in it, having little fluidity and heat, becomes cooled while the fluid is evaporating together with the hot, and is formed into parts that are hard and earthy, e.g., nails, horns, hoofs, and bills; that is why these can be softened, but none of them can be melted by fire; though some, e.g., eggshell, can be melted by fluids.

It is by the internal heat that both the sinews and the bones are formed as the moisture dries up. That is in fact why bones, like pottery, are indissolvable; for they have been baked by the heat present in their generation, as though in an oven. But this <heat> does not make any chance thing flesh or bone, nor in any chance place nor at any chance time; rather <it makes into flesh or bone> that which is naturally so <made> and where it is naturally so <made> and when it is naturally so <made>. For neither will that which is potentially *be* by the agency of a motive source that doesn't have the actuality, nor will that which has the actuality produce <a thing> out of any chance thing, just as the carpenter would not make a chest out of something other than wood and just as there will not be a chest <made> out of the wood without him.

The heat is present in the spermatic residue having its movement and actuality in an amount and character suitably proportioned (*summetros*) to each of the parts. To the extent to which it is deficient or excessive, the thing being formed will come out inferior or deformed, similarly to what happens with things that are constituted outside the body, via boiling, for culinary enjoyment or any other use. But there we supply a suitable proportion (*summetria*) of heat to the movement, while here it is provided by the nature of the generator. In spontaneous generations the cause is a seasonal movement and heat.

Cold is a deprivation of heat. The nature uses both [sc. cold and heat], as having a potentiality to make, of necessity, one thing this and another that; in the generations <under discussion>, however, their cooling and heating occur for the sake of something, that is, each of the parts is formed such <as it is>—the flesh soft, the sinew dry and elastic, the bone dry and crackable—in one respect of necessity, in another respect for the sake of something.[27]

The formation of parts subsequent to the heart is described in this passage in the following order (which may or may not be the chronological order of their formation):

[27] My thanks to Devin Henry for reminding me some time ago of the importance of this passage for the issues of this chapter, and to Alan Bowen and Jim Lennox for valuable advice on the passage's translation. The Greek of this important passage reads (beginning with T8):

(T8) (743a1) *Ἐκ δὲ τῆς καρδίας αἱ φλέβες διατεταμέναι, καθάπερ οἱ τοὺς κανάβους γράφοντες ἐν τοῖς τοίχοις· τὰ γὰρ μέρη περὶ ταύτας ἐστίν, ἅτε γιγνόμενα ἐκ τούτων. ἡ δὲ γένεσίς ἐστιν [ἐκ] τῶν ὁμοιομερῶν ὑπὸ ψύξεως καὶ θερμότητος·* (a5) *συνίσταται γὰρ καὶ πήγνυται τὰ μὲν ψυχρῷ τὰ δὲ θερμῷ.* (T10) (743a6) *περὶ δὲ τῆς τούτων διαφορᾶς εἴρηται πρότερον ἐν ἑτέροις, ποῖα λυτὰ ὑγῷ καὶ πυρὶ καὶ ποῖα ἄλυτα ὑγρῷ καὶ ἄτηκτα πυρί. διὰ μὲν οὖν τῶν φλεβῶν καὶ τῶν ἐν ἑκάστοις πόρων διαπιδύουσα ἡ τροφή, καθάπερ ἐν τοῖς ὠμοῖς κεραμίοις* (a10) *τὸ ὕδωρ, γίγνονται σάρκες ἢ τὸ ταύταις ἀνάλογον ὑπὸ τοῦ ψυχροῦ συνιστάμεναι, διὸ καὶ λύονται ὑπὸ πυρός. ὅσα δὲ γεηρὰ λίαν τῶν ἀνατελλόντων, ὀλίγην ἔχοντα ὑγρότητα καὶ θερμότητα, ταῦτα δὲ ψυχόμενα ἐξατμίζοντος τοῦ ὑγροῦ μετὰ τοῦ θερμοῦ γίγνεται σκληρὰ καὶ γεώδη τὴν μορφήν, οἷον* (a15) *ὄνυχες καὶ κέρατα καὶ ὁπλαὶ καὶ ῥύγχη· διὸ μαλάττεται μὲν πυρί, τήκεται δ' οὐθέν, ἀλλ' ἔνια τοῖς ὑγροῖς, οἷον τὰ κελύφη τῶν ᾠῶν.*

ὑπὸ δὲ τῆς ἐντὸς θερμότητος τά τε νεῦρα καὶ τὰ ὀστᾶ γίγνεται ξηραινομένης τῆς ὑγρότητος. διὸ καὶ ἄλυτά ἐστι τὰ ὀστᾶ ὑπὸ τοῦ πυρὸς καθάπερ κέραμος· οἷον (a20) *γὰρ ἐν καμίνῳ ὠπτημένα ἐστὶν ὑπὸ τῆς ἐν τῇ γενέσει θερμότητος. αὕτη δὲ οὔτε ὅ τι ἔτυχε ποιεῖ σάρκα ἢ ὀστοῦν οὔθ' ὅπου ἔτυχεν οὔθ' ὁπότ' ἔτυχεν, ἀλλὰ τὸ πεφυκὸς καὶ οὗ πέφυκε καὶ ὅτε πέφυκεν. οὔτε γὰρ τὸ δυνάμει ὂν ὑπὸ τοῦ μὴ τὴν ἐνέργειαν ἔχοντος κινητικοῦ ἔσται, οὔτε τὸ τὴν ἐνέργειαν ἔχον ποιήσει ἐκ τοῦ τυχόντος,* (a25) *ὥσπερ οὔτε κιβωτὸν μὴ ἐκ ξύλου ὁ τέκτων ποιήσειεν οὔτ' ἄνευ τούτου κιβωτὸς ἔσται ἐκ τῶν ξύλων.*

ἡ δὲ θερμότης ἐνυπάρχει ἐν τῷ σπερματικῷ περιττώματι τοσαύτην καὶ τοιαύτην ἔχουσα τὴν κίνησιν καὶ τὴν ἐνέργειαν ὅση σύμμετρος εἰς ἕκαστον τῶν μορίων. καθ' ὅσον δ' ἂν ἐλλείπῃ ἢ (a30) *ὑπερβάλλῃ ἢ χεῖρον ἀποτελεῖ ἢ ἀνάπηρον τὸ γιγνόμενον, παραπλησίως τοῖς ἔξω συνισταμένοις διὰ τῆς ἑψήσεως πρὸς τροφῆς ἀπόλαυσιν ἤ τινα ἄλλην ἐργασίαν. ἀλλ' ἐνταῦθα μὲν ἡμεῖς τὴν τῆς θερμότητος συμμετρίαν εἰς τὴν κίνησιν παρασκευάζομεν, ἐκεῖ δὲ δίδωσιν ἡ φύσις ἡ τοῦ γεννῶντος.* (a35) *τοῖς δὲ αὐτομάτως γιγνομένοις ἡ τῆς ὥρας αἰτία κίνησις καὶ θερμότης.*

ἡ δὲ ψύξις στέρησις θερμότητός ἐστιν. χρῆται δ' ἀμφοτέροις ἡ φύσις ἔχουσι μὲν δύναμιν ἐξ ἀνάγκης ὥστε (743b1) *τὸ μὲν τοδὶ τὸ δὲ τοδὶ ποιεῖν, ἐν μέντοι τοῖς γιγνομένοις ἕνεκά τινος συμβαίνει τὸ μὲν ψύχειν αὐτῶν τὸ δὲ θερμαίνειν καὶ γίγνεσθαι τῶν μορίων ἕκαστον, τὴν μὲν σάρκα μαλακὴν τῇ μὲν ἐξ ἀνάγκης ποιούντων τοιαύτην τῇ δ' ἕνεκά τινος, τὸ δὲ* (b5) *νεῦρον ξηρὸν καὶ ἑλκτὸν τὸ δ' ὀστοῦν ξηρὸν καὶ θραυστόν.*

(i) blood-vessels;[28]
(ii) flesh;
(iii) parts that are hard and earthy, e.g., nails, horns, hoofs, and bills;
(iv) sinews and bones.

This remark about the agency by which sinews and bones are formed is brought home with an apparently more general example that speaks of

(v) flesh or bone,

and then elaborates on the agency involved, and the conditions of that agency; the final lines of the passage then take as their subject flesh, sinew, *and* bone. This suggests that when Aristotle switches to flesh or bone he is switching to commenting on the agency involved in the formation of *all* uniform parts. I will provide confirmation of this in a moment. But let us return to the start of this list to note the agency to which each of these formations is ascribed:

(i) The blood-vessels are not clearly assigned to any agency, but the process is analogized to the sketching out by artists of preliminary figures on walls, and it is said that on that framework certain (unspecified) parts form 'around' the blood vessels; the text goes on to speak about 'constitution' and 'solidification' of the uniform parts, and this is ascribed to the agency of heat and cold. There is no mention in this discussion of the action of a formal nature, though this *might* be implied by the craft analogy, and by the fact that the blood-vessels are formed first after the heart, which is the *seat* (*archē*) of the nutritive soul and formal nature.[29]

(ii) and (iii) Flesh is 'constituted' by the agency of the cold as the blood 'wells up' through the walls of the blood vessels; the cold also produces hard and earthy parts, such as nails and horns; in both cases there are frequent references to 'chemical' facts that support the assignment of material agency, with no (immediate) reference to a formal nature.[30]

(iv) Sinews and bones likewise have 'chemical' properties that show that they are formed by some sort of heating; the heating is assigned to the 'internal heat'.[31]

[28] I put aside for the moment those parts, whatever they are, that are said there to be generated 'out of the blood-vessels' (see n. 23 above).

[29] Indeed, as Mariska Leunissen has suggested to me, the agent of this process has been mentioned earlier in 742a37-b3 and 742b33-35: it is the heart—the part that 'has the source and end of the whole nature' (742b1).

[30] By 'chemical facts' I mean the sort of material causal connections that may be found in *Meteor.* IV.1-11. These have to do with the connection between a material body's perceptible properties and the elementary powers from which the body is constituted. Compare, e.g., 743a11-17 with *Meteor.* IV.6, esp. 383a26-32 (n.b. *keras* [in sing. or pl.] at 743a15 and 383a32).

[31] It is not clear whether the mention here of *internal* heat (*hupo d'entos thermotētos*) brings in a reference to formal powers, as 'vital-heat' or 'soul-heat' does. Although I suspect it may (cf. 742a14-16, and Peck *ad loc.,* on the internal formation of fresh *pneuma*), I will not appeal in my argument to this mention.

(v) Since 'flesh or bone' is introduced as subject in the midst of a statement about the formation of 'sinews and bones', and since 'flesh or bone' becomes the subject for the rest of what is said, I think it is clear that 'flesh or bone' is a stand-in for 'all the uniform parts'. The immediate agent for the formation of flesh or bone is 'the heat present in the spermatic residue';[32] this heat does not make into flesh or bone 'any chance thing', 'at any chance place' and 'at any chance 'time', but only a 'natural' material at a 'natural' place and a 'natural' time (*to pephukos kai hou pephuke kai hote pephuken*). The remainder of the passage then makes the following four points about that heat:

(a) It must be such as to 'have its (full) actuality', at least in the way that the semen and *katamēnia* have it in full actuality;[33]
(b) This heat, which was said to be present in the spermatic residue, is now said to be due to 'the nature of the generator';[34]
(c) The heat must be there 'in an amount and character suitably proportioned (*summetros*) to' (the movement needed to produce) each particular part;[35] and
(d) The heat (and cold) which supplies the already mentioned movement and (thereby) material agency is *made use of* by the (formal) nature that is at work,[36] in such a way that, while the immediate effects of the material

[32] As Mariska Leunissen has also pointed out to me, this 'spermatic residue' must be what remains of the mother's contribution, since the father's residue is no part of the *kuēma* (fetus), while the mother's residue has already been identified (at 744b28-38) as the first, and best, nutriment, used for the most important parts. On the other hand, see the end of the next note.

[33] 743a28. Matters here are a bit tricky: as my previous note suggests, the 'actuality' mentioned in (a) is ascribed to the heat in the mother's residue, i.e., the *katamēnia*. By contrast, T3 above, the first of our passages to speak of actuality, does not mention heat, or the mother, but says that the semen, qua tool of 'the nature in the male', contains its movement 'in actuality' (unlike a craftsman's tool which, when not in touch with the craftsman, has the movement of the art only potentially). T4, to be cited again momentarily, likewise speaks only of movement when it speaks of actuality, this time of the movement of 'the generator', which is said to have that movement 'in full actuality' (*entelecheiāi*). T7 also speaks of the movements imparted to the craftsman's tools as containing the actuality of the art, but, in speaking of animal generation, analogizes to those tools the heat and cold which serve as agents of the nutritive soul. What links our claim in T10 with those of T3, T4, and T7 even more closely than it might otherwise seem to be linked is (i) the fact that both T10 and T7 speak of the movement as carrying, and using as a tool, an appropriate heat; and (ii) the fact that the 'spermatic residue' in T10, though originally the mother's *katamēnia*, is *katamēnia* that has been 'set' or 'constituted' by the heat from the *sperma*; the *kuēma*'s own heat is thus closer in character or quantity to the father's heat than that heat was before the *katamēnia* was 'set'. This, I take it, helps explain why Aristotle can go on immediately to speak of the 'spermatic heat' as the heat of 'the generator'.

[34] 743a34. See the previous note.

[35] The fact that Aristotle now speaks of a *summetria* for 'each of the parts' (743a29) further supports my suggestion that he is now speaking of the formation of uniform parts in general. For more on this *summetria*, see below, pp. 108-109.

[36] 743a36-37.

agency—'the flesh soft, the sinew dry and elastic, the bone dry and crackable'—are *of necessity*, the formal nature making use of it is producing that result (under some description) *for the sake of something.*

Our question is what the respective roles are of the formal and material natures in the formation of the uniform parts, as this is described in this passage. A close look makes it clear that those roles are at least very close to what they are said to be in passages T4 and T7 above. Here, again, are the operative parts of T4 and T7;[37] to show the parallels in their accounts of the roles of heat and cold both in their own right and as tools of the formal nature, I identify via the same letters in square brackets the portions of T4 corresponding to the portions of T10 that are given letters in round brackets in the summary just above.

(T4) Now, hard and soft and supple and brittle, and whatever other such conditions (*pathē*) that belong to the parts containing soul—[d] heat and cold would make those, but would not go so far as to make the definition (*logos*) in virtue of which the one is now flesh and the other bone: that is due to [b] the movement (*kinēsis*) from the generator who is [a] in *full actuality* (*entelecheiāi*) what the thing out of which the product comes is *potentially*, just as is the case with the things produced according to an art; for [d] the hot and the cold make the iron hard or soft, but what makes it a *sword* is the movement of the tools which contains a definition (*logos*) belonging to the art.

(T7) And just as things which are produced by the agency of an art are produced by means of tools—it's truer in fact to say [d] 'by the *movement* of the tools' (this <movement> being [a] the actuality of the art, and the art being the shape of that which comes to be in another)—in this way [d] the nutritive soul-capacity, just as, in the animals and plants themselves later on, it produces growth from the food using as tools heat and cold (for [a] the soul-capacity's movement is *in* these, and each thing is produced according to a certain definition [*logōi tini*]), in this way also at the very beginning the thing produced by a nature is constituted. . . . If then this is the nutritive soul, [b] it is also that which generates, and this is the nature of each <organism> and present in all plants and animals.

(a), (b), and (d) show clearly that the same causal theory that is outlined in T4 and T7 lies behind T10—and that the primary generative agent of the uniform parts in T10 is the formal nature of the embryo, aimed at the production of an offspring one in form with the parents. T4 attributed to heat and cold as agents in their own right only the *pathē* of the ensouled parts, giving as examples their hard or soft and their supple or brittle character. The last lines of T10 provide as examples of what is produced of necessity *pathē* characteristic of the uniform parts under discussion: 'the flesh soft, the sinew dry and elastic, the bone dry and crackable'. (c), which we find in T10 but not in T4 and T7, introduces a concept we have not seen before in *GA*: the *summetria* (which I have translated 'suitable proportion') between the 'amount and character' of the 'movement and actuality' of the internal generative heat that

[37] For the full passages, see above, pp. 95 and 98.

embodies the formal nature and the parts being produced. This is a difficult notion, made more difficult to pin down by its long and varied history from ancient Greece to the present day.[38]

I take it, however, that this requirement is a specification, or narrowing, of the requirement conveyed in T10 at 743a21-26: the immediate agent for the formation of flesh or bone is 'the heat present in the spermatic residue'[39] this heat does not make into flesh or bone 'any chance thing', 'at any chance place' and 'at any chance 'time', but only a 'natural' material at a 'natural' place, and a 'natural' time (*to pephukos kai hou pephuke kai hote pephuken*). (c) presumably stresses that the heat is 'targeted' to the specific material, place, and time needed. Such targeting is a function of the formal nature, so once again we have T10 paralleling T4 and T7: the formings of the uniform parts are ascribed by Aristotle to the formal nature. Only the *pathē* of the non-uniform parts, as such, are due to heat and cold, as such.[40]

This does not mean, as already said, that II.6 is concerned only with the formation of uniform parts. In fact, this will be confirmed if we, now, widen our focus to the entire chapter, and note that the discussion of uniform parts in T10 is embedded in a larger, single, partly chronological narrative of the order of the formation of the entire embryo and its parts. This narrative begins very early in II.6. It sets as its primary focus 'what parts come into being after what' (742a16-17), and though it names very few non-uniform parts, it speaks of, for example, the upper and lower parts of the body, the head, the eyes, and the thighs. And it is in this context that attention is given, in the passage we have already discussed (T10), to the formation of the uniform parts, such as flesh, bone, and sinew, and, in particular, to the formation of their *pathē*.

This narrative is structured overall by an extended passage, very early in the chapter, that aims to connect various priority relationships of the parts, in difference senses of priority, with the order of their coming to be. The passage is surprisingly

[38] See Hon and Goldstein 2010. Their discussion of *summetria* in Aristotle is not of great help for our purposes, although I do make use of their English translation for the relevant meaning of the word. Oddly, they do not discuss these two *GA* occurrences.

[39] On this 'spermatic residue', see nn. 33 and 34 above. This is the female residue *after* it has been acted upon by the heat and movements of the male *sperma*.

[40] This reading is confirmed by the concluding lines of T10, 743a37-b5: 'Cold is a deprivation of heat. The nature uses both [sc. cold and heat], as having a potentiality to make, of necessity, one thing this and another that; in the generations <under discussion>, however, their cooling and heating occur for the sake of something, that is, each of the parts is formed such <as it is>—the flesh soft, the sinew dry and elastic, the bone dry and crackable—in one respect of necessity, in another respect for the sake of something.' Heat and cold, *acting as such, from material necessity*, produces the *pathē* of the parts—'the flesh soft, the sinew dry and elastic, the bone crackable'. They have no larger direction, as it were. But, in another respect—i.e., (I supply) *qua used by the formal nature*—they are directed *heneka tou*—for the sake of the functional parts they are tools in creating.

cumbersome, repetitive, and over-long (so far as I can tell). I will quote its first two-thirds or so in full, in part to show that the narrative includes the wide range of parts I have mentioned, and in part to allow us to see how II.6 carries out a good part of what we identified at the beginning of this chapter as the overall aim of *GA*—the integration of the efficient and final causes of generation. The portion I will quote is 742a16-b17.

(T11) Some of the early students of nature made an attempt to state what comes to be after what among parts, but had little experience of what actually happens. For among parts as in other things one is naturally prior to another, and there are many sorts of priority, for the thing for the sake of which and the thing for the sake of this differ, and the latter is prior in generation but the former is so in being. The thing for the sake of this also has two differentiae—the thing from which motion originates *and* the thing which that for the sake of which uses. I mean, for example, the thing which can generate and the thing which is instrumental to the thing generated; for one of these must be present earlier—the one that makes (e.g., the *aulos* teacher *before* the learner and the *auloi after* the one learning to play the *aulos*, since it is superfluous that there be *auloi* present to those who do not know how to play them).Given that there are three [*sc.* sorts of relata in the various priority relations we have distinguished[41]]—one being the end which we say is the thing for the sake of which, a second (belonging to things for the sake of this) being that which initiates movement and is generative (for that which makes and is generative, insofar as they are such, are relative to the thing made and generated), and third being useful, that is, being the thing the end uses—it is necessary that there be present first the part in which is the source of movement (for from the start this is one part of the end and indeed the most governing part), then, after this, is the whole and the end; and third, and last, the parts instrumental to these in relation to some uses.

As a result, if there is any such thing which must be present in animals, in that it contains a source and end of its whole nature, then this must come to be in so far as it is first a mover and in so far as it is part of the end, along with the whole. As a result, all of the instrumental parts that can generate the nature must themselves always be present earlier (for they are for the sake of another *as a source*), but all the instrumental parts that are not of this sort must be present later than the ones for the sake of another. That is why it is not easy to distinguish which of the parts are prior, all those which are for the sake of another or that for the sake of which these are. For the parts that initiate movement complicate things, in that they are

[41] I add this phrase to bring out the progression of Aristotle's argument. He starts from a general account of different priority relationships in a wide class of comings-to-be—cf. 'among parts *as in other things*' (742a18-19). These 'other things' include craft comings-to-be, such as in the case of the coming to be of a trained music student presented at a26-28. Having indicated the range of comings-to-be he will consider, he then (a20-23) identifies the different sorts of priority relationships present in these comings-to-be and, incidentally, the different relata encompassed by these relationships. At a28-32, in a genitive absolute construction, he focuses in on those three distinct relata, still at the general level of comings-to-be. Only then, at a32-33 does he return to animals parts in particular, to cash out this analysis of priority relationships that determine *ti meta ti ginetai*. This manner of progression helps to some extent to explain the slow, lumbering, and to my mind repetitive progression of the text here and especially in what follows—a progression which is not even completed at the point where I leave off in my quotation.

prior in the generation of the end, and it is not easy to distinguish the movement-initiating parts from the instrumental.

And yet in accordance with this method one must seek what comes to be after what, for the end is later than some parts but earlier than others. And on account of this the part which has the source comes to be first, then following it the upper body. That is why the parts about the head and the eyes in embryos appear largest, while those below the umbilicus, e.g., the thighs, appear small. The reason is that the lower parts are for the sake of the upper, and are neither parts of the end nor generative of it.[42]

As I mentioned, this passage aims at providing a framework for creating an account of the order in which the parts of an embryo come to be, an account which will include both uniform and non-uniform parts. In identifying the generative parts as coming to be first, and the source of these—the heart or its analogue—as that which comes to be first of all,[43] this passage identifies an efficient causal sequence, though only in the broadest strokes. It is only late in the passage, and then again later in the chapter, that there is any specificity in the efficient causal account of

[42] I thank Alan Bowen for his help with the translation of this passage. Here is the Greek: (742a16) *τῶν δ' ἀρχαίων τινὲς φυσιολόγων τί μετὰ τί γίγνεται τῶν μορίων ἐπειράθησαν λέγειν, οὐ λίαν ἐμπειρικῶς ἔχοντες τῶν συμβαινόντων. τῶν γὰρ μορίων ὥσπερ καὶἐπὶ τῶν ἄλλων πέφυκεν ἕτερον ἑτέρου πρότερον. τὸ δὲ πρότερον* (a20) *ἤδη πολλαχῶς ἐστιν· τό τε γὰρ οὗ ἕνεκα καὶ τὸ τούτου ἕνεκα διαφέρει, καὶ τὸ μὲν τῇ γενέσει πρότερον αὐτῶν ἐστιτὸ δὲ τῇ οὐσίᾳ. δύο δὲ διαφορὰς ἔχει καὶ τὸ τούτου ἕνεκα· τὸμὲν γάρ ἐστιν ὅθεν ἡ κίνησις, τὸ δὲ ᾧ χρῆται τὸ οὗ ἕνεκα. λέγω δ' οἷον τό τε γεννητικὸν καὶ τὸ ὀργανικὸν τῷ γεννωμένῳ·* (a25) *τούτων γὰρ τὸ μὲν ὑπάρχειν δεῖ πρότερον, τὸ ποιητικόν, οἷον τὸ διδάξαν τοῦ μανθάνοντος, τοὺς δ' αὐλοὺς ὕστερον τοῦ μανθάνοντος αὐλεῖν· περίεργον γὰρ μὴ ἐπισταμένοις αὐλεῖν ὑπάρχειν αὐλούς· τριῶν δ' ὄντων—ἑνὸς μὲν τοῦ τέλους ὃ λέγομεν εἶναι οὗ ἕνεκα, δευτέρου δὲ τῶν τούτου ἕνεκα τῆς ἀρχῆς* (a30) *τῆς κινητικῆς καὶ γεννητικῆς (τὸ γὰρ ποιητικὸν καὶ γεννητικόν, ᾗ τοιαῦτα, πρὸς τὸ ποιούμενόν ἐστι καὶ γεννώμενον), τρίτου δὲ τοῦ χρησίμου καὶ ᾧ χρῆται τὸ τέλος—πρῶτον μὲν ὑπάρχειν ἀναγκαῖόν τι μόριον ἐν ᾧ ἡ ἀρχὴ τῆς κινήσεως (καὶ γὰρ εὐθὺς τοῦτο τὸ μόριόν ἐστι τοῦ τέλους ἓν καὶ κυριώτατον),* (a35) *ἔπειτα μετὰ τοῦτο τὸ ὅλον καὶ τὸ τέλος, τρίτον δὲ καὶ τελευταῖον τὰ ὀργανικὰ τούτοις μέρη πρὸς ἐνίας χρήσεις. ὥστ' εἴ τι τοιοῦτόν ἐστιν ὅπερ ἀναγκαῖον ὑπάρχειν ἐν τοῖς* (742b1) *ζῴοις, τὸ πάσης ἔχον τῆς φύσεως ἀρχὴν καὶ τέλος, τοῦτο γίγνεσθαι πρῶτον ἀναγκαῖον, ᾗ μὲν κινητικὸν πρῶτον, ᾗ δὲ μόριον τοῦ τέλους μετὰ τοῦ ὅλου. ὥστε τῶν μορίων τῶν ὀργανικῶν ὅσα μέν ἐστι γεννητικὰ τὴν φύσιν, ἀεὶ πρότερον* (b5) *δεῖ ὑπάρχειν αὐτά (ἄλλου γὰρ ἕνεκά ἐστιν ὡς ἀρχή), ὅσα δὲ μὴ τοιαῦτα τῶν ἄλλου ἕνεκα ὕστερον. διὸ οὐ ῥᾴδιον διελεῖν πότερα πρότερα τῶν μορίων, ὅσα ἄλλου ἕνεκα ἢ οὗ ἕνεκα ταῦτα. παρεμπίπτει γὰρ τὰ κινητικὰ τῶν μορίων πρότερον ὄντα τῇ γενέσει τοῦ τέλους, τὰ δὲ κινητικὰ πρὸς* (b10) *τὰ ὀργανικὰ διελεῖν οὐ ῥᾴδιον. καίτοι κατὰ ταύτην τὴν μέθοδον δεῖ ζητεῖν τί γίγνεται μετὰ τί· τὸ γὰρ τέλος ἐνίων μὲν ὕστερον ἐνίων δὲ πρότερον. καὶ διὰ τοῦτο πρῶτον μὲν τὸ ἔχον τὴν ἀρχὴν γίγνεται μόριον, εἶτ' ἐχόμενον τὸ ἄνω κύτος. διὸ τὰ περὶ τὴν κεφαλὴν καὶ τὰ ὄμματα μέγιστα* (b15) *κατ' ἀρχὰς φαίνεται τοῖς ἐμβρύοις, τὰ δὲ κάτω τοῦ ὀμφαλοῦ, οἷον τὰ κῶλα, μικρά. τοῦ γὰρ ἄνω τὰ κάτω ἕνεκεν καὶ οὔτε μόρια τοῦ τέλους οὔτε γεννητικὰ αὐτοῦ.*

For the balance of this passage see first below, chapter 7, p. 157, and then the next note.

[43] 'Now with those things that do not change the origin or source (*archē*) is the essence (*ti esti*); but in things that come into being, immediately there are several [sources], and another manner [of source] and not all the same. Among them, one in number, is the source from which the movement comes, and that is why the heart is the first part which all blooded animals have, as we have said at the beginning; in others the analogue to the heart comes to be first' (742b33–743a1).

the formation of parts, and even there, as we have seen, primarily in the account of the uniform parts' coming to have the *pathē* they do. The focus of this passage is rather on the logical movement from the *end* of the development—that for the sake of which it takes place—to that which comes to be *for that*: first the source or origin of the development—the parts which form first and facilitate the formation of subsequent parts. This is a movement via *conditional necessity*—and it mirrors surprising well what is called for in the *PA* I.1 statement of the aims of a proper treatment of generation that we looked at, at the beginning of this chapter:

(T2) And because he is such a thing, his coming to be necessarily happens so and is such. And that is why this part comes to be first, and then this.

In (T11) the conditionally necessary inference starts from 'the thing for the sake of which'—the *end* (*telos*) of generation, which is, roughly, the integrated set of parts of the organism which are essential to its functioning, all of which are found in the upper part of the body. The inference then moves to two sorts of parts which are for the sake of that end—the generative parts, including the heart (or its analogue), and the instrumental parts, which serve the more fundamental parts and the organism as a whole. Those include the parts found in the lower part of the body. The situation is complicated, as the passage says (742b6-10), for several reasons; the one we will stop over is that the generative *source* (*archē*) of the embryonic development, the heart (or its analogue), is also the *source* for the organism as such at each of its stages—the seat of its soul—through the rest of its life. As such it is part of the end—the thing for the sake of which—as well as, in its generative capacity, a key part that comes to be and is for the sake of that end. But, that understood—and the relationship between the lower, instrumental parts and the upper, more defining parts understood as well—our passage is easily interpreted as satisfying the first sentence of T2: 'And because he is such a thing, his coming to be happens so and is such.' The passage likewise satisfies the second sentence of T2: 'And that is why this part comes to be first, and then this'—as the opening words of (T11) make clear: our aim is 'to state what comes to be after what among parts'. T11, then, not only gives us a method (*methodos*, 742b10-11) for identifying the order in which the parts come to be, it also gives us the teleological tools for understanding *why* they come to be in the order they do.

It is within this framework that Aristotle goes on, in the rest of the chapter, to assign agency for the formation of the parts in the embryo, and to assign it primarily to a formal-efficient cause and only secondarily to material-efficient causes, as per such earlier passages as T4 and T7. What T11 does is to place that complex efficient-causal account in the framework of a broader teleological account, thereby completing, in principle anyway, the full account of generation promised in the opening passage of *GA* itself.[44]

Size and shape: indirect confirmation for our reading of *GA* II.6

I have said I could not find any place in II.6 where Aristotle attempts to give an account, at least in outline, in terms of material agency, of the formation of a heterogeneous non-uniform part. I would like to close with a look at one possible exception, since its ultimate failure to work is instructive as well. We might consider that the first move from uniform to non-uniform involves adding shape, and there is a case to be made that Aristotle thinks that the generation of shape can be shown to be necessitated at the material level. For, in animals which possess bones (and the same can be said for other supporting materials), the shape is significantly (if not fully) determined by the differing lengths of the bones, and Aristotle *does* in this chapter talk about length of bones. Here is what he says, at 745a4–10:

(T12) . . . and because of this the bones continue growing until a certain point; for all animals have a certain limit to their size, for which reason there is one as well for the growth of their bones. For if they continued growing forever, then whatever animals contain bone or its analogue would go on growing as long as they lived, for these are the determinant of the size in these animals.[45]

[44] In saying this, I don't mean to imply, of course, that the presentation of the theory of generation is completed at *GA* II.6. As Andrea Falcon and I will show in the paper mentioned in n. 1 above, *GA* III–V are integral to the overall project of a theory of generation described in T2 and in the opening lines of *GA* I.1.

This is perhaps a good place to take up again my objection to one key aspect of the influential account of Aristotelian natural teleology given by Susan Sauvé Meyer in Meyer 1992. As I explained in the preceding chapter (above, p. 80ff.), Meyer claims that for Aristotle the intrinsic efficient cause of the generation of animals and their parts is a 'reliable mechanism' operating in the generative process so as to 'overdetermine' that result. Meyer, I noted, simply assumes this to be the case, and doesn't cite any texts from *GA* to support it. I complained that she effectively begged the question against my own view that the relevant intrinsic efficient cause is precisely the formal nature, the *dunamis* for form, that underwrites my own interpretation of Aristotle's teleology (for *it* is responsible for the regularity of the outcome), and that she was being anachronistic in taking for granted the presence, in Aristotle's view, of any sort of material mechanism. We see the soundness of both of these charges more clearly, now that we have looked more closely at these II.6 passages. There simply is no such 'mechanism' and indeed no material necessitation of the generative outcome. Aristotle's account in this chapter of the generation of animals and their parts is unequivocally based primarily on the action of the formal nature, what I have called the animal's *potential for form*, a *dunamis* which, once set up in the newly formed embryo, is the embryo's *soul*, some of it still potential but its initial part—the nutritive capacity—actual; and this is the formal-efficient cause of the embryo's subsequent development, as we have seen. There is no material-efficient cause of that development *per se*.

[45] . . . *καὶ διὰ τοῦτο τὰ μὲν ὀστᾶ μέχρι τινὸς λαμβάνει τὴν αὔξησιν· ἔστι γάρ τι πᾶσι τοῖς ζῴοις πέρας τοῦ μεγέθους, διὸ καὶ τῆς τῶν ὀστῶν αὐξήσεως. εἰ γὰρ ταῦτ᾽ εἶχεν αὔξησιν ἀεὶ καὶ*

In this passage's first line, it is not clear what the referent of the 'this' (*touto*) in 'because of this' (*dia touto*) is. Perhaps Aristotle is thinking, while discussing in the preceding lines the special ('seminal' or 'nutritive') nutriment from which new bone growth must come, that there is only a *fixed amount* of such nutriment available; and if limiting material conditions fixed the length of each bone, then indeed they would fix the size—and thus, presumably, the shape—of the animal.

I would stress in response, however, that this account is at best speculative: we see no actual hint of it in the text of II.6, nor any general mathematical perspective on Aristotle's part from which it would be natural to project, D'Arcy-Thompson-like,[46] that radial lines mapping skeletons would fix sufficiently specific shapes for Aristotle's purposes here. Nor must we forget that an insistence on quantitative limit is not necessarily understood by Aristotle fully quantitatively: he may be thinking of the limit in question not as a number or measurable amount, but irreducibly in terms of *form*, the bones growing to *just the limit called for by* the form.[47] And, indeed, although T12 doesn't mention the shape of the whole animal, it does mention size, and says that animals have a limit to the growth of their bones *because* they have a limit to their size.[48] Given that, we should expect him to say the same for shape: shape as well as size is inherent in the nature of an animal,[49] and it is shape along with size that is responsible for the length

τῶν ζῴων ὅσα ἔχει ὀστοῦν ἢ τὸ ἀνάλογον ηὐξάνετ' ἂν ἕως ἔζη· τοῦ γὰρ μεγέθους ὅρος ἐστὶ ταῦτα τοῖς ζῴοις.

[46] Cf. Thompson's magisterial *On Growth and Form* (Thompson 1942).

[47] Even if one takes T12 to be saying that the amount of material flowing in a definite direction and solidifying is fixed by material factors alone, is it clear that this would for Aristotle fix the precise situating of bones and flesh, and so forth needed for the highly specific shapes, densities, and so forth that are required by the form and function of the part in question? That seems entirely unlikely, as does the idea that material factors could set the bones at the angles they need to be relative to each other, and the curvature of the ribs, and so forth. As I and others have noted a long time ago, in most of the many passages in which Aristotle speaks of a nature's diverting (*katachrētai, parakatachrētai*) materials to some teleological use, he specifically does *not* say that that diversion itself occurs of material necessity: cf., e.g., *PA* III.2 663b20-35: 'So then, what the nature of the horns is for the sake of has been stated, and owing to what cause some have such things while others do not; but, since there is a necessary nature, we must say how the nature according to the account makes use of things present of necessity for the sake of something.... For the residual surplus of this sort of body, being present in the larger of the animals, is used by <the> nature for protection and advantage, and the surplus, which flows of necessity to the upper region, in some cases it distributes to teeth and tusks, in other cases to horns' (663b20-24, 31-35; tr. Lennox 2001a, with one modification, in pointed brackets). See also Lennox 2001a ad loc., and my own discussions below, pp. 133 (with Mariska Leunissen) and 390.

[48] Mariska Leunissen reminds me that Aristotle identifies size as something that is part of the definition of the substantial being of animals in the following passages: *PA* IV.9, 685b12-15 (cf. *IA* 8, 708a9-20); *PA* IV.6, 683a18-19; *IA* 8, 708a9-20; and *Pol.* V.9, 1309b18-35.

[49] See chapter 10 below, pp. 232-33, 235-38.

the bones grow to (as well as their directions, something crucial for shape). Finally, advocates of any such account will have to remember the general fact about Aristotelian matter that, as one moves gradually downward from organisms to their non-uniform and then uniform constituents and then on down to the elements and then elemental powers out of which the uniform parts are built, one finds less and less structure and fewer and fewer resources to explain precise, complex structures and other mathematical properties at the level of organisms.[50]

Returning to Aristotle's text, it is important also to remember that there is no sign in II.6 of any in-general concern with explaining the sorts of features of non-uniform parts that distinguish them from uniform parts as such.[51] This makes my projection as to how one might try to read the passage about bone size all the less likely. One would expect to be told that one was being given a paradigm for the necessary formation of a vast swath of animal features, if that were the case.

Finally, the proposed way of taking T12 would appear to run against the grain of *An.* II.4 415b8-416a19, where in the context of insisting that *soul* is the formal, final, and efficient cause of functioning living bodies (with the matter *qua* efficient cause only a sort of *contributory* cause [*sunaition pōs*, 416a14]), Aristotle insists that Empedocles' view that fire is the cause *simpliciter* of growth cannot be right,

> for the growth of fire is unlimited while there is something to be burnt, but in all things which are naturally constituted there is a limit and a *logos* both for size and for growth; and these belong to soul, but not to fire, and to *logos* rather than to matter.[52]

Here, once again, a feature of an animal that is more than just a *pathos* of one of its parts is attributed not to material but to formal agency.

[50] e.g., Waterlow [Broadie] 1982, 82-88; Cooper 1987, 270-71; Cohen 1989, 258; cf. Furth 1987, esp. 24-25, 42-46.

[51] For a list of such features, see Furth 1987, 46-49.

[52] 416a15-18, tr. Hamlyn 1993. Hamlyn gives 'proportion' for the first '*logos*' and 'principle' for the second. I think rather that the first '*logos*' points to the fact that an animal kind's size (range) is inherent in its definition, so that 'a limit and a *logos*' could loosely be translated as 'a definable limit'. Hamlyn does better with the second '*logos*', since this is clearly that familiar usage in which '*logos*', though it carries the sense of 'definition', tends to signify what is identified in the definition, something here like 'definable form'. Since the line of thought in this *An.* passage is fully consonant with the different causal roles the *GA* texts we are studying assign to material factors on the one hand and form, *logos*, (formal) nature and so forth, on the other, it is not unreasonable to look for consonance also between the *An.* passage and our T11 and thus not unreasonable to reject an interpretation of T12 which sees it as assigning to material factors the same causal responsibility for embryonic shape as it does for embryonic *pathē*.

On the basis of our review of T1–T12 and of *GA* I–II at large, then, I conclude that Aristotle's discussion of embryogenesis in *GA* II.6 is entirely compatible with the interpretation of his natural teleological theory that I have developed in the first three chapters of this book—and indeed lends strong support to that interpretation.

5

'What's Teleology Got to Do with It?'

A Reinterpretation of Aristotle's *Generation of Animals* V

Co-authored with Mariska Leunissen★

Introduction: the problem of *Generation of Animals* V

Despite the renewed scholarly interest in Aristotle's *Generation of Animals* (*GA*) in recent years,[1] the subject matter of *GA* V, its preferred mode(s) of explanation, and its place in the treatise as a whole remain misunderstood.

Scholars typically focus on the first four books of *GA*, which contain Aristotle's key theories of reproduction, embryogenesis, and heredity; by contrast, they treat book V as a mere appendix,[2] since it discusses 'the attributes by which the parts of

This chapter is a very slightly revised version of Leunissen and Gotthelf 2010.

★ The fourth Pittsburgh/London (Ont.) Workshop on Aristotle's *Generation of Animals*, organized by Allan Gotthelf and Devin Henry, and held at the University of Pittsburgh, April 26-27, 2008, was devoted to *GA* Book V. At this workshop we each separately offered new readings of (i) the general statement in the book's first chapter of the mode(s) of explanation appropriate to the attributes under discussion in the book as a whole (Gotthelf) and (ii) the manner in which, in the book's last chapter, the explanations pertaining to teeth appeal to both material-efficient and teleological causation (Leunissen). Both of us noticed the light each of our accounts shed on the other's, and how the two together pointed the direction to a new interpretation of the aims and the unity of *GA* V, and the continuity of its investigation with the investigation of animal generation in the first four books of *GA*. We were helped in these reflections by outstanding, sometimes electric, discussion at the workshop, most especially with Alan Code, Devin Henry, and Jim Lennox, but with other participants as well. We are grateful to all who participated in the workshop, and to Devin for his role in its organization. Following the workshop, Leunissen did a draft of the bulk of the chapter, to which Gotthelf added the section on the V.1 passage discussing the modes of explanation and the Appendix. Extensive revision by the two of us in concert, followed by incorporation of useful written comments on an earlier version provided by Devin Henry, Robin Smith, and an anonymous referee for OUP, has resulted in the chapter you have before you.

[1] See in particular Balme 1987a; Bolton 1987; Code 1987; Coles 1995; Henry 2006a; 2006b; 2007a; 2009; Liatsi 2000; and Witt 1985.

[2] See Liatsi 2000, 13, 25, and Morsink 1982, 148.

animals differ' (*GA* V.1 778a16-17), such as differences in eye-color and pitch of voice. Whereas the first four books explain the coming to be of entire organisms and their parts in terms of efficient-final causation, book V is thought to concern (a) *individual, accidental differences* among animals, which are (b) purely *materially necessitated* and thus have nothing to do with the operation of teleology, and (c) are only *tangentially related* to the topics discussed in the earlier books.[3] Within book V, scholars consider Aristotle's discussion of the differences of teeth in chapter 8 to be especially problematic. There, Aristotle 'paradoxically' combines references to material necessity with teleology, which has led a recent commentator to postulate that chapter 8 does not belong to book V properly speaking, but that it is instead a second, separate appendix to the first four books.[4]

The purpose of this chapter is to defend an alternative and more integrated account of the subject matter of book V, its preferred mode(s) of explanation, and its relation to the first four books by closely examining two key passages: Aristotle's methodological introduction in chapter 1 and his discussion of the differences of teeth in chapter 8.

Our argument consists of two parts. In the first section, we shall offer a detailed interpretation of Aristotle's introduction to the project of book V in *GA* V.1 778a16-b19. First, we show that what unifies the differences singled out for discussion in book V, and characterized in 778a16-29, is that they are all 'differences of the more and the less' that come to be due to changes during the development of the animal *after its birth*. Under this interpretation, book V continues the investigation of the first four books into the causes of animal generation operative during embryogenesis and surrounding the animal's birth. Second, we give a reading of *GA* V.1 778a29-b19 according to which the passage's careful distinction between teleological and material-efficient modes of explanation does not require *either* that (i) everything universal to a subject must be explained teleologically *or* that (ii) the subject matter of book V is not in any way subject to teleological explanation. We offer a short analysis of the explanations in use in *GA* V.1-7 in further support of this reading.

In the second section, we turn to Aristotle's explanation of the differences of teeth in chapter 8. Though we grant that the primary cause of the generation of all the differences studied in book V, including the ones of teeth in *GA* V.8, is material necessity, we argue that this mode of explanation is fully compatible with *secondary* forms of teleology. The differences of teeth, as do teeth themselves according to Aristotle's explanation in *GA* II.6, come to be of material necessity after the animal's birth, from the residues of matter produced for the growth of bones.

[3] Interpretations that restrict the scope of book V to accidental, non-teleological differences at the subspecies level are defended by Balme 1987a, 11; 1987d, 305-306, 312; and 1990, 51; Cooper 1990, 81-83; Johnson 2005, 59 and 197; Liatsi 2000, 14-19, 23, 25; Lloyd 1990, 20-21, 23-24; and Pellegrin 1986, 157.

[4] Liatsi 2000, 19, 195.

The developing animal's formal nature then uses what has come to be of material necessity for something good, e.g., to enhance the function of nutrition. We use this notion of secondary teleology to show that there is a continuity not only in theme but also in the modes of explanation used, both within book V as a whole and between that book and the first four books of *GA*.

If our interpretation is right, book V forms an intrinsic part of Aristotle's investigation into the generation of animals.

The scope and nature of book V

The nature of 'the attributes by which the parts of animals differ'

The methodological introduction to book V consists of two parts: Aristotle first introduces the subject matter of the investigation to be undertaken (in *GA* V.1 778a16-28) as 'the attributes by which the parts of animals differ' and then explicates which mode(s) of explanation he deems appropriate for these attributes (in 778a29-b19). In our discussion below, we shall closely follow Aristotle's text.

In his introduction of the subject matter, Aristotle provides two sets of specifications from which we might glean the nature of these 'attributes'. First, he gives examples:

> I mean such attributes of parts as blueness or blackness of eyes, and height and depth of voice, and differences of color in hairs and feathers.[5] (778a17-20)

We take it that the group of attributes Aristotle is singling out for discussion here by means of these examples is a subset of the category of 'differences of the more and the less', a category of difference which Aristotle carefully distinguishes in the delineation of types of sameness and difference of animal parts at the opening of *History of Animals* (among other places).[6] He there includes in this category differences of parts (i) 'by oppositions of their attributes, e.g., color and shape', and then (expanding the category to include all 'differences by excess and defect') (ii) 'by <the parts> being more or fewer, and larger and smaller'. These differences always belong to animals that are part of one and the same kind (*genos*), as opposed to differences (and samenesses) by analogy, which are found across different animal kinds.[7]

[5] *λέγω δὲ τὰ τοιαῦτα παθήματα τῶν μορίων οἷον γλαυκότητα ὀμμάτων καὶ μελανίαν, καὶ φωνῆς ὀξύτητα καὶ βαρύτητα, καὶ χρώματος [ἢ σώματος] καὶ τριχῶν ἢ πτερῶν διαφοράς.*

[6] *HA* I.1 486a25-b8; cf. *PA* I.4 644a16-21, b8-15, and *Metaph. H*.2 1042b29-35. While the account at the opening of *HA* illustrates the different types of sameness and difference by reference to animal parts, later, at the start of book VII, Aristotle explicitly applies that account to differences in *ēthē* (which include (i) animal traits such as fierceness and timidity, (ii) enmities between animals, and (iii) degrees of *phronēsis*, intelligence, all discussed in VIII). Differences of the more and the less are clearly to be found in every category of animal 'difference' (or attribute). See *HA* VII.1 588a18-b3. On differences of the more and the less in general, cf. Lennox 2001a, 160-81.

[7] See the previous note. It is unclear in which type of sameness and difference Aristotle places differences in the position of the same part (his example is the mammaries), a type which he tacks

In the methodologically prior *Parts of Animals*, Aristotle had already discussed many differences of the more and the less. The reason why he singles out *these particular differences* of the more and the less (such as in color and pitch) for discussion here in *GA* V is made clear in the subsequent section of his introduction, where he notes the distinctive features of this group of attributes that make them worthy of study in the current context:

> It turns out that *some* of such <attributes> belong to whole kinds, but to *some* <kinds> they belong at random, such as happens especially with regard to human beings. In addition, following the changes of life *some* <changes> belong equally to all living beings, while *others* <belong> in the opposite way, such as <the changes> concerning voice and the color of hair. For *some* do not grow grey visibly in old age, while man suffers from this most among all animals. And *some* immediately follow coming into being, *others* become plain as age advances or in old age.[8] (778a20-28)

The reason for discussing this group of attributes separately in the final book of the *Generation of Animals* is obviously not because Aristotle thinks these are all non-generic, individual features (by contrast with the ones discussed in *Parts of Animals*, which all belong to entire species, or even wider kinds). What makes these attributes listed here interesting is exactly the fact that for some species the attributes are possessed unvaryingly as well as universally, whereas in others, and particularly in human beings, the attributes are variable and occur randomly. In other words, the attributes of eyes, hair, and voice are *not realized uniformly* in all living beings that possess them and sometimes not even within one and the same species. A second characteristic is that these attributes come to be not during embryogenesis, but as a result of changes during the *later development* of the animal, either immediately after its birth, or in later life, or in old age. Again, some of these changes belong in the same way to all living beings that possess them, while others do not. A third point of interest in Aristotle's characterization of these differences is their *prominence in human beings*: for reasons to be specified in the remainder of book V, differences in eyes, hair, and pitch of voice are most visible in human beings.

on at the end of this passage. The word *παθήμα* is standardly used in all these contexts to pick out the qualities of parts by which they differ in the more and the less (cf. *PA* I.4 644b13: *τοῖς σωματικοῖς πάθεσιν*; *Metaph.* *H*.2 1042b21-22: *τοῖς τῶν αἰσθητῶν πάθεσιν*), and color is explicitly mentioned as such a difference (*HA* I.1 486b6; *GA* V.1 779b33-34: *τὰ δὲ μεταξὺ τῶν ὀμμάτων τούτων τῷ μᾶλλον ἤδη διαφέρει καὶ ἧττον*). Differences in the pitch of voice are to our knowledge nowhere explicitly mentioned as differences of the more and the less, but see *HA* IV.9 for Aristotle's discussion of differences in pitch of voice due to locality and *PA* II.17 660a14-b4 for his explanation of differences of the more and the less in vocalization.

[8] *τυγχάνει δὲ τῶν τοιούτων ἔνια μὲν ὅλοις ὑπάρχοντα τοῖς γένεσιν, ἐνίοις δ' ὅπως ἔτυχεν, οἷον μάλιστ' ἐπὶ τῶν ἀνθρώπων τοῦτο συμβέβηκεν. ἔτι δὲ κατὰ τὰς τῶν ἡλικιῶν μεταβολὰς τὰ μὲν ὁμοίως πᾶσιν ὑπάρχει τοῖς ζῴοις τὰ δ' ὑπεναντίως, ὥσπερ περί τε φωνὰς καὶ περὶ τριχῶν χρόας τὰ μὲν γὰρ οὐ πολιοῦται πρὸς τὸ γῆρας ἐπιδήλως, ὁ δ' ἄνθρωπος μάλιστα τοῦτο πάσχει τῶν ἄλλων ζῴων. καὶ τὰ μὲν εὐθὺς ἀκολουθεῖ γενομένοις, τὰ δὲ προϊούσης τῆς ἡλικίας γίγνεται δῆλα καὶ γηρασκόντων.*

What unites these attributes, then, is that their changes are especially variable and noticeable in the maturation and aging of human beings: as often in the biological works, Aristotle takes the human (male) body as a starting point of explanation and as a guide for the organization of his exposition (see, e.g., *HA* I.6 491a19-26, *PA* II.10 656a9-14, and *GA* II.4 737b25-27). The change of eye-color in babies, growing grey or bald in men, and the changes in pitch of voice of humans (who have voice more than any other kind of animal) are the primary explananda in this book, which in their turn prompt the discussions of the (often different and very variable) causes of the changes of similar attributes in other animals.

Aristotle's subsequent accounts of these attributes in the remainder of chapters 1-7 follow this programmatic introduction rather closely (we postpone the question of how chapter 8 fits into the larger structure of book V to section two below). First of all, Aristotle's concern with attributes that come to be due to changes *after* the birth of an animal explains why the starting point of his discussion is the moment of birth (778b19-21).[9] Before turning to the attributes, Aristotle first needs to solve some problems that pertain to the first development of animals (778b23-779a26; cf. b23-24: ἔχει δ' ἀπορίαν περὶ τῆς ἐξ ἀρχῆς γενέσεως). From then on, he explains the causes of the coming to be of the three kinds of attributes mentioned in the introduction: changes in eye-color in human beings are the topic of the remainder of the first chapter (followed by a general discussion of differences in eyes in 779b12ff. and of hearing and smell in chapter 2); changes and differences in the structure and color of hair and feathers are discussed in chapters 3-6 (starting off with discussions of baldness and greyness in men); and differences in voice are explained in chapter 7.

Second, preceding his explanations, Aristotle states whether an attribute belongs to a whole *genos* or not; whether the attribute is a result of a later change and if so, when exactly during the later development after an animal's birth the attribute comes to be; and finally, how the occurrence and distribution of the attribute is different in human beings as opposed to its realization in other animals.[10]

And third, the actual explanations Aristotle provides in the first seven chapters (about which we shall say more below) are sometimes remarkably precise in classifying both the nature of the attribute to be explained and the kinds of causes that are responsible for the coming to be of that attribute.[11] In doing so, Aristotle

[9] *Pace* Liatsi 2000, 19.

[10] See in particular *GA* V.1 778b20-779b6, V.3 781b30-782a19, V.4 784a22-25, V.6 785b16-26, and V.7 786b6-22.

[11] For instance, in his explanation of balding, Aristotle specifies that even though balding is most visible in human beings, it is a generally occurring attribute that also holds of some plants and birds (*GA* V.3 783b8-13; b9-10: ἔστι δέ τι καθόλου τὸ τοιοῦτον πάθος; cf. V.1 779b12-13). Many explanations differentiate explicitly between natural causes and affections such as diseases, aging, or environmental influences (see, e.g., *GA* V.5 785b1-2, V.6 785b30-34, and 786a8-14), and one even differentiates between *per se* causes and accidental ones (see V.3 783a33-b2).

provides detailed accounts—not just for why a particular difference of the more of the less belongs to or comes to be in a particular kind of animal—but also for why one and the same attribute can be realized so differently among different species, and especially among human beings.

The focus on generation we find both in the introduction and in the actual explanations of book V is of course typical for the whole of *GA* (see especially *GA* I.1 715a1-18 and *GA* II.1 731b20-22). This perhaps also explains why Aristotle does not discuss these differences in the *Parts of Animals*, in which the focus is primarily on the *being* of animals, and on their generation so far as it is controlled primarily by teleological causes; in those cases the complete, mature animal is taken as a starting point of explanation. Differences that are not realized uniformly within a whole species and/or that change over the animal's life-time do not appear as explananda in that context.

These facts about the focus of book V already make clear the book's close relationship to the first four books of *GA*. Books I–III discuss the mechanisms of reproduction and embryogenesis and the natural order of the generation of parts. An appropriate next step might well be a discussion of the coming to be of *differences* of parts—especially of those that occur after birth (cf. *GA* V.1 778b20). What is arguably needed first, however, is an account of the mechanisms of heredity and other phenomena related to pregnancy and birth itself, and this book IV provides.[12] The concluding words of book IV and the opening of book V suggest that book V is in fact an immediate follow-up from the preceding discussion:

> We have spoken about the nourishment of animals within <the mother> and of their birth into the outside <world>, both of each kind separately and of all in common. || We must now investigate the attributes by which the parts of animals differ.[13] (*GA* IV.10 778a10-V.1 778a17)

The *δέ* in the opening sentence of book V picks up the *μὲν* in the concluding remarks of book IV: where book IV concludes with a discussion of phenomena related to the growth of animals within the womb and to their birth, book V continues 'chronologically' with a discussion of the attributes by which their parts differ that arise during their development after birth.

Under this interpretation, Aristotle continues his investigation of the coming to be of animals by shifting from identifying the causes of what happens during embryogenesis and surrounding pregnancy to identifying the causes of what

[12] For instance, in book IV, Aristotle already briefly touched upon attributes such as excessive hair-growth in animals rich in semen (*GA* IV.5 774a30-b4), the re-growth of eyes in very young birds (*GA* IV.6 774b27-34), and changes of voice during sexual maturation in male animals and most prominently in humans (*GA* IV.8 776b13-25): these are all differences that are immediately related to gestation and reproduction.

[13] *περὶ μὲν οὖν τῆς ἔσωθεν τροφῆς τῶν ζῴων καὶ τῆς θύραζε γενέσεως εἴρηται, καὶ χωρὶς περὶ ἑκάστου καὶ κοινῇ περὶ πάντων. || περὶ δὲ τῶν παθημάτων οἷς διαφέρουσι τὰ μόρια τῶν ζῴων θεωρητέον νῦν.*

happens to some particularly variable parts of animals after their birth. Thematically speaking, there is no reason to treat book V as an appendix.

The appropriate mode of explanation of the attributes in GA *V: the general statement in* GA *V.1 778a29-b19*

Although the broad outlines of Aristotle's discussion, in *GA* V.1 778a29-b19, of the mode(s) of causation operative in the generation of the attributes which are the subject matter of book V are clear enough, much in it is difficult to understand, not least its opening sentence. The discussion starts as follows:

> About these <attributes> and all the things of this sort, one must no longer think that there is the same manner of explanation. For all those which are neither the products of a nature in common nor a distinctive feature of each kind—of those sorts of things, none either is or comes to be for the sake of something. For, while an eye is for the sake of something, its being blue is not for the sake of something—unless this attribute is a distinctive feature of the kind. In some cases <the attribute> does not reach to the definition of the substantial being, but the explanations must be referred back to the matter and the source of motion as coming to be from necessity.[14] (778a29-b1)

The expression 'one must no longer think that there is the same manner of explanation' (τὸν αὐτὸν τρόπον...τῆς αἰτίας) might mean either (a) 'one must no longer think that there is here the same manner of explanation as we used previously' or (b) 'one must no longer think that there is one and the same manner of explanation for every type of attribute, as one might have thought previously'. We will argue for a certain version of (a).

Whatever is meant by the first sentence is explained in the next sentence by reference to a contrast between two manners of explanation, one that refers to teleological causation and one that refers to a non-teleological one. The contrast is first given in just those terms (in one set of attributes, 'none . . . for the sake of something') and then exemplified (eyes, when present, are for the sake of something, blue eyes normally are not). Attributes not possessed in common across some wider or narrower kind are said not to be present for the sake of something. Presumably, those that are common in one of those ways *may* be present for the sake of something.[15]

[14] περὶ δὲ τούτων καὶ τῶν τοιούτων πάντων οὐκέτι τὸν αὐτὸν τρόπον δεῖ νομίζειν εἶναι τῆς αἰτίας. ὅσα γὰρ μὴ τῆς φύσεως ἔργα κοινῇ μηδ' ἴδια τοῦ γένους ἑκάστου, τούτων οὐθὲν ἕνεκά του τοιοῦτον οὔτ' ἔστιν οὔτε γίγνεται. ὀφθαλμὸς μὲν γὰρ ἕνεκά του, γλαυκὸς δ' οὐχ ἕνεκά του πλὴν ἂν ἴδιον ᾖ τοῦ γένους τοῦτο τὸ πάθος. οὔτε δ' ἐπ' ἐνίων πρὸς τὸν λόγον συντείνει τὸν τῆς οὐσίας, ἀλλ' ὡς ἐξ ἀνάγκης γιγνομένων εἰς τὴν ὕλην καὶ τὴν κινήσασαν ἀρχὴν ἀνακτέον τὰς αἰτίας.

[15] Though interpreters regularly take the sentence to make the stronger claim, that universality is sufficient for the causation to be teleological, the sentence actually commits Aristotle only to the view that universality is *necessary* for the causation to be teleological. Likewise in the example which follows (a33-34), Aristotle's point may well be that blue eyes are non-teleologically caused, unless they are characteristic of the entire kind in question, in which case they *may* be teleologically caused (rather than *must be* teleologically caused).

Only in the fourth sentence is the non-teleological mode of causation described positively (though still by contrast with a corresponding feature of the teleological case): the explanation of an attribute will not start from the definition of the substantial being (λόγος τῆς οὐσίας) of its possessor, but from its 'matter and the source of motion' showing the attribute 'as coming to be from necessity'.

This is a non-teleological, non-conditional necessity, as the last sentence of the methodological discussion makes clear: 'For while <an animal> will have an eye from necessity (for an animal is posited as being this sort of thing),[16] it will have a particular sort of eye from necessity, though not that sort of necessity but another kind, because it is its nature to act and be acted upon in this or that way' (778b16-19).[17] As the sentence just prior to this one says, in regard to differences such as the blueness of an eye, (in most cases) 'the cause must be sought in the motion and in the generation as having acquired these differences during its process of constitution (ἐν αὐτῇ τῇ συστάσει)' (778b14-15).

In the methodological passage under discussion, then, we are told first that, in the case of the attributes which are the subject of book V, we should no longer think that the manner of explanation is the same. Then we are introduced, from various angles, to a distinction between two manners of explanation, one that refers to (an apparently less familiar) non-teleological material-efficient causation, the other that refers to (the apparently more familiar) teleological mode of causation.

The traditional reading of this passage takes the opening sentence to be saying, in effect, that one mode of explanation has been used in books I–IV, but that a different mode will be needed here in book V.[18] The balance of the passage is then understood to be introducing the non-teleological mode by carefully contrasting it with the teleological mode used in the previous books. We think this cannot be right for several reasons.

[16] Cf. *PA* I.1 640a33-b1, and note that the lines in *PA* I that Aristotle refers to in the course of our *GA* V.1 passage (at 778b2) appear just before this, at 640a16-19.

[17] *ὅτι τοιονδὶ ἢ τοιονδὶ ποιεῖν πέφυκε καὶ πάσχειν*. Note the similar wording at *PA* I.1 642a34-35, where it is explained that when one says that '*this* comes to be of necessity because of *those*', 'necessity' signifies not conditional necessity but that *ἔστιν οὕτως ἔχοντα καὶ πεφυκότα* ('this is their state and nature'; tr. Balme 1992).

[18] This reading, which focuses on the difference between books I–IV and book V in the modes of explanation that are used, goes back to Michael of Ephesus, *CAG* 14.3 (ed. Hayduck), 211.14-212.28 (see especially 211.19-20: *καὶ ἐπὶ τῶν ὁμοίων τὸ μὲν ὑλικὸν καὶ ποιητικὸν αἴτιον ζητητέον καὶ ἀποδοτέον, τὸ δὲ ἕνεκά του οὐ ζητητέον*), and has been defended explicitly most recently by Liatsi 2000, 22. This reading is often asserted without argument, e.g., in Peck 1961, lxxvii and Louis 1961, *ad* 778a20 (176 n. 1). Others have pointed to a break in thematic continuity between books I–IV and V, and argued that since book V is only concerned with the non-teleological generation of non-functional features, it must be treated as an appendix: see, e.g., St.-Hilaire 1887, cclix-cclxvi, 418, and 424-25; Zeller 1897, 92 n. 2; and Morsink 1982, 148. On both aspects (or variants) of the traditional reading, see also n. 3 above.

First, and most important, it is simply not true either that books I–IV make no use, in the explanations they provide, of material-efficient causation operating on its own *or* that book V makes no use of teleological causation. For instance, material-efficient causation operating on its own is appealed to several times in book IV.[19] And, second, as we will show below, teleological causation, quite apart from its appearance in chapter 8 of book V, is never excluded as a possible mode of explanation of the attributes under discussion in the first seven chapters, even though its actual use in those chapters is entirely marginal.

This leaves us, we think, with a different reading. The opening sentence is to be taken as saying that in previous books the *dominant* (not necessarily *only*) mode of explanation was the teleological one, while in the present book the *dominant* (and not necessarily *only*) mode of explanation is a non-teleological one appealing primarily to material-efficient causes.

In favor of our reading is the fact, already mentioned, that distinct modes of explanation have already been used in book IV, and will be used, as we will show, in book V. Nowhere in his introduction does Aristotle claim that we should rely exclusively on just one mode of explanation for all of the attributes at hand. For instance, as the example concerning eye-color indicates, Aristotle believes that if an attribute is proper to an animal kind, it may well have to be explained teleologically.[20] Similarly, the use of 'in some cases' in *GA* V.1 778a34 may suggest that in *other* cases attributes by which parts of animals differ do 'reach to the definition of the substantial being' and are thus essential to, or at least *per se* of, those animals. (These attributes, we would submit, are the ones already explained teleologically in the *Parts of Animals*; see, e.g., *PA* II.2 648a13-19, II.13 657b22-27, IV.12 693b28-694a9). And, a few lines later, at 778b7-10, Aristotle criticizes his predecessors for not seeing that natural phenomena, including the attributes under discussion, might involve formal-final causation *as well as* material-efficient causation. It is wrong, he says, to explain all such phenomena in terms of (not very well distinguished) material-efficient causation *alone* (778b8).

However, Aristotle's delineation of the two distinct modes of explanation that follow presents the material-efficient mode as less familiar (to Aristotle's audience if

[19] For example, in *GA* IV Aristotle explains monstrosities solely as a result of material necessity (see IV.3-4, especially IV.4 770a6-7) and the production of milk as being due to a combination of material necessity and teleology (see IV.8 776a15-b3).

[20] This point is already well made by Lennox 2001b, 176. Based on our reading of the text at a30-32 (and 32-34) we are a bit more cautious: though Lennox may indeed be right that Aristotle thinks that the appropriate sort of universality provides 'prima facie evidence' that teleological explanation is in order, we don't think the passage quite says that, as we explained in n. 15 above. (It is no part of Aristotle's claim here, incidentally, that *having eyes* is among the attributes that *GA* V is meant to provide explanations of; surely it is not, even though *having blue eyes* of course is. Its point is simply to exemplify, strikingly in this context, a mode of explanation which *some* attributes that are the subject of *GA* V may require.)

not to his predecessors) than the teleological one; so it is natural to take it as introducing a mode which is now to be used more than previously. The new subject matter at hand makes it no longer (οὐκέτι) right to think what one had been right to think regarding the prior subject matter, namely that the biological phenomena in question are to be explained predominantly (although not exclusively) by reference to teleology.[21]

Our reading thus differs from the traditional reading in taking Aristotle to be saying, not that teleological explanation is present *only* in books I–IV and explanation via material-efficient causes present *only* in book V. It takes him to be speaking rather of the type of explanation that is *dominant* in the two places, and dominant in two senses. In the first sense (as we will explain further below when we distinguish 'secondary teleology' from 'primary teleology'), while both teleological and material-efficient causes may be at work together in each of the two places, in explanations in books I–IV, even if material-efficient causes are involved, the teleological cause is primary; but in explanations in book V the material-efficient cause is primary, even if (as we show is indeed sometimes the case) teleological causation is involved. In a second sense of 'dominant', our reading acknowledges that there may be exceptions to its claims about the modes of explanation present in the different books;[22] its claim is rather that explanations in which material-efficient causes play the only or the primary role are *rare* in books I–IV and that explanations in which teleological causes are primary are rare in book V.

In the section below, we shall provide further support for this reading of the primary methodological passage in *GA* V, showing that Aristotle does not from the outset restrict the explanations he provides in V.1-7 to material-efficient causation alone. Then we will turn to V.8.[23]

The appropriate mode of explanation of the attributes in GA *V: the explanations in use in* GA *V.1-7*

The explanations in use in *GA* V.1-7 (which discuss the specific kinds of attributes listed in *GA* V.1 778a17-20) offer further support for our readings.

As should be expected from Aristotle's methodological remarks (cf. also V.3 782a20-24 and V.7 786b23-25), the vast majority of the explanations offered in V.1-7 are of a material nature. The most common type of explanation refers to the material constitution of a part and its specific potentials in each kind of animal. For

[21] The γάρ ('For') immediately following the first sentence (778a20) might be thought to provide further support for this reading.

[22] For example, in the joint explanation of milk in *GA* IV (above, n. 19), material necessity plays the primary role: its genesis is an example of what, in pp. 130-32 below, we call 'secondary teleology'.

[23] For a detailed outline-analysis of our reading of the whole of 778a29-b19, see the Appendix below, 139-41.

example, differences in degrees of the liquidity of the eye (where medium liquidity is optimal for sight) determine what particular eye-color animals have (see *GA* V.1 779b7-12, 779b14-33, and 780b3-11) as well as how discriminative their eyesight is (V.1 779b33-80a13, 780b12-781a11; cf. V.3 782a24-b11 and V.7 787a6-8). (Occasionally, Aristotle also picks out the location of a part or its size as the cause for a particular difference.)[24] Others pick out external or internal efficient causes that affect the material constitution of parts and thereby bring about certain changes (that are not for the sake of something). Such causes are: (a) aging;[25] (b) diseases;[26] (c) the material potentials of the environment;[27] and (d) the kind of food an animal digests.[28] And finally, Aristotle sometimes refers more generally to what may be the material nature of the whole animal kind.[29]

Teleology is mostly absent in these chapters, but Aristotle does on occasion seem to entertain the possibility that some of these attributes might need to be explained teleologically. For instance, there are a few times when Aristotle refers to the goal-directed actions of formal natures,[30] which suggests that he must have assumed there

[24] See *GA* V.1 780b12-781a11, V.2 781b5-16, and V.3 783b35-784a3.

[25] Aging is a form of decay that causes a deficiency of heat and moisture, which makes the body go cold and dry, which then causes the coming to be of various attributes; see, e.g., *GA* V.1 780a30-32, V.3 783b2-7 and V.4 784a26-34.

[26] Diseases are a (temporary) deficiency of natural heat (just as aging is), which can cause the coming to be of attributes; see, e.g., *GA* V.1 780a14-21 and V.4 784a25-30.

[27] e.g., rainy weather or a damp climate affects the movement of air in the ears and deteriorates hearing (*GA* V.2 781a33-34), and moist environments cause hair to grow straight whereas hot and dry environments cause curly hair (V.3 782b32-783a1); see also *GA* V.3 783a11-32, 784a12-20, V.6 786a30-34, and V.7 788a16-20.

[28] e.g., more omnivorous animals are more vari-colored, because 'if foods are the causes of change (*εἰ γὰρ αἱ τροφαὶ αἴτιαι τῆς μεταβολῆς*), it is reasonable that more varied foods make the motions and the residues of the food more varied, from which hairs and feathers and skins come into being' (*GA* V.6 786a34-b4); see also V.6 786a2-20 and IV.2 767a28-35.

[29] e.g., some animals are multi-colored 'because not having a single color belongs to the nature of the whole kind' (*GA* V.6 785b16-786a2; cf. V.1 779b2, V.3 784a4-5, and V.6 786a2-4). Aristotle nowhere offers a teleological explanation for the specific colorations of hairs, but refers their causes in all non-human animals back to the material nature of the skin (V.4 784a23-24: *τῶν δὲ χρωμάτων αἴτιον τοῖς μὲν ἄλλοις ζῴοις καὶ τοῦ μονόχροα εἶναι καὶ τοῦ ποικίλα ἡ τοῦ δέρματος φύσις*), which seems to suggest that the references to nature must be to the animal's *material* nature, rather than to its formal nature. For an alternative understanding of 'the nature of the whole kind' in this passage, however, see Henry 2008, 67-70.

[30] For instance, in *GA* V.1 780b9-12 Aristotle explains differently colored eyes within one individual by reference to the formal nature of the animal *failing* (*ὅταν οὖν μὴ δύνηται ἀπαρτίσαι ἡ φύσις* . . .) to make the eyes correspond by concoction. There are two other references in *GA* V.1-7 to the goal-directed actions of the formal nature: in *GA* V.3 782a20-22 Aristotle states that he has explained earlier for the sake of what nature has made hairs, but he nowhere provides a teleological explanation of the attributes of hair; in *GA* V.7 786b19-22 he explains why the difference in pitch of voice between male and female is especially clear in human beings, by stating that 'nature has given this faculty to them in the highest degree, because they alone of animals make use of speech, and voice is the material of speech.' The difference is thus especially

to be a potential role for teleology to play in the explanation of the attributes under discussion, even though in practice this role turns out to be marginal at the least. For instance, in discussing the attributes of the organs of sight, smell, and hearing (in V.1-2) Aristotle indicates what effects the presence of these attributes has on the performance of the functions of perception (e.g., in V.1 779b33-780a13), although he never embeds this into a teleological explanation or even uses teleological language. It seems that, even though the attributes under discussion here change the sense-organs for the better or the worse, Aristotle does not believe that they come to be for the sake of this functional difference (as he does in the case of the more-and-less differences in sense organs he considers in the *Parts of Animals*: e.g., *PA* II.2 648a13-19, II.13 657a30-b4, 657b30-658a10, and IV.11 691a19-27).

Within *GA* V.1-7, there is only one example of an explanation that is unequivocally teleological in nature (V.2 781b22-28):

> Nature acts reasonably also with regard to the things concerning the seal. For, although it is four-footed and live-bearing, it does not have ears but only passages. The cause is that its way of life is in the water. For the part <that is the> ears is attached to the passages for preserving the motion of air from afar: now this is not useful for it, but it would accomplish the opposite by receiving a great amount of water in them [i.e. in the passages].[31]

This explanation of the absence of outer ears in seals follows a discussion of how longer passages in ears and noses produce greater accuracy in perception from afar. It is not clear what the passage is doing here. It is possible to read the absence of outer ears in seals as another example of a difference 'of the less', but one would rather expect to find such a teleological explanation of why a type of animal lacks an entire (type of) part in the *Parts of Animals*. In fact, the explanation given here complements the short remark about the absence of outer ears in seals that Aristotle makes in *PA* II.12 657a23-25 rather well: contrasting seals with birds and egg-laying four-footers who lack outer ears because of a lack of the kinds of materials from which outer ears are typically formed, Aristotle states that seals lack outer ears because they are 'a deformed four-footed animal' (on the deformity of seals, cf. *HA* II.1 498a32-b4 and *IA* 19 714b12-13). Seals are 'dualizers', and like other dualizers, they lack the features that are universally present in one of the natures to which they tend (here: outer ears in four-footed animals) if the presence of those features would be harmful or useless to them (cf. *PA* IV.10 689b31-34, IV.11 690b19-24, IV.13 697b1-13 and IV.14). Owing to the seal's aquatic way of life the presence of

audible as a result of there being more voice in human beings than there is in other animals, but this in itself is not for the sake of something.

[31] *εὐλόγως δ' ἀπείργασται ἡ φύσις καὶ τὰ περὶ τὴν φώκην· τετράπουν γὰρ ὂν καὶ ζῳοτόκον οὐκ ἔχει ὦτα ἀλλὰ πόρους μόνον. αἴτιον δ' ὅτι ἐν ὑγρῷ αὐτῇ ὁ βίος. τὸ γὰρ τῶν ὤτων μόριον πρόσκειται τοῖς πόροις πρὸς τὸ σώζειν τὴν τοῦ πόρρωθεν ἀέρος κίνησιν· οὐθὲν οὖν χρήσιμόν ἐστιν αὐτῇ, ἀλλὰ καὶ τοὐναντίον ἀπεργάζοιτ' ἂν δεχόμενα εἰς αὐτὰ ὑγροῦ πλῆθος.*

outer ears would impair rather than improve their hearing from afar, and—since 'nature does nothing in vain'—this is why these outer ears are absent. The seal-passage in *GA* V.2 thus elaborates on the rather elliptical remark in the *Parts of Animals*, and so the passage might well be an interpolation.

It is not until Aristotle's discussion of the differences of teeth in *GA* V.8, that we find actual, fully fledged teleological explanations for the *presence* of attributes, and, as we shall show below, these explanations go hand in hand with references to material necessity.

The appropriate mode of explanation of the differences of teeth in *GA* V.8

The connection between GA *V.8 and* GA *II.6*

Having treated in *GA* V.1-7 the three sorts of attribute mentioned at the opening of chapter 1 as examples of the book's subject matter (differences of eyes, voice, and color of hair and feathers), Aristotle turns in the book's final chapter, chapter 8, to a fourth sort—differences of teeth (*GA* V.8 788b3-9):

> About teeth, it has been stated previously that animals do not have them for one thing, nor <do they have them> all for the sake of the same thing, but some because of nutrition, others also for defense and for speech in sound. Why the front teeth come to be earlier and the grinders later, and <why> the latter do not fall out, but the former fall out and grow back again—we should consider the explanation <of this> to be of a similar kind to the accounts concerning <their> generation.[32]

The explananda of *GA* V.8 are thus (a) why front teeth come to be first and the grinders later, which is a difference in the *timing* of their growth; and (b) why front teeth fall out and grow back again, whereas grinders do not fall out, which is a difference in *shedding*.

The attributes at stake here are similar to the ones explained in *GA* V.1-7 in that the differences of teeth also come to be during the later development of animals and are especially visible in human beings (see *HA* II.1 501a8-II.5 and *HA* IX.10 587b13-16). Aristotle did not mention these differences in his introduction in chapter 1, but the reference to his account of teeth in *Parts of Animals* (in which he identifies the final causes of their presence)[33] is similar to the start of the explana-

[32] *περὶ δὲ ὀδόντων, ὅτι μὲν οὐχ ἑνὸς χάριν οὐδὲ πάντα τοῦ αὐτοῦ ἕνεκα τὰ ζῷα ἔχουσιν ἀλλὰ τὰ μὲν διὰ τὴν τροφὴν τὰ δὲ καὶ πρὸς ἀλκὴν καὶ πρὸς τὸν ἐν τῇ φωνῇ λόγον, εἴρηται πρότερον· διότι δ' οἱ μὲν πρόσθιοι γίγνονται πρότερον οἱ δὲ γόμφιοι ὕστερον, καὶ οὗτοι μὲν οὐκ ἐκπίπτουσιν ἐκεῖνοι δ' ἐκπίπτουσι καὶ φύονται πάλιν, τοῖς περὶ γενέσεως λόγοις τὴν αἰτίαν συγγενῆ δεῖ νομίζειν.*

[33] Teeth are present for the sake of nutrition, and sometimes also for defense and speech; see *PA* II.9 655b2-15, III.1 661a34-662a15, and IV.5 678b6-26.

tions of the attributes of hair and voice in *GA* V.3 782a20-24 and 786b23-25. Additionally, he explicitly links the present investigation to the accounts about generation in the earlier chapters of *GA*, because it is 'congenial'[34] to them.

In fact, the question Aristotle sets out to answer in *GA* V.8 was already announced in *GA* II.6 (745b13-15: 'The explanation for why some teeth come to be and fall out but others do not fall out *will be stated later*'), following an account of the causes of the coming to be of teeth. Since this account forms the background of the causal explanation of the difference of teeth in chapter 8, we shall discuss this first.

Aristotle's account of the coming to be of teeth in GA *II.6*

The context of the account of teeth in *GA* II.6 is Aristotle's attempt to track and explain the natural order of generation of the various animal parts within the embryo (see *GA* II.1 733a32, 734a16-34, and II.4 740a2-16). In II.6 744b11-27, he uses an analogy comparing a nature to 'a good housekeeper',[35] to illustrate this order: in the allocation of resources, a nature first creates the most important parts from the best materials, and then—provided that there are enough leftovers to spare—uses the extra materials to make parts that perform subsidiary or luxury functions.

We believe that these two categories of parts are the result of two different patterns of causation. Parts that are of vital or essential importance to the animal in question are all the result of what we call *primary teleology*: these parts are the necessary realizations of a pre-existing, internal potential for form, specified by the definition of the substantial being of the animal.[36] Because of their importance to the animal for living, these parts come to be first and are generated by that form, or formal nature, through conditional necessity: if there is to be an animal of that form, it *must* have these very parts. Parts that serve subsidiary functions are the result of what we call *secondary teleology*: the materials constitutive of these parts come to be as materially necessary by-products of prior teleological processes and are then *used* by the animal's formal nature, which acts as 'a good housekeeper',[37] for the

[34] For the use of συγγενῆ as referring to things that belong to the same scientific genus, cf. *APo*. I.9 76a1, a9, a30, I.28 87b4, and I.32 88b24. Cf. also Aristotle's introduction to the *Generation of Animals* in *GA* I.1 715a11-18.

[35] For a discussion of this image of nature as 'a good housekeeper', the hierarchy of parts, and the distinction between primary and secondary teleology, see Leunissen 2010b, ch. 3 and 2010c. (Gotthelf would like it stressed that the account we develop here of the primary/secondary teleology distinction was originated by Leunissen.)

[36] This characterization of 'primary teleology' builds on the understanding of teleological causation in the case of animal generation provided in Gotthelf 1987b (= chapter 1, above).

[37] In a fuller study of what we are calling 'secondary teleology' we would include those cases where a nature uses already formed parts for a second function, without incorporating any additional material. A familiar case is an elephant-nature's use of its already flexible trunk for grasping (*PA* II.16 659a20-36; cf. a21-22: ἡ φύσις παρακαταχρῆται, καθάπερ εἴωθεν, ἐπὶ πλείονα τοῖς αὐτοῖς μορίοις). Two more cases are the use in some animals of (i) the air they already breathe for voice as well as respiration and (ii) the tongue already there for taste for articulation. See *An*. II.8

production of parts that while they are not directly necessary for the form, are necessary for an animal of that form to live or serve the animal's well-being.[38] This second category of parts comes into being *after* the first category, since the nature produces parts of the second category only if there is enough residue available from productions of the first category *or* if those productions result in the formation, by material necessity alone, as by-products, of extra materials. If this is not the case, the nature will not be able to produce (all of the) subsidiary parts. Every animal will still have all the parts that are the necessary prerequisites for the performance of its necessary functions (for instance, as Aristotle points out in *GA* I.4 717a11-31, *all* sexually reproducing males need ducts for semen; without this part, reproduction would be impossible for them), but it might lack other features which merely perform subsidiary functions (for instance, not all sexually reproducing males will have testes, but only those in which having testes is possible and in which their presence is 'for the better'). Subsidiary parts are parts an animal—hypothetically speaking—could do without (i.e., the animal *could have been* designed in a way such that it would not rely on the presence of subsidiary parts for its survival); they are not conditionally necessary given the animal's form.[39]

Both types of parts are present because of the function they perform, so both are caused teleologically. But whereas the causation of the first category of parts—the vital and essential ones—is *wholly* due to form (for even the material used in that process is produced of conditional necessity by the organism's—or its mother's—form), the causation of the subsidiary parts *begins* from material by-products of the former process, which are formed at a later stage and then act according to their own natures. The function is imposed on these materials only secondarily to their independent production. That is, the nature did not produce these materials for the sake of (ἕνεκα) realizing a necessary function as in the case of primary teleology, but uses (καταχρῆται) what is available of material necessity in a way that is (a) consistent with the potentials the available material has of necessity and (b) in all or most cases adds just to the well-being of the animal. In these cases Aristotle characterizes the resulting part as being 'for the better' or 'for' (πρός, εἰς) something.

To sharpen further this distinction between primary and secondary teleology, we would like to offer a *non-metaphorical* characterization of secondary teleology, along the lines of the non-metaphorical account we draw on of primary teleology, as the realization of an internal pre-existent potential for form (see n. 36). We start

420b13-22 (Aristotle refers to his account of voice in *De Anima* in *GA* V.7 786b24-26): 'it is reasonable that the only creatures to have voice should be those which take in air. For nature then uses the air already breathed in for two functions (καταχρῆται ἡ φύσις ἐπὶ δύο ἔργα); just as it uses the tongue for both tasting and articulation . . . ' In the present context, however, the form of secondary teleology specified in our text here is what is at work.

[38] Cf. *PA* I.1 640a33-b1.

[39] Cf. Gotthelf 1997a, 90 (= chapter 8, this volume, p. 177).

from two premises: (i) that there is, throughout the generation of an animal, *one* formal nature, i.e., one 'potential for form', in operation, and (ii) that material by-products of the primary operation of that nature are not part of what that formal nature or potential is a potential to produce, in just the way that the aim (and desire) of the carpenter is not to produce a table *plus* (e.g.) the excess shavings. However, just as the art of woodworking includes the capacity to make good use of leftovers or unexpected extra materials, so too we think it is the case for Aristotle that the one nature, the one potential for form, *includes* the potential to act for the best on (excess) material of a certain sort (should such material happen to be around), at some point in the process *later* than the initial activation of that potential.[40]

In the discussion that follows the analogy, which largely pertains to the nature and the coming to be of bones (744b27-745a18), Aristotle suggests that there are different kinds of material building blocks and residues available to the formal nature of an animal, which all play different roles in the formation of parts. First, there is spermatic residue (S), i.e., the menses of the female, which are from the very beginning available for the development of the embryo in the body of the female. Second, there is natural food (F), of which there are two kinds: natural food that is 'nutritious' (F1) and which is used for the formation of complete parts, and natural food that is 'conductive to growth' (F2) and which is used for the quantitative increase of parts that are already complete (i.e. they are fully developed, but may still grow in size). Finally, there are residues of these two kinds of materials (RS, RF1, RF2). Aristotle uses these distinctions among the available materials to differentiate between bones and sinews on the one hand, which come to be in the first formation of parts and are made from the spermatic residue (RS), and parts such as 'nails, hair, hoofs, horns, beaks, the spurs of birds, and any other such part' (745a1-3), on the other hand, as being made 'from the food that is taken in later and is concerned with growth (F2), which is acquired from the mother and from the outer world' (745a3).

Teeth are not mentioned until after Aristotle completes his discussion of the distinction between these two groups of subsidiary parts, apparently because teeth present a problem (745a18: περὶ δ' ὀδόντων ἀπορήσειεν ἄν τις). The problem pertains to the categorization of teeth (745a19; 745b2-4): are teeth like bones

[40] Note that there is a certain complexity involved even in the characterization in Gotthelf 1987b of how parts are brought about by *primary* teleology: 'On this picture, the fact that something, *A*, is necessary for some end, *B*, will *explain* the presence of that *A* because (i) a potential for *B* is being realized and (ii) if a potential for *B* is being realized, anything necessary for *B* is being realized. (That is part of what it *is* for a potential for *B* to be being realized.)' (232 [= above, p. 31]). To achieve a one-sentence characterization of how parts are brought about teleologically, *some of them* via primary teleology and *others* via secondary teleology that already complex formulation will require at least one additional clause (to be labeled '(iii)') along the lines of the sentence in the text to which this note is attached. We don't see the point of giving such a characterization, just as we wouldn't think it necessary to give such a one-sentence formulation of the rational teleological powers of *actual* 'good housekeepers' that incorporates their skilled secondary teleological capacities in order to understand the character of such powers.

(and sinews) or not? That teeth are subsidiary parts (they are either present for the preparation of food or for defense, or both: see n. 33 above),[41] rather than vital or essential ones, is already implied in an explanation of the coming to be of horns Aristotle offers in *PA*, where he mentions teeth as being among those parts that are made from residues (*PA* III.2 663b22-35):

> But we must say what the character of the necessary nature is, and, how nature according to the account *makes use* of things present of necessity for the sake of something. . . . For the residual surplus of this sort of body [i.e. earthen], being present in the larger of the animals, is *used* by nature *for* protection and advantage, and the surplus, which flows of necessity to the upper region, in some cases it *distributes* to *teeth* and tusks, in other cases to horns.[42]

The image of nature using materials available of necessity indicates that teeth are a product of secondary teleology (as are, we would argue, hair and voice: on hair, see *GA* II.6 744b25 and 745a1; on voice, see n. 37 above). The question Aristotle sets out to settle in *GA* II.6 745a18 is rather *what kind* of subsidiary part teeth are. As he points out in the next lines, teeth are of the same nature as bones (cf. *HA* I.11 493a2: ὀδόντες ὀστέινοι), and they thus belong to the first group of subsidiary parts, but not without qualification: the coming to be of teeth is intrinsically connected to the growth of bones, but in a way that makes their coming to be dependent on and secondary in time to the growth of bones (745b2-5):

> And it turns out that teeth do not have the same nature as the other bones: for they all come to be in the first formation and none of them later, whereas teeth come to be later.[43]

[41] Strictly speaking, and at the most general level, the function of nutrition in animals requires—i.e., conditionally necessitates—only the presence of a part for the intake of nourishment and for the excretion of residues (*PA* II.10 655b29-31) and a part in which the incoming food is changed (*PA* III.14 674a12-16; cf. 674a21-22: the contributory (συντελεῖς) parts for the stomach referred to likely include teeth); teeth are not necessary in this strict sense, as is evidenced by the fact that some kinds of animals, such as birds, lack teeth entirely and are nevertheless able to perform the function of nutrition.

This subsidiary status of teeth, however, does not necessarily pose problems for Aristotle's reliance on the example of the growth of teeth in his defense of natural teleology in *Phys.* II.8 198b16-199a8: the three examples (growth of crops, teeth, and entire organisms) presented by the materialist opponent represent for Aristotle *increasingly stronger cases of teleology*, starting with a kind of artificial production that is most natural (cf. ps.-Arist. *Econ.* I.2 1343a24-b1), and ending with a kind of natural phenomenon that can only be explained by reference to primary teleology. The example of teeth presents a case in between, as being due to secondary teleology. For a defense of this reading, see Leunissen 2010b, ch. 1. (Gotthelf agrees re the teeth, but thinks that crop growth is not a case of teleology in this passage; see above, p. 23 n. 41.)

[42] πῶς δὲ τῆς ἀναγκαίας φύσεως ἐχούσης τοῖς ὑπάρχουσιν ἐξ ἀνάγκης ἡ κατὰ τὸν λόγον φύσις ἕνεκά του κατακέχρηται, λέγωμεν. . . . τὴν γοῦν τοιούτου σώματος περισσωματικὴν ὑπερβολὴν ἐν τοῖς μείζοσι τῶν ζῴων ὑπάρχουσαν ἐπὶ βοήθειαν καὶ τὸ συμφέρον καταχρῆται ἡ φύσις, καὶ τὴν ῥέουσαν ἐξ ἀνάγκης εἰς τὸν ἄνω τόπον τοῖς μὲν εἰς ὀδόντας καὶ χαυλιόδοντας ἀπένειμε, τοῖς δ᾽ εἰς κέρατα.

[43] συμβαίνει δὲ μηδὲ τὴν αὐτὴν ἔχειν φύσιν τοῖς ἄλλοις ὀστοῖς τοὺς ὀδόντας· τὰ μὲν γὰρ ἐν τῇ πρώτῃ συστάσει γίγνεται πάντα καὶ οὐθὲν ὕστερον, οἱ δ᾽ ὀδόντες ὕστερον.

The explanation is that, whereas bones come to be from the spermatic residue (RS) and take their growth from the residues of the natural food (RF1) that is not used up for the production of the 'most important parts' (744b27-38), teeth come to be from the bones (745a19-20). More precisely, teeth come to be from the food distributed to the bones (RF1) but that is not used up by the bones for their growth—once the bones 'have reached their number', teeth come to be (745b7-9). In other words, teeth come to be from the leftovers of materials that were conditionally necessitated for some other use.

The particular connection of teeth to the growth of bones also accounts for two of their distinctive features. First, unlike the other bones, teeth can keep growing throughout the animal's lifetime (745a25), as long as there are residues available from the food distributed to the bones. (As Aristotle is quick to point out in 745a27-b2, the continuous growth of teeth also serves a final cause: they would blunt fast, if there were not some flow of materials towards them.) Second, because teeth are not grown together with bones and come to be later, they can also grow back again (only once in human beings, but repeatedly over their lifetime in some of the other animals with teeth, such as in sharks and reptiles) after they have fallen out. As we shall see shortly, Aristotle's explanation of the differences of teeth in book V, chapter 8 builds upon these two distinctive features.

Aristotle's account of the differences of teeth in GA *V.8*

Aristotle's explanations in book V, chapter 8 also exhibit the pattern of secondary teleology, involving references both to an initial material necessity and to a subsequent final causation. Aristotle mixes his explanations with criticisms of Democritus' account of the differences of teeth; for the sake of clarity, we shall reconstruct Aristotle's explanations separately and then turn to his critique of Democritus.

Aristotle starts his explanation of the difference in the timing of the growth of the front teeth and the grinders by applying the teleological principle that 'nature does nothing in vain, but always, given the possibilities, does what is best for the substantial being of each kind of animal' (*GA* V.8 788b20-24):[44]

> But since we suppose—making suppositions based on the things we observe—that nature neither falls short nor produces anything pointless among the possibilities in each case, it is necessary for animals that are to take in nourishment after suckling to have instruments for the chewing of food.[45]

Usually Aristotle invokes this principle to identify the causally basic feature of some animal that explains why it has the specific part it has, when observation of the

[44] For formulations of the complete principle, see *IA* 2 704b12-18, 8 708a9-12, and 12 711a18-29. See also Lennox 2001b, 205-23.

[45] *ἐπεὶ δὲ τὴν φύσιν ὑποτιθέμεθα, ἐξ ὧν ὁρῶμεν ὑποτιθέμενοι, οὔτ' ἐλλείπουσαν οὔτε μάταιον οὐθὲν ποιοῦσαν τῶν ἐνδεχομένων περὶ ἕκαστον, ἀνάγκη δὲ τοῖς μέλλουσι λαμβάνειν τροφὴν μετὰ τὴν [τοῦ γάλακτος] ἀπογαλάκτισιν ἔχειν ὄργανα πρὸς τὴν ἐργασίαν τῆς τροφῆς.*

wider kind shows that there are multiple ways in which its nature could have realized the function the part in question fulfills (see, e.g., Aristotle's explanation of why humans have hands instead of forelimbs in *PA* IV.10 687a15-18). Here he uses the principle to identify the causally basic feature that explains why teeth arise *at the time* they do: among the possibilities, it is best for teeth to arise after suckling (and not when approaching maturity, as Democritus claimed: 788b25), because *that* is when animals need instruments for chewing. If teeth arose earlier, they would be pointless; if they arose later, 'nature would have failed to do something that is among the possibilities for her' (788b25-26; cf. *GA* II.6 744a35-b1 and *An.* III.12 434a30-b8). In *GA* II.6 Aristotle explained that the growth of teeth occurs after 'the first generation of parts' when there are residues available from the food assigned to the growth of bones. The exact timing, however, of *when* formal natures use these residues for the production of teeth is determined by the animals' functional needs (including the needs of the process by which they are produced).

The full explanation of the difference in timing appeals both to teleology and material necessity (788b29-789a4): front teeth arise before grinders because their function is needed earlier than the function performed by grinders *and* because 'what is smaller—*even* when set in motion at the same time—by nature comes to be faster than the bigger'. Grinders are bigger than front teeth, because the jaw-bone is narrower towards the front of the mouth, and bigger and flat where the grinders grow, and *of necessity* there is a bigger outflow of food to the surface from what is bigger. The remark 'even when set in motion at the same time' (*κἂν ἅμα ὁρμηθῇ*) is significant: even if formal natures did not divert the residues first to front teeth and later to grinders, the *same* beneficial result would have occurred of material necessity alone.

The explanation of the differences of shedding also refers to both teleology and necessity (789a8-14): front teeth blunt fast, so they need to be replaced if animals are to remain able to cut food. However, there is no such functional need for grinders (they do not blunt, but get smooth): it is therefore 'for the better' (*τοῦ μὲν βελτίονος χάριν*) that front teeth shed, and not grinders, which do not need replacement. Front teeth also shed of material necessity (*ἐξ ἀνάγκης δ'*): whereas grinders are fixed firmly in the broad, strong parts of the jaws, front teeth grow in the thinnest parts and are therefore easily moved.[46]

Finally, the availability of constitutive materials is given as the one and only explanation why front teeth grow back after shedding (789a14-17): they grow back 'because the shedding occurs while the bone is still growing and there is still time for teeth to grow'. This explanation builds upon Aristotle's discussion of teeth

[46] In the biological works, Aristotle typically provides such double-barreled explanations for subsidiary parts or differences: the 'for the better' picks out the operation of secondary teleology, the 'of necessity' the operation of material necessity. For examples, see *GA* I.4 717a12-21 and 26-31; *PA* II.14 658b2-10, III.9 672a1-21, IV.3 677b22-32, and IV.12 694a22-b11; cf. also Leunissen 2007, 173-74.

in *GA* II.6: front teeth are shed while there is still a supply of food assigned to the growth of bones, which leaves enough residue for a second set of front teeth to grow (and for wisdom teeth to form: *GA* V.8 789a16-20).

In both explanations of the differences of teeth, teleology and material necessity are identified as operating alongside and even as complementing each other, just as teeth themselves come to be both of material necessity and due to (secondary) teleology.

For Aristotle's views about the interaction of these two forms of causation, we need to turn to his second criticism of the explanations provided by Democritus (*GA* V.8 789b2-15):[47]

> Democritus, neglecting to state that for the sake of which, brings back all things of which nature makes use to necessity—they are such, but on the other hand they are for the sake of something and for the best concerning each. Thus nothing prevents [them] from coming to be this way and being shed, but it is not because of those but because of the end. For those things are causes in the sense of being the movers and the instruments and the material, since it is also likely that the operation of *pneuma* is as an instrument for many things—as for instance some [instruments] are of many uses to those involved in arts, such as the hammer and anvil in the art of the smith, so too is *pneuma* in those things which are put together by nature. It seems that to say that the causes are of necessity is similar to if somebody would think that it is only because of the surgeon's knife that the water has been drawn off in those who are dropsical, but not because of being healthy, for the sake of which the knife made the cut.[48]

The basic mistake Democritus made according to Aristotle is not the fact that he explained *all* natural phenomena in terms of material necessity, but that he *failed to state* the final cause of all those things *which nature uses*. We take it that what Aristotle is saying here is that in cases where the formal nature of an animal makes use of things that have come to be of material necessity, one should include a reference to the function for which those things are used in the explanation *as well*.[49] As Aristotle points out (in *GA* V.8 789b3-4): the coming to be of such things *is* materially necessitated, but that is not the whole story—they also serve a

[47] In his first critique of Democritus, Aristotle faults Democritus for providing a general explanation without having examined all cases (*GA* V.8 788b9-21).

[48] *Δημόκριτος δὲ τὸ οὗ ἕνεκεν ἀφεὶς λέγειν πάντα ἀνάγει εἰς ἀνάγκην οἷς χρῆται ἡ φύσις—οὖσι μὲν τοιούτοις, οὐ μὴν ἀλλ' ἕνεκά τινος οὖσι καὶ τοῦ περὶ ἕκαστον βελτίονος χάριν. ὥστε γίγνεσθαι μὲν οὐθὲν κωλύει οὕτω καὶ ἐκπίπτειν, ἀλλ' οὐ διὰ ταῦτα ἀλλὰ διὰ τὸ τέλος· ταῦτα δ' ὡς κινοῦντα καὶ ὡς ὄργανα καὶ ὡς ὕλη αἴτια, ἐπεὶ καὶ τὸ τῷ πνεύματι ἐργάζεσθαι τὰ πολλὰ εἰκὸς ὡς ὀργάνῳ—οἷον γὰρ ἔνια πολύχρηστά ἐστι τῶν περὶ τὰς τέχνας, ὥσπερ ἐν τῇ χαλκευτικῇ ἡ σφύρα καὶ ὁ ἄκμων, οὕτω καὶ τὸ πνεῦμα ἐν τοῖς φύσει συνεστῶσιν. ὅμοιον δ' ἔοικε τὸ λέγειν τὰ αἴτια ἐξ ἀνάγκης κἂν εἴ τις διὰ τὸ μαχαίριον οἴοιτο τὸ ὕδωρ ἐξεληλυθέναι μόνον τοῖς ὑδρωπιῶσιν, ἀλλ' οὐ διὰ τὸ ὑγιαίνειν οὗ ἕνεκα τὸ μαχαίριον ἔτεμεν.*

[49] See also Aristotle's final example of water-drawing, where he equates 'stating that the causes are of necessity' with believing that the water draws off *only* because of the knife's cut, and not because of health, which is that for the sake of which the cut was made. The suggestion is again that some phenomena are due to the operation of both material necessity and (secondary) teleology, and their explanations thus need to pick out both causes.

purpose in the animals that have those features because of the use nature has made of them.

In his application of this critique to the example of the shedding of teeth, Aristotle claims that front teeth come to be and are shed of material necessity, but that ultimately phenomena such as these should be explained by reference to (a form of) teleology, because it is ultimately the formal nature of an animal that is responsible for using those 'materials' and 'instruments'—such as *pneuma*—in a goal-directed way within the animal. It is usually assumed that Aristotle here offers a general defense of teleology and speaks of the operation of material necessity only within the framework of an operation of conditional necessity.[50] However, if we limit the scope of Aristotle's critique to Democritus' explanation of teeth and of any other such feature that is the result of nature *using* what is present of material necessity, such an interpretation might well be too strong. Aristotle's own explanations of the differences of teeth, alongside many other examples throughout the biological works and especially the ones provided in chapters 1-7, offer sufficient evidence that material necessity can operate independently of teleology, and that formal natures often, though not always,[51] make use of what happens of material necessity for something good, where this good is not already specified in the definition of the animal's substantial being.

Aristotle refers to this distinction between the attributes of teeth that are used for something and the attributes discussed in the first seven chapters that do not serve any teleological purposes in his summary of book V at the end of the chapter V.8 789b16-20:

> We have thus spoken of teeth, saying why some fall out and grow again, and others not, and generally because of what cause they come to be. And we have spoken also of the other attributes of parts, as many as turn out to come to be not for the sake of something but of necessity and because of the moving cause.[52]

Aristotle first sums up the immediately preceding account of teeth: it has now been stated (in chapter 8) why there is a difference in the shedding of teeth, and why they come to be in general (in 788b3-6, summing up the accounts provided in the *Parts of Animals*). The next section sums up the earlier accounts of book V and *contrasts* them with the account of the differences of teeth: he has also spoken of the *other* attributes (in chapters 1-7), which are characterized by the fact that they do *not*

[50] e.g. Cooper 1987, 259, 268.

[51] Cf. *PA* IV.2 677a15-18: 'Sometimes a nature makes use even of residues for some benefit (*κατάχρηται μὲν οὖν ἐνίοτε ἡ φύσις εἰς τὸ ὠφέλιμον καὶ τοῖς περιττώμασιν*), yet one should not on this account search for what something is for in every case; on the contrary, when certain things are such as they are, many other such things happen from necessity because of these.'

[52] *περὶ μὲν οὖν ὀδόντων, διότι οἱ μὲν ἐκπίπτουσι καὶ γίγνονται πάλιν οἱ δ' οὔ, καὶ ὅλως διὰ τίν' αἰτίαν γίγνονται εἴρηται. εἴρηται δὲ καὶ περὶ τῶν ἄλλων τῶν κατὰ τὰ μόρια παθημάτων ὅσα γίγνεσθαι συμβαίνει μὴ ἕνεκά του ἀλλ' ἐξ ἀνάγκης καὶ διὰ τὴν αἰτίαν τὴν κινητικήν.*

come to be for the sake of something, but arise due to material and efficient causes alone.

Aristotle thus signals a discontinuity between the first seven chapters of book V and chapter 8, but if our interpretation of chapter 8 is correct, this discontinuity does not run very deep: the dominant type of causation in the coming to be of all attributes of parts discussed in book V is material-efficient causation, but the differences of teeth discussed in chapter 8 are different in that they, unlike the others, are *also* (though secondarily) for the sake of something.

Concluding remarks

The upshot of the preceding analysis of Aristotle's explanation of the differences of teeth for our assessment of the scope and nature of book V is as follows.

First of all, we found that the differences of teeth discussed in chapter 8 are similar to the particular attributes of eyes, hair, and voice discussed in chapters 1-7 in that all are differences of the more and the less that come to be due to material and efficient causes during the later development of the animals that have them, and most particularly human beings. The scope of book V is thus not limited to the explanation of accidental, individual differences among animals, but encompasses some of the most variable attributes that come to be, in one variant or another, in either whole species or in individuals during their later life, and that are very conspicuous in human beings. What unites all explanations of the attributes discussed in book V is thus that they pick out material necessity as the primary cause for the later development or changes responsible for these attributes. The differences of teeth are different, however, in that the formal natures of animals with teeth use or co-opt the differences of teeth that come to be of material necessity for something good, whereas this is evidently not (explicitly) the case for the other attributes.

Second, Aristotle's concern for the appropriate mode of explanation to be used for the explanation of the differences of teeth (expressed in his critique of Democritus) is consistent with his advice in his introduction in the first chapter of book V: in both cases, Aristotle warns us not to think that the manner of explanation is the same for all explananda. As is evidenced also by the actual explanations provided across book V, as we have seen, Aristotle never bans the use of teleology for the explanation of attributes in book V, but only insists that it should be applied to the appropriate phenomena and in the appropriate way.

Finally, as for the place of book V in the *Generation of Animals* as a whole, there is sufficient evidence that the particular explanations of the differences of teeth provided in chapter 8, just as the explanations of the other attributes discussed in chapters 1-7, presuppose and build upon Aristotle's earlier accounts of the coming to be of the parts themselves. Together with the explicit cross-references between book

V and the earlier books, this suggests that Aristotle considered all these books as part of one and the same investigation into the coming to be of animals and their parts. Also, the continuity in content and method between chapters 1-7 and chapter 8 in our opinion outweighs its discontinuity: the hypothesis that chapter 8 is unrelated to the preceding seven chapters and forms instead a separate appendix to books I–IV ought to be rejected.

Appendix: GA *V.1 778a29-b19*

The manner of explanation appropriate to the attributes now under study

1. (a29-30) Introduction via a contrast: the manner of explanation is 'no longer the same' as was appropriate previously (i.e., in *GA* I–IV)

2. (a30-32) Distinction between two manners of explanation, the primary one to be used in *GA* V, distinguished from the primary one used previously, by reference to the dominant mode of causation employed in each case:

   [i] The subject neither is nor comes to be for the sake of something. <Of this sort are> things that are:
   neither (α) products of a nature in common (*sc.* across several kinds) *nor* (ß) distinctive features (ἴδια) of any particular kind

   [NOTE: This statement only commits Aristotle to holding that satisfying (α) or (ß) is *necessary* for such an attribute to be for the sake of something, *not* that satisfying (α) or (ß) is *sufficient* for that. The statement does not guarantee, then, that *every* feature common to a kind must be for the sake of something.]

   [ii] The subject is <and comes to be?> for the sake of something.

3. (a32-34) Examples of each of the two dominant modes of causation in 2.:

   [ii] An eye *is* present for the sake of something

   [i] Its being blue *is not* (unless this attribute is distinctive (ἴδιον) to the kind of animal its possessors belong to)

   [NOTE: Here, too, the conditional (in this case, the 'unless'-clause) may only be stating a *necessary* condition for the attribute's, the blueness's, being for the sake of something, and not a sufficient condition; in that case what is being said is that if the attribute is distinctive to the kind, it *might* be for the sake of something, not that it necessarily *will be* for the sake of something.]

4. (a34–b1) Fuller statement of the first of the two manners of explanation just distinguished, the one which is appropriate to (at least the majority of) the attributes under discussion in *GA* V:

In those cases where the attribute being explained does not (as [ii] would require) reach back to (*συντείνει*) the account of the substantial being (*λόγος τῆς οὐσίας*; *sc.* of the animal possessing it), then the attribute must (qua case of [i]) be referred to 'the matter and the source of motion, as coming to be of necessity'.

5. (b1–10) Reminder of what we know of things explained that come to be for the sake of something (= [ii]) and critique of predecessors:

[a] As has been said in the opening *λόγοι* (*sc. PA* I; cf. 640a10ff.), in the case of 'the ordered and determinate products of <a> nature', coming to be follows upon and is for the sake of substantial being (*οὐσία*—and so it's not because a thing comes to be of a certain character that it is that character, but rather because it is of this particular character that it comes to be such as it does).

[b] Earlier natural philosophers thought the opposite, for one reason because they did not realize that the causes are many but saw *only* the cause that belongs to the matter and the one that belongs to the motion (and did not distinguish these).

[Note: Here, arguably, Aristotle distinguishes material and efficient components of what is often called the 'material-efficient' cause. Later (b11–12), he will, refer to the efficient aspect of the formal cause. In short, it should be kept in mind that, for Aristotle, both matter and form can, and do at different points in *GA*, serve as efficient causes.]

6. (b10–19) Fullest statement of the contrast between the two modes of causation (and their respective sorts of necessity) that are dominant, respectively, in the two manners of explanation of the coming to be of animal differences previously distinguished:

[ii] The teleological mode (essential to teleological explanations of the primary sort): The difference whose coming to be is being explained is formed both for the sake of the animal which is to possess it and through the other causes, and the necessity involved is *conditional necessity*. (b10–14)

[Note: Such differences, Aristotle reports here, include:

(*α*) those which are present in the account of each (*sc.* of these 'ordered and determinate products of nature');

(ß) those which are for the sake of something (*sc.* i.e., for the sake of differences in (*α*) or (*γ*)); and (γ) those (*sc.* other than the ones in (α)) which are that for the sake of which other differences (*sc.* those in (ß)) are.]

[i] The non-teleological mode (essential to explanations that are primarily material-efficient): The difference being explained is formed (*sc.* not because the animal's substantial being (*οὐσία*) requires it, as in [ii], but) in the very constituting (*σύστασις: sc.* of the animal), as a result of the movement *(κίνησις)* and formation (*γένεσις*) going on itself, via *the (non-conditional) necessity of* 'its being its nature to act or be acted upon in this or that way' (*ὅτι τοιουδὶ ἢ τοιουδὶ ποιεῖν πέφυκε καὶ πάσχειν*). (b14-19)

6

Teleology and Spontaneous Generation in Aristotle

A Discussion

The problem[1]

Aristotle's mature theory of spontaneous generation and his evident denial that spontaneous generations have a teleological character pose an intriguing challenge to certain interpretations of his teleology, my own included.

I have argued in chapter 1 that Aristotle's natural teleology rests in part on a denial that the generation and development of living organisms is wholly due to, and thus can wholly be explained by reference to, the simple natures and potentials of the elements which constitute these organisms. These generations and developments, I have claimed, are due rather to an 'irreducible potential for form', which is borne by the heat and/or 'movements' transmitted by the male, either via the *pneuma* in its semen (if it emits semen) or directly through contact (if it does not).[2]

The challenge from spontaneous generations can be put as a dilemma. *Either* a spontaneous generation is due wholly to the simple natures and potentials of the

This essay was written for a volume in memory of Joan Kung, who died of cancer in 1987 at the age of 48 (Penner and Kraut 1989).

[1] Joan Kung and I often discussed issues in and around Aristotle's teleology and my interpretation of it, and I drew comfort from her unstinting encouragement. She was a dear friend, open with her observations and judgments, whether professional or personal, as well as sensitive, perceptive, and challenging. The topic of the present essay is one we discussed, so it seems an appropriate, if small, memorial to her, and to her contributions to our field, especially as she and I and Jim Lennox, whose views I criticize (and who cared for her very much as well), talked jointly about teleology on several occasions. I miss Joan a great deal.

[2] Strictly, the 'active' potential is embodied in the male heat or 'movements', the 'passive' potential in the female material. The union of these passive and active *potentials* results in the offspring's *nature*, which is its budding *soul* (first a mostly potential soul, then, as the capacities get formed, an increasingly actual one). This way of speaking follows Aristotle's definitions of *nature* and *potential* more precisely than does my own tendency to speak of both the capacity in the semen and the capacity in the developing stages of the organism itself as 'potentials for form', though Aristotle himself licenses this broader use of 'potential' at *Metaph.* Θ.8 1049b5–10.

elements (as Aristotle's remarks about spontaneous generation in the *Metaphysics* would seem to support[3]), in which case it is unclear how Aristotle could have been so certain about irreducibility in the non-spontaneously generated cases, *or* a spontaneous generation is due to the 'capture' from the environment of pneumatic heat in 'the exact amount and strength' that would be present if it were a sexual generation (as the mature theory of spontaneous generation in *GA* III.11 might be thought to have it), in which case it isn't clear how, consistent with my interpretation, Aristotle could ascribe teleology to sexual generations but deny it to spontaneous ones.

The second horn of the dilemma was brought to the fore by James Lennox, who in a 1982 paper argued for an interpretation of the *GA* III.11 account of spontaneous generations along the lines I have summarized.[4] Lennox offered a reading of a key passage in that important chapter of *GA*, based an account of Aristotle's theory of spontaneous generation on it, and then argued for an interpretation of Aristotelian teleology (the 'formal replication model') built around the contrast drawn elsewhere in the corpus between spontaneous and sexual generations. As he wrote in a footnote, after kindly acknowledging the debt his own interpretation owed to mine, 'I was led to the formal replication model of teleology originally by troubling

[3] *Metaph.* Z.9 1034a33-b1, 4-7: 'It is this way too with things constituted by nature. For their seed produces just as in the case of things that are made from art—it has the form potentially, and that from which the seed comes is in a way the same in name.... Those can come to be spontaneously like the ones <from art> considered above, of which the matter can be moved and changed by itself in the way in which the seed <characteristically> moves and changes it; but those things which have not such matter cannot be produced except from <parents>.' Translation after Lennox (1982, 226 [2001b, 234]) and Ross; the passage is discussed also in Ross 1924, ad loc. and Balme 1962b, 96-97. Balme argued that the view of spontaneous generation behind this passage is both different from and earlier than that in *GA* III.11. For instance, according to the account of spontaneity in this and surrounding passages in *Metaph.* Z.8-9, as well as the accounts in *Phys.* II.4-6 and *PA* I.1 640a27-33 (cf. *APo.* II.11 95a3-6), most of the members of species in which there are spontaneous generations should be generated sexually, but that is precisely not so according to *GA* III.11, which insists that virtually all of the testacea, all of the eels, all of certain species of river fish, and all of certain species of insects are generated spontaneously. In fact, wherever an animal is generated spontaneously the whole of at least its species are also. Now it is just possible that the *GA* account is compatible with the earlier one, *if* (i) the 'neither always nor for the most part' clause applies here only to generations as such, across the animal world (cf. Lennox 1982, 235 [2001b, 241-42]) and (ii) the action of the pneumatic heat in the *GA* III.11 theory counts as the matter moving by itself in just the way the seed would move it. I am disinclined to accept this view, however, both because the deflation of the 'neither always or for the most part' condition does not seem in the spirit of the *Phys.* II.4-6 discussion, and because of the way in which the second of the two *GA* III.11 passages I discuss below stresses the existence in spontaneous generations of an analogue to the male contribution *in addition to* the analogue to the female contribution of the 'material source'. (On this, see further below, pp. 147ff.)

[4] Lennox 1982 (= Lennox 2001b, 229-49). The words in quotations marks in the previous paragraph are Lennox's 1982, 225 (2001b, 233).

over three difficulties in Gotthelf's paper. First, it is not clear to me how that account can distinguish between non-teleological, spontaneous processes and teleological, reproductive ones.'[5] For if the heat in spontaneous generations is species-specific, as Lennox's account entails, then the same grounds for ascribing an irreducible potential for form to the species-specific vital heat in sexual generations would seem to apply to spontaneous generations as well.[6]

In what follows, I would like to explore, somewhat briefly because of space considerations, the *GA* III.11 theory of spontaneous generation. The theory is of course important in its own right as part of a notorious and central chapter in the history of biology; as Lennox's paper shows, it is important too for the implications it might have (or not have) for the interpretation of central aspects of Aristotle's philosophic thought. Building on a close look at the passage in *GA* III.11 on which Lennox based his interpretation of the theory of spontaneous generation, I will aim to show:

(1) that, contrary to Lennox's claim, the pneumatic heat enclosed at the beginning stages of a spontaneous generation is *not* species-specific, and there are thus no grounds for ascribing to it an irreducible potential for the form of the outcome;

[5] Lennox 1982, 237 n. 42 (2001b, 248 n. 41). Lennox's second reservation concerns the number of irreducible potentials at work and would need more elaboration before I can deal with it. His third reservation, that 'it remains unclear to me how "the end for the sake of which" is *responsible for* the end-directed process on Gotthelf's account', stems, I think, from the same error that vitiates his own 'formal replication model'. Lennox claims that what generates teleological explanation in sexual and craft productions is the identity of the form that initiates the process with the form that results. The form at the end is responsible for the process which leads to it because it is the form which initiates that process (1982, 222-23 [2001b, 231]). But this, notwithstanding Lennox's protestation (1982, 227 n. 18 [2001b, 245 n. 18]), is to confuse efficient and final cause. (It is also based on a less satisfactory view of the sameness in form that is involved than Lennox himself presented in Lennox 1987b.) What one needs to explain is not how the form at the end might serve to initiate the process (e.g. by also being there at the beginning), but how it is responsible *as end* (*Phys.* II.3, 194b32) and this Lennox's view simply doesn't do. My own view, and its implications for Lennox's concern, is perhaps more clearly put now in my Postscript 1986 in chapter 1, pp. 31-32.

[6] In an important passage in the exposition of his theory of sexual generation (*GA* II.3, 736b27-737a7), Aristotle ascribes differences in value among souls, and thus, by implication, among animals, to differences in pneumatic heat, and I have used this passage in support of the view that vital heat in sexual generation is species- (or at least kind-) specific (chapter 1 above, pp. 17-18). I use the term 'species' here as shorthand because Lennox did in his 1982 paper. Both of us would now prefer to resist translating *eidos* as 'species', as is explained in Lennox 1987b, 339 asterisked note (2001b, 160n.); see also Balme 1987c, 306 and Pellegrin 1987 *passim*, with references.

Balme has argued, I think successfully, that the 'movements' that are responsible for generation (and thus, in some respect, the heats which carry them) are not only kind-specific but also parent-specific (Balme 1987c, 293 n. 14); that conclusion, however, will not be relevant to the present discussion.

(2) that a spontaneous generation is nonetheless probably *not* explainable wholly by reference to the simple natures and potentials of the elements involved in the process, but probably involves a non-species-specific irreducible potentiality; and

(3) that, in any case, it is unwise to rest interpretations of Aristotelian teleology on Aristotle's discussions of (what he took to be) *actual* spontaneous generations.

Spontaneous generation

The key passage on spontaneous generation in *GA* III.11 is 762a18-27. In his 1982 paper, Lennox translated it as follows:

> Animals and plants come into being in earth and in liquid because there is water in earth, and air [*pneuma*] in water, and in all air there is vital heat [*thermotēta psuchikēn*]; so that in a sense all things are full of soul. Therefore <living things> form quickly whenever <this air and vital heat> are enclosed <in anything>. When they are so enclosed, the corporeal liquid's being heated, there arises as it were a frothy bubble. The differentiae which determine whether the kind is more or less honorable are <determined by> the organization of the vital principle in the enclosed. And both the places and the enclosed material are causes of this <organization>.[7]

The crucial sentences are the last two, the Greek of which runs:

> *αἱ μὲν οὖν διαφοραὶ τοῦ τιμιώτερον εἶναι τὸ γένος καὶ ἀτιμότερον τὸ συνιστάμενον ἐν τῇ περιλήψει τῆς ἀρχῆς τῆς ψυχικῆς ἐστιν. τούτου δὲ καὶ οἱ τόποι αἴτιοι καὶ τὸ σῶμα τὸ περιλαμβανόμενον.*

Lennox's rendering of *to sunistamenon* in a25 as 'the organization' is what allowed him to see the process as 'organizing' or, as he later puts it, 'capturing' vital heat in an 'exact amount and strength'. This is Lennox's only textual basis for the view that the vital heat in spontaneous generations is species-specific. *to sunistamenon* cannot, however, mean 'organization', and cannot here take *tēs archēs tēs psuchikēs* as

[7] Words in pointed brackets are marked by Lennox as not in the Greek but to be supplied from the sense. Two small points on this translation, before going on to my major criticism: (i) *pneuma* is not just any air, but air infused with a special heat (*GA* II.2 736a1; cf. 735b32-34, II.3, 736b33-37 and the references in n. 19 below) and the term is best simply transliterated (as Lennox himself did in his discussion of the passage); (ii) '<in anything>' is both too broad and unnecessary: the enclosing material will always be an earth/water mixture, but the range of possible mixtures is an empirical question which apparently cannot be predicted at this level of theory (for references to the variety of materials, cf. Balme 1962b, 100 n. 2).

genitive object. The present participle here and elsewhere clearly refers to the outcome of a process and is a passive, not a middle;[8] it means 'that which is (or: is being) constituted'.[9] David Balme has suggested[10] that in the first of these two sentences *to sunistamenon* is best taken as the subject of *einai*, and *to genos* as an accusative of respect. 'This puts *estin* with the plural subject *diaphorai* and no complement, but that is not unusual—the verb "to be" is as roughly handled in Greek as in English . . .'[11] As a literal rendering, he proposed:

> The differences of more and less value in respect of kind, in that which is being constituted, lie in that which encloses the soul-beginning.[12]

An even more literal rendering of both sentences together might run:

> The differences in regard to that which is constituted being more or less valuable in respect of kind are found in the enclosure of the soul-beginning. Of this both the places and the enclosed body are causes.[13]

The theory in 762a18–27 is thus to be reconstructed as follows. A bit of earthy water containing *pneuma*, which itself contains an '*undifferentiated* [vital heat]'[14] is warmed by the (ordinary) heat of the surrounding air or water (= 'places', a26). This warming forms a bubble around the earthy water and thickens the earthy water around the portion of *pneuma* it contains.[15] In the case of testacea, the heated earthy water becomes the shell (a27–32). There is, one may note, no explicit reference to any thinner earthy/watery material inside the thickening material,

[8] LSJ s.v. *sunistēmi*; Bonitz s.v. *sunistanai*.

[9] Cf. *GA* III.11 762b1.

[10] In correspondence, in 1984.

[11] Professor Balme noted that Drossaart Lulofs 'swallows' this as well in his OCT edition (1965). Peck conjectures *eisin*, but like Balme I prefer not to 'rewrite' the text when there is no MS authority.

[12] My anonymous reader complained that Balme's construction requires us to take the *tou* in a24 with *timiōteron*, but of course that's wrong: the *tou* goes with *einai*, the articular infinitive being 'used to limit the meaning of' (Smyth, §2032) *diaphorai*; *timiōteron* (along with *kai atimoteron*) is the predicate compliment of *einai*. My own translation, which follows, attempts to bring that out more clearly.

[13] Lennox may have based his interpretation in part on the ending of *perilēpsis*, viewing the differences in kind as due to the enclos*ing* of (a specific 'amount and strength' of) the soul-beginning, rather than as due to, as Balme translated, 'that which encloses'. I think the context rules firmly against Lennox's way of taking the sentence, and in view of the number of nearby participles derived from forms of *lambanō*, have translated *perilēpsis* with the more neutral 'enclosure', as does Lennox. One might be tempted to take the *sōma* in a27 to be the *pneuma* or even the vital heat, but against that see n. 15 below.

[14] Balme 1962b, 99 (my emphasis).

[15] The same material which does the enclosing of the soul-source is itself enclosed by the frothy bubble (cf. *emperilambanetai*, a22). By the time the material hardens, the bubble is long gone.

such as would become the inner, fleshy part of the testacean; but nothing in the description rules out there being such.

The referent of *toutou* in a26, then, is not any organizing of vital heat (for the existence of which there is no independent evidence), but *either* the nature of that which encloses the vital heat and is responsible for the kind of animal that is produced from the process (if one takes, with Lennox, *to sunistamenon* as antecedent), *or* the production of that very kind of animal itself rather than some other (if one takes the entire sentence as antecedent). In either case, the 'differences of more or less value' lie not in differences in the soul-source, in 'amounts and strengths' of vital heat, but in the differences in the enclosing material, e.g. in the different sorts of shells of the testacea.[16] Nor does Aristotle cite as cause of these differences anything about the vital heat. As Lennox himself notes, these differences are due to 'the places and the body which is enclosed'—i.e., to the environment (including climatic conditions and, generally, whether the immediate surroundings are land or sea or river, etc.) and to the particular proportion and character of the earth and water that was separated off (and 'enclosed' by the bubble) in the first place.

This account of the process by which the spontaneously generated organism is produced is confirmed by the longer discussion at 762a35-b18. Here Aristotle clearly describes a two-stage process, on analogy to the two-stage process in sexual generation in which, first, the 'material principle', a 'residue', is formed in the female and, then, this material is acted upon by 'the male principle'. In the spontaneous cases the climatic heat first produces something like a female residue, in which the pneumatic heat then 'implants movement'. As the passage is very difficult, I offer a literal translation, with elucidatory and interpretative comments in pointed and square brackets respectively:

Someone wanting to explore this in the right way might explore what it is that is being worked up (*to sunistamenon*), in the sense of the *material source*, in such [*sc.* spontaneously generated] cases. For, in females this is a certain residue of the animal which, when the source from the male imparts movement <to it>, <it> being potentially such [*sc.* an animal of the same kind] as the very thing [*sc.* the female] from which it came, results in the animal [*sc.* offspring]. What corresponds to this <material source> here, and whence and what is the movement-imparting source corresponding to the male?

We need to understand that even in the generating animals the heat in the <female> animal, by separating off and concocting the incoming nutriment, forms the residue which is the <material> source of the embryo. Similarly also in plants, except that in them, as well as in some of the animals, there is no additional need of the source from the male (for they

[16] The same view is maintained in *HA* (cf. 528b13). None of the passages from *HA* that Lennox cited (1982, 225 [2001b, 233]) suggests differences in the character or quantity of vital heat, and 547b18-23, which he does not mention, says just what the *GA* passage does. As Balme points out (1962b, 100-101), the *HA* discussions of spontaneous generation do not mention *pneuma* or vital heat at all (cf. Lennox 1982, 224 n. 13 [2001b, 245 n. 13]).

have this already mixed in), while in most of the animals there is need of something beyond the residue. Now in the case of these [*sc.* plants] the nutriment is water and earth, while in the case of those [*sc.* the generating animals] it is things made of earth and water, so that that very thing [*sc.* the residue] which the heat in the <female> animals produces out of its nutriment, this the heat of the season (*ōra*) in the surrounding environment (*to periechon*), by a process of concoction, composes and constitutes out of sea-water and earth.

That <portion> of the soul-source which has been enclosed or separated off in the *pneuma* produces [*sc.* out of this material source produced by the climatic heat operating on the earth/water mixture] an embryo, and implants movement in it.

Though nothing is said here as to which of the factors is responsible for the nature of the outcome, the passage should, taken by itself, make one wary of ascribing too much to the role of the pneumatic heat and not enough to that of other factors; and it is thoroughly in harmony with the account of Aristotle's theory extracted from 762b18-27.[17]

In sum, then, there is nothing in Aristotle's discussions of spontaneous generation that suggests that the vital heat in an organism in process of spontaneous generation is specific to the kind of organism being generated. The vital heat 'captured' in early stages of spontaneous generation cannot therefore be said to have a specific capacity to produce in the relevant material an offspring of a given form, and the disanalogy with the sexually generated case is clear. The second horn of the dilemma described above is thus disarmed.

It is significant, however, that Aristotle's term for what I have mostly translated as 'vital heat' is *thermotēs psuchikē*—'*soul*-heat',[18] and that it, or the *pneuma* which contains it, is referred to as the *psuchikē archē*—'*soul*-beginning' or '*soul*-source'. It is the sort of heat appropriate for producing (a potential) soul in its product—a special heat, having a different *dunamis* from the heat of fire, as Aristotle explains at 736b29-737a6.[19] This suggests that the heat that actually brings life to the spontaneous generation[20] has an undifferentiated, but distinct and irreducible, potential for life. As the language of 'matter' in the second of the two passages we translated (762a35-b18) suggests, it is this 'soul-heat' that imparts the potential for form in the spontaneously generated organism, even if it doesn't impart a potential for a specific form (the specifics of the organism's form being, in spontaneous generations, due to the material mixture in which it is embedded). And many of the very

[17] A statement made in passing in an earlier discussion (743a35-36) positively suggests that, among heats, it is in fact the climatic heat that makes the difference, though the sentence is too loose for much weight to be put on it.

[18] e.g. 762a20 (for other references, cf. Bonitz s.v. *thermotēs*, esp. 327a24-30).

[19] See my discussion of this passage, and cf. the parallel passage in Theophrastus' *On Fire*, in Gotthelf 1975, 144ff., more accessible in a briefer version in chapter 1 above, pp. 17-18, with n. 30. 737a4-6 is most likely a reference to the heat in putrefying material from which some spontaneous generations begin.

[20] 762a32, as distinct from the heat in the environment.

same arguments which support the irreducibility of the potential for a specific form in the sexually generated cases support the irreducibility of this undifferentiated potential for life. Thus, at the time Aristotle developed his mature theory of spontaneous generation, he probably did *not* think that a spontaneous generation was explainable wholly from the simple natures and potentials of the elemental materials involved in the process.

In fact, I suspect that it was precisely the conviction that even spontaneous generations were not due to element-potentials alone, but involved the same sort of special heat as is involved in sexual generation, that motivated the mature theory of spontaneous generation. In *Metaph.* Z.9 and elsewhere (cf. n. 3 above) Aristotle characteristically explained why only some products of nature and craft come about spontaneously, by distinguishing those in which the matter could be moved by itself and its surroundings in the way the form would move it (e.g., a medicinal heating, a simple movement) and those in which it could not (e.g., a house or statue, a dancing movement); he must have thought that those animals which could be spontaneously generated were simpler and required simpler movements in their generation than those which did not. Very likely, as Aristotle studied the generative process in detail, he came to think that it was too complicated, too balanced and precise—that even the spontaneously generated cases were rather like the dancing movements—especially as he became convinced, on observational grounds, that even some blooded animals, such as the eel and some species of river fish, are generated spontaneously.[21] A form, and the characteristic heat that bears it, he probably thought, must be involved even in spontaneous generations, even if only generically, in an undifferentiated potential for life, carried by environmental *pneuma*, which *pneuma* must be a lot more widespread than previously suspected.

Teleology

What does this undifferentiated potential for life mean for teleology? On my own account of Aristotle's natural teleology,[22] one might expect some of the organ formations in the spontaneously generated case—or some aspects of them—to be explained teleologically, even if it isn't entirely clear how that would go. Do we find any evidence of such explanations? This question brings us to the last of the three claims of this chapter: that it is unwise to rest interpretations of Aristotelian teleology on Aristotle's discussions of (what he took to be) *actual* spontaneous generations. Though it would be enjoyable to speculate on an Aristotelian answer to the question with which I began this paragraph, the corpus does not give any evidence that Aristotle himself reflected on it, or on any of the implications of his

[21] See Zwier (unpublished).

[22] See the previous chapters in this Part of the present book, esp. chapter 1, 'Postscript 1986', sec. I.

mature theory of spontaneous generation either for his natural teleology or for much else in his natural philosophy and metaphysics.[23]

More generally, there does not even seem to be any evidence that Aristotle *developed* his teleology by reflecting on the difference between sexual and (what he, at the time he was developing it, took to be *actual* cases) of spontaneous generation. Nowhere that I can think of does Aristotle explicitly remark on the non-teleological character of spontaneous generations as he conceived them actually to take place. As Lennox's discussion shows, in the *Metaph.* passages in which Aristotle contrasts the two types of causation involved, respectively, in sexual and spontaneous generations, it is the different sorts of *efficient* cause involved that occupy his attention. And when, in *Phys.* II, he does contrast teleological causation with cases he calls 'spontaneous', it is either the imaginary, Empedoclean cases, or cases not obviously similar to spontaneous generation, that he has in mind.[24] Similarly, *GA* never mentions teleology in connection with spontaneous generation (or vice versa). One might, as I say, speculate with much pleasure about such connections (and I do), but I cannot see that in so doing one would be recapturing either the process by which Aristotle was led to, or the essential content of, his conception of final causality or his doctrine of natural teleology.

[23] *GA* I.1 715b7-16 is a small and only partial exception. (For a contrasting view, see Stavrianeas 2008.) There are many other questions one could raise about the connection between teleology and spontaneous generation. For instance, I have argued elsewhere that, for Aristotle, a part's being for the sake of something presupposes its coming to be for the sake of that thing (chapter 1 above, 'Postscript 1986', sec. II). Could that be why, as James Lennox, speaking of *PA* IV, has noted (2001a, 293), 'The chapters dealing with [the parts of] the bloodless animals are entirely devoid of Aristotle's usual language of teleology—where one would expect *heneka* or *charin* Aristotle uses other prepositions in ways that encourage their translation with a teleological nuance (*pros*, *eis*, *dia*)'? Does that hesitancy reflect a concern with the spontaneously generated cases? On the other hand, this stretch of text includes bloodless animals that are *sexually* generated, and *PA* II-IV says nothing explicitly either about spontaneous generation in general or about its bearing on the functional explanation of parts so generated in particular. Again, the more important point is that Aristotle *says* nothing about this, or about the many other very interesting questions (e.g. about the souls and forms of spontaneously generated organisms) that one might raise as one attempts to integrate the mature theory of spontaneous generation with the rest of his thought.

[24] The reference to Empedocles is, of course, to *Phys.* II.8. Ross (1936, 524) argues that *Phys.* II.5 197b32-37 refers to spontaneous generation, but he is not convincing: see Charlton 1970, 110-11.

PART II

First Principles and Explanatory Structure in the *Parts of Animals* (*PA*)

7

First Principles in Aristotle's *Parts of Animals*

It is often maintained that Aristotle's practice in such explanatory treatises as the *Parts of Animals* does not correspond to the theory of science presented in the *Posterior Analytics*. Two major respects in which this is so, it is held, are (i) the absence in *PA* of the explicit syllogisms in which, according to *APo.*, the explanations are to be cast, and (ii) the absence in *PA* of the axiomatic structure of explanation *APo.* calls for.

Writing of the second point in his little book on *Aristotle* (1982), Jonathan Barnes explains that

> Aristotle's scientific treatises are never presented in axiomatic fashion. The prescriptions of the *Posterior Analytics* are not followed in, say, the *Meteorology* or the *Parts of Animals*. These treatises do not lay down axioms and then proceed to deduce theorems; rather, they present, and attempt to answer, a connected series of problems. (37)

Barnes's formulation makes it clear that he takes the requirement that proper science have an axiomatic structure to mean that it is to be *presented* in a certain way—with its first principles presented first, and labeled as such, and its deductions following sequentially, 'in the geometrical manner'. Similarly, the requirement of the full theory of *APo.* that the deductions be syllogistic in form, Barnes and others take to be a requirement that the deductions be presented in explicit syllogisms. The absence of explicit syllogisms and of an explicit axiomatic structure is the basis of the twofold discrepancy between the *APo.* theory and the *PA* practice. How is this discrepancy to be explained?

In 'Aristotle's Theory of Demonstration' (Barnes 1975b), Barnes proposed that while the *APo.* theory was a theory of how to present and impart a completed science, the biological (and other scientific) treatises were a record of 'research',

On the place of this chapter in my work, see the Preface above, pp. vii–viii. The chapter appeared in print originally in *Philosophical Issues in Aristotle's Biology* (Gotthelf and Lennox 1987). It is an appropriate bridge between Part I on teleology and Part II on first principles and explanatory structure, since it treats, among other topics, the way teleological principles enter into the structure of explanation. Chapter 8, which follows immediately, supplements the present chapter in important ways.

'progress-reports', 'the sharing of tentative . . . explorations'. As such they had a 'rich and variegated' methodology of their own, neither syllogistic in form nor axiomatic in structure.[1] Since the scientific treatises did not aim to be the sort of thing *APo.* described, there was no conflict. One problem with this view, which scholars familiar with the biological treatises noticed, was that these treatises did not seem very tentative, and that they were less the presentation of 'research' into facts than of explanations of facts previously gathered, and that as explanations they invited, after all, comparison with the theory of explanation in *APo.*

In 'Proof and the Syllogism' (1981), Barnes argued that a pre-syllogistic theory of demonstration could be detected in *APo.*, and that the lateness of the development of the syllogistic would explain its absence in the biological works; this invited once again consideration of the relationship between the theory of demonstration (now shorn of the syllogism) and the scientific practice.[2] This pre-syllogistic theory would of course require of Aristotelian science an axiomatic structure, but on the apparent absence of this in the scientific treatises Barnes said nothing in this paper. However, in his *Aristotle*, Barnes offered a new solution:

> Why are the *Meteorology* and the *Parts of Animals*, say, not presented axiomatically? The answer is disconcertingly simple. Aristotle's system is a grand design for finished or completed science . . . Since the treatises are not the final presentations of an achieved science, we should not expect to find in them an orderly succession of axioms and deductions. Since the treatises are intended, in the end, to convey a systematic science, we should expect them to indicate how that science is to be achieved. And that is exactly what we do find: Aristotle was a systematic thinker; his surviving treatises present a partial and unfinished sketch of his system. (38-39)

Barnes offers no argument for this view of *PA* (and other scientific treatises), but he does raise for us a question rarely given much attention: *Is* there in fact an implicit axiomatic structure, or the outlines of one, in treatises such as *PA*, which might reflect movement towards a system of the sort *APo.* calls for? If the solution to the apparent discrepancy between theory and practice in Aristotelian science is to be found in the relationship between an achieved ideal and its partial realization, we will need a very full picture of the practice, to see if in fact it is a realization, even if only partial, of that ideal. We will need to take a close look at the actual explanations in *PA*, at their internal structure and their premises, and at their

[1] Barnes 1975b, 77, 84.

[2] 'For centuries, learned Aristotelians sought and found the influence of the *Posterior Analytics* on the scientific πραγματεῖαι; more recently, impressed by the Syllogistic content of the *Posterior Analytics*, scholars have tried to divorce that work from the rest of the Aristotelian corpus. If, instead, we liberate Apodeictic from the shackles of Syllogistic, we may again look for the influence of the *Analytics* on Aristotle's scientific and philosophical practice . . . I do not pretend to know how much we shall find; but I now think, as I once did not, that the search is worth undertaking' (Barnes 1981, 58).

connections to each other. We will need to ask if there is, even implicit in the treatise, anything like an axiomatic structure to these explanations, and anything like the principles which such a structure demands. It is these questions I would like to address in this chapter.

This is a large inquiry, and I can only make a first entry into it. I will attempt to show that there *is* an implicit axiomatic structure to the explanations in the *Parts of Animals*, and will attempt to identify the sorts of first principles embedded in this structure. Such an examination will inevitably use many of the concepts and categories of the *Posterior Analytics*, but I want so far as possible to avoid, at least in the bulk of this chapter, the many interpretative controversies that burden the study of that difficult treatise, and to focus instead on *PA* itself, and on the explanatory structure to be found there. In the last part of this chapter I will make some general remarks about possible connections between the two treatises, focusing especially on the question of the absence of both explicit syllogisms and an explicit axiomatic structure in *PA*, but these will be brief and largely independent of the analysis I offer of the structure of *PA* itself.

In asking whether there is an implicit axiomatic structure to the explanations in *PA*, we need to start with a quite general notion of such a structure, so as to be open to the variety of sorts of 'axioms' we might find. In general, we are asking, roughly, whether the explanations to be found in *PA* are based on first principles, i.e., on facts about the animal kinds whose possession of certain features are being explained, which facts are not themselves explained by reference to more basic facts. To take one possible variety, are there, for instance, features of an animal kind that explain other features of that kind without themselves being explained by features of that kind? Aristotle's theory of science would call for such features, and in *APo.* he treats statements expressing the possession by entities of such features, when properly formulated, as definitions, and includes such definitions among the first principles of a science. But are there animal definitions in *PA*, and do they play such roles? It has often been claimed, and not without reason, that there are no definitions of animal kinds in *PA*, or anywhere in the biological treatises. That the theory of science calls for such definitions does not entail that they will be present in the scientific practice. Aristotle might, for instance, have come to believe after doing his biological work that explanations are not linear and directed in the way just described but, say, circular, animal features being explained ultimately in terms of each other, or in a highly localized and contextual way. I have already implied that I do not myself believe this, but that is the sort of alternative our argument will have to exclude.

Biology as demonstrative science: *PA* I and *GA* II.6

Though I have referred to *PA* as a whole, it is strictly only the last three books, *PA* II–IV, that offer explanations of biological phenomena, and can be counted as 'biology proper'. *PA* I is, as David Balme has called it, 'a string of papers', a series of methodological discussions, probably put together as an introduction to the zoology as a whole.[3] Its use cannot be a substitute for a close look at the actual workings of II–IV. At this stage, I want to draw on *PA* I only for a picture of how Aristotle himself conceived the nature and structure of his enterprise in II–IV and the rest of the biology proper, and then turn immediately to II–IV themselves.

It seems clear from *PA* I, we may note first, that Aristotle himself saw *PA* II–IV as offering (and not just aimed at) *epistēmē*, in the sense of a systematic knowledge of causes, and as containing demonstrations.

The famous motivating passage of 1.5, for instance, recommends attention to the lectures that are immediately to come—such as the *PA* series, one would presume—first of all on grounds of the greater *epistēmē* it offers (645a2) in providing what is an immeasurably pleasant knowledge of *causes* (a9-10). Though the heavenly bodies are much greater in value than the living things 'which grow beside us', these living things, because 'the information about them is better and more plentiful, take the advantage in *epistēmē*' (645a1-2).

The enterprise being introduced and characterized in I.1 is called *phusikē epistēmē* at 640a2,[4] where the mode (*tropos*) of demonstration (*apodeixis*) and of necessity which its contents possess are distinguished from that of the *theōrētikai epistēmai* (probably, or primarily, the mathematical sciences[5]). The immediately preceding paragraphs, and in fact most of I.1, make clear that this *epistēmē* embraces a knowledge of causes, and the fact that the contrast between biology and mathematics comes parenthetically within a discussion of the necessity that obtains in nature, and the way the contrast is characterized, ensure that the *apodeixeis* of *phusikē* are no less productive of *epistēmē* than those of mathematics (even if they are somewhat different in character).[6] And we can connect these characterizations of the explanations of natural science to *PA* II–IV via the specification at 640a33ff. of the modes of explanation of the parts of animals required by their hypothetically necessary (or generally teleological) character. That these are the proper modes is inferred (*dio*: 640a33) from the arguments for teleology which surround the discussion of necessity in nature; and as we will see later, these are in fact the major modes of explanation used in *PA* II–IV.[7]

[3] Balme 1972, v, 69. (James Lennox has since shown, in Lennox 2010b, that there is a greater unity to *PA* I than Balme had recognized. See also my discussion of the *aporia* across *PA* I.1, 4, and 5 in chapter 9 below, pp. 186-89, though Lennox makes a much fuller case for the book's unity.) Because of its methodological character, *PA* I can serve as a bridge between the general theory of *APo.* and the practice in both *PA* II–IV and the rest of the biology proper. See, e.g., Lennox 1987a, 114-15 (2001b, 29-30) and now 2010b.

[4] Only 'in effect', if *epistēmōn* does not as noun carry back all the way to *phusikēs*, but the enterprise is in any case explicitly called *epistēmē* at 639a3.

[5] Cf. Grene 1985.

[6] See 'Additional note on *PA* I.1 639b21-640a9' below, pp. 182-85.

[7] See below, p. 175.

Both Wolfgang Kullmann and James Lennox have pointed us to the serious use of *apodeixis* (with its *peri hōn* and its *ex hōn*) in *HA* I.6.[8] We might also note a most interesting passage in *Generation of Animals* II.6 (beginning at 742b17), where Aristotle criticizes people 'like Democritus' who accept the observation that 'this is how they are always formed' as a sufficient explanation of certain matters on the grounds that they are eternal and thus can have no prior principle:

[b23] On this line of argument . . . there will be no demonstration (*apodeixis*) of any single one of the 'eternal' things. It is obvious, however, that demonstrations of many of these (some of them things which *always* come to be, some things which *always* are) do in fact exist. For instance the angles of a triangle are *always* equal to two right angles, and the diagonal of a square is *always* incommensurable with the side; in both of these cases we have something 'eternal', yet there is a cause for them and they are demonstrable (*estin autōn aition ti kai apodeixis*). Thus it is right to say that we cannot undertake to try to discover a starting-point [Peck: first principle] in all things and everything; but it is not right to deny the possibility in the case of all things that *always* are and that *always* come to be; it is impossible only with the first principles of the eternal things, for of course the first principle does not admit of demonstration (*apodeixis*) but is apprehended by another mode of cognition. Now with those things that are 'immutable' (*en tois akinētois*) the first principle is the essence (*ti esti*); but in the things which come into being, immediately there are several, and another manner of principle and not all the same.[9] Among them the source whence the movement comes must be reckoned as one, and that is why the heart is the first part which all blooded animals have . . .[10]

This is an important passage for our purposes. First of all, Aristotle speaks clearly and self-consciously of the presence of demonstration (*apodeixis*) not only in mathematics but also in natural science, and in the study of animals in particular ('things which always come to be': 742b30-31). Furthermore, he makes clear that the demonstrations in both cases have *first principles* (*archai*); it must be stressed that the reference is not only to *archai* of the subject matter but also to *archai* of the

[8] 491a13-14. Cf. Kullmann 1974, 81 n. 8 and Lennox 1987a, 101-102.

[9] Here I have softened Peck's translation, which reads: 'but as soon as we begin to deal with those things that come into being though a process of formation, we find there are several first principles . . . principles, however, of a different kind and not all of the same kind'. The Greek of the last clause is *τρόπον δ' ἄλλον καὶ οὐ πᾶσαι τὸν αὐτόν*, and means that the *additional* principles in this case are of another manner from the *ti esti*, and that these additional ones are different from each other. There is no reason to exclude the *ti esti* from being one of the principles of *gignomena*.

[10] 742b23-36: tr. Peck, with one change, noted. The Greek in Drossaart Lulofs's OCT edition reads: *καίτοι κατὰ τοῦτον τὸν λόγον . . . οὐθενὸς* (b25) *ἀπόδειξις ἔσται τῶν ἀϊδίων· φαίνεται δ' οὖσα πολλῶν, τῶν μὲν γιγνομένων ἀεὶ τῶν δ' ὄντων, ἐπεὶ καὶ τὸ τρίγωνον ἔχειν δυσὶν ὀρθαῖς ἴσας ἀεὶ καὶ τὸ τὴν διάμετρον ἀσύμμετρον εἶναι πρὸς τὴν πλευρὰν ἀΐδιον, ἀλλ' ὅμως ἐστὶν αὐτῶν αἴτιόν τι καὶ ἀπόδειξις. τὸ μὲν οὖν μὴ πάντων ἀξιοῦν ζητεῖν* (b30) *ἀρχὴν λέγεται καλῶς, τὸ δὲ τῶν ὄντων ἀεὶ καὶ γιγνομένων πάντων οὐ καλῶς, ἀλλ' ὅσαι τῶν ἀϊδίων ἀρχαὶ τυγχάνουσιν οὖσαι· τῆς γὰρ ἀρχῆς ἄλλη γνῶσις καὶ οὐκ ἀπόδειξις. ἀρχὴ δ' ἐν μὲν τοῖς ἀκινήτοις τὸ τί ἐστιν, ἐν δὲ τοῖς γιγνομένοις ἤδη πλείους—τρόπον δ' ἄλλον καὶ οὐ πᾶσαι* (b35) *τὸν αὐτόν—ὧν μία τὸν ἀριθμὸν ὅθεν ἡ κίνησίς ἐστιν. διὸ πάντα τὰ ἔναιμα καρδίαν ἔχει πρῶτον . . .*

demonstrations: as Peck notes ad 742b33, there is a clear reference in the words *tēs gar archēs allē gnōsis kai ouk apodeixis* to standard Aristotelian doctrine about the first principles of demonstration in the *Posterior Analytics, Metaphysics*, and *Nicomachean Ethics*. Finally, the passage suggests (as one might well expect) that propositions about the *archai* in the sense of the *causes* of a subject matter will be *archai* of the demonstrations regarding that subject matter, and that this is so as much in biology, where several causes are operative, as in mathematics, where only one is.[11]

In sum, it seems clear that Aristotle himself is happy to speak the language of *epistēmē, apodeixis*, and first principles in talking of his existing biological work.

This language is too pervasive and self-conscious to be dismissed, but we are entitled to ask with how much precision and technical richness it is being used. And it is here that we must turn to the biology itself, exploring the reference of the language directly as our best guide to its sense.

The overall structure of *PA* II–IV

Let us begin with a descriptive overview of the whole of *PA* II–IV, so that we may see features of the structure of the whole, and place our discussions of the individual passages and explanations properly within that structure.

PA II begins with the three *suntheseis* (Peck: 'levels of composition'): of elementary powers into simple compounds, of these into the uniform parts of animals, of those into the non-uniform parts.[12] After some arguments (reminiscent of I.1) that

[11] This passage is interesting for at least one other reason as well. As discussed in my 'Additional Note on *PA* 1.1 639b21-640a9' (below, pp. 182-84), it seems to reveal Aristotle's sensitivity to the issue of how non-eternal objects could be the subject of a demonstrative science, and suggests, as I propose in that Note, his solution: though we do not have objects which 'always are', we do have objects which 'always come to be', and that will somehow be sufficient to guarantee the reference of the subject term in the propositions of the science. For details, see below, p. 184.

There are other places in the biology where Aristotle betrays sensitivity to concerns expressed in the *Posterior Analytics*. For example, at 747b27-748a14 (and ff.) he considers and rejects an explanation of the sterility of mules on grounds that it does not proceed from the principles proper to the subject matter. (The explanation also rests on premises that are false, but that is a *second* difficulty.) Here too the term *apodeixis* is used, for what this proffered explanation would be if successful.

[12] The immediate beginning places *PA* II–IV relative to the *historiai* concerning the animals (probably, though not necessarily, our *HA*), as investigating (*episkepteon*) 'on account of what causes' the parts of each animal are as they had been said there to be. The passage has the ring of the lecture room, and may well be placing II–IV relative to a previous course based on (something like) our *HA*. But it is interesting that we are asked to put aside (*chorisantas kath' hauta*) what was said there, since as David Balme has observed, II–IV quotes all the data which it explains (referring us to the *historia* or the *historiai* only for additional descriptive information not relevant to the present explanatory task). See in the Coda below, p. 397, on the relative dates of these two treatises.

each level is for the sake of the succeeding, Aristotle turns to the last two levels, as the basic division between parts which controls the structure of the three books. He first explains why animals must have uniform parts at all, and non-uniform parts at all (cf. 646b35ff.), then explains why sense-organs must be uniform, and why viscera (especially the heart) must be in one way uniform and in another non-uniform, thus distinguishing them from the true organs (*organika*), which are simply non-uniform.

At this point (our[13] chapter 2) begins the study of uniform parts, which are immediately divided into the soft and fluid, and the hard and firm. The various sorts of cause the uniform parts may have (as matter or as nutriment for the non-uniform, or as residue of nutriment) are listed, and the existence of differences within each type noted, with the observation that we should expect (*hupolepteon*) them to be explained as required for the functions and in general the *ousia* of the animals that have them, or as making their working better than it would be without them.[14] The part-by-part review then begins, and blood is first, but Aristotle is not long into the discussion when he announces the need to distinguish the various ways in which things can be hotter or colder, and more solid (or dry) or fluid, for the reason (to be discussed below) that 'the nature of many things is referred to these'.

Finally the review actually commences. Returning to blood, Aristotle discusses one by one across chapters 3-9 first the more fluid, then the more solid uniform parts, stating 'what each is and on account of what causes' (e.g., 651b13-19, 652a19-21). From II.10 until the end the focus is on the non-uniform parts, with the sequence as follows. Aristotle begins by identifying the 'most necessary' (non-uniform) parts in all animals[15]—that which receives food, that which expels residue, and that in between, 'in which is the *archē* of its life' (655b37). The increasing complexity of animals, as we compare them with plants and move up

[13] I say 'our' because the chapter divisions have no ancient authority, and are not found in the manuscripts, nor in any printed edition before the second Basel edition of 1539. The Basel divisions, which were taken over by Bekker, are identical to those in the 1492 Venice edition of Theodorus Gaza's Latin translation of the biology, *De Animalibus* (= *HA, PA, GA*), in which we are told that 'Sebastianus Manilius Romanus recognouit & per capita disposuit'. The great majority of these mirror the unnumbered divisions produced via enlarged initials in the first printed edition (Venice 1476), prepared just before Gaza's death, though Manilius redid a significant minority. Whether Gaza himself produced the 1476 division, however, is unclear. (My thanks to the British Library for making all these wonderful editions so easily available.) It is worth remembering that our chapter divisions are the construct of late medieval or Renaissance editors, and not Aristotle's own, and reading the text right through them to see if its import changes. Sometimes it does. (For more details on the origin of the chapter divisions, see now Beullens and Gotthelf 2007, 477-83, 503-508.)

[14] . . . *ta men pros ta erga kai tēn ousian hekastōi tōn zōiōn, ta de pros to beltion ē cheiron* . . . (648a15-16; cf. 647b29-30).

[15] The subject of the sentence is actually all 'complete' (*teleia*) animals, but see below nn. 40, 45.

the scale to man, is noted, and man's sharing in the divine, the relative familiarity of his external parts, and the correspondence of his 'up' and 'down' with the normal 'up' and 'down' of spatial direction are given as reasons for beginning with him. (It is not that all of man is done first, but that the list is taken from man, and for each part man's is dealt with first, then others, accordingly as they are essentially the same, or different from man's.[16]) Though he does not say he is doing so, Aristotle considers for the time being only the blooded animals, leaving the bloodless for much later.

The review of non-uniform parts starts at the top and works down. Aristotle takes a portion of the body at a time, first the head, then the neck, and looks outside then inside.[17] Within each portion the order is mostly, but as we will see later, by no means completely, from top down. Having finished the internal parts in the neck area, he moves not to the chest, or to the trunk generally, but directly to the internal parts in that area—the viscera. In each case, the tendency, as James Lennox has shown,[18] is to start with the common features, and with the widest class possessing them, then move to the differences—e.g., first with the possession of horns as such in the class of animals that have them (carefully delineated), then with why some have one horn, some two, etc. Existing animal groups are made use of to facilitate the identification of the widest classes possessing each feature. For instance, the classification into *megista genē* (cf. *HA* I.6 *init.*), and the broad divisions between live-bearing animals and egg-bearing animals, and between land and water animals, are in evidence, but the review within a part is not rigorously by any one of these divisions. Rather, the animals are, as David Balme has said (1987b, 88), 'grouped and regrouped', according to the needs of widest-class generalization for each difference of that part.

After completing the viscera and related internal parts, Aristotle moves to the bloodless animals, introducing them by their lack of viscera. He offers an explanation of that lack, deriving it from their nature as bloodless, notes the four main kinds, and begins a review of their internal parts part by part (or area by area), apparently using the cephalopods as the standard (comparable to man in the blooded), and perhaps going top down, but looking first at the parts concerned with nutrition (681b13-15), then at those that serve as sense organs. At 682a30-34 we are told that we have now completed our discussion of the internal parts of animals, and are ready for the external, first in the bloodless, then in the blooded, picking up where we left off with the latter only after the shorter discussion

[16] Cf. *HA* I. 491a19-25 on doing man first and going top to bottom. The 'up' and 'down' of spatial direction is given, of course, by the outermost sphere and the center of the earth. Biological 'up' and 'down' is given functionally (*ergōi*): the 'up' is the direction of 'the part from which is derived the distribution of nutriment and [thus] the grown of each thing', the 'down' 'the part in which it finally ends'. In animals, then, the 'up' is where the mouth is, in plants, where the roots are. Cf. *IA* 4 705a27-b8 and Lennox 1985b.

[17] In the Loeb edition (Peck and Forster 1961, 15-18) Peck tabulates the parts discussed.

[18] Lennox 1987a, 109-14 (2001b, 25-29); cf. now chapter 9 below, and chapter 14, pp. 297-99.

required of the bloodless.[19] That is what we get, in a general review, largely across *megista genē* (plus man), but often moving down to differences within large kinds, of head and neck, overall layout and degree of uprightness, limbs, breasts, trunk features, excretory organs, rear parts and feet. After overall layout and degree of uprightness, the discussion was restricted to the live-bearing, so it now proceeds to the egg-bearing animals, focusing first on the quadrupeds and snakes, then birds, then fish. The cetaceans are introduced as animals that, with respect to being water or land animals, 'tend to both sides',[20] and that invites a brief explanatory review of the external parts of some others—seals, bats, and the Libyan ostrich. *PA* II–IV then closes with the notice that 'We have discussed, concerning the parts of all the animals individually (*kath' hekaston*; Peck: '[taken] severally'), on account of what cause each is present to them. Having determined that, the next thing in order is to deal with the things that pertain to their modes of generation.'

This is a complicated structure, but in essence it is: part by part, first uniform, then non-uniform; within the non-uniform, separating the blooded and the bloodless animals, then proceeding top to bottom (as measured by man), moving from external to internal and back. Within a part it moves from the general explanation of why its possessors have it at all to an explanation of why possessors of one type have that type and those of another that other. There is nothing prima facie axiomatic about this structure, but I would like us to take a closer look at some of its features.

1. If explanations in a (subbranch of a) science depend on premises drawn from a prior (branch of that) science, it would be no surprise if those premises were presented first, especially if the scientist were sensitive to the logical order of his explanations. There is some reason to think this is why Aristotle distinguishes the three *suntheseis* as early as he does, and why he discusses first the elementary powers, then uniform parts, then non-uniform parts. As we saw, he says explicitly that his reason for discussing the elementary powers is that 'the nature of many things is referred to these'. And that is the case in his own explanations. Thus, at 649b20ff. he makes use of his discussion of the various ways things can be hotter, colder, more fluid or solid, by stating in what way blood is hot; this enables the establishment of (part of) the definition of blood, which is part of his aim. The establishment of this partial definition is not of course a demonstration, but the definition plays a role in the explanation of some of the features of blood,[21] and makes

[19] This may be lecture-preparer's glue, as may 685b27-28, introducing the resumption of discussion of externals in the blooded, since the latter discussion has the feel of a self-contained essay, with striking echoes of the *Timaeus*, and our *PA* II–IV could well be an amalgam, as I + II–IV surely is, but this does not matter for our purposes.

[20] David Balme's rendering of ἐπαμφοτερίζειν. Peck's 'dualizing' is misleading, as the animals do not in fact fall into both of the two contrary groups. See now Balme 1991 *ad* 589a21.

[21] e.g., that it congeals when separated from the body (649b28-9). At 651a5ff. basic facts about hot and cold help to explain how congealing takes place, and why some blood (i.e., the

possible the explanation of certain psychic differences (e.g., in intelligence, and readiness to anger) in terms of differences of fluidity or heat in the blood.[22]

When a sense of 'hot' or 'fluid' plays a role in explanation, it does so by way of a premise regarding the actions of the power conceived in that sense. Thus certain fluids when deprived of heat congeal, while other thicker ones when deprived of heat thin out; moist earth is in the former category: 'the cold expels the heat and makes the fluid evaporate'. 'Passion produces heat; and solids, when they have been heated, give off more heat than fluids.'[23] Premises about the elementary powers also occur in explanations of the presence of features of several of the uniform parts besides blood. Thus, suet congeals because it is earthy, lard does not because it isn't.[24] Things that are brittle or fluid cannot pass without a break right through something like a whole spine.[25] Appropriate amounts of hot and cold will balance in an intermediate position.[26] A certain process occurs according to the principles involved in rain-showers.[27] Heat does not pass as easily through thick earthy material as it does through thinner fleshy material.[28] And so on.[29]

2. The order of discussion of the uniform parts is not without explanatory purpose. Blood is surely discussed first because it is an *archē* for the rest. It is explained as present to provide nourishment to the other parts, being the *eschatē trophē* (Peck: final form which the nourishment assumes).[30] Lard and suet are blood that has gone through certain types of concoction, as is marrow, and the explanations of marrow's presence draws explicitly on the prior explanation of blood's.[31]

blood of some animals) congeals more than (that of) others. On how complete this account of blood is, see the next chapter, n. 18.

[22] II.4 *passim.*

[23] 649b28ff.; 651a7-10 (and surroundings); 650b35-36 (and surroundings).

[24] 651a26-29, 33-36.

[25] 651b32-36.

[26] 652a31-33, and II.7 *passim.*

[27] 652b34-653a8.

[28] 654a3-8.

[29] As I discuss below (pp. 174-75), in most cases in *PA* II–IV, premises about the matter function in explanation only in conjunction with premises about the final cause. But, as John Cooper has shown, all explanations that appeal to hypothetical necessity, and most teleological explanations generally, do make use of distinct premises about the nature of the matter. (Cf. esp. Cooper 1987, 255-64, and Balme 1987c, 283-85.)

[30] 650b2-3, 651a15. In the latter passage Aristotle says this makes the blood the matter of the (blooded) living thing. For an interesting discussion of this see Freeland 1987, 398-404. The argument that concludes with 650b2-3 seems an example of one that establishes blood's function rather than one that uses that knowledge to explain its presence: surely blood isn't the final form of nourishment *because* (explanatorily) it is the only fluid in the blood-vessels. That is what helps us to *know* that blood must be nourishment, but does not *explain* it. The demonstration itself is probably signaled by the *heneka* in b2, but is not actually given. Cf. below, pp. 164-65 and n. 35.

[31] Lard and suet: 651a20-24; cf. 651b28-30, 652a7-10. Marrow, 651b20, 652a2-23. Note the definition of marrow which results (a19-23) from the investigation: it mentions material, final,

Bone and certain other hard materials are for the sake of supporting flesh or its analogue; flesh or its analogue is itself necessary to all animals as the organ of, or associated intimately with, the defining sense of touch.[32]

3. Surely at least part of the reason that non-uniform parts are discussed after uniform is that explanations of the former sometimes make use of the facts about the latter. Thus, the explanation of blood-vessels and of the heart presupposes that of blood, as does one part of the explanation of the liver and spleen.[33]

4. The reason the bloodless are treated separately from the blooded is probably that most of their parts are the same as those of the blooded only by analogy. They are treated after the blooded probably because the blooded are better known and understood (compare the explanation for treating man first among the blooded), as evidenced by how many of their parts have no name of their own but are called '*x*-analogue', where '*x*' is the name in the blooded case.

5. Finally there is surely an explanatory basis for treating the common features of a part (including its presence at all) before treating the differences in that part. For instance, horns are present in animals which have them because they have no other suitable defense and there is earthy material available to make them. The explanation why some animals have one horn, others two, presupposes that horns as such come from earthy material, and refers to the amount of such material available. Where there can be only one, its placement in the center is explained by reference to the generic defensive function.[34] (This pattern of explanation, explaining the

efficient, and (as some combination of the preceding) formal causes. This is a fairly rare paradigm of the sort of definition of a part that each of the discussions of the uniform parts, and probably of the non-uniform parts as well, is aimed at. It presupposes, and can probably be seen as embedded in, the explanation of the presence of marrow that precedes, and as such has at least an affinity with the second of the three types of definition in *APo.* II.8–10. Cf. Robert Bolton's discussion of the three forms of definitions in Bolton 1987, 142–46, and cf. below, pp. 164–65 (with n. 35) and pp. 167–68.

[32] Flesh and its analogue: 653b20–30 (cf. b22–23: *ton logon . . . horizometha*). Bone and its analogues: 653b30–36 and ff. (to end of II.8), 654b27–29 (and most of II.9).

[33] Hair and blood-vessels: 665b9–16, and III.4–5 *passim*. Liver and spleen: 670a20–21. Here too, many facts about the presence of uniform parts will be explained by reference to the needs of the non-uniform parts, and someone might therefore expect discussion of the latter to precede discussion of the former. But, as noted above (n. 29), these explanations via hypothetical necessity presuppose certain facts about the uniform parts, and it is certainly reasonable that these facts should be stated first. (Certain facts about the matter will only be explained teleologically, by reference to the needs of the whole organism and the irreducible *dunameis* for the fulfillment of those needs, but again those explanations will presuppose the sort of facts about the matter given in the early parts of *PA* II.)

[34] III.2, esp. 662b27, b30–663a8. Aristotle seems to me not to have progressed as far in his explanation of horns as he has with most other organs, since he does not seem to be able to isolate the class of horn possessors, and so has to describe it as including animals that have no means of defense other than horns. It may be that he has independent explanations for why these others have the means of defense they do, and why horn-possessors cannot have any of those, and believes that horns are the only possible means left. But there is no indication of that, and we

variations of a certain general feature by reference to variations in the cause(s) of the general feature, is explored more fully in chapter 9 below.)

In sum, then, *PA* II–IV has a structure which is at least amenable to exhibiting relations of priority among demonstrations, and may well have been in part designed for such a purpose.

A representative explanation: the multiple stomachs of ruminants

To see more clearly the different sorts of premises that may appear in the explanations in these books, and to observe the structure more microscopically, it will be useful to look at a particular stretch of text. The discussion of the multiple stomachs in ruminants (in the first part of III.14) will be an interesting test case, as the explanation is non-obvious, and referred to in *APo.* (II.14). Before embarking on this, however, I need to make an important general observation about *PA* II–IV.

Though I have been arguing, and will continue to argue, that *PA* II–IV has a broadly axiomatic structure, with demonstrations based on first principles, I should make it clear that I do not think that that is all these books contain. A fair amount of the time, what we get are systematic discussions—arguments, actually—aimed at *establishing* what the cause (generally, the function) of the part in question is. Both of these are tasks of Aristotelian science—to establish non-demonstratively the causes and then to demonstrate that they are the causes—and *PA* II–IV takes them on together, sometimes doing one, sometimes the other.[35] I mention this in

probably have an example of an obvious way in which *PA* II–IV is incomplete (and was probably thought to be so). But we can see how this sort of incompleteness does not in the least stop the treatise from having the structure which it would have if it were complete in this sense. There is just a gap, and a temporary filler, where the ultimate explanatory differentia(e) of horned animals qua horned will go. But it is worth noting that the filler is not without explanatory value. At its present stage the demonstration has the form 'Horns belong to all live-bearing animals that are not polydactylous and do not have another means of defense; being live-bearing and not polydactylous with no non-horn means of defense belongs to all horn-possessors.' ('Horn-possessors': where there isn't a ready-made name for the widest class that possesses a feature *x*, Aristotle often describes the group as 'those which have x', and often summarizes by saying 'we have now discussed on account of what causes those which have *x* have x'.) As for the useless horns in the deer, Theophrastus was puzzled too: cf. his *Metaphysics* 9 10b11-14 (Ross and Fobes 1929), discussed in Lennox 1985b, 148-49 [Lennox 2001b, 264-45]. (G. E. R. Lloyd is surely unfair in suggesting that the 'temporary filler' I have provided is meant as a paradigm of what I claim is the demonstrative character of Aristotle's explanations [Lloyd 1990, 33 n. 85; 1991, 396 n. 85]. See Lennox 1997a for more on the syllogistic character of Aristotle's explanations in *PA*.)

[35] The arguments aimed at *establishing* what the cause of the part in question is, amount, I believe, to attempts to establish the definitions of these parts, in the spirit of *APo.* II.8-10, much as Robert Bolton explains it in Bolton 1987. According to *APo.*, these arguments should not, of course, be demonstrations of the definitions they establish, and it would be interesting to explore

part for its place in a picture of the structure of the whole of *PA* II–IV, and in part because it will help us to understand certain portions of the passage in which Aristotle explains why some animals have one stomach and others more than one, to which we now turn.[36]

It is a long passage, but with apologies I quote in full, in Peck's translation (and paragraphing), so that we can have it before us, and I can refer to it in numbered portions.

([1] 674a9-12) Below the diaphragm is the stomach, which is placed where the esophagus ends (if there is an esophagus; if not, immediately next to the mouth). Next after the stomach and continuous with it is what is called the 'gut'.

([2] a12-16) It must be obvious to everyone why all animals have these <two> parts. It is necessary for them to have some receptacle for the food they take in, and to expel it again when its moisture has been extracted from it; and there must be two different places for these two things—the unconcocted food and the residue, and there must be a certain place in which <the food> changes. ([(3] a16-21) Two receptacles, then, one for the incoming food, one for the residue which is <of> no more use—as there is a separate time for these so there must be a separate place. However, it will be more appropriate to go into these matters in our work on generation and nutrition. ([4] a21-27) At the present we must consider the variations (*diaphorai*) that are to be found in the stomach and its subsidiary parts.

The stomach differs both in size and <in> appearance in different animals. Those of the blooded live-bearing animals which have front teeth in both jaws have one stomach; e.g., man, the dog, the lion, and the other polydactyls; so also those that have solid hoofs, e.g., the horse, the mule, the ass; and those which although they are cloven-hoofed have front teeth in both jaws, e.g., the pig—([5] a27-b7) unless, because of a large body and the character of its food, being, say, not easy to concoct but thorny and woody, it has several, such as the camel. So also have the horned animals, as they have not front teeth in both jaws. Thus also the camel has not the two rows of front teeth either, although it has no horns; this is because it is more necessary for the camel to have several stomachs than to have all these

their form in *PA* II–IV, as Bolton has done in Bolton 1987 for some of the arguments in *GA* I–II. Bolton argues, with respect to the theory of *APo.* and the practice of *GA* I–II, that the definition-establishing arguments are not essentially dialectical, though they may contain dialectical parts, because they rely on empirical data in ways not permitted or countenanced by dialectic proper, and appeal to explanatory relations not properly introduced in dialectic. *Endoxa*, according to Bolton, are often brought in as alternative hypotheses to be examined against, and refuted by, the empirical data, and I think this is right for the cause-establishing arguments in *PA* II–IV. (Cf. n. 42 below.) In the course of Freeland 1990, Cynthia Freeland has attempted to show the same thing for *Meteorology* I–III. (Malcolm Wilson is preparing a full-length study of *Meteor.* I–III; for now see Wilson 2009.)

[36] Our passage will exhibit one more dimension of the exposition in *PA* II–IV. In addition to arguments establishing the causes, and demonstrations that exhibit them, there will be stretches of text expounding the facts whose explanations are to be given. (Cf. n. 12 above.) Finally, in understanding what is going on in *PA* II–IV, it must not be forgotten that what we have are lecture notes, or memoranda for students, and not really finished 'treatises', as we are wont to call them.

front teeth. So, as it resembles the animals which lack the upper front teeth in that it has several stomachs, therefore the arrangement of its teeth is that which normally accompanies the multiple stomachs: in other words, it lacks these front teeth, as they would be no use to it. And also, as its food is thorny, and as the tongue has of necessity to be of a fleshy character, the nature has made use of the earthy matter saved from the missing teeth to make the roof of the mouth hard. Again, the camel ruminates as the horned animals do, because it has stomachs that resemble theirs. ([6] b7-15) Every one of the horned animals (such as the sheep, the ox, the goat, the deer, and the like) has several stomachs; and the purpose of them is this (*hopōs*): Since the mouth is deficient in teeth, the service which it performs upon the food is deficient; and as one stomach after another receives the food, which is quite untreated when it enters the first stomach, more treated in the next, completely treated in the next, and a smooth pulp in the next. And that is why these animals have several such places or parts, the names of which are [i] the paunch (*rumen*), [ii] the net or honey-comb bag (*reticulum*), [iii]) the manyplies (*omasum*), [iv] the reed (*abomasum*). ([7] b15-17) How these stand in relation to each other as regards position and appearance must be studied both from the *historia* dealing with animals and from the dissections.[37]

The beginning portion, labeled [1], names the next two subjects in the roughly top-to-bottom survey of the viscera in blooded animals, but also notes their

[37] 674a9-b17; tr. Peck, occasionally modified. Peck's Greek (with my numbering inserted) reads as follows: ([1] 674a9-12) *ὑπὸ δὲ τὸ ὑπόζωμα κεῖται ἡ κοιλία τοῖς ζῴοις, τοῖς μὲν ἔχουσιν οἰσοφάγον ᾗ τελευτᾷ τοῦτο τὸ μόριον, τοῖς δὲ μὴ ἔχουσιν εὐθὺς πρὸς τῷ στόματι· τῆς δὲ κοιλίας ἐχόμενον τὸ καλούμενον ἔντερον.* ([2] a12-16) *δι' ἣν δ' αἰτίαν ἔχει ταῦτα τὰ μόρια τῶν ζῴων ἕκαστον, φανερὸν πᾶσιν. καὶ γὰρ δέξασθαι τὴν εἰσελθοῦσαν τροφὴν καὶ τὴν ἐξικμασμένην ἀναγκαῖον ἐκπέμψαι, καὶ μὴ τὸν αὐτὸν τόπον εἶναι τῆς τ' ἀπέπτου καὶ τοῦ περιττώματος, εἶναί τέ τινα δεῖ τόπον ἐν ᾧ μεταβάλλει.* ([3] a16-21) *τὸ μὲν γὰρ τὴν εἰσελθοῦσαν ἕξει μόριον, τὸ δὲ τὸ περίττωμα τὸ ἄχρηστον· ὥσπερ δὲ χρόνος ἕτερος ἑκατέρου τούτων, ἀναγκαῖον διειλῆφθαι καὶ τοῖς τόποις. ἀλλὰ περὶ μὲν τούτων ἐν τοῖς περὶ τὴν γένεσιν καὶ τὴν τροφὴν οἰκειότερός ἐστιν ὁ διορισμός·* ([4] a21-27) *περὶ δὲ τῆς διαφορᾶς τῆς κοιλίας καὶ τῶν συντελῶν μορίων νῦν ἐπισκεπτέον. οὔτε γὰρ τοῖς μεγέθεσιν οὔτε τοῖς εἴδεσιν ὁμοίας ἔχουσιν ἀλλήλοις τὰ ζῷα· ἀλλ' ὅσα μέν ἐστιν αὐτῶν ἀμφώδοντα τῶν ἐναίμων καὶ τῶν ζῳοτόκων, μίαν ἔχει κοιλίαν, οἷον ἄνθρωπος καὶ κύων καὶ λέων καὶ τἆλλα ὅσα πολυδάκτυλα, καὶ ὅσα μώνυχα, οἷον ἵππος, ὀρεὺς, ὄνος, καὶ ὅσα διχαλὰ μὲν ἀμφώδοντα δέ, οἷον ὗς,* ([5] a27-b7) *πλὴν εἴ τι διὰ μέγεθος τοῦ σώματος καὶ τὴν τῆς τροφῆς δύναμιν, οὖσαν οὐκ εὔπεπτον ἀλλ' ἀκανθώδη καὶ ξυλικήν, ἔχει πλείους, οἷον κάμηλος, ὥσπερ καὶ τὰ κερατοφόρα. τὰ γὰρ κερατοφόρα οὐκ ἔστιν ἀμφώδοντα. διὰ τοῦτο δὲ καὶ ἡ κάμηλος οὐ τῶν ἀμφωδόντων ἐστίν, ἀκέρατος οὖσα, διὰ τὸ ἀναγκαιότερον εἶναι αὐτῇ τὴν κοιλίαν ἔχειν τοιαύτην ἢ τοὺς προσθίους ὀδόντας. ὥστ' ἐπεὶ ταύτην ὁμοίαν ἔχει τοῖς μὴ ἀμφώδουσι, καὶ τὰ περὶ τοὺς ὀδόντας ὁμοίως ἔχει αὐτῇ, ὡς οὐδὲν ὄντας προέργου. ἅμα δὲ καὶ ἐπεὶ ἡ τροφὴ ἀκανθώδης, τὴν δὲ γλῶτταν ἀνάγκη σαρκώδη εἶναι, πρὸς σκληρότητα τοῦ οὐρανοῦ κατακέχρηται τῷ ἐκ τῶν ὀδόντων γεώδει ἡ φύσις. καὶ μηρυκάζει δ' ἡ κάμηλος ὥσπερ τὰ κερατοφόρα, διὰ τὸ τὰς κοιλίας ὁμοίας ἔχειν τοῖς κερατοφόροις.* ([6] b7-15) *τούτων δ' ἕκαστον πλείους ἔχει κοιλίας, οἷον πρόβατον, βοῦς, αἴξ, ἔλαφος, καὶ τἆλλα τὰ τοιαῦτα τῶν ζῴων, ὅπως ἐπειδὴ τῆς ἐργασίας ἐλλείπει περὶ τὴν τροφὴν ἡ λειτουργία ἡ τοῦ στόματος διὰ τὴν ἔνδειαν τῶν ὀδόντων, ἡ τῶν κοιλιῶν ἑτέρα πρὸς ἑτέρας δέχηται τὴν τροφήν, ἡ μὲν ἀκατέργαστον, ἡ δὲ κατειργασμένην μᾶλλον, ἡ δὲ πάμπαν, ἡ δὲ λείαν. διὸ τὰ τοιαῦτα τῶν ζῴων πλείους ἔχει τόπους καὶ μόρια. καλοῦνται δὲ ταῦτα κοιλία καὶ κεκρύφαλος καὶ ἐχῖνος καὶ ἤνυστρον.* ([7] b15-17) *ὃν δ' ἔχει τρόπον ταῦτα πρὸς ἄλληλα τῇ θέσει καὶ τοῖς εἴδεσιν, ἔκ τε τῆς ἱστορίας τῆς περὶ τὰ ζῷα δεῖ θεωρεῖν καὶ ἐκ τῶν ἀνατομῶν.*

physical connections, to each other and to other organs already discussed.[38] This is relevant because such connections are often evidence of causal relation, and of function, or their lack.[39]

The general explanation for the presence of stomach and gut follows immediately, in portions [2] and [3]: [2] explains why there is a need for the organs to do what they do, and [3] explains why there are two separate ones. What these organs do, Aristotle says, is quite obvious to everyone, and the explanation, informally expressed, amounts only to one sentence, which we may express as: 'All animals must have a place where unconcocted food is received for concocting and a place for the residue of concoction prior to its expulsion.'[40] But this statement, simple as it sounds, packs in a great deal, and it will be worth taking some time to see that. Actually, what Aristotle says is 'obvious to all' is not this statement, but an assumed one (or two) which we could supply as follows: 'The stomach is the place for unconcocted food ready for concoction; the gut is the place for the residue of concoction prior to its expulsion.' This is the (joint) fact apparent to all which we are being asked to explain, and we can note that it is a sort of definition of its subject, which serves also, in one form or another, as the conclusion of the explanatory deduction, much as Aristotle says of 'Thunder is noise in the clouds' in *APo*. II.8-10.[41]

I say 'in one form or another' because, as the conclusion in the latter case is sometimes given as 'There is noise in the clouds' (or perhaps: 'Noise belongs to certain clouds'), so in our text it is given first as 'All animals have a stomach and intestines', then as 'All animals have a place for unconcocted food ready for concoction and a place for the residue of concoction prior to expulsion.' But in our case the second is a scientific advance on the first. The second feeds into the conclusion the identification of what the stomach and intestines do, so that the

[38] The esophagus has already been said to be continuous with the mouth at 650a15 and again at 664a21-25.

[39] For an example, cf. n. 43 below.

[40] Peck's own translation of *εἶναί τέ τινα δεῖ τόπον ἐν ᾧ μεταβάλλει* in a16—'and there must also be *another* place where the change is effected' (my emphasis)—makes it sound as if we are getting an explanation for the presence of the esophagus as well as for that of the stomach and intestines. That of course can't be what Aristotle is saying, and the word Peck translates 'another' is *τινα*—'a certain'. It's the *same* place—the stomach—where unconcocted food is first received and then concocted. Aristotle's reason for describing that one place in two ways may lie in the premises he thinks he needs to establish that that same place can't also be used to store the residue of concoction prior to expulsion. (Lennox's translation of the relevant clause at a16, as usual when his and Peck's differ, is noticeably truer to Aristotle's text.) Though the subject of this sentence is the unrestricted 'all animals' (*τῶν ζῴων ἕκαστον*, a13), what Aristotle says elsewhere may imply a restriction to 'complete animals' (*τελεία ζῷα*)—cf. n. 45 below. Since the discussion from II.10 to IV.4 is in any case focused on the blooded animals, these too may be the domain under consideration in the present statement.

[41] e.g., 94a7-9. On this see, again, Bolton 1987, 143-44, and Bolton 1976.

explanation can go on to show what about the animals that have them (i.e., have organs that do what they do) *necessitates* that they have them, and thus explains *why* they have them.[42]

If it were not 'apparent to all' what stomachs and intestines do, we would probably have been given an argument for this being its function, an argument which would not itself be a demonstration, and which, as indicated above, would combine dialectical examination of alternative hypotheses and appeals to empirical evidence, including the physical connections, or absences thereof, mentioned above.[43] This discovery of what these organs do would at that point have been fed into the original 'nominal'[44] definitions of stomach and intestines, to give the fuller statement that appears in the text, although (and this is important) without any mention of necessity. The assertion of necessity is what makes the statement as we have it part of the explanation.

Interestingly, this sentence we have been discussing is still strictly the conclusion of the explanation but now in proper, modal form. The actual explanation—i.e., of why animals must have such places—is simply assumed, because, I want to suggest, getting back finally to our main purpose, the various premises have been given (and demonstrated) earlier for it to be simple to construct it. For, what is the full explanation? Why must all animals have a place for unconcocted food ready for concoction and a place for the residue of concoction prior to expulsion? The answer clearly must be first that: 'It is not possible to be (*einai*) or to grow (*auxanesthai*) without food' (655b30-32). The discussion of non-uniform parts, as we noted in our overview, opens, in II.10, with one of Aristotle's usual 'fresh starts', 'beginning first with the things that are first'. And what is first about non-uniform parts is the question of what parts are most necessary (*anankaiotata*) to all animals.[45] The first two mentioned are 'that by which food is received and that by

[42] A lesson of this is that identifying what an organ does is not sufficient to explain its presence. One must go on to show why the organisms which have it *need* something that does what it does or are otherwise better off for having it rather than not. It seems too that for Aristotle to call what some organ does its *function* (*ergon*) is for him to say both that it does that, *and* that it is necessary (or better) for the organism that it do that. This has interesting parallels with the contemporary account of functions given by Larry Wright, on which see section III of my 'Postscript 1986' in chapter 1 above.

[43] For an example see the discussion of the function of the windpipe at 664a35-b19 (and note yet another of Aristotle's humorous plays on words in his concluding sentence).

[44] If Bolton is right to view the 'nominal' definition as involving an assertion of existence: cf. Bolton 1987, 158 n. 55 and Bolton 1976. (For a contrasting view see Charles 2000, pt. I.)

[45] As noted above in n. 15, the subject of the sentence is actually all 'complete' (*teleia*) animals. It is hard to see why there is the restriction. In his note to 655b29-30, Peck suggests, with references to *GA*, that only the live-bearing animals are *teleia*. At 682a34, however, all of the blooded seem to be so called. But then most even of the bloodless have these three 'most necessary' parts, with soul-source in the middle, and even most of the testacea (cf. 681b31-682a2). One has to go, it seems, to the lowest of the low, the poor ascidian (*tethua*), to find an animal which, though it has distinct openings by which it takes in fluid matter for food and lets

which residues are eliminated',[46] and they get the explanation just quoted. (The third, 'intermediate between these two, is the part in which is the *archē* of its soul'.)

The surrounding context of this premise (viz., that it is not possible to be or to grow without food) makes clear that its intended application is to all living things. Plants we are immediately told get their food from the soil, already concocted, and thus have no place for useless residue. Nor do they have any stomach. (The earth, in effect, is their stomach, he says at 650a22.) So the fact about animals which distinguishes them from plants, and is central to their mode of nutrition (their 'threptic' soul), is that they concoct their food inside themselves. The argument thus starts with a basic fact about all living things, then adds a basic fact about (in effect) animal soul, then derives the need for certain organ-systems. All this is a general framework within which existence of the stomach and associated organs is explained, as the organs which perform these necessary functions.

But we should note also another twofold assumption that is made here, and taken over in the explanation of stomachs and intestines, about the metabolic process, namely that food must be *processed* (and in particular 'concocted'), and that the processing leaves a *residue*. These are physiological facts, as it were, having to do (at least in part) with the nature of the uniform parts of which the non-uniform parts are constituted, and with the parts of *phusikē* wider than the study of animals, the parts dealt with in *Generation and Corruption* and in *Meteorology* IV, two treatises in fact referred to early in *PA* II as sources on such issues.[47]

So the initial, general explanation of the presence of stomach and intestines in the animals that have them makes use of premises (i) about the basic nature of all living things, (ii) about the interactions of elementary powers, and these with

out the surplus moisture, 'has no evident residue like the other testacea'. These, and some equally (or more) primitive creatures, the sea-anemones ('the ones called *knides* or *akalephai*, which are not testacea but fall outside of the *genē* which have been divided off', 681a36-b2), Aristotle says, are barely animals, having as they do several features that make them plant-like, including the lack of residue. The sea-anemones 'tend towards the plants because their nature is incomplete (*ateles*) and they quickly attach themselves to rocks; and further, because they have no apparent residue though they have a mouth' (681b6-8). Could Aristotle have had (only) *these* creatures in mind when he qualified his statement in II.10?

[46] At this level Aristotle is probably talking more about 'systems', e.g., 'mouth-esophagus (if present)-stomach-liver-etc.' and 'intestines-kidney-bladder-etc.' than specifically and only about stomach and intestines.

[47] See below, p. 172. Part of the physiological story had been noted already in *PA* II–IV, in connection with the nutritive function of blood, at 650a2ff.: 'Everything that grows must of necessity take food. This food is always supplied by fluid and solid matter, and the concoction and transformation of these is effected by the agency (*dunamis*) of heat' (ἐπεὶ δ' ἀνάγκη πᾶν τὸ αὐξανόμενον λαμβάνειν τροφήν, ἡ δὲ τροφὴ πᾶσιν ἐξ ὑγροῦ καὶ ξηροῦ, καὶ τούτων ἡ πέψις γίνεται καὶ ἡ μεταβολὴ διὰ τῆς τοῦ θερμοῦ δυνάμεως). This is in line with the already noted frequent use of the facts about the elementary powers and their interaction throughout the discussion of uniform parts.

foodstuffs and with the uniform parts of animals, which are themselves conclusions of explanations in prior parts of the study of nature.

The part of our passage from III.14 marked [3] attempts to explain why there are two separate organs doing what stomachs and intestines respectively do. It is a brief and cryptic argument, based on the premises that (1) the storing of unconcocted food ready for concoction, and the forming of the residue ready for expulsion, take place at different times, and (2) different times require (or render more advisable) different places. There is some spelling out to be done here, at least of the second premise. But let us pass over that for now, to move on to the explanation of the difference in number of stomachs among animals.

The part of our passage marked [4] identifies two types of difference, then peculiarly moves on to discuss a third; the first two are not discussed until much later in the chapter. The third has to do with the number of stomachs. The general explanation spoke as if stomachs come in ones (*hē koilia*); now we learn that while many animals do indeed have only one, many others have several. The question is why, but the first question is *which*: what correlates with each possibility. First Aristotle catalogues those animals that have one stomach (I have marked that part [4]), then (at the start of [5]) those that have several, noting what each have in common and how they differ. He looks first at how the number of stomachs correlates with the number of rows of teeth and the presence of horns, then at how it correlates with the type of feet. Those with one stomach have teeth in both jaws. Those with several stomachs have teeth in only one jaw, and with one exception are all horned. Those with one stomach include all polydactyls, and all the solid-hoofed; and all those with several stomachs are cloven-hoofed, but the pig is cloven-hoofed and yet has only one stomach, so the decisive factor must be not 'being cloven-hoofed' but 'having one row of teeth'. All we need now is an explanation of the connection between having only one row of teeth and having several stomachs, and that would be easy, since Aristotle has already (III.1) noted that teeth are for mastication, and less mastication should mean that more concoction is required than the normal one stomach can handle. But instead of going on to give that explanation (which he does, finally, in [6]), he spends a good deal of time with the camel.

The reason seems clear: the camel does not have horns, and we know from our discussion of horns (III.2) that it is when an animal needs the limited upper earthy material for horns that is loses one of the two rows of teeth. Yet here is the camel, which does not need horns because its large size provides sufficient defense, and thus it does not have any horns, yet it is missing that row of teeth. It seems as if we have got to understand why the camel has those several stomachs *before* we explain why the rest do, in order to insure the soundness of the latter explanation. And so Aristotle proceeds. The camel, Aristotle seems to infer, would need several stomachs even if it had two rows of teeth, for its food is thorny and woody and thus hard to concoct. That is the source of its need of four stomachs, not its teeth. If it has one row of teeth, this is because, given the four stomachs, it no longer needs the second row of teeth—and one row tends to go with four stomachs,[48] and anyway, given that its food is hard, we could use some hard containing-material in the mouth, and the tongue has got to be fleshy, so let us put it on the roof of the mouth. 'A nature does nothing without a point but acts always for the best within what is possible to the being of the thing.' Now we can go on to the several stomachs in the horned animals, and in [6] we finally get that explanation. For details, now, on the other differences among the stomachs of live-bearing animals, we are referred to something like our *HA*, although after some discussion of features of the stomachs of birds and fish which have related explanations to that just given, Aristotle actually turns to the other differences among live-bearing animals.[49]

Three types of first principles

Instead of following Aristotle any further in these explanations, let us step back and take a look at the premises that have been mentioned, alluded to, or otherwise made use of in the explanation of multiple stomachs, and note their provenance. In the general explanation, Aristotle draws on the connection between horns and teeth, demonstrated earlier. That earlier explanation itself drew on the fact that the larger live-bearing animals, though they alone have *some* earthy material available in the developmental process for the construction of hard parts in the upper area, have only a limited amount of such material. The explanation of the camel's stomachs depends on the fact of what it eats, which is a given for this explanation (and not derived, for example, from what an antecedently given stomach might be able to process). And the explanation of *its* one row of teeth, and of the hardness of the roof of its mouth, seems to make use of the principle that nature always operates for the best among what is possible in the construction of a living thing.

Since the 'nature' which operates for the best is just the natures of particular organisms, the application of that premise to the camel presupposes that a camel nature exists. That is to say, teleological explanation of features of camels requires that we posit the existence of camels as ends, on the basis of which we can establish the necessity (or 'at least best'-ness) of those features which are explained by reference to this end. (I will elaborate on this shortly.)

[48] It would not be amiss to see a vague development hypothesis in the background: if they are correlated in being, their production in the generative process is probably connected.

[49] If someone were to argue that a separately written discussion of birds and fish, and the reference to 'the animal *historia* and the dissections', were both inserted between a continuous discussion of stomachs in live-bearing animals, he would get no objection from me, though these things are always hard to decide. (On 'dissections' cf. Lennox 1987a, 97 n. 16 [2001b, 35 n. 16].)

There seem to me to be several groups of premises at work here, which bear on our theme.

1. There are premises about the elementary powers and their interactions with various uniform compounds of them—and these appear drawn from prior parts of the study of nature. Early in the presentation in *PA* II, as mentioned in our overview above, Aristotle announces the need to distinguish the various ways in which things can be hotter or colder, and more solid (or dry) or fluid, explaining that these are *archai* 'to which the nature of many things is referred'.[50]

There is almost the identical wording in Theophrastus' *De Causis Plantarum* at I.21.4. In both Aristotle and Theophrastus these words introduce a review of the different ways in which things can be said to be, or known to be, hotter or colder than each other. To motivate further his discussion of the elementary powers, Aristotle writes, a few lines later, that it will be important to make this review (and a comparable one for 'dry' and 'fluid'):

> Since it seems clear that indeed these (but not roughness and smoothness, or heaviness and lightness, or any such things) are pretty much causes (*aitia*) of death and life, and also of sleep and waking, prime and old age, sickness and health. And this makes good sense, for, as was said previously elsewhere (*en heterois*), these—hot, cold, dry, fluid—are *archai* of the natural elements (*phusikōn stoicheiōn*).[51]

The 'elsewhere' is probably *Generation and Corruption*, though the thesis is made use of in *Meteorology* I–III, and IV. (Cf. *GC* II.2, where we are given an argument for the priority of these contraries over the others, a thesis only asserted here.) The reference to another work at 649a33 is, as Peck notes, almost certainly to *Meteor.* IV. Here we have Aristotle drawing on facts established in a prior portion of the general study of nature, in just the way one might expect of a linear, axiomatic-like structure of explanations.[52]

[50] *πολλῶν γὰρ ἡ φύσις ἀνάγεται πρὸς ταύτας τὰς ἀρχάς* (648a23).

[51] 648b4-10; tr. mine, with help from Peck. Peck's Greek reads: *ἐπεὶ ὅτι γ' αἴτια ταῦτα σχεδὸν καὶ θανάτου καὶ ζωῆς ἔοικεν εἶναι φανερόν, ἔτι δ' ὕπνου καὶ ἐγρηγόρσεως καὶ ἀκμῆς καὶ γήρως καὶ νόσου καὶ ὑγιείας, ἀλλ' οὐ τραχύτητες καὶ λειότητες οὐδὲ βαρύτητες καὶ κουφότητες οὐδ' ἄλλο τῶν τοιούτων οὐδὲν ὡς εἰπεῖν. καὶ τοῦτ' εὐλόγως συμβέβηκεν· καθάπερ γὰρ ἐν ἑτέροις εἴρηται πρότερον, ἀρχαὶ τῶν φυσικῶν στοιχείων αὗταί εἰσι, θερμὸν καὶ ψυχρὸν καὶ ξηρὸν καὶ ὑγρόν.*

[52] When one begins to make close comparisons between the practice in the biology and the theory in *Posterior Analytics*, one will want to ask whether the study of *animals* (or, for that matter, animal *parts*, animal *generation*, etc.) is the unit *epistēmē*, or rather the whole study of nature (*phusikē*, 'second philosophy'), or whether in the end it doesn't much matter. At that point one will want to consider the relationship between the studies of the elements in such treatises as *GC* and *Meteor.* IV *and* the biology, and perhaps to compare that relationship with the *APo.* account(s) of the relationship between 'superordinate' and 'subordinate' sciences. The biology clearly draws causes from the prior (portion of the) science, though here the borrowed (material) causes are secondary to the (final) causes which belong to the biology itself. See also Bolton 1987, 158 n. 56 (and now Lennox 1997a).

2. There is also here, as I have said, a general principle that for anything with a (formal) nature, its nature does nothing without a point but always acts for the best within what is possible to its being—and there are the posits of the individual natures for all the kinds of animals there are (known to be). Since explanation can take place at various levels of generality, these natures too can be posited at various levels. Thus, the explanation of beaks, for example, depends on the positing of a bird nature; this would not of course exist separately, but only as a component of the natures of the individual bird-forms. Nonetheless, explanation at the level of birds would require the positing specifically of that generic aspect of these individual forms. We thus have the positing not only of the *genos* of animal as such (since some explanation takes place at that level), but also of all the sub-kinds of animals the parts of which get explanations at the level of those sub-kinds by reference to the natures of those sub-kinds. Such posits, even if not given explicitly, are embedded in the use of teleological explanations for the kind in question, and are thus central to the enterprise of *PA* II–IV.[53]

I mentioned that in addition to the posit of the individual formal nature there is a general posit to the effect that these natures do nothing without a point but always for the best. Aristotle himself makes this very observation in a rare explicit identification of first principles in the study of nature, in *IA* 2. At 704b12-17, the principle that 'a nature makes nothing without a point but always the best from among the things possible to the *ousia* for each kind of animal' is said to be among

[53] I sidestep here some complicated, if fascinating, questions about these posits. First of all, there are many explanations in *PA* where the class of animals whose possession of a feature is being explained does not constitute a *genos* and does not have a name. One example is the class of animals whose possession of four stomachs is being explained in our *PA* III.14 passage; another (a subset of that class) is the widest class of possessors of four stomachs which gets a single explanation, the horned animals. They are identified by a description, and not a name, a common practice in biology (Balme 1987b, 81-85). I will need to make clear what sort of nature is being posted.

Now, it may be that the fact that explanation takes place at the level of horned animals is grounds for viewing them as a *genos*, and that raises the question generally of how *genē* are fixed in Aristotle. Here *PA* I.4 is of utmost importance, stating as it does that one should take as one's *genē* those classes which 'have been satisfactorily marked off by popular usage and possess a single nature (*mia phusis*) in common and forms not far separated in them—bird and fish and any other that is unnamed but like the kind embraces the forms that are in it' (644b1-7, tr. Balme, with 'kind' and 'forms' in place of 'genus' and 'species'). But there is a question of whether *PA* I.4 refers primarily to the initial stages of a science, before causes have been systematically identified, or refers equally to the developed stages. Lying behind this are questions about what counts as a 'single nature', how one grasps these natures initially, and how one establishes that what one took as single natures really are. [Some of these issues are discussed in Charles 1990 and 2000, and in Lennox 1990 and 2004; cf. also Balme 1987b, 78-89. I review their positions and develop my own in chapter 12 below.]

the things we must here posit (*hupothemena*) as we customarily do in *phusikē* when studying 'the works of nature'.[54]

Teleological explanation is licensed jointly, then, by two posits: the existence of formal natures (really a set of posits), and the fact that these natures make nothing without a point but always for the best among what is possible to the organisms of that form.

That such formal natures are posited in teleological explanation is made very clear by Aristotle's language in *PA* I. The material necessity which operates in the living world, Aristotle explains at two places, is not simple but *ex hupotheseōs*.[55] As John Cooper explains in his discussion of hypothetical necessity in Cooper 1987, 'the hypothesis relative to which a hypothetical necessity, in Aristotle's usage, is necessary is always a goal posited or set up (*hupotethen*) as something to be achieved' (243). Since Aristotle believes that biological natures act for an end, Cooper explains, 'the concept of hypothetical necessity will apply also to natural formations' (244). The application of hypothetical necessity to the explanation of the parts of animals thus depends *both* on the existence of formal natures *and* on their acting for the sake of something, i.e., for the best from among what is possible to things of that formal nature.[56]

We thus have, at the basis of the explanations in *PA* II–IV, principles pertaining to the matter of animals, drawn from the prior study of that matter, and principles pertaining to their form and final cause, embedded in posits of the formal natures of individual animals (plus a general teleological principle about formal natures). Cooper's study points the way to an understanding of how these two sorts of principles interact in these explanations, and thus how these explanations may be understood as demonstrations in a fairly strong sense.

Though, as we have seen, Aristotle is happy to acknowledge, with respect to the elementary powers which constitute the first level of the matter of animals, that

[54] *ἀρχὴ δὲ τῆς σκέψεως ὑποθεμένοις οἷς εἰώθαμεν χρῆσθαι πολλάκις πρὸς τὴν μέθοδον τὴν φυσικήν* . . . Aristotle names two additional sets of *hupothemena*: the six directions, and the fact that the *archai* of change of place are 'thrusting and pulling' (Forster). In the discussion of family resemblance in *GA* IV.3 Aristotle identifies three *katholou hupotheseis* (768b5-6) required for the discussion, two of which derive from the general theory of interaction presented in *GC* (as the apparent reference at 768b23-24 suggests).

[55] 639b24, 642a9; cf. 642a33 and *Phys.* II.9 199b34ff.

[56] Cooper rightly makes a distinction between strict hypothetical necessity, which explains some feature of an animal as *necessitated* by the being or essence of the kind in question, and explanation by reference to what is better for the kind from among several alternatives all of which are possible to it. I take it, though, that in the *ouden matēn* principle quoted in its fullest form in *IA* 2 but mentioned frequently elsewhere (cf. Bonitz, *Ind. Arist.* 836b28ff.), Aristotle means to collapse the two sorts of explanation into one. If a nature always makes what is best from among what is possible to the being of a kind of animal, then surely if there is only one possibility available to that being, it will choose that one (rather than, say, some alternative which would make the animal's being impossible). Cf. chapter 10 below, pp. 235-36, for an occurrence of the principle in which this may be the case. On the principle generally, see Lennox 1997b. [*Added 2011:* In the follwing chapter, I separate out more clearly three premises here: that formal natures exist, that they are ends, and that they are best-organized. See pp. 191-92, and in the Coda, p. 393.]

'the nature of many things is to be referred (*anagetai*) to these *archai*', he is very unhappy with those of his predecessors who referred *all* things to them.[57] For material natures by themselves are insufficient to generate the complex plant and animal forms we find around us; the existence of form is a separate principle, and one which acts for the sake of something.[58] Explanations of parts must thus (*dio*, 640a33) begin not from the material natures of the elements which constitute these parts, and the animals of which they are parts, but from the natures of the animal forms themselves:

> Hence we should if possible say that because this is what it is to be a man, therefore he has these things: for he cannot be without these parts. Failing that, we should get as near as possible to it: we should say altogether that it cannot be otherwise, or that it is at least good thus. And these follow . . .[59]

I substantially follow David Balme (1972, 87) and Cooper (1987, 254) in finding three teleological and one non-teleological mode of explanation in this passage.[60] The three teleological modes are: (i) of a part as necessary, given an essential function, because it is *the* organ which performs it; (ii) of a part as necessary, given an essential function, because it makes some necessary contribution to the performance of that function; and (iii) of a part as not necessary but 'at least' best, given an essential function, because it is one among the possible contributors to that function, and the best of that group. The non-teleological mode, derivative from these, is of a part (usually a residue or other byproduct) as a materially necessary consequence of the production of parts explained in one of the prior ways.[61]

3. The passage just quoted, in which Aristotle outlines the various modes of explanation of the parts of animals (640a33ff.), starts not only from a posit of the existence of an animal kind (in Aristotle's example, man), but from a statement of (some aspect of) *what it is to be* a member of the kind in question (*epeidē tout' ēn to anthrōpōi einai*). This cannot but bring to mind the claim of the *Posterior Analytics* that among the first principles of a demonstrative science are *definitions*. Putting aside again for now the various controversies concerning the *APo.* prescription, let us ask if the explanations of *PA* II–IV rest on definitions.

[57] 639b22, 640b4ff., *GA* V.8 789b2-4, *Phys.* II.8 198b12ff., and see chapters 1-4 above.

[58] Cf. Furth 1987, 24-25 (with n. 3) and 33 [= Furth 1988, 69-70 (with n.4) and 79), and, of course, Part I of the present volume, *passim*.

[59] 640a33-b1, tr. Balme. Peck's Greek reads: *διὸ μάλιστα μὲν λεκτέον ὡς ἐπειδὴ τοῦτ' ἦν τὸ ἀνθρώπῳ εἶναι, διὰ τοῦτο ταῦτ' ἔχει· οὐ γὰρ ἐνδέχεται εἶναι ἄνευ τῶν μορίων τούτων. εἰ δὲ μή, ὅ τι ἐγγύτατα τούτου, καὶ ἢ ὅτι ὅλως ἀδύνατον ἄλλως, ἢ καλῶς γε οὕτως. ταῦτα δ' ἕπεται . . .*

[60] For further details see below, chapter 10, pp. 219-20 with n. 4.

[61] Other views of this passage may be found in Kullmann 1974, 37 and Sorabji 1980, 155-56. The Balme–Cooper alternative seems supported by the rough parallel at I.5 645b 28-33, with modes two and three collapsed into this passage's second way, as well as by Aristotle's practice in distinguishing modes (i) and (ii), as Cooper argues. I give some further evidence for this in chapter 10 below, esp. n. 4.

In our 'stomachs' passage we saw explanations going back to what I loosely called 'the basic nature' of this or that. But does Aristotle in his biology in fact believe that there are features that can be called basic, i.e. features of a kind that explain other features of that kind without themselves being explained by features of that kind? As I mentioned towards the beginning of this chapter, he might, for instance, have come to believe after doing his biological work that explanations are not linear and directed in that way but are, say, circular, features being explained in terms of each other, or highly localized and contextual. I think in fact that this is not the case, and that the biology speaks firmly to the existence of such basic features for each authentic kind.

This is a large subject, and there is not space to deal adequately with it here. In outline, the evidence is this. First, there are some eight passages in *PA* II–IV, and one more in *IA*, in which Aristotle explicitly says that some feature of an animal kind is in the *ousia*, or in the *logos ousias*, etc. of that kind, and in every case that feature so identified is at work explaining some other feature of that kind. I have discussed these passages in detail in chapter 10 below, where I have also provided the Greek; here I just present them for your examination, in a literal translation.

(1) **II.8 653b19-27:** We must now examine the other uniform parts, and first flesh in those that have flesh, and its analogue in the others; for this [*sc.* flesh-or-its-analogue] is of animals an origin and a body in itself. This is clear also from **the *logos*, for we define an animal by its having sensation, and first the primary sense, which is touch**, and such a part is the sense organ of this—either its primary organ, as the pupil is of sight, or the organ taken together with that-through-which, as if someone took the whole of the transparent together with the pupil.

(2) **III.6 669b8-12:** Generally, then, the lung is for the sake of respiration, though it will be bloodless and such [as has just been described] for the sake of a certain animal kind [sc. the egg-bearing animals]. But there is no common name in the case of those [which have a lung], not like 'bird' is applied in the case of a certain kind. This is why, just as the being (*einai*) for a bird is out of something, also in these cases **having a lung is present in the *ousia*.**

(3) **IV.5 678a26-34:** The commonly called 'softies' [cephalopods] and 'soft-shells' [crustacea] are very different from those [*sc.* the blooded animals], since they have nothing at all of the nature of viscera. Nor, likewise, do *any* of the bloodless animals. (There are two kinds of the bloodless remaining, the 'potsherd-skins' [testacea] and the insect-kind.) For, none of them has that out of which viscera-nature is composed—blood—because **some such condition belongs to their *ousia*,** for **that some animals are blooded, while others are bloodless, will be present in the *logos* defining their *ousia*.**

(4) **IV.6 682a35-b32:** Now, then, the insects are all many-footed, in order to make their motion quicker and to counteract their natural slowness and coldness, and those that are especially cold on account of their length are very-many-footed, such as the centipede-kind. Furthermore, because they have many origins [of life-activity], they have their 'insections' and are correspondingly many-footed.

They are insected both on account of the aforementioned causes and in order that they may protect themselves by curling up to avoid injury. For those of them that are long coil themselves up, and this could not happen if they were not insected. Those that do not coil up make themselves harder by pulling their sections together. And it is necessary for them to be insected: for **having many origins is present in their *ousia*** and in this they are similar to plants. For, just as plants, they also can live when cut up, though the latter live only a while, whereas the former become complete in nature and two or more in number from one.

(5) [**IV.8 684a32-b1:** In lobsters only, whether male or female, it is a matter of chance which claw is bigger. The cause of their having claws is that **they are in the kind that has claws**; but they have them irregularly because they are deformed [Balme: 'imperfect'] and do not employ them for their natural use but for walking.[62]]

(6) **IV.9 685b12-16:** Now, the others [that have suckers] have two rows of suckers, but one kind of octopus has one row of suckers. The cause of this is **the length and slimness of their nature**, for a narrow one will necessarily have only one row. Thus, it is this way not as being the best but as necessary because of **the particular *logos* of its *ousia*.**

(7) **IV.10 686a25-31:** Now, man, instead of forelegs and forefeet, has, as we call them, arms and hands. For he alone of the animals stands upright, on account of **his nature and *ousia* being divine, and the function of that which is most divine is to think and reason**; and this would not be easy if there were a great deal of the body at the top weighing it down, for weight hampers the motion of the intellect and the common sense.

(8) **IV.12 693b2-13:** Two legs like man, bent inward like quadrupeds and not outwards like man; and the wings, as the forelegs of quadrupeds, with the convex side outward. They are each biped of necessity, for **the *ousia* of a bird is among the blooded**, and at the same time also winged, and the blooded do not move on more than four points. So birds have four attached parts, just as the other [blooded] ones, i.e., the ones that live and move on the ground; but whereas these have arms and legs (quadrupeds having forelegs), birds instead of forelegs or arms have—uniquely—wings, for they are stretchable out with these, and **being a flier is in the *ousia* of a bird.**

(9) **IV.13 695b17-25:** Fishes do not have attached limbs because **their nature according to the *logos* of their *ousia* is being a swimmer**, since nature makes nothing either superfluous or pointless. And since **they are according to their *ousia* blooded**, because they are swimmers they have fins, while because they do not walk they do not have feet, for the addition of feet is useful for moving on the ground. At the same time it is not possible for them to have four fins and feet as well, or any other such limbs, since they are blooded.

(10) ***IA* 8 708a9-20:** The cause of the footlessness of snakes is *both* that nature makes nothing pointless, but always with a view towards the best for each thing from among the things which are possible to it, preserving **the particular *ousia* and essence of each** (*to ti ēn autōi einai*), *and* what we have said earlier, that it is not possible for blooded animals to move with more than four points. For, it is evident from these things that **all blooded animals which are disproportionate in length relative to the rest of the bodily nature, as are the snakes**, cannot possibly be footed. For it is not possible for them to have more than

[62] I argue in chapter 10 below (p. 232) that this is *not* a generic portion of the lobster's definition, so I bracket the passage, but include it for completeness.

four feet (since then they would be bloodless), and if they had two or four feet they would be pretty much unable to move at all; this way their movement would necessarily be slow and useless.

These are of course only *partial* definitions: David Balme and Pierre Pellegrin are certainly right that there are no full definitions of animal kinds offered in the biology.[63] But partial definitions are all that *APo.*itself requires.[64] Interestingly they appear at all levels of generality, from the (unnamed) blooded and bloodless animals down to a certain kind of octopus. The suggestion seems to be that for every level at which there is explanation of features of some animals, there are features basic (causally primary) at that level, features which define *at that level* one part of the being of the animals that have the features explained at that level. What this implies for the existence of kinds (*genē*) at levels identified only descriptively, and for the posits at those levels, is not clear (cf. n. 53 above), but the metaphysics of this is less relevant than the fact that we have in the biology definitions that serve as principles of demonstration.

Once one notes the existence of some partial definitions, one is tempted to look for features that play the same role but are not labeled as so doing. I call these features the 'givens' (for any kind)—features which, though not identified as such, in fact serve to explain features of their possessors without themselves being explained. This is to some extent speculative, of course, since the fact that Aristotle has not given an explanation of some features does not establish that he thought there was none to be given, but as one tries to isolate such features from a close look at the actual explanations in *PA* II–IV, a picture begins to emerge which looks plausible to me at least. For, it includes all the soul-faculties, in their various differentiated forms, that *De Anima* identifies as part of the *ousia* of animals generally, and also those (primarily) non-functional features that appear in the nine *ousia* passages, such as size, and being blooded (or bloodless—i.e., possessive of that other, nameless, nutritive fluid). Also frequently serving as 'givens' in *PA* II–IV are the *bioi* of the animals kinds—not just their capacities but their capacities in relation to some environment. They are given by such predicates as 'land-animal', 'water-animal', 'marsh animal'.[65] One component of a *bios* seems often to be the

[63] Balme 1987d, 310; Pellegrin 1985, 99.

[64] Cf., e.g., Barnes 1975a, 247.

[65] The following is a possibly complete list of places in *PA* and *IA* where something labeled the '*bios*' of some animal kind seems to serve as a given: 657b29, 660b33, 662b6, 662b15, 663b19, 665b3, 682b7, 682b13, 684a5, 691a26, 693a4, 693a5, 693a11, 693a15, 694a2, 694a6 (text disputed), 694a7, 694b7, 694b12, 694b16, 710a27, 713b21, 713b28, 714a10, 714a21, 714a23. Cf. the detailed and elaborate division of *bioi* in *HA* I.1 (487a10-488b10), and the general discussion of *bioi* and *praxeis* in *HA* VII.1-2 (588a16-590a18) and at the beginning of VII.12 (596b20-28). (See now also Lennox 2010a, though I don't think the paper's main argument is successful.)

type of food an animal eats, as in the case of our camel, with its thorny and woody food. Aristotle does not think that animals eat the food they do because they have organs that can process that sort of food—'nature makes the organ to suit the function not the function to suit the organ' (694b13-14).

One might take food as a given, but in *HA* VII.1-2 he states that animals eat food with an elemental blend like their own. The cetaceans, though land animals (*peza*) so far as lungs and internal structure is concerned, have a more watery blend than land animals standardly do, and require food that is more watery. Such food being found only in the water (for the same reason, I suppose), they live in the water.[66] The same doctrine is alluded to in *PA* III.6 in the discussion of lungs, to explain why cetaceans violate the correlation of lungs with living on land, and again in *Resp.* and in *GA* and even in the discussion of growth in *GC*.[67] This suggests that an animal's 'elemental blend', whatever that is exactly, may have been thought by Aristotle to be part of the *ousia* of that animal. While this opens intriguing metaphysical doors, generating reflections on the unity of form and matter, and on their relationship in teleological explanations, and so forth—issues I hope to pursue elsewhere—what is relevant to the present discussion is the clear existence of a structure of essential, explaining features on the one hand, and explained ones on the other.

The explanations in *PA* II–IV rest, then, on at least these three sorts of principles—premises about the elementary matter of which animals are composed, premises about the existence of formal natures and their character as ends, and definitions of these formal natures. For this reason, I am strongly inclined to take Aristotle at his word when he asserts, in some of his introductory and methodological discussions, that his biological works offer an *epistēmē* of animals, full of *apodeixeis*, based on a variety of *archai*. Based on our examinations of the overall explanatory structure of *PA* II–IV, and the detailed explanation of the presence of one or more stomachs in animals, and the frequent dependence of later explanations on premises established in earlier explanations, I am equally inclined to see *PA* II–IV as exhibiting the sort of interconnected structure of demonstrations which one typically associates with the Aristotelian idea of science.

This structure may be disguised in part by the fact that the material so structured is unfamiliar to us (both as students of biology and as scholars of Aristotle), and in part by the fact that Aristotle does not in *PA* II–IV appear interested in showing that this material exhibits that structure. He is working not against the background of those who might wonder if his biological material is amenable to the sort of structure that is outlined in *APo.*, but against the background of those who have

[66] 589a5-8, 589a29-590a12.

[67] 669a8-13. Cf. *Resp.* 14, *GA* IV.2 767a30-4, and *GC* II.8 335a10. The explanation of the differences in mode of reproduction among types of animals in *GA* II.1 (732a25-733b16) also makes use of differences in elemental blend.

thought that one could explain the nature and generation of animals and their parts by reference to low-level material principles alone. Aristotle's claim is that final causes as well as the necessary natures of the elementary materials are operative in nature, and that their operation is prior to these necessary natures in the sense that it is only on the hypothesis that an animal's formal nature is to be realized as an end that these material natures are operative in the genesis of that animal or any one of its parts. These theses are the background for the systematic study of animal parts which is exhibited in *PA* II–IV, and generate much more talk about causes than about premises, and about reasons why than about demonstrations. But the structure is there, and the language is present enough, to warrant our speaking of an axiomatic structure for the theory of these books.

Need the syllogistic and axiomatic structures be explicit?

In making these claims about the epistemic character of *PA* II–IV, I have purposely avoided giving anything like rigorous definitions of such notions as *epistēmē*, demonstration, *hupothesis*, definition. Our major source for these is in *APo.*, and I have thought it important in our exploration of *PA* II–IV not to entangle us in the many controversies that surround the interpretation of that terribly difficult text. The attempt to establish the precise relationship between the structure exhibited in a general way in this chapter, and the precise theory of the structure that *APo.* requires of a properly developed science, is a project in itself. But it is a project which requires as groundwork the sort of examination of *PA* II–IV begun here.

I would like to close, however, with some brief remarks on the relationship between the practice in *PA* II–IV and the theory in *APo.* which do arise directly out of our study of *PA* II–IV as a whole, and which bear on the question of the significance of the absence of syllogistic in *PA* II–IV.

The fact that *PA* II–IV can have the sort of structure I have argued it has, and be declared by Aristotle to be a body of demonstrative understanding, in the face of its clearly (logically) *informal* character, should perhaps give pause to the notion that *APo.* calls for science to be formal.

In his study of 'Aristotle's theory of demonstration' mentioned above, Jonathan Barnes claimed that according to *APo.*, a demonstrative science must be presented in the language and form of the syllogistic. By that criterion, of course, *PA* II–IV fails miserably. When *PA* II–IV was thought to be 'tentative', 'exploratory', or 'dialectical', its informal character was no surprise. But what are we to make of a highly structured *explanatory science* that is presented without syllogistic form? Is it awaiting the stage of final science (which it clearly isn't in its present form) before being cast into syllogisms?

In a more recent paper (1981) also referred to above, Barnes argued for the view, maintained years ago by Solmsen, that the central core of the theory of demonstration is independent of the syllogistic, and earlier. It follows from this, of course, as Barnes observes, that scientific treatises written prior to the discovery, or the implementation into apodeictic, of the syllogism, could be demonstrative without containing any syllogisms. Barnes did not, however, say much about the requirements that the pre-syllogistic theory of demonstration places on the logical form of demonstrative premises, to help us judge whether *PA* II–IV would on that earlier theory count as demonstrative science. Nor did he suggest that the pre-syllogistic theory of demonstration does not require proper science to be presented in *explicit* axiomatic form. But I want to suggest that there is reason to think that *PA* II–IV would count as demonstrative science even on the post-syllogistic theory, or would at least not be debarred from that status by virtue of its lack of an explicitly syllogistic form or of an explicit axiomatic presentation. This has to do with the character of the *Posterior Analytics* itself as a presentation of a theory of science, and it is a point that was first brought to my attention by Aryeh Kosman, in discussion at the 1983 Williamstown conference.

Kosman's thesis is that *APo.* should be understood as offering a formal description of proper science, not a requirement that proper science itself be formal. As he put it in that discussion, it is much like getting the quantifiers in the right place, though here it is the 'formal' operator: a formal description of science not a description of formal science. On this view, the *APo.* theory requires of the natural philosopher only that his exposition be putt*able* into the appropriate form, not that it actually be so *put*; and it relieves proper science of a self-conscious concern with that form, allowing it to satisfy its various informal purposes—though it of course does not give the scientist a blank check, since the formalization of a stretch of informal discourse is not an arbitrary matter.

This view finds support, I think, from the analysis of the syllogistic itself in Jonathan Lear's valuable study of *Aristotle and Logical Theory* (1980). Lear argues that 'Aristotle's project is to provide a formal analysis of the non-formal deductions with which he was familiar.' I take it this implies that there is no suggestion that speakers, to get their logic fully right, should reason in formally explicit syllogisms. Lear reminds us of the ambiguity of the term *sullogismos*, and its match with the ambiguity of our word 'deduction', and observes that 'This ambiguity is tolerable since the value of the formal syllogistic is supposed to derive from the fact that a syllogism in the broad sense can be represented as a syllogism or chain of syllogisms in the narrow sense.[68] Since the study of the syllogism is announced at the opening of the *Analytics* as a part of the study of demonstration, one might expect Lear's point to transfer over to the theory of science. It would be an interesting question to consider whether there is any explicitly support from *APo.* itself for this view of it, but I pass by that for now.

[68] Lear 1980, 10-11.

On such a view, which as Kosman has suggested is certainly an intuitively plausible one for Aristotle to take, one would expect *PA* II–IV to have much the logical form it has now. For, whether or not the explanations can in fact be cast into syllogistic form, there is much reason to think that Aristotle himself thought they could be so cast.[69]

But even if this line of approach to *APo.* is not right, and that treatise does call for proper science to be presented in syllogistic form, one will want to ask what constraints syllogistic theory places on the character of preliminary, informal science. For not every form of explanation at the informal level will make later formalization of the appropriate sort possible. And the same can be said for its premises. An informal structure without at least the prototypes of axioms, *hupotheseis*, definitions, etc., or with an essentially circular structure, or with explanations that are highly contextual or piecemeal, will not be axiomatizable in the way required.

Regardless of whether proper and final science is to be formal or informal, then, we might expect close study of the explanatory structure of such biological treatises as *PA* II–IV to be of help in resolving some of the interpretative logjams that plague the study of *APo.*, as we might expect close study of *APo.* to illuminate further the explanatory structure of *PA* II–IV. Towards those ends the present chapter is only a first step, but I hope a useful one.

Additional note on *PA* I.1 639b21-640a9[70]

Aristotle's use of *tropos* in *Metaph.* Δ and elsewhere (cf. Bonitz, *Ind. Arist.* 774b2ff.) suggests that he would not speak of another *manner* of demonstration unless the explanations in natural science referred to *were* demonstrations in no merely homonymous way, and this is borne out by the fuller parallel passage in *Phys.* II.9 (esp. 200a15ff.), most probably the 'elsewhere' referred to in 640a2. For, in that passage, while noting the differences between explanations in the case of things that

[69] The clearest piece of evidence is, perhaps, Aristotle's actually doing so, for some sample explanations concerned with plants, in the later chapters of the *Posterior Analytics*. For references and discussion, cf. Lennox 1987a, 95-96. (I owe the reference to Lear 1980 to Lennox's briefer discussion of this issue at the close of that paper.) As Alan Code pointed out in the discussion of this paper at the 1985 Cambridge/Trenton conference, the fact that Aristotle does not present his biological explanations in syllogisms, or even that they cannot be so presented (if that too is a fact), should not count against the demonstrative character of *PA* II–IV. After all, Greek mathematics cannot be formalized within the unaugmented syllogistic Aristotle presents, and Aristotle does little to try to show that it can, but that doesn't stop him from using mathematics as an example (if not the paradigm) of demonstrative science. There is, of course, the question of whether and, if so, how Aristotle thought that the facts of natural science, which are most of them true only 'for the most part', could be syllogized. This is a difficult (but not, I think, insuperable) matter which must await another occasion.

[70] See n. 6 above.

come to be and explanations in mathematics, he focuses as well on their *similarities*, and these can show us why he counts the former as demonstrations clearly productive of *epistēmē*.

Here, as in *PA* I.1 (200a13, 639b24), Aristotle begins by insisting that the necessity in things which come to be is only hypothetical, and moves immediately to compare it with the necessity in mathematics. In *Phys.* II.9 he starts with the similarity—necessity appears in both 'in a parallel fashion' (*paraplēsiōs*, 200a16):

> Because the straight is so and so, it is necessary that a triangle should have angles together equal to two right angles, and not the other way round.... With things which come to be for something the case [sc. though parallel] is reversed: if the end will be or is, that which comes before will be or is...[71]

The parallel is that in both cases (if in different ways) the necessity is one-directional, and from *what* some primary subject is to some other feature of the subject whose presence follows from the subject's being that. (The geometrical explanation assumes that straight sides are essential to triangles.) 200a34-b4 brings this home:

> And the end is that for which, and the start is from the definition and the account;... so if this is what a man is, then so and so, and if so and so, then such and such.

Though parallel in this way, the cases are 'reversed' (*anapalin*, 200a19). In mathematics, the starting points are (in some sense) eternal (cf. 639b24), and thus *are* and are *of (unqualified) necessity*, so that whatever they necessitate is also of unqualified necessity. But with things which come to be, the starting-point of explanation, the end, is *not* necessary (200a13), and thus what the starting point necessitates is necessary only *on a hypothesis*: '*if* the end will be or is, that which comes before will be or is' (a19-20).

Interpreters have made much of this use of the future tense, especially in the abbreviated *PA* I.1 version, where only the *contrast* between the mathematical and natural-coming to be cases is noted, and where it is drawn in terms of tense: 'For the latter begin from what is, the former from that will be' (640a3-4). But these interpreters fail to notice both that the portion of the *Phys.* II.9 passage just quoted that corresponds to 640a3-4 reads 'if the end will be *or is*' (200a21), and that the sample demonstration that in abbreviated form immediately follows 640a3-4 is entirely in the present tense: 'Because health or man is such, it is necessary that this be or come to be' (a4-5). This illustration makes clear that the demonstrations *both* in the mathematical *and* in the natural cases will typically have the same outward form, and when formalized will be indistinguishable. (Cf. Lennox

[71] a16-22; tr. Charlton. Ross's Greek text reads: *ἐπεὶ γὰρ τὸ εὐθὺ τοδί ἐστιν, ἀνάγκη τὸ τρίγωνον δύο ὀρθαῖς ἴσας ἔχειν· ἀλλ' οὐκ ἐπεὶ τοῦτο, ἐκεῖνο ... ἐν δὲ τοῖς γιγνομένοις ἕνεκά του ἀνάπαλιν, εἰ τὸ τέλος ἔσται ἢ ἔστι, καὶ τὸ ἔμπροσθεν ἔσται ἢ ἔστιν.*

1987a, 111 n. 40 [= Lennox 2001b, 36 n. 37].) Aristotle's use of the future serves rather (and only) to reflect the contrast in the necessary status of the starting-points in the two cases, and the temporal order of necessitating and necessitated factors in explanations of coming to be. It also permits an economical way of contrasting his own view of the latter case with the view of his predecessors that complex organic features come to be of unqualified necessity from their elemental antecedents (640a6–9).

Thus, in sum, though the difference between demonstrations in mathematics and those in natural science (in the case of things that come to be) is of great significance, it does not lead Aristotle to deny that there *are* demonstrations in natural science, productive of *epistēmē*, and our *Phys.* II.9 passage helps us to see why. As the passage explains, in both the mathematical and the biological cases the explanation proceeds from *what the subject is* to some less essential attribute of it, and in both cases it proceeds by necessity. These are central features of demonstration for Aristotle, and evidently central enough for their presence in biological explanations to entitle these explanations to be called 'demonstrations'. Their *mode* of demonstration is different from that in the case of mathematical explanations, since the mode of necessity is different (Balme's 'i.e.' for *kai* in 640a1 seems right), but both are demonstrations—and productive of *epistēmē* nonetheless.

(That the individuals subsumed by the minor terms in these biological demonstrations do not exist of necessity will have to be dealt with eventually. To start, we will have to distinguish between the necessary existence (or not) of the subject and predicate terms in demonstrative premises, and the necessity of these premises themselves. Aristotle's concern in the *PA* I.1 and *Phys.* II.9 passages is with the former; the questions just raised have to do with the latter. On this question I imagine he would say that while there is no necessity that any particular man, say, should exist, it *is* necessary that some man or other exist, and that this is sufficient to render the relevant universal minor premises necessary in the way appropriate for demonstration, even though their referents are not necessary existents. The passage from *GA* II.6 quoted above (p. 157) seems to support this, suggesting perhaps that the amenability of things which come to be to demonstration lies in the fact that there is an eternal coming to be of things of the relevant kind. This is a large problem, but the question will be to understand *why* at *PA* I.1 640a1 Aristotle allows demonstration in natural science; *that* he does is clear from that text and its parallel in *Phys.* II.9.)

In the preface to his *Aristotelian Explorations* (Lloyd 1996a), G. E. R. Lloyd reflects on our methodological differences, referring to me by name and then speaking of the 'lumpers' and the 'splitters'. But as the foregoing discussion shows, I don't *lump together* essentially different conceptions and practices of demonstration, rather I attempt to *relate* or even *connect* apparently different ones, by searching out essential *similarities*. (Lloyd's view is developed in his 1996b; see Wians 1992 for criticism of the original version. I provide some detailed replies to Lloyd myself in chapter 8 below, in nn. 1, 15, and 18.) The division is not between 'lumpers' and 'splitters', but between *integrators* and *fragmenters*. And an increasing number of scholars, approaching the relationship between the *Analytics* and the biological works with (to adapt Nietzsche's words) a tuning fork and not a hammer (or a wedge), have found much consonance. Among the scholarship in this vein I may mention Lennox 1997a and 1997b (and cf. his additional remarks in Lennox 2001b, where Lennox 1997a are 1997b and reprinted), and Leunissen 2010a (my largely favorable comments on which appear in the same volume as does her chapter, as Gotthelf 2010); and see Charles 2000, ch. 12 for a nuanced intermediate view, though one with important integrationist sympathies.[72]

[72] Some of the main ideas of this chapter were developed in an undergraduate seminar at Trenton State College in Spring 1985, which began with nine students. To the three noble souls who survived to the end, Don Angellini, Tom Church, and Laura Italiano, my gratitude for their stimulating attention and suggestions. Some of these ideas were anticipated in Kullmann 1974. An earlier draft of the chapter was read as a paper at the Cambridge/Trenton Conference on Aristotle's Philosophy of Biology, at King's College, Cambridge, in June 1985. I would like to thank all those who participated in an exceptionally valuable discussion, but especially David Balme, Myles Burnyeat, Alan Code, Jim Lennox, Geoffrey Lloyd, and Malcolm Schofield. An initially skeptical response by David Balme to an early outline of the first draft, and his suggestion that I try my thesis out on the 'stomachs' passage, helped shape some of the direction of this chapter. I am grateful to him for those comments. I owe thanks, too, to David Charles and Martha Nussbaum, for important discussions outside the Cambridge session, and to Cynthia Freeland, Joan Kung, Mario Mignucci, and, especially, Jim Lennox for very helpful written comments.

8

The Elephant's Nose

Further Reflections on the Axiomatic Structure of Biological Explanation in Aristotle

In the previous chapter I argued that the explanations across *PA* II–IV have an implicit axiomatic structure. In the present chapter, I would like to illustrate further and to some extent develop that basic claim, via an examination of the explanation in *PA* II.16 of the elephant's unique 'nose-passage', its trunk (as we call it in English). I will not here be concerned with the syllogistic character of the explanation, but with its *premises*, and with the complex logical *structure* across the explanations in *PA*.[1]

This chapter is based on a paper read to a 1995 conference on Aristotelian Biology in Bad Homburg organized by Wolfgang Kullmann and Sabine Föllinger, and was published in the proceedings of that conference (Kullmann and Föllinger 1997). For more details of its background see n. 26 below.

[1] In several publications Geoffrey Lloyd has raised various questions about my thesis and its associated claims; I will respond to several of them in the course of this chapter. A fuller discussion will have to await another occasion, on which I will also want to discuss a fundamental difference in methodology that is worth mentioning here. Lloyd's focus has been on trying to identify *differences* between Aristotle's explanatory theory and practice. In some cases (as I argue below) the differences are illusory, and a more sensitive reading either of the *Analytics* or of the biology can show that. In others there are indeed differences, but the question remains how to *weight* them in judging the closeness of the practice to the theory. By contrast, my approach has been to look for *similarities* between the practice and the theory, especially the unevident ones. My own view, to put it loosely (and to be further specified and argued, as I say, on another occasion), is that the similarities far outweigh the differences. Lloyd's efforts are salutary (and I have learned from them), but only, I think, if one does not ignore the fundamental similarities that are there to be found. For example, one common concern, expressed by many (myself included), is that while the *Analytics* aims (in part, anyway) to specify the structure of a completed science, the body of science in *PA* is not complete, nor was thought by Aristotle to be so. But as (e.g.) *APo.* II.8 93b12–14 makes clear, even if we have not yet discovered, with respect to any property or phenomenon, the *ultimate* middle term or cause, we do have, at any stage of explanatory science, features that are *epistemologically* basic—i.e., basic *at that level of knowledge*—and the explanations we are able to provide at that level will have many features of a full axiomatic structure, even if the epistemologically-first principles are not the ultimate ones in that case. Aristotle has said less than he might about the form(s) a science takes as it develops, but as the passage from *APo.* II cited above makes clear, there are rich hints in his extant work that we can develop. (See, e.g., the Coda below, on the revisability of Aristotelian science, pp. 394–97). A focus on the *similarities*

In the previous chapter and elsewhere I have used the term 'linear branching structure' to describe the structure of explanation across *PA*, but I now think this expression is, or can be, misleading, for it may suggest the following pattern (Fig. 1).

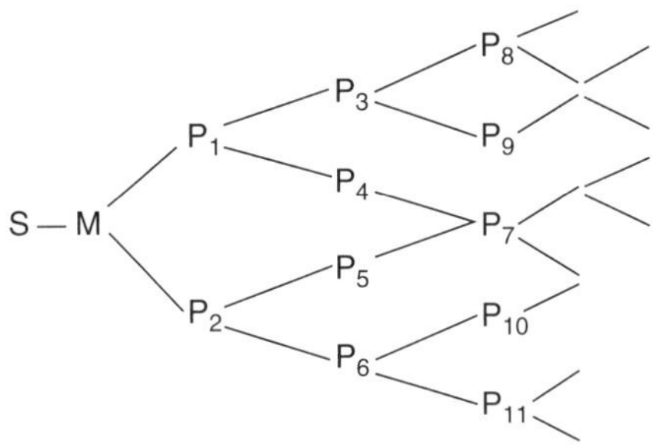

Figure 1

I use 'S' for the subject kind (e.g., elephant), 'M' (from 'middle term') for a single feature that is explanatorily basic, and 'P_n' for features which belong to elephants *kath' hauta*, 'in virtue of themselves'.

On this model there is one *essential* feature of the kind, and one only, and all other features are explained from it. Here, the structure does indeed branch outward, so that the right-hand end is much wider (or taller) than the left hand beginning.

But this is not the structure I meant to impute to *PA*. The structure I described in the previous chapter differs from this on several counts. Let me start with two of them.

(1) As John Cooper argued convincingly in his study of hypothetical necessity,[2] and as I made clear at several places in the previous chapter, many explanations draw on premises about the *matter*, and do so in various ways.[3] The paradigm Cooper offered is the move from eyes to eye lids via certain facts about material natures such as the necessary fluidity and thus easy penetrability of the only transparent materials available to an organism (viz., water and air), which are the only sort of materials that can serve as the organ or vehicle of sight.[4] We will see other examples of premises concerning material natures shortly. In Figure 1, while some of the links between terms (e.g., between S and M) may be definitional, others (e.g., between M and P_1) will depend on distinct premises about material

between explanatory theory and practice (with a proper appreciation for apparent, and real, differences) might help point us to a richer exploitation of those hints. (On the *APo*. II.8 passage, see Bolton 1987, 144-45.) For more on this general methodological difference with Lloyd, see above, pp. 184-85.

[2] Cooper 1987.

[3] See also Lennox 1997a, 163ff.

[4] Cooper 1987, 255.

natures; neither these premises, nor all of their terms, are represented in this diagram.

(2) Also, although the diagram in Figure 1 presumes that one and only one feature of the subject kind S is essential ('M'), in fact many explanations in Aristotle's biology proceed from *multiple* features of the animal kind in question, all of which are 'givens' for those explanations,[5] none of which are explained anywhere else in *PA*, and for each of which there is good theoretical or intuitive reason to think that Aristotle considered them explanatorily basic and thus essential to the subject kind. My favorite example of such an explanation is that of the *elephant*, and it will be worth our time to stop over Aristotle's discussion of the animal in *PA* II.16.

The discussion comes in the course of Aristotle's account of the various parts around the head, in the portion of that account in which he has been discussing the nature and placement of the parts where the organs of sight, hearing, and smell may be found, viz., the eyes, ears, and nose (*muktēr*, 'nostrils' in the plural), and the associated parts such as eyelids and eyelashes. He mentions that the four-footed live-bearing animals do not differ much in their organ of smell, except for (i) those that have a 'snout' (*rugchos*) in which jaws and nostrils coalesce—and (ii) the elephant, in which the 'nose' is absolutely unique (*idiaitaton . . . tōn allōn zōiōn*, 658b33-34) in respect of its extraordinary length and ability (*dunamis*, viz., its ability to grasp and transport objects, such as food to its mouth, and its strength—it can, e.g., uproot trees). Its length and ability need explanation, and the explanation proceeds from (what appears to be thought of as) the essential nature of the elephant. In fact, it proceeds arguably from perhaps as many as *seven* basic facts (certainly at least five) about the elephant (which I will label M_1–M_7), in a sequence of two premise steps. Of course the truth of Aristotle's attributions is not at issue, so we will not stop over that.[6]

First, the elephant is a swamp or marsh animal, in that it gets its food in water that is in some places deeper than the elephant is high (M_1, 659a2-4, 14). Second, the elephant is blooded (M_2, a4-5); it thus has a degree of heat which requires an internal organ of cooling (cf. III. 668b34-669a3); but it is a *pezon* (M_3, 659a3,4), which literally means 'walker' or 'mover-on-land' but in this context effectively means 'air-cooler'.[7] So, it has to breathe (that's P_1, 659a4, from M_2 and M_3).[8] But

[5] I explain the term 'givens' in the previous chapter, p. 178.

[6] To follow the next three paragraphs, readers should consult Fig. 2 on p. 190 below.

[7] Cooling by air usually correlates with moving on land when the animal is blooded, although not always since cetaceans are air-coolers and thus in this sense *peza*—cf. 668b32-669a13, esp. a9-13. See also below, chapter 13, p. 302 n. 20.

[8] In some cases in my analysis of this passage it will be disputable what exactly should count as a feature of the elephant for purpose of scientific explanation, and thus what term exactly we should designate by 'P_i'. The claims I make about the *structure* of the explanation Aristotle offers do not, however, depend on how any such disputes are resolved, but only on it being the case that Aristotle thought that such disputes could be objectively resolved.

because it must spend much time in deep water (M_1), it must be able to reach air quickly (I count that as P_2, even though it is not entirely explicit in the text [but cf. a5]; P_2, then, follows from P_1 and M_1). Yet because of its great size (M_4, a7), it cannot do that in the standard way, by moving quickly out of the water onto land (P_3: cannot move quickly from deep water to land, a5–7; this follows from M_4 and certain basic facts about movement in general regardless of whether the moving thing is living or not. I label premises that derive from the material nature of the animal, $h_1, \ldots, h_n$).[9] As a consequence of that privative feature, P_3, together with P_2 (its need to reach air quickly), and material facts which entail what dispositional properties various structures of various materials will have, the elephant must have, or is best served by having, or is well-served by having, a breathing-instrument long enough to reach above the surface (P_4, a9–15). Such an instrument is by definition a nose, and so the elephant has its distinctive long nose, its trunk. (More about the teleological principles that license these explanations in a moment.)

But that extraordinary length would interfere with the elephant's activities on land unless the trunk were made of soft and pliable material, and so it is made of such material (P_5, a16–21, from P_4 and various material principles). Were Aristotle to go on to the constitution of such material, as he does of uniform parts elsewhere, he would probably point out that such a soft pliable material would have to be both earthy and watery, given certain necessary facts about the material dispositions that various material natures in combination give rise to. Such an explanation of the material elements that ultimately constitute a trunk would make use as premises of certain facts about the necessary natures of materials, facts developed quite systematically in *Meteor.* IV and elsewhere.[10]

So, four different basic features of the elephant, M_1–M_4, are at work in the explanation of the extraordinary length and character of its nose. But Aristotle is not finished yet: the elephant's nose has not only a distinctive length but also a distinctive capacity to function as a strong grasping and moving instrument. At 659a23–25, Aristotle notes that in four-footed animals (M_5) that are many-toed (many-'digited', *poludaktula*), the front feet are typically there not only for the sake of supporting their weight but also 'in place of hands', i.e. for grasping and moving objects, such as food. I am not sure if being many-toed is *explained by* that fact, even if it facilitates it; if not, it probably counts as M_6. In any case, the elephant cannot use its front feet that way: because of its great size (M_4) and weight (which I take to follow directly from its size and the kind of materials needed to constitute a four-footed live-bearing animal of such size, for which reason I label its great weight P_6), its feet have to be devoted fully to the task of supporting that weight (which privative property we could label P_7). (Incidentally, as I have just suggested, the

[9] 'h' from *hulē*, since 'M' is already taken for 'middle'.

[10] See Gill 1997, 145ff. and Lennox 1997a, 163ff.

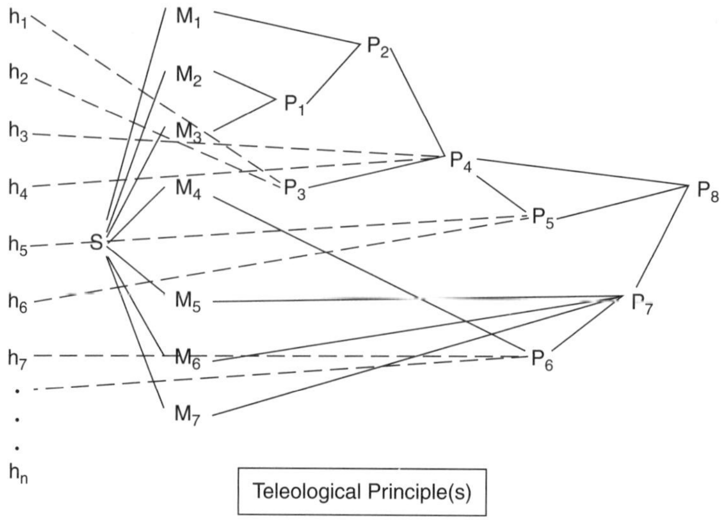

Figure 2

term 'four-footed' at 659a23-24 may in fact be short for 'four-footed live-bearing',[11] as is often the case when the context is live-bearing animals alone, and insofar as 'live-bearing' plays any role in the explanation, which is not entirely clear, it would be my seventh basic feature, M_7.) So the front feet cannot be put to that very important use, but a soft and pliable breathing instrument of the kind that can be generated in an elephant *can* be so put, with no change or minimal change in its constitution, and so the elephant's nature puts it to that additional use (*parakatachrētai*, a21), as natures often do—which is another version of a teleological principle. I label the trunk's use for grasping and moving P_8—the final feature of the elephant's 'nose' discussed, and explained, in II.16.

So, the diagram looks rather like the one given in Figure 2, and not like the simple one given above in Figure 1.

As this diagram suggests, we have (i) material principles drawn from a prior study of the elemental powers, the four elements they constitute, and the uniform combinations they form. We have (ii) the postulation of the existence of an elephant kind ('S') with a distinctive formal nature. And we have (iii) the postulation of the existence of certain features of the elephant as basic—multiple features. From these we establish the existence of other features of the elephant as either necessary or better for it. While the diagrams required to exhibit the comparable structure of the explanations of other features of the elephant (or of some other

[11] I translate *zōotokos* as 'live-bearing', to capture in English the not-yet-technical feel of Aristotle's Greek.

animal kind) may be less (or more) complicated than this one, they will not be significantly different. The expression I have used, 'linear, branching structure', is perhaps not a very good one for this sort of structure. Although it is linear, in the sense of one-directional, it is not clear that the number of final features (P_is) on the right-hand side must be significantly greater than the number of primitive features (M_is) on the left. Of course there will be many more features that are explained than that do the explaining, but the bulk of the features may be intermediate ones, so that the figure does not strictly speaking branch out. In addition, we must not forget the material principles, and we must not forget one more thing—the role of the teleological principle or principles involved. I have claimed that there are at least two sorts of teleological principle involved, one which asserts that formal natures are ends and one which asserts that those natures operate for the best. Since the first of these principles is pervasive and the second frequent, and since they do not affect the layout of the *terms* of the explanation, but only the manner of their *connection*, I place them in a horizontal box at the bottom, to indicate their role in underwriting the explanations, without indication as to how exactly they do that.

I said in the previous chapter that explanations which move from an animal's essential nature to what is necessary for the animal presuppose the postulation of these formal natures *as ends*,[12] since the necessary features of the animal are necessitated conditionally on the formal nature's status as end, and not simply.[13] The move to what is better (sc. than any other available option) requires, I have suggested, a stronger teleological principle than that, to the effect that a nature always produces what is best from among what is possible to the being of the kind of animal in question (*IA* 704b15-17, *PA* 687a15-16). But if one adds this premise to the already existing premises that the formal nature in question exists and that some feature would indeed be best for it from among all those possible (given the nature and materials available), then it would appear that this feature too is present of (conditional) necessity. But I have been cautioned by James Lennox that Aristotle seems at 640a35 to say precisely that these features do *not* belong of necessity to their possessors, and by David Charles that, as is the case on one standard model of practical reasoning, Aristotle may mean no more than that this is *one good* way to do it. In part my answer to Lennox is that these features are not necessitated by the formal nature or essence *alone* (which is the context of 640a33ff.); this does not rule out their being necessitated by the formal nature of essence *together with an additional principle*. As for Charles's hypothesis, none of the relevant passages in *PA* II–IV *clearly* supports it: 'better' often seems to mean 'better than any other option' (cf. the equation of 'better' and 'best', in connection with the general statement of the stronger principle, at *IA* 704b17), and 'useful' often

[12] Above, pp. 173-74.

[13] That is one difference between natural and mathematical demonstrations. Cf. *PA* I.1 639b21-640a9 and the Additional Note to chapter 7, pp. 182-84.

equates with *heneka tou* and does not as such imply options.[14] So I am still inclined to think that the result does follow of necessity from the teleological principle at work. In any case, teleological principles of both sorts are certainly at work, and count as first principles of Aristotelian natural science.[15]

Now, we must note that the entire diagram we produced, just as the whole of II.16, is devoted to only *one* distinctive feature of the elephant. There will be some such diagram, of greater or lesser complexity, for *each* of the attributes, the *sumbebēkota kath' hauta*, of elephants recognized in *PA*. Each such explanation will start from one or more of the basic features of elephants, which include M_1–M_7 and possibly some others. So the diagram for features other than the trunk will to some extent overlap the one for the trunk, starting on the left with some of the same features—and in some cases going even part of the way along, as would be the case if some other feature, say some property of the elephant's lung, follows from the fact that it must breathe (P_1 in our diagram).

And to complicate the matter yet further, let us remember that the elephant is only one of many indivisible animal kinds, so that the elephant kind must be labeled not just S but S_i, and that there will be for the whole of *PA* II–IV as many S_is, and as many such diagrams, as there are indivisible animal kinds recognized in those books.

So, the structure of explanation differs from the simple one-starting-point model (Fig. 1) that arises from thinking only of one kind, and from thinking of there being only one feature of that kind which is essential, and from thinking that each of its attributes follows from the essential attribute alone. The actual structure differs from that simple one in including (i) basic premises about material natures, (ii) multiple essential features, and (iii) teleological principles of various sorts. It differs in one other very important respect as well, which I have discussed in the previous

[14] At a minimum we must remember that Aristotelian natures are stellar craftsmen of the highest standards, not merely minimally rational agents (cf. 639b19)—but the many passages in which Aristotle appeals to what is better or useful (etc.) do indeed need closer study before this issue can be fully resolved.

[15] Geoffrey Lloyd has protested that such teleological principles are for Aristotle only regulative and not axiomatic (Lloyd 1990, 32). Now these principles are of course regulative of inquiry, but they also play a role in *explanation*, and Lloyd says nothing about that role. As for their status as first principles of explanation, Lloyd seems to expect that first principles must always have the initial self-evidence (to us) of the axiom of non-contradiction, or the equality axiom in mathematics, and he contrasts the elaborate series of arguments that Aristotle gives for teleology in *Phys*. II.8 (and the character of the opponents' denial) with the elenctic demonstration in *Metaph*. *Γ*.4. But non-contradiction must be known before *anything* can be known (or understood). The teleological principles are first principles of the operation only of *formal natures*, and neither the existence of these natures nor their first principles need be self-evident 'to us'. The arguments in *Phys*. II.8 (and those across book II culminating in the arguments of chs. 8-9) are clearly dialectical (or in any case move *from* the *phainomena*). They thus qualify as arguments to first principles. On the role in explanations of the 'nature does nothing without a point' principle, see also Lennox 1997b.

chapter and elsewhere, but warrants more attention, and which wreaks havoc on the attempt to diagram the structure. It is the fact that Aristotle insists that (iv) explanation be at *the proper level of generality*, which is often well above that of the subject kinds he counts as the *ousiai* of the world, such as elephant.[16]

This requirement on explanation, that it be at the right level of generality, is obviously central to Aristotle. The very first issue of method in biological explanation that he addresses in *PA* I.1 (at 639a15, presented there as an *aporia*) is: should one 'take each *ousia* singly and clarify its nature independently, making individual studies of, say, man or lion or ox and so on, or should one first posit the attributes common to all in respect of something common'? The solution, developed across *PA* I.4-5, is to opt for the latter, and indeed the explanations across *PA* II–IV are mostly given at the widest level of generality.[17] In many cases, however, there are significant differences within a generic feature. So Aristotle outlines in *PA* I.5 a related pattern of explanation, which involves moving from those explanations of the generic feature to explanations of differences in those features, and this pattern is also common in *PA* II–IV.[18]

[16] In speaking of what we often call 'species' as the *ousiai* of the world, I do not mean to prejudice the question of the existence of particular or individual forms and/or essences, nor to suggest that species and *not* their particular members are the *prōtai ousiai* of the world. Nor, I think, does Aristotle. For discussion of how this language in *PA* I could be compatible with a metaphysics that treats particulars as *prōtai ousiai* ('primary substances'), see Balme 1987c and Lennox 1987b, and see Balme's revised note to *PA* I.4 644a24 in Balme 1992 ad loc.

[17] For discussion of the *aporia* and its solution, and for examples of the *PA* practice, see my chapter 9, below. For further examples from *PA*, see Lennox 1987a. Lennox has also observed that it is a requirement of *APo*. I.4-5, II.14, and *passim* that explanation be at the level of the widest class that possesses the feature—the primitive universal (whether that class is an authentic kind or merely the class of all the animals that have some other feature(s) in common). On *APo*. II.14 see also Balme's seminal Second Symposium Aristotelicum paper in its revised form, Balme 1987b.

[18] I discuss this pattern of explanation at some length in chapter 9. In a bit more detail than I have given in the text above, the pattern goes like this: for each feature, identify first the widest class of animals that possess it. Then identify the variations (or differences, *diaphorai*) in which it comes. Then identify the widest class of animals that possess each of the differences. Then explain why all animals that have the feature at all do have it. Finally, explain the variations in the feature by reference, wherever possible, to variations in the cause of the feature. In the simplest case we deal with kinds and their subkinds. In those cases we explain the *differences* in some generic attribute (e.g. feathers) across subkinds of the large kind (e.g. Bird) by reference to the *differences*, across those subkinds, in the features which explain the presence of that generic attribute in the large kind.

This pattern is not spelled out in *PA* I.4, but on one reading of the first sentence in ch. 5 that follows the inserted exhortation to biological study, it is spelled out right there. At 645a36-b3 Aristotle writes: *ἀναγκαίον δὲ πρώτον τά συμβεβηκότα διελεῖν περὶ ἑκαστον γένος, ὅσα καθ' αὐτὰ πάσιν ὑπάρχει τοῖς ζώοις, μετὰ δὲ ταύτα τὰς αἰτίας αὐτών πειράσθαι διελεῖν*, which I translate: 'It is necessary first to divide the attributes pertaining to each kind which are present in all the animals <of that kind> in virtue of themselves, and after that to try to divide their causes.' This injunction is immediately followed with the observation that animal features can be the same not only by being the same in kind but differing in the more and the less, which is the

For example, observe how Aristotle proceeds in his explanation of eye-coverings in II.13. He first (657a25) gives the widest class of animals that have eye-coverings; then briefly divides the eye-coverings, identifying the different types and indicating the widest class of animals that possess each type (a27); then gives the general cause for the presence of eye-coverings (a31); then divides the causes, showing how differences in the kind of protection needed or in the materials available to form a protective covering explain the differences in the eye covering found.[19] Likewise, in III.14 Aristotle explains the presence of stomachs in all animals

standard way important features within a broad animal kind are related, but also by being the same by analogy, because they have, even if not a similar material constitution and structure, still the same *dunamis* (645b9). Aristotle's point seems to be that the method of 'first divide the attributes, then try to divide their causes' can and should be extended to predicate unities even wider than those that differ only in the more and the less—just as, by the way, *APo.* II.14 enjoins (cf. 98a20-23). For an example of analogical unity in action in *PA* II–IV, see the next note.

The evident match between explanatory theory and practice, with regard to this pattern of explanation in general, and the scientific acceptability of analogical unity in particular, should at least raise some doubts about Geoffrey Lloyd's oft-voiced concern that the 'univocity' requirement of *Analytics* is often not met in the biological corpus (and that Aristotle is probably aware of this). (See, e.g., Lloyd 1987b, ch. 4 and Lloyd 1990, 29-31.) Two of Lloyd's examples, the difference between two-leggedness in birds and man and the difference in the blood of animals different in kind where blood is part of their essence, are clear cases of differing 'in the more-and-the-less' and thus involve a sameness in kind paradigmatic of one type of *Analytics* univocity. (For the examples and some discussion, see Balme 1992, 110.) A third example, heat, is a trickier one, since Aristotle indeed does not appear to have come up with an essential characterization of heat that will do everything he wants it to in biology. But it does not follow from this, nor is there any good reason to think, that he could not ever, with his existing apparatus of types of unity, do so. Nor, *a fortiori*, is there any good reason to think that Aristotle *thought* he could not ever, nor had any interest in whether he could ever, come up with such an essential characterization of heat. And even without a satisfactory essential definition of heat, Aristotle is able, in biological explanation throughout the corpus (especially, but not only, in *PA* II.4-9 and *GA* II.2 and 6), to draw on general propositions established in *Meteor.* IV as to what heat will do to various kinds of materials. Lloyd's discussion of a fourth example, concoction, also treats evidence that the science is unfinished as evidence that it cannot be finished, in a manner consistent with the full explanatory theory of the *Analytics*. (For an extended argument that Aristotle's theory of the nature and biological roles of heat, and his conception of the nature of concoction, are both more developed than has been thought, see Freudenthal 1995.)

A different question of compatibility between the *Analytics* theory, on the one hand, and both the claim of *PA* I.3 that biological definitions require multiple differentiae and the *PA* II–IV practice of indeed explaining from multiple differentiae (illustrated above in the case of the elephant), on the other, is discussed in Charles 1997. I hope to address this very interesting issue on another occasion. (For an approach to Charles's problem that I do not think succeeds, see Lennox 2010a.)

[19] 'Eye-covering' appears to be a good example of an analogical unity, with a role in explanation such as is described at the end of the previous note. For Aristotle unites under that heading eyelids, nictating membranes, and hard outer eyes. I will make use of that fact in my discussion below of the way in which the demand that explanation be at the proper level of generality makes a proper 'diagram' of the overall structure of explanation significantly more complex than that given in Fig. 2 above.

from two basic facts about animals;[20] then he explains differences of stomachs by reference to differences in the general cause of stomachs as such.[21]

How, then, are such explanations linked to the *ousia*, the indivisible kinds such as man, lion, ox, elephant? Aristotle does not spell that out in *PA* I.5, but his practice, and the theory in the *Analytics*, makes clear that the answer is: by a type of covering explanation which James Lennox has called 'A-type explanations'[22] and which Richard McKirahan calls 'application arguments'.[23] All Ks must have (or go better by having) P; Ss are Ks; Ss must have (or go better by having) P.

But what does explanation at the generic level do to our diagram? Notice that first we have to add to the elephant all the indivisible kinds there are, creating a long vertical sequence of S_is each of which has its own very complicated structure, with its various P_is. And then we have to move to a *third dimension* in which we have features $R_1, \ldots, R_n$, which are various generic and analogical unities[24] of the P_is, and these have to be put on various levels in that dimension. For example, there is an R_i, eyelid, which has various forms of eyelids at the first level (P_es), and is itself united with R_j, nictating membrane (itself, say, having various types), and R_k, hard outer eye, into a higher level unity, R_m, eye-covering. Then, for each of those features at each level in the third dimension, there will be one or (usually) more other features, perhaps each equally a generic or analogical unity, from which the former feature is ultimately explained. These basic explanatory features will each be a generic part of the essential nature of each indivisible kind of animal which has the feature being explained (as, e.g., being living and an animal is part of the essential nature of each stomach possessor, including the elephant).[25] I am at a loss

[20] All animals (in fact all living things) must have food to be and to grow, and all animals must concoct their food inside themselves. (These facts were laid down much earlier, at the beginning of II.10, as among the first and most necessary facts about animals.) Given these facts, and the material nature of the process, concoction requires a receptacle (674a13-15). For a much fuller discussion of this explanation, and for exploration of various issues connected with the theme of this chapter, see (in conjunction with pp. 168-69 of the previous chapter) Detel 1997.

[21] Thus, different numbers of stomachs (one or four) are explained by differences in the way concoction must go in the stomach, depending on whether the animals in question have *two* rows of teeth to aid concoction or only one. The explanation here is more complicated than in the previous example, involving the demands on limited materials of several functions (e.g., defense), but the pattern under discussion remains a central component of the full explanation. See here too the fuller discussions referred to in the previous note.

[22] Lennox 1987a, 92-99, 109-14. Lennox gives examples from *PA*; my only caveat regarding his discussion is that (as Lennox now agrees) those count as explanations only *after* the explanation at the proper level has been established; before that, the pattern exemplifies rather the move, prior to starting explanation of a feature, to the widest class of possessors of that feature, in order to have the explanandum at the proper level of generality, and is not as such a type of explanation. See now Lennox 1991, 273-74 with n. 16, where this point is taken up.

[23] McKirahan 1992.

[24] See nn. 19 and 20 above.

[25] See the previous chapter, p. 178.

as to how one could put all of this into a diagram of any sort. If one could, or could produce a three-dimensional physical model of the structure I have described, one would come a good deal closer to capturing that structure than does the diagram in Figure 2. Nevertheless, that diagram, together with the accompanying explanation, may give a fuller indication of what I have in mind, and had outlined in my earlier papers. And the diagram shows how hopelessly inadequate the first, simple model, and even the verbal image of a 'linear branching structure', is for capturing the three-dimensional structure actually there in *PA* II–IV. But it also shows how incredibly rich the logical structure of explanation is in Aristotle's biology, and how worth systematic study, from a philosophical point of view, the biological treatises are.[26]

[26] Earlier versions of this chapter were presented in 1994 in graduate courses I taught at Oxford University (jointly with David Charles and James Lennox) and Tokyo Metropolitan University (see above, Preface, p. x). I would like to thank all those who participated in discussion on those occasions and at Bad Homburg, especially David Charles, Jim Lennox, Shinro Kato, Kei Chiba, Alan Code, Wolfgang Detel, and Wolfgang Kullmann (to whom gratitude is owed also for the organization of a very successful conference).

9

Division and Explanation in Aristotle's *Parts of Animals*

The nature and proper methods of logical division, *diairesis*, are important concerns in at least four of Plato's later dialogues, and reflections on the process may even have led Plato to a revised conception of the very nature of a Form. The place of division in Aristotle's *Topics*—both in the organization of some of its materials and as the method for seeking definitions which many of the *topoi* are designed to regulate—suggests the importance *diairesis* must have had in the Academy,[1] as does the Epicrates fragment and Aristotle's criticism of alternative views of the nature or purpose of division in *Prior Analytics* I.31, *Posterior Analytics* II.5 and II.13, and *Parts of Animals* I.2-3.

But, as David Balme has argued,[2] Aristotle criticizes these alternative views not in order to reject division but to *reform* it. When he criticizes those who think the production of a definition by division constitutes a deduction, or even demonstration, of that definition, he reminds us that division is nonetheless *useful*, in the establishment of definitions, and appears to devote most of a long chapter (*APo.* II.13) to the central role of division in 'hunt[ing] out what is predicated in what a thing is' (96a22), which *may* be a reference to the definitions that feature as first principles in demonstrations.[3] And when Aristotle rejects dichotomous division as a way of defining *eidē* in *PA* I.2-3, he replaces it with division by multiple

This chapter was first written for presentation at a meeting of the Society for Ancient Greek Philosophy in San Francisco in 1991, and was revised for inclusion in the 1997 Festschrift for Wolfgang Kullmann (Günther and Rengakos 1997). The questions of the significance of the maxim the chapter discusses—'first divide the attributes, then try to divide the causes'—and of how widely it is applied in *PA* had been on my mind at least since the mid-1980s (not long after Jim Lennox gradually convinced me of the importance of division in Aristotle). These questions, especially the first, were the subject of a session I led at the 1988 NEH Summer Institute on Aristotle's Metaphysics, Biology, and Ethics which I directed with Michael Frede and John Cooper. See further n. 28 below.

[1] Cf. Cherniss 1944 and Tarán 1981.

[2] Balme 1987b; cf. Pellegrin 1986, ch. 1.

[3] Useful: *APr.* I.31 46a31-32, *APo.* II.13 96b25-26 (cf. II.5 91b28-32). In fact, it is 'only' (96b35) with division that one can ensure completeness. Alternatively, II.13 may be concerned with pre-causal definitions, as is suggested by Ross (ad loc.) and argued for in Bolton 1993.

differentiae simultaneously, and argues that that procedure successfully avoids the criticisms dichotomous division falls prey to. Finally, in *Metaphysics* Z.12, he indicates how one of his reforms of division—the rule that one must divide 'by the differentia of a differentia',[4] e.g., *footed* by *split-footed*[5] and not by *feathered*—allows at least a partial resolution of the problem of the unity of a definition.

Balme has even argued—controversially but not entirely implausibly—that definitions by division are involved in the solution to be found in *Metaph. H*.6 of the unity problem for the *object* of definition. And Pierre Pellegrin has insisted that the very concepts of *genos, eidos* and *diaphora* which feature so centrally in the discussion of definitions and much else throughout the corpus derive their precise logical relationship to each other from their place in the divisional process. These interpretative claims have been challenged, and alternative readings of some of the passages I've mentioned have been given, but these controversies, currently unresolved as they are, only underscore the fact that the topic of the place of division in Aristotle's logic, metaphysics, and biology is an important one, and warrants more study.

The biological case is perhaps the most perplexing. For the critique of dichotomous division in *PA* I.2-3 ends, as I've said, with the firm endorsement of a method argued to *work* where dichotomous division does not. In seeking a definition of an *atomon eidos* (such as sparrow or crane), Aristotle instructs us to divide, not the *genos* Animal by one differentia at a time, but a commonly recognized *genos* of animals (in this case Bird), marked out by multiple differentiae, and then divide these differentiae simultaneously, each as many times as necessary to reach the definition sought. In *PA* I.4 he even goes on to explain how to select the *genē* from which to start the process.

Thus the method, and now to work, one might think. But, as Balme put it in 1976: 'The puzzle is, what happened to *diaeresis* after that? Aristotle makes no further use of it in biology; and no *genos* or *eidos* is actually defined.' Or, to put the question more neutrally, *does* division have any role in the biology and, if so, is that role connected in any way with the reforms of division presented in *PA* I.2-4, and more generally, with the discussions of division in the logical and metaphysical writings?

Now, I don't pretend to have the answers to these larger questions. I don't think anyone does, yet. But work done since 1976, especially by Balme, Pellegrin, and James Lennox, has moved us closer to solving the puzzle, and I would like, in this small chapter, to add one more piece to that eventual solution.

[4] 1038a9-10; cf. *PA* I.3 643b17, and Balme 1972, 102-103, 117 and 1987b, 69, 73.

[5] 'For split-footedness is a certain sort (*tis*) of footedness' (1038a15).

Some recent work

In a recent essay on 'The Reforms of *Diaeresis*' (part I of Balme 1987b), Balme suggested that the reforms summarized in *PA* I.2-4, with their focus on sameness, difference, and subdivision of differentiae, require a precise grasp of the differentiae that might appear in a divisional definition of animal *eidē* prior to the production of those definitions, and that the *Historia Animalium* (chronologically the last of the biological treatises[6]) was conceived as the vehicle for achieving that grasp. *HA* would thus not itself contain definitions by division, but would make such definitions possible. Aristotle never reached the stage of providing such definitions, and 'we may indeed doubt whether [he] maintained this aim when his experience as a naturalist increased', since he would have realized that 'there is no end to the recognition of fresh significant differentiae'. But the 'unfinishable' character of the collection and analysis of differentiae would not preclude the use of 'the correct identification and grouping' of differentiae collected, which would '[lead] directly to the fundamental causes of animals' attributes and differences. This might well have seemed the best way towards a methodical *apodeixis* of living nature' (80).

But if Balme is right, it would seem that the reforms of division have now become useless—*unless* division so reformed plays a role in the search for and/or organization of scientific *explanations*. In two recent papers, James Lennox has argued that division indeed plays such a role, specifically in the *pre-explanatory* stage of Aristotelian science. In a paper significantly titled 'Divide and Explain' (1987a), Lennox, as he himself summarized:

> develops Balme's suggestion that *APo*. II.14 is a key text in solving the puzzle of the relationship between division and demonstration in Aristotle's scientific method. Central to the theory of the *APo*. is the articulation of a distinction between incidental and unqualified understanding. Unqualified understanding requires locating the widest kinds to which differentiae belong universally: this allows explanations of why subkinds of these widest kinds have the differentiae in question, and identifies those features which are either essential to the kind, or explicable as consequences of its essence.[7]

To use an example from *APo*. I.5, isosceles triangles have an angle sum equal to two right angles because they are triangles; we understand the former connection fully only when we understand why triangles *as such* have that attribute. Thus *APo*. II.14 instructs one to seek, for any differentia to be explained, the widest class to which it belongs, and to do so by 'selecting from . . . the divisions' (98a1-2). This procedure, as Lennox puts it, 'assumes a set of divisions of a subject domain ready at hand, and divisions, the examples suggest, organized along Aristotelian lines' (98). Those divisions will be the source both of the predicates of the propositions to be

[6] Cf. Introduction to Balme 1991.

[7] Gotthelf and Lennox 1987, Introduction to Part II, 66-67.

demonstrated and of their subjects, as we look up and down the relevant divisions for the widest class to which each predicate applies universally. Division thus helps 'direct, or redirect, inquiry to the appropriate level for... explanations of... primitive predications' (99). Division as such neither supplies the true propositions nor identifies the actual causes—further empirical inquiry of specific sorts is required for that—but it organizes the former and facilitates the latter. Research in *HA* which Lennox and I did in collaboration, he goes on to argue, shows that this 'downward division/correlation methodology' is much 'like what we find in' that treatise (99).[8]

In a subsequent paper,[9] Lennox connects both the prescription of *APo.* II.14 and the practice of *HA* with the account in *APr.* I.27-31 of 'how we may ourselves always be supplied with deductions about what is set up, and the route by which we may obtain the principles concerning any particular subject'.[10] Although, as Aristotle puts it, 'division by means of kinds is only a small part of the procedure that has been described' (45a31-32), Lennox is surely right that it is an indispensable part.

Division thus plays an important role in the pre-explanatory stage of science. But what about the explanatory stage itself? In his 1987 paper Lennox suggests that division is involved in explanation of why sub-kinds of a kind that possess a feature primitively also possess that feature, but as he notes this is at best 'partial demonstration'.[11] What of unqualified demonstration, proper explanation? Is division involved there? That is our question.

Dividing the causes

Once we have the connections to be explained at the proper level of generality, and are ready to search for the causes of these connections, is division of any help? There is at least one passage in *PA* I which suggests that it might be:

ἀναγκαῖον δὲ πρῶτον τὰ συμβεβηκότα διελεῖν περὶ ἕκαστον γένος, ὅσα καθ' αὑτὰ πᾶσιν ὑπάρχει τοῖς ζῴοις, μετὰ δὲ ταῦτα τὰς αἰτίας αὐτῶν πειρᾶσθαι διελεῖν.[12]

It is necessary first to divide the attributes pertaining to each kind which are present in all the animals <of that kind> in virtue of themselves, and after that to try to divide their causes.

[8] For discussion of the aims of *HA* see, in addition to Lennox 1987a: Balme 1991, Lennox 1991, and chapters 12 and 14 below.

[9] Lennox 1991.

[10] 43a20-22, tr. Smith.

[11] Demonstration *kata meros*. For references, and a discussion of my reservations about Lennox's view of this, see Lennox 1991, 273-79 with nn. (Lennox 2001b, 48-53 with nn.).

[12] 645a36-b3. Here and throughout this chapter, I use Bekker's text of *PA*, unless otherwise noted, and Balme's 2002 text of *HA*.

Now, I have translated *dielein* as 'divide', but the word also could be translated 'distinguish', in the sense of 'set out separately'. Peck has 'describe', Louis has 'analyser' and 'discerner'; Balme, on the other hand, has 'divide off', and Pellegrin, in his discussion of this passage,[13] insists on 'divide'. If 'divide' is not translated, this precept ('the *PA* I.5 precept', as I'll call it) would then be pretty much equivalent in content to the general methodological remark at *HA* I.6 491a9-11, with which it has some striking verbal similarities:

δι' ἀκριβείας δ' ὕστερον ἐροῦμεν, ἵνα πρῶτον τὰς ὑπαρχούσας διαφορὰς καὶ τὰ συμβεβηκότα πᾶσι λαβωμεν. μετὰ δὲ τοῦτο τὰς αἰτίας τούτων πειρατέον εὑρεῖν.

We will speak in detail later, in order that we may first grasp the differences and attributes which are present in all. After that we must try to find their causes.

I think, however, that the stronger reading is the more likely one, for two reasons. First, the precept appears as part of a discussion of an issue that spans three passages, one each in chs. 1, 4, and 5 of *PA* I; these passages together make a case for beginning explanation at a generic level and moving downward. Aristotelian division of differentiae, as we'll see, is the natural mechanism for that downward movement. Second, the organization of much of the explanation across *PA* II–IV is consistent with the precept, *and is illuminated by it*, if διελεῖν is translated as 'divide'. In what follows I'd like to develop these points, and make a case for the view that Aristotelian division does play a certain role in Aristotelian explanation.

Starting explanation at the generic level: *PA* I

The three *PA* I passages I have in mind are: ch.1, 639a15-b5; ch.4, 644a12-b15 (most of the chapter); and ch. 5 645b1-14 (which begins with our precept). These three passages hang together. The I.4 passage continues an *aporia* introduced in the I.1 passage; the I.5 passage resolves that *aporia*, referring in the process back to the two preceding passages. Let us look at them in order.

PA I.1 announces itself as concerned with the methodological principles (*horoi*, 639a13) that ought to be used in evaluating 'the method of demonstration' in inquiries about nature.[14] At first we don't actually get principles, we get questions, the proper answers to which will be the principles. I reproduce Balme's translation of the first question (with *eidos* however translated 'form' rather than 'species'):

[13] Pellegrin 1986, 43, 156.

[14] The verb *deiknumi* is general enough to permit the list to include principles that pertain to the organization of information that occurs at the *hoti* stage of inquiry, as well as principles at the *dihoti* stage, pertaining to the type of explanation that ought to be given. And the noun *historia*, used as it is by Aristotle either for the *hoti* stage alone or for the entire scientific enterprise from start to finish, positively suggests it.

Should one take each *being* (*ousia*) singly and clarify its nature independently, making individual studies of, say, man or lion or ox and so on, or should one first posit the attributes common to all in respect of something common? For many of the same attributes belong to many different kinds of animals, for example sleep, breathing, growth, wasting, death, and any other affections and conditions of this sort (for at present we are not in a position to speak of them with clarity and precision). If we do speak of the animals severally, it is plain that we shall often be saying the same things about many of them. For each of the above attributes belongs to both horses and dogs and men, so that if one refers to each of their attributes one will have to speak repeatedly about the same ones—all that are the same in different forms of animal which have no differentia themselves. On the other hand there are no doubt others which, although they have the same designation, differ by the formal differentia. Animal locomotion, for example, is evidently not one in form, for there are differences between flying, swimming, walking, and creeping. Therefore we must not overlook the question how the examination should be made, that is whether one should first survey common general attributes and then later the peculiar ones, or take them individually straight away. At present this has not been clarified.[15]

It has often been pointed out that the first side of the *aporia*—the case for treating common attributes across various kinds—cannot merely be recommending brevity.[16] The kind of repetition Aristotle complains of is just the sort discussed in *APo.* I.5, as a case of failing to have grasped the *actual* cause, and to that extent failing to give proper demonstration: to give separate proofs, for instance, that isosceles, equilateral, and scalene triangles, respectively, have an angle sum equal to two right angles is not to see that it is in virtue of the features which make them *triangles*, and not of the equality or non-equality of their sides, that they have that angle sum. But Aristotle does not make the connection with the *APo.* theory explicit. Nor does he develop clearly the argument on the other side, for starting with the *ousia*; the one consideration he brings to bear, that some common attributes come in different forms, pushes us to lower levels of generality, but not necessarily all the way to

[15] (639a15-b6). The Greek text is: *λέγω δ' οἷον πότερον δεῖ λαμβάνοντας μίαν ἑκάστην οὐσίαν περὶ ταύτης διορίζειν καθ' αὑτήν, οἷον περὶ ἀνθρώπου φύσεως ἢ λέοντος ἢ βοὸς ἢ καί τινος ἄλλου καθ' ἕκαστον προχειριζομένους, ἢ τὰ κοινῇ συμβεβηκότα πᾶσι κατά τι κοινὸν ὑποθεμένους. πολλὰ γὰρ ὑπάρχει ταὐτὰ πολλοῖς γένεσιν ἑτέροις οὖσιν ἀλλήλων, οἷον ὕπνος, ἀναπνοή, αὔξησις, φθίσις, θάνατος, καὶ πρὸς τούτοις ὅσα τοιαῦτα τῶν λειπομένων παθῶν τε καὶ διαθέσεων· ἄδηλον γὰρ καὶ ἀδιόριστόν ἐστι λέγειν νῦν περὶ τούτων. φανερὸν δ' ὅτι καὶ κατὰ μέρος μὲν λέγοντες περὶ πολλῶν ἐροῦμεν πολλάκις ταὐτά·καὶ γὰρ ἵπποις καὶ κυσὶ καὶ ἀνθρώποις ὑπάρχει τῶν εἰρημένων ἕκαστον, ὥστε ἐὰν καθ' ἕκαστον τῶν συμβεβηκότων λέγῃ τις, πολλάκις ἀναγκασθήσεται περὶ τῶν αὐτῶν λέγειν, ὅσα ταὐτὰ μὲν ὑπάρχει τοῖς εἴδει διαφέρουσι τῶν ζῴων, αὐτὰ δὲ μηδεμίαν ἔχει διαφοράν. ἕτερα δὲ ἴσως ἐστὶν οἷς συμβαίνει τὴν μὲν κατηγορίαν ἔχειν τὴν αὐτήν, διαφέρειν δὲ τῇ κατ' εἶδος διαφορᾷ, οἷον ἡ τῶν ζῴων πορεία· οὐ γὰρ φαίνεται μία τῷ εἴδει· διαφέρει γὰρ πτῆσις καὶ νεῦσις καὶ βάδισις καὶ ἕρψις. διὸ δεῖ μὴ διαλεληθέναι πῶς ἐπισκεπτέον, λέγω δὲ πότερον κοινῇ κατὰ γένος πρῶτον, εἶτα ὕστερον περὶ τῶν ἰδίων θεωρητέον, ἢ καθ' ἕκαστον εὐθύς. νῦν γὰρ οὐ διώρισται περὶ αὐτοῦ . . .*

[16] e.g. Balme 1972, Kullmann 1974, Lennox 1987a. For an interesting discussion of the historical background to the question, see Kullmann 1974, 12.

ousia. Interestingly, in I.1 the *aporia* is left unresolved, apparently because he hasn't yet presented the resources needed for dealing with it.

In I.4 the *aporia* is taken up again. Here the case for starting not with common attributes of the broad kinds (such as Bird), but straightaway with the attributes of the 'indivisible forms' (such as sparrow and crane), is given a metaphysical basis: doing it that way, we get our subjects as close as scientifically possible to the actual *ousiai*—the metaphysically *proper* subjects (644a29-33[17]). Again, the counter-argument is repetition, now not only its length but its 'virtual absurdity' (*hupatopon*), which may be an allusion to explanatory inadequacy. Here he proposes a solution:

Perhaps then the right course is to speak of some affections in common by kinds (*genē*), whenever the kinds have been satisfactorily marked off by popular usage and possess both a single nature in common and forms not far separated in them—bird and fish and any other that is unnamed but like the kind embraces the forms that are in it; but whenever <the affections> are not like this [i.e. common,] to speak of particulars, for example about man or any other such.[18]

The presentation of the *aporia* in I.1 ended with a question. Here, the alternative of starting with attributes possessed in common across *genē* is endorsed, although tentatively (*isōs*, b1), and an apparatus for dealing with the I.1 objection to going this route is introduced. The I.1 objection was that some attributes which are possessed in common exist in the animals in different forms. When Aristotle, at the opening of chapter 4, explained the notion of common attributes that differ in the more and the less, he introduced a way of dealing with at least a set of cases that meet that description without abandoning the virtues of starting at the generic level.

For all kinds that differ by degree and by *the more and the less* have been linked under one kind, while all that are analogous have been separated. I mean, for example, that bird differs from bird by *the more* or by degree (one is long-feathered, another is short-feathered), but fishes differ from bird by analogy (what is feather in one is scale in the other).[19]

[17] For Balme's revision of his 1972 reading, translation and note for this passage, see Longrigg 1977. The revised translation and note have been incorporated into the more recent edition (Balme 1992).

[18] 644b1-7. Readers of the Loeb *PA* should note that Peck rewrites the text of this passage. Balme translates the MSS and I have followed him. I take ὅσα δὲ μὴ τοιαῦτα (b6) to modify the same subject the coordinate τὰ μὲν κατὰ γένη κοινῇ (b2) modifies, namely the *attributes*, referred to as ὑπάρχοντα at a26 and as πάθη at a34, and so interpolate as I do; but it *could* be modifying *genē*. The entire passage reads: ἴσως μὲν οὖν ὀρθῶς ἔχει τὰ μὲν κατὰ γένη κοινῇ λέγειν, ὅσα λέγεται καλῶς ὡρισμένων τῶν ἀνθρώπων, καὶ ἔχει τε μίαν φύσιν κοινὴν καὶ εἴδη ἐν αὐτοῖς μὴ πολὺ διεστῶτα, ὄρνις καὶ ἰχθῦς, καὶ εἴ τι ἄλλο ἐστὶν ἀνώνυμον μέν, τῷ γένει δ' ὁμοίως περιέχει τὰ ἐν αὐτῷ εἴδη· ὅσα δὲ μὴ τοιαῦτα, καθ' ἕκαστον, οἷον περὶ ἀνθρώπου καὶ εἴ τι τοιοῦτον ἕτερόν ἐστιν.

[19] 644a16-22 ὅσα μὲν γὰρ διαφέρει τῶν γενῶν καθ' ὑπεροχὴν καὶ τὸ μᾶλλον καὶ τὸ ἧττον, ταῦτα ὑπέζευκται ἑνὶ γένει, ὅσα δ' ἔχει τὸ ἀνάλογον, χωρίς· λέγω δ' οἷον ὄρνις ὄρνιθος διαφέρει

But now let us consider why starting at the generic level is a virtue in these cases: what, exactly, would be repeated if we only gave separate explanations for each variation of the common attribute? The only answer can be that the common aspect of the varying attributes is explained by the same thing in each case, a thing which would be said many times if it isn't said all at once for the entire kind. We could, for instance, explain why long-feathered birds have long feathers, and short-feathered birds short feathers. In both cases what would be repeated is the explanation of why they each have feathers. Feathers are something they have in virtue of being birds, not in virtue of being the sort of birds they are. To that extent, it is the same argument as in the case of the angle sum of the three types of triangles. But the difference is also instructive, and spells out a pattern of explanation Aristotle must have had in mind. In the triangle case, the feature in question, having an angle sum of two right angles, is possessed identically by the three types of triangles, and nothing is learned by moving down to that level. But, in the case of birds, while we must start at the generic level, something further is learned when one moves from understanding why all birds have feathers (say because they're needed to lift the bird's weight in flying) to understanding why e.g. eagles have long feathers but sparrows have short ones (perhaps because of their *particular* weights and the *type* of flying their *particular* feeding needs require).

So, the pattern I am suggesting is as follows: explain the *differences* in some generic attribute (e.g., feathers) across sub-kinds of a large kind (e.g., Bird) by reference to the *differences*, across these sub-kinds, in the features which explain the presence of that generic attribute in the large kind. Now, this pattern is not actually spelled out here, since the focus of the chapter is not on explanation. And in principle nothing prevents there being cases where this isn't so: certain variations in wings, for example, might just be used say for defense instead of for a distinct type of flying. But it would certainly be reasonable to expect our pattern to be the predominant one, and that, as we'll see, is most often the case across *PA* II–IV.[20]

τῷ μᾶλλον ἢ καθ' ὑπεροχήν (τὸ μὲν γὰρ μακρόπτερον τὸ δὲ βραχύπτερον), ἰχθύες δ' ὄρνιθος τῷ ἀνάλογον (ὃ γὰρ ἐκείνῳ πτερόν, θατέρῳ λεπίς).

Cf. *HA* I.1 486a15-487a10 for a fuller discussion of the sorts of samenesses and differences in animal features, and the sorts of 'more and less' relationships; the parallel examples for 'same by analogy' at 486b19-21 and 'differ by analogy' at 644a21-22 make clear that they are the same relationship (described from two different standpoints).

[20] I referred two sentences earlier to the focus or aims of *the PA* chapter under discussion, *PA* I.4. In fact, there seem to be two aims, and it would be useful to connect them, as I have not yet been able to do. One aim, which I noted at the start of this chapter (of the present book), is to explain how to select the kinds from which the reformed division just described in *PA* I.3 should proceed; the second is to resolve the *aporia* developed in *PA* I.1. The two aims are obviously connected, and since one bears on division and the other, in the end, on explanation, understanding the connection should contribute importantly to the wider aims of this chapter of the present book.

This discussion in *PA* I.4 is followed by the famous exhortation to biological study which begins in chapter 5, evidently inserted by some editor (it has no connecting particle). Our precept comes next but the sequence is not a logical one and the precept cannot have been written to follow the exhortation. The lines immediately following the precept, however, *are* directly connected with the precept, and tie it right back to the I.4 passage, referring to that passage twice as it does:

It is necessary first to divide the attributes pertaining to each kind which are present in all the animals <of that kind> in virtue of themselves, and after that to try to divide their causes. Now *we have said before* that many belong in common to many animals, some simply (for example feet, feathers, scales, and affections too in the same way), but others analogously. (By 'analogously' I mean that some have lungs while others have not lungs but something else instead which is to them what lungs are to the former; and some have blood while others have the analogous part that possesses the same capability [*dunamis*] as blood does for the blooded.) To speak separately about each particular will, *as we said before*, often result in repetition when we speak of every attribute: the same ones belong to many. Let this then be determined so.[21]

Here too we are instructed to start with common features and to move downward to more specific ones. There is one difference, which we will return to later, namely that analogy is now explicitly identified as a type of commonality—blood and its counterpart in the bloodless animals (or lungs and gills), although not common *haplōs, are* common by analogy, because they possess the same *dunamis* (at least at some level of generality). This does not contradict the spirit of the early discussions, although *they* were restricted to the paradigm case of simply common attributes, those possessed uniquely across a single *genos*; for we can either divide the common function directly (e.g., in the case of lungs and gills, divide cooling into *air-cooling, water-cooling*), or divide the analogous group of simply common features (e.g., *lung-or-gill*) into the individual *cases* (*lung, gill*), and then divide each of them down in the way required. But the addition of features common only by analogy will prove very important in understanding the discussions in *PA* II–IV, where sometimes different parts which perform the same function (e.g., eyelids,

[21] 645a36-b14; tr. Balme, with opening sentence modified and emphases added. The Greek reads: *ἀναγκαῖον δὲ πρῶτον τὰ συμβεβηκότα διελεῖν περὶ ἕκαστον γένος, ὅσα καθ' αὑτὰ πᾶσιν ὑπάρχει τοῖς ζῴοις, μετὰ δὲ ταῦτα τὰς αἰτίας αὐτῶν πειρᾶσθαι διελεῖν. εἴρηται μὲν οὖν καὶ πρότερον ὅτι πολλὰ κοινὰ πολλοῖς ὑπάρχει τῶν ζῴων, τὰ μὲν ἁπλῶς, οἷον πόδες πτερὰ λεπίδες, καὶ πάθη δὴ τὸν αὐτὸν τρόπον τούτοις, τὰ δ' ἀνάλογον. λέγω δ' ἀνάλογον, ὅτι τοῖς μὲν ὑπάρχει πλεύμων, τοῖς δὲ πλεύμων μὲν οὔ, ὃδὲ τοῖς ἔχουσι πλεύμονα, ἐκείνοις ἕτερον ἀντὶ τούτου· καὶ τοῖς μὲν αἷμα, τοῖς δὲ τὸ ἀνάλογον τὴν αὐτὴν ἔχον δύναμιν ἥνπερ τοῖς ἐναίμοις τὸ αἷμα. τὸ δὲ λέγειν χωρὶς περὶ ἑκάστων τῶν καθ' ἕκαστα, καὶ ἔμπροσθεν εἴπομεν ὅτι πολλάκις συμβήσεται ταὐτὰ λέγειν, ἐπειδὰν λέγωμεν περὶ πάντων τῶν ὑπαρχόντων· ὑπάρχει δὲ πολλοῖς ταὐτά. ταῦτα μὲν οὖν ταύτῃ διωρίσθω.*

nictating membranes, and an unusually hard eye-material) will be grouped and explained under one common heading: eye-protection (657a26).

This passage in I.5 may also remind one of the passage in *APo.* II.14 where Aristotle explains that in looking for the proper subject of propositions to be explained—the widest class possessing a given feature—that subject will sometimes be a *genos* (e.g., Bird), sometimes a group (such as the horned animals) that is not a *genos* but nonetheless shares common features, and sometimes a group which shares only analogous features (such as pounce [a rigid structure in some cephalopods], spine and bone); they should be treated together because the common function they all perform will either explain, or be explained by, other features they all possess in common.[22] In fact, that is exactly what is done in *PA* II, at 654a20ff., where bone, spine, pounce (and 'pen', a structure in certain other cephalopods) are identified as analogues of each other and explained by a unified cause, a common function they serve. Their unified cause is the need, in the bodies of the organisms which possess them, for a certain degree of straightness and inflexibility, which need is best served in their cases by a hard internal supporting organ. So, the inclusion of analogous unity as a kind of unity or commonality in I.5 significantly expands the range of cases that fall under our precept, and allows our precept more fully to satisfy the concern that generated the initial *aporia* in the first place—the concern that explanation be at the proper level of generality.

We began the survey of these three passages in *PA* I in the attempt to understand better the I.5 precept first to διελεῖν the essential attributes then try to διελεῖν their causes. On the reading of it that arises from considering it in connection with the related passages in I.1 and I.4, it says, in effect:

> One should start with larger divisible kinds and their common, divisible attributes. For each common attribute, state its (divisible) cause. Then divide the common attribute, identifying for each of its subdivisions the widest subclass (of the large kind) which possesses it. Then, wherever possible, seek the causes of these more specific attributes by dividing the cause of the common attribute. Since many divisible attributes *are* possessed in common across large kinds (and sometimes even more widely), we mustn't in those cases, for the reasons we indicated in an earlier chapter, start with the indivisible forms. Let this then be determined so.

If this reading is right, we now have a way in which division is used in explanation—not just at the pre-explanatory stage, but in the laying out, perhaps in some cases even in the very discovery, of the explanations themselves. There are various ways to test this reading. One way is to see if the pattern described can be found in the actual explanations offered in the biological corpus.

[22] Perhaps only analogously in common, if *APo.* II.17 99a15 applies: 'And things which are the same by analogy will have their middle terms the same by analogy too'.

The organization of explanations in *PA* II–IV

I think in fact that the pattern recommended by our I.5 precept, so interpreted, pervades *PA* II–IV. I've suggested that the precept as worded focuses on a paradigm case but is meant to be taken more broadly. The paradigm case is one in which we have a proper kind, such as Bird, and a set of attributes possessed by animals of that kind in virtue of itself (feathers, beaks, inward-bending two-leggedness, etc.). Each such attribute is given a causal account (involving, typically, both final and material factors). Each attribute is divided into its various forms (typically according to more-and-less variations), and each distinct form of attribute is explained from a corresponding variation in the generic causal factors (i.e., from the *particular* nature and needs of, and the *particular* materials available in the construction of, the subgroup of birds that possess it). But, as we've seen, the immediately following lines (as well as the related passage in I.4, and the discussion in *APo.* II.14), allow that the feature being explained may only be analogously common, and the subject class need not be a proper kind, but any grouping which correlates with a common attribute.

These extensions are important in understanding *PA* II–IV, because that treatise is organized not primarily by animal kinds but by animal parts. The subject class for each part tends simply to be the widest class of animals which possess that part, and that class may or may not form a kind. To illustrate my claim that the pattern is pervasive, I've chosen three chapters from different parts of the study, covering a variety of cases: the discussion of marrow, a uniform part, in II.6; the discussion of modes of eye-protection, usually eye-coverings, a non-uniform part (or set of analogous parts), in II.13; and the discussion of external parts of birds, in IV.12.

The chapter on marrow forms part of a series of discussions of uniform parts in II.2-9. By the time marrow is reached, there have already been discussions of blood, lard, and suet (two fatty materials, one earthier than the other). The chapter opens with an argument for the material nature (and material cause) of marrow: it is concocted blood. Differences in the degree of fattiness of marrow across different animals are noted, and correlated with the presence of lard or suet in these animals. Then differences in quantity of marrow from one animal to the next are noted, and correlated with different strengths and densities of bone. Then we get causal explanation. The function of marrow (as evidenced by its source in blood and its location in bones) is established: it is the nourishment of bones. Since smallness of quantity of marrow correlates with strength and density of bone, the explanation is simple: a difference in function—'the nourishing of strong, dense bones'—explains the difference in quantity—'small amount of marrow'. Similarly, the correlation of degree of fattiness of marrow with presence of lard or presence of suet points to the cause of that difference: since lard and suet are themselves concocted blood, and their relative presence has already been established to be a function of the density of the blood being concocted, differences in density of the blood being concocted explain the different degrees of fattiness of marrow.

Aristotle then turns, interestingly, to the marrow found in animals with spine rather than bone. Here the material cause is generically the same—the marrow is still concocted out of blood, but it is somewhat different (*alloioteros*) from marrow in animals with bones, being more sticky and sinewy, because it serves here a fastening and stretching as well as a nutritive role.

The pattern? Identify the part, determine its general cause (involving both material and final factors), divide the part along the different more-and-less dimensions it is found to vary in. Correlate with those variations variations in the causal factors, and establish that variation in the causal factors is in fact responsible for variation in the part, thereby 'dividing the causes'. Be prepared, however, to find on occasion that some variation in the part may be due, not to a variation in the cause, but to an additional causal role which that modified part serves (although expect that to happen more often when the correlated features are only analogously the same, as with bone and spine in the marrow case).

And the modes of eye-protection in II.13? The pattern is much the same. I translate at length so that we may more easily see the pattern at work.

> 657a25 Men and birds and four-footed animals, both live-bearing and egg-bearing, have protection for their eyes. The live-bearing ones have two eyelids, with which they also blink. The heavy birds and some others, and the egg-bearing four-footed animals, close their eyes with the lower eyelid, but birds blink with a membrane from the corner of the eye.
>
> 30 Now the cause of their having this protection is that eyes are moist in order that they may naturally see sharply. For, if they were hard-skinned, they would be less subject to damage by outside things striking them, but then they would not be sharp-sighted. So, for the sake of this, the skin around the pupil is fine, and for safety there are eyelids. And because of this they all blink, and humans most of all. They all do so in order that the eyelids may prevent things from striking the eyes (and this is not due to choice but produced naturally); humans do so most often because they are the most fine-skinned.
>
> b2 The eyelid is enveloped with skin; that is why neither the eyelid nor the foreskin will grow back together—because the skins are lacking in flesh.
>
> 5 Birds which close their eyes with the lower eyelid, and the egg-bearing four-footed animals, close them in this way because of the hardness of the skin around the head. For in the heavy, feathered animals, because they are not fliers, growth of wings has been turned into thickness of skin. That is why they too close their eyes with their lower eyelid, while pigeons and the like do so with both. The four-footed egg-bearing animals, on the other hand, are horny-scaled, and these scales are all harder than hair; so their skins are also harder than ordinary skin. The skin around their heads, then, is hard, and that is why they do not have an eyelid there; but the skin lower down is fleshy, so that there they have an eyelid that is fine and extensible. The heavy birds, though, blink not with this eyelid but with their membrane, because the

former's movement is slow while quick is what's needed and the membrane is that. And they blink from the corner of the eye alongside the nostrils because it is better for things of their nature to be from one source, and they have a source where they attach to the nostrils, and the front is more a source than the side.

22 The four-footed egg-bearing animals do not blink in the same way because it is not necessary for them, being earthbound, to have their eyes moist and their sight accurate; but for birds it is necessary, for they use their sight for seeing at great distances. That is why the crook-taloned birds are also sharp-sighted (for they sight their food from above, and that is why they, most of all birds, fly to great heights). Birds that are earthbound and not able to fly, such as domestic fowl and the like, are not sharp-sighted, for nothing in their mode of life demands it.

30 Fishes and insects and hard-skinned animals have different sorts of eyes, but none of them has an eyelid. For the hard-skinned animals can't have it at all: the eyelid's use requires action that's quick and thus by skin. But instead of this sort of protection, they are all hard-eyed; it's as if they were seeing through an eyelid that was directly attached. Since they must see less sharply because of the hardness, the nature made the eyes for insects movable, and for hard-skinned animals even more so (just as some of the four-footed animals have movable ears), in order that they might see more sharply by turning towards the light and receiving its rays. Fishes, on the other hand, are moist-eyed, for, moving around as much as they do, they must use their sight from afar. Now, in the case of terrestrial animals, the air is easy to see through. But in the case of fishes, since water is a hindrance to seeing sharply and does not, as the air does, contain many things that can hit the eyes—because of this they do not have an eyelid (for nature makes nothing without a point); but for dealing with the water's thickness they are moist-eyed.[23]

[23] 657a25-658a10, my translation. Bekker's Greek reads:

(657a25) καὶ οἱ μὲν ἄνθρωποι καὶ οἱ ὄρνιθες καὶ τὰ ζῳοτόκα καὶ τὰ ᾠοτόκα τῶν τετραπόδων φυλακὴν ἔχουσι τῆς ὄψεως, τὰ μὲν ζῳοτόκα βλέφαρα δύο, οἷς καὶ σκαρδαμύττουσι, τῶν δ' ὀρνίθων ἄλλοι τε καὶ οἱ βαρεῖς καὶ τὰ ᾠοτόκα τῶν τετραπόδων τῇ κάτω βλεφαρίδι μύουσιν· σκαρδαμύττουσι δ' (a30) οἱ ὄρνιθες ἐκ τῶν κανθῶν ὑμένι. τοῦ μὲν οὖν φυλακὴν ἔχειν αἴτιον τὸ ὑγρὰ τὰ ὄμματα εἶναι, ἵνα ὀξὺ βλέπωσι τοῦτον τὸν τρόπον ὑπὸ τῆς φύσεως. σκληρόδερμα γὰρ ὄντα ἀβλαβέστερα μὲν ἂν ἦν ὑπὸ τῶν ἔξωθεν προσπιπτόντων, οὐκ ὀξυωπὰ δέ. τούτου μὲν οὖν ἕνεκα λεπτὸν δέρμα τὸ περὶ τὴν κόρην (a35) ἐστί, τῆς δὲ σωτηρίας χάριν τὰ βλέφαρα· καὶ διὰ τοῦτο σκαρδαμύσσει τε πάντα καὶ μάλιστ' ἄνθρωπος, πάντα μὲν ὅπως τὰ προσπίπτοντα τοῖς βλεφάροις κωλύωσι (καὶ τοῦτο (b1) οὐκ ἐκ προαιρέσεως, ἀλλ' ἡ φύσις ἐποίησε), πλειστάκις δ' ὁ ἄνθρωπος διὰ τὸ λεπτοδερμότατος εἶναι. ἡ δὲ βλεφαρίς ἐστι δέρματι περιειλημμένη· διὸ καὶ οὐ συμφύεται οὔτε βλεφαρὶς οὔτε ἀκροποσθία, ὅτι ἄνευ σαρκὸς δέρματά ἐστιν. τῶν (b5) δ' ὀρνίθων ὅσοι τῇ κάτω βλεφαρίδι μύουσι, καὶ τὰ ᾠοτόκα τῶν τετραπόδων, διὰ τὴν σκληρότητα τοῦ δέρματος τοῦ περὶ τὴν κεφαλὴν οὕτω μύουσιν. οἱ μὲν γὰρ βαρεῖς τῶν πτερωτῶν διὰ τὸ μὴ πτητικοὶ εἶναι τὴν τῶν πτερῶν αὔξησιν εἰς τὴν τοῦ δέρματος παχύτητα τετραμμένην ἔχουσιν. διὸ καὶ οὗτοι μὲν (b10) τῷ κάτω βλεφάρῳ μύουσι, περιστεραὶ δὲ καὶ τὰ τοιαῦτα ἀμφοῖν. τὰ δὲ τετράποδα τῶν ᾠοτόκων φολιδωτά ἐστιν· ταῦτα δὲ σκληρότερα πάντα τριχός, ὥστε καὶ τὰ δέρματα τοῦ δέρματος. τὸ μὲν οὖν περὶ τὴν κεφαλὴν σκληρόν ἐστιν αὐτοῖς, διόπερ οὐκ ἔχει βλέφαρον ἐκεῖθεν, τὸ δὲ κάτωθεν σαρκῶδες, (b15) ὥστ' ἔχειν τὸ βλέφαρον λεπτότητα καὶ τάσιν. σκαρδαμύττουσι δ' οἱ βαρεῖς ὄρνιθες τούτῳ μὲν οὔ, τῷ δ' ὑμένι, διὰ τὸ βραδεῖαν εἶναι τὴν τούτου κίνησιν, δεῖν δὲ ταχεῖαν γίνεσθαι· ὁ δ' ὑμὴν τοιοῦτον. ἐκ δὲ τοῦ κανθοῦ τοῦ παρὰ τοὺς μυκτῆρας

The feature being explained is the having of eye-protection. This feature, as we'll see later, is one by analogy. In most cases eye-protection is provided by a part that is one in kind, an 'eye-covering', and Aristotle starts by focusing on that general attribute. The I.5 precept is applied to that general attribute. He later broadens the attributes under consideration from eye-covering to protection provided by the character of the eye-material itself, after he explains why eye-*coverings* are impossible in those cases; he shows that the protective function is served here by the distinctive eye-material. (He concludes by identifying why some animals do not need eye-protection at all and so don't have it.)

Thus, Aristotle begins (a25) by identifying the widest class of animals which possess any sort of eye-covering. He then (a27) briefly divides the types of eye-coverings. In the second paragraph (a30) he gives the cause of the possession of eye-coverings as such, identifying both final and material factors. This cause explains blinking, and he immediately notes a difference in blinking between man and other animals and explains it by reference to a difference in the causal factors (b1). He then returns, in the fourth paragraph (b5), to the differences in eye-covering and goes on to 'divide the causes', explaining the *differences* in the eye-coverings across groups of animals by reference to the *differences*, among these groups, in the general features which explain the presence of eye-covering as such. Along the way additional facts about eye-coverings (e.g., that they don't grow back when cut) are explained from the basic ones, and Aristotle concludes (b30), as I noted just above, with an account of how animals which don't

σκαρδαμύττουσιν, ὅτι βέλτιον ἀπ' ἀρχῆς μιᾶς (b20) τὴν φύσιν εἶναι αὐτῶν, οὗτοι δ' ἔχουσιν ἀρχὴν τὴν πρὸς τὸν μυκτῆρα πρόσφυσιν· καὶ τὸ πρόσθιον ἀρχὴ τοῦ πλαγίου μᾶλλον. τὰ δὲ τετράποδα καὶ ᾠοτόκα οὐ σκαρδαμύττει ὁμοίως, ὅτι οὐδ' ὑγρὰν αὐτοῖς ἀναγκαῖον ἔχειν καὶ ἀκριβῆ τὴν ὄψιν ἐπιγείοις οὖσιν. τοῖς δ' ὄρνισιν ἀναγκαῖον· πόρρωθεν (b25) γὰρ ἡ χρῆσις τῆς ὄψεως. διὸ καὶ τὰ γαμψώνυχα μὲν ὀξυωπά (ἄνωθεν γὰρ αὐτοῖς ἡ θεωρία τῆς τροφῆς, διὸ καὶ ἀναπέτονται ταῦτα μάλιστα τῶν ὀρνέων εἰς ὕψος), τὰ δὲ ἐπίγεια καὶ μὴ πτητικά, οἷον ἀλεκτρυόνες καὶ τὰ τοιαῦτα, οὐκ ὀξυωπά· οὐδὲν γὰρ αὐτὰ κατεπείγει πρὸς τὸν βίον. οἱ (b30) δ' ἰχθύες καὶ τὰ ἔντομα καὶ τὰ σκληρόδερμα διαφέροντα μὲν ἔχουσι τὰ ὄμματα, βλέφαρον δ' οὐδὲν αὐτῶν ἔχει. τὰ μὲν γὰρ σκληρόδερμα ὅλως οὐκ ἔχει· ἡ δὲ τοῦ βλεφάρου χρῆσις ταχεῖαν καὶ δερματικὴν ἔχει τὴν ἐργασίαν· ἀλλ' ἀντὶ ταύτης τῆς φυλακῆς πάντα σκληρόφθαλμά ἐστιν, οἷον (b35) βλέποντα διὰ τοῦ βλεφάρου προσπεφυκότος. ἐπεὶ δ' ἀναγκαῖον διὰ τὴν σκληρότητα ἀμβλύτερον βλέπειν, κινουμένους ἐποίησεν ἡ φύσις τοὺς ὀφθαλμοὺς τοῖς ἐντόμοις, καὶ μᾶλλον (658a1) ἔτι τοῖς σκληροδέρμοις, ὥσπερ ἔνια τῶν τετραπόδων τὰ ὦτα, ὅπως ὀξύτερον βλέπῃ στρέφοντα πρὸς τὸ φῶς καὶ δεχόμενα τὴν αὐγήν. οἱ δ' ἰχθύες ὑγρόφθαλμοι μέν εἰσιν. ἀναγκαία γὰρ τοῖς πολλὴν ποιουμένοις κίνησιν ἡ τῆς ὄψεως (a5) ἐκ πολλοῦ χρῆσις. τοῖς μὲν οὖν πεζοῖς ὁ ἀὴρ εὐδίοπτος· ἐκείνοις δ' ἐπεὶ τὸ ὕδωρ πρὸς μὲν τὸ ὀξὺ βλέπειν ἐναντίον, οὐκ ἔχει δὲ πολλὰ τὰ προσκρούσματα πρὸς τὴν ὄψιν ὥσπερ ὁ ἀήρ, διὰ μὲν τοῦτ' οὐκ ἔχει βλέφαρον (οὐδὲν γὰρ ἡ φύσις ποιεῖ μάτην), πρὸς δὲ τὴν παχύτητα τοῦ ὕδατος ὑγρόφθαλμοί (a10) εἰσιν.

have eye-coverings deal with the need which give rises to them in those that do. 'Divide the attribute . . . then try to divide its cause.'

The external parts of birds are grouped and studied together in IV.12. Birds do, of course, share certain generic external parts with other animals; some of these have already been mentioned in that connection, and others are noted in the course of this chapter, but birds do have a distinctive outer appearance and shape, which is the source of their initial grouping as a kind (644b7–15; cf. Balme 1972 *ad loc.*), and a single common nature associated with aspects of that appearance which validates that initial grouping (644b3). In fact, Aristotle begins the chapter by making the case for studying the birds separately:

In the birds, the differentiation is in the excess and deficiency, and with respect to the more and less, of their parts; for some of them are long-legged, others short-legged, and some have a broad tongue, others a narrow one, and similarly for the other parts. And they have a few parts among themselves which differ <significantly>, but when compared with the other animals they do differ <significantly> in the form (*morphē*) of their parts. (692b3–9)

Having said this, Aristotle lists a number of external parts distinctive (*idia*) to birds: their feathers (and in general their characteristic wings), their beak, their inward-bending two-leggedness, and their distinctive breast, along with features they have in common with other animals, such as 'a neck which sticks up' (692b20) and other generic features of their 'trunk'. Some of these distinctive parts are explained in terms of the basic nature of a bird ('the *ousia* in the case of the bird includes being blooded . . . and being a flier is in the *ousia* of a bird'[24]). Differences among the different types of bird (grouped by correlated features, not by indivisible bird forms such as sparrow and crane) are then explained by reference to differences in, e.g., mode of flight, type of food, mass of body, i.e., by differences in the broad general factors in the nature of a bird, both material and final, which explain the distinctive features of a bird. Here is just one example. Earlier on, in the discussion of mouths of blooded animals in III.1, Aristotle identified the functions that the beak, which substitutes for both lips and teeth, performs: feeding and defense. Now, in IV.12, having explained how variations in neck size are explained by variations in the animal's mode of life (*bios*, roughly its habitat and type of food), he says:

Their beaks differ in accordance with their modes of life. For some have a straight beak, others a curved one; straight, those that have it for the sake of nourishment, curved, those that are carnivores; for such a beak is useful for overpowering, and it is necessary that their nourishment must be procured from animals. But all birds whose mode of life

[24] 696b6, 13; cf. chapter 10 below, p. 234.

is marsh-based and are plant-eating have a flat beak, for such a beak is useful both for digging up and cropping off their nourishment. And in some cases, the beak of such animals is also long, as is the neck, because they take their nourishment from the depths.[25]

This chapter is not as simply organized as the others we have discussed, perhaps because of the complexity of the material, perhaps also because, as some peculiar features of the last four chapters of book IV suggest, it may be very early; but the pattern is still prevalent. One can open *PA* II–IV more or less at random and light immediately upon such a pattern of argument.[26]

But are they actual *divisions*?

Now, granted that differences in generic attributes are explained, for the most part, by differences in the causes of those attributes, why should we speak of these hierarchical relationships as *divisions*? For the I.5 precept to be recommending this pattern of explanation, as I have claimed it does, I will have to show that these subdifferentiations, of both the features that turn out to be explained, and those that turn out to do the explaining, count as divisions. I think we have two criteria to go on. First, we must ensure that the sequence of attributes down which the explanatory progression proceeds fits, where relevant, the various Aristotelian reforms of Academic division, so that it would count as a division for Aristotle. Second, we can see if any of the reasons Aristotle gives for the usefulness of division in other contexts (I am thinking of the various remarks in the *Analytics* cited above) apply in this case. On the first criterion, I think we get a positive answer. Although the reforms of division are focused on definition, they apply more generally to the sorts of pre-definitional divisions we have evidence Aristotle himself prepared.[27] Balme (1987b) isolated three reforms: (i) distinguishing the categories of *genos, eidos*, and *diaphora* from each other, viewing a *genos* not as a separate entity but as a potentiality that exists only in its different *eidē*, as a determinable or abstraction; (ii) insisting on *successive* differentiation—dividing only by 'the differentia of a differentia'; and (iii) requiring that division be (where necessary) by multiple differentiae simultaneously. The first two reforms obviously apply to the cases of

[25] 693a11-19. Here I borrow Lennox's 2001a translation, with some modifications. Bekker's Greek reads: *διαφορὰν δ' ἔχει καὶ τὰ ῥύγχη κατὰ τοὺς βίους. τὰ μὲν γὰρ εὐθὺ ἔχει τὰ δὲ γαμψόν, εὐθὺ μὲν ὅσα τροφῆς ἕνεκεν, γαμψὸν δὲ τὸ ὠμοφάγον· χρήσιμον γὰρ πρὸς τὸ κρατεῖν τὸ τοιοῦτον, τὴν δὲ τροφὴν ἀναγκαῖον ἀπὸ ζῴων πορίζεσθαι, καὶ τὰ πολλὰ βιαζομένοις. ὅσων δὲ ἕλειος ὁ βίος καὶ ποοφάγος, πλατὺ τὸ ῥύγχος ἔχουσιν· πρός τε γὰρ τὴν ὄρυξιν χρήσιμον τὸ τοιοῦτον καὶ πρὸς τὴν τῆς τροφῆς σπάσιν καὶ κουράν. ἔνια δὲ καὶ μακρὸν ἔχει τὸ ῥύγχος τῶν τοιούτων, ὥσπερ καὶ τὸν αὐχένα, διὰ τὸ λαμβάνειν τὴν τροφὴν ἐκ τοῦ βάθους.*

[26] See now Lennox 2001a, 125, 132, 146, 176.

[27] Cf. Barnes 1975a, 240.

differentiation we have looked at. The subject of study is the immense diversity of animals in the natural world, not some independent realm of kinds that are to be 'interwoven' in some logical manner. The ontological distinction between generic potentialities or determinables, and formal actualities or determinations, is fully realized in the doctrine of the-more-and-the-less (and the functional unity behind only analogous similarity) which permeates the biology. And the explanatory pattern I've described requires that the division be by 'differentia of a differentia', since there has to be entailment from the bottom up if the cause of the lower differentia is to be a form of the cause of the higher differentia. The multiple-differentiae reform is meant to apply primarily to definitions of whole animals, but as Pellegrin has observed, it should apply as well to all those animal parts that require more than one line of division to characterize them. In fact, as I've indicated, the 'cause' which needs to be divided in biological cases is itself usually a combination of material and final factors, and the divisions of the materials and of the functions are of course distinct axes of division. So the sequences of attributes Aristotle uses in his explanatory progression fit the Aristotelian reforms of division.

But are they drawn from, or do they represent, actual *divisions*? Are these lists actually organized according to these divisional rules and into groups of 'divisions'? Is there something *PA* II–IV is based on, or embodies, which could count as the 'Divisions' (capital D) of which *APo.* II.14 speaks? (You recall my references earlier to Lennox's citation of 98a1-2, where Aristotle instructs the scientist to seek the proper subject class for explanatory propositions, 'selecting from . . . the divisions'.) The answer here too seems to be yes, even if the exact form in which it was recorded is unclear, if we consider what the usefulness of these divisions is. We have already commented on the entailment relationship, from the bottom up, that is crucial to make the explanatory scheme work; the same relationship is commented on in *Metaph.* Z.12 in connection with definition and said to be ensured by taking 'the differentia of a differentia'. One can do that without pre-existing division tables, but they facilitate the process. When *APo.* II.13 speaks of the usefulness of division, it speaks of its role in ensuring (in fact it is the 'only' thing which ensures) 'completeness'. That has a certain meaning in definition; but in explanation, I take it that the value of a complete division, reflecting the whole of one's survey of the attributes in one's realm of study, is to give one an exhaustive supply of attributes to be explained and attributes that do the explaining. The all-and-only correlations Aristotle draws on in the course of producing his explanations depend on completeness of a certain sort, a completeness which the organization of attributes *by division* best ensures.

For all these reasons I am inclined to think that *PA* II–IV embodies a use of Aristotelian division at the *explanatory* level, not previously discussed, a use which is reasonably encapsulated in the I.5 precept that 'it is necessary first to divide the attributes pertaining to each kind which are present in all the animals <of that kind> in virtue of themselves, and after that to try to divide their causes.'[28]

[28] I dedicate this chapter to Wolfgang Kullmann, with gratitude for his groundbreaking work on the philosophical significance of Aristotle's biological treatises. Drafts were read at a Society for Ancient Greek Philosophy meeting in San Francisco in 1991 and at the University of Tokyo in 1994. I would like to thank all those who participated in discussion on those occasions. For helpful discussions during the preparation of the original paper, I would like to thank in particular Robert Bolton, Catherine Culver, David Depew, and James Lennox. Professor Kullmann's *Wissenschaft und Methode* (1974) provides a valuable background to this study.

PART III

Metaphysical Themes in *PA* and *GA*

10

Notes towards a Study of Substance and Essence in Aristotle's *Parts of Animals* II–IV

Readers of Aristotle's *Metaphysics* puzzle over such issues as whether the *aporiai* of book *Z* are resolved within *Z*, or not until *H*.6 or beyond; whether its primary substances are (in the case of animals) animal species, the forms of animal species, individual animals, or *their* forms; whether the essence of an animal is its *form* or some unity of form and matter (or whether animal form itself *includes* matter), and if either of the latter two, then *which* material aspects are included in the form; whether there are *individual* forms and essences as well as (or instead of) species-forms, and if so, whether forms of individuals of a species are qualitatively or only numerically distinct, and so on.

David Balme has taught us that there is much to be learned about Aristotle's philosophical thought from his biological works, and in the spirit of that lesson it may be useful to ask what light these treatises can shed on those dark puzzles and in general on Aristotle's conceptions of substance and essence. As a small contribution to our celebration of this large lesson and of its gentle author, I should like to describe a research project for the study of substance and essence in Aristotle's biological works and to present a few preliminary results from my own pursuit of that project in one such work, *Parts of Animals*.

In undertaking a study of the lessons of the biological treatises for our understanding of substance and essence in Aristotle, we need at the outset to distinguish two questions:

(a) With what conceptions of substance and essence, if any, does Aristotle operate in the biology?

(b) What can we learn from our study of these conceptions in the biology about the conceptions of substance and essence in the *Metaphysics*?

This essay, written in 1982, was my contribution to the Festschrift for David Balme (Gotthelf 1985a). I retain, in this reprinting, the use of Peck's text, translation, and (occasionally) apparatus for the *PA* II–IV passages.

The need for the distinction is clear: we cannot say in advance that the notions at work in the biology are the very same ones under scrutiny in the *Metaphysics*. Perhaps the biological treatises were written much later (or much earlier) than *Metaph. ZHΘ*; or perhaps they were written without careful reflection on the metaphysical concepts being used. Or perhaps the biology and the *Metaphysics* diverge for some other reason. While I do not myself believe any of that, we must not prejudge the issue and should thus keep the two questions separate. Question (a) is of course the prior question and will be our concern in this chapter; Question (b) is for another time.[1]

The project

Our goal, then, is to understand the conceptions of substance and essence with which Aristotle works in the biological treatises. The research project I propose has two parts. The first is simple, yet quite powerful: collect and study each occurrence in these treatises of the terms *ousia, to ti ēn einai, to einai* + dative, and so on. As it turns out, these fall naturally into two groups: (i) those in which Aristotle states explicitly that some feature of an organism is 'in the *ousia*' or 'in the *logos tēs ousias*', and so on of that (kind of) organism; and (ii) those others in which the term appears as part of a methodological, or otherwise theoretical, statement.[2]

Occurrences of type (i) are more frequent in the innards of the biological works and are, from our standpoint at least, the more exciting, as they promise us a look at just that sort of definition (or at some portion of just that sort of definition) of which the *Metaphysics* worries over the possibility—and possible content: definitions of living organisms. There are roughly ten such passages, all from *PA* II–IV

[1] I touch on these issues in the next chapter, but see Balme 1987d for a much fuller discussion.

[2] I have made a working distinction between those biological texts which have traditionally been considered of philosophical significance, and thus already gotten at least some discussion in the literature—viz., *PA* I.1, 2-4 and *GA* V.1—and the less explicitly philosophical or methodological 'innards' of the biology, and will devote my attention in this project to the latter. A full study must of course incorporate the former, though a fresher look might just be had if that study be done *after* close examination of the 'innards'. Examples of *ousia*-containing methodological or otherwise theoretical statements of the innards include: *PA* 647b25, 648a14; *GA* 742a22, 767b34 (in the very important IV.3, on which see Balme 1987d and Cooper 1997); *Juv.* 469a30. On the *GA* V.1 passage see now chapter 5 above.

There are very few passages outside of *PA* and *IA* which fall into group (i). I have only been able to find *GA* 715b18 (on testacea), 731a25 (on plants generally), and *Resp.* 477a23 (paralleling our passage 3 below, and remarked on in n. 14 below). Wolfgang Kullmann, as part of a discussion with similar motives to the present study (which I became aware of only after completing the bulk of the work reported here), includes passages in which some feature is said to be part of the *nature* (*phusis*) of the animal in question (e.g., *PA* 664a3-8, on deer's horns). I refrain from doing so here, however, out of a worry that 'nature' (even in the sense of 'formal nature') might incorporate more than *ousia* does. The issue is worth more attention, however. Cf. Kullmann 1974, ch. 4, sec. 3, esp. 308-25.

and its 'appendix', *Progression of Animals*. In the next section of this chapter I shall quote (with surrounding context) and briefly discuss each one. I will not here discuss the few occurrences of types (ii) in these works (see *PA* 647a16, 652b18),[3] as they fit more naturally into the second part of the overall project.

The project's second part is a more complicated undertaking, and a report of its results will require a separate occasion (see n. 2 above). But I should like at least to describe this part of the project itself, both as an invitation to readers to join me in this study and as a suggestion of the promise the overall enterprise might have.

At *PA* I. 640a33-b1, Aristotle draws, from his case for the priority of the final cause, conclusions regarding the proper method for expounding the causes of the parts of animals:

> Hence we should if possible say that because this is what it is to be a man, therefore he has these things; for he cannot be without these parts. Failing that we should get as near as possible to it: we should either say altogether that it cannot be otherwise, or that it is at least good thus . . .[4]

[3] *PA* 646a25, 26, and b1, which mirror passages in *PA* I.1, are too general to be of any fresh use. For *ti ēn einai*, see 649b22 (and n. 12 below).

[4] tr. Balme. Sorabji (1980, 155-56), commenting on the alternative described in the first half of the second sentence, claims that 'it may well be simplest to suppose that (A′) [= " . . . altogether . . . it cannot be otherwise"] simply repeats (A) [= " . . . he cannot be without these parts"]'. He mentions as an acceptable alternative an interpretation of Kullmann's, but strangely does not mention the clearly preferable (if not entirely clear) one suggested by Balme (1972, 87): 'The best explanation is to be able to say (1) that because a man is such (e.g. perceptive) therefore he must have these parts (sense organs). If we cannot argue so, we may be able to show (2) that he cannot survive without this part (e.g. heart or liver, *P.A.* III. 670a23).' One might take the distinction Balme has in mind to be something like the following: the better explanation is by way of showing an organ to be *the* organ that performs a defining function, while the second-best one shows an organ to make only some necessary (and perhaps causal) contribution through its functioning to some defining function, but not to be *the* organ for that function. (For example, man is nutritive; but there is no single organ which absorbs and metabolizes food, flushing waste products, and so on, as there is a single organ for, say, sight. Organs such as the liver contribute to nutrition but do not by themselves 'perform' nutrition. The eye thus gets the first sort of explanation, the liver the second. [I avoid any complications introduced by the role of the heart as seat of all soul-functions.])

Cooper 1987 makes important use of this passage. He claims to be following Balme's interpretation of the passage, though he does not identify the sort of features he thinks will be the subject of the first mode of explanation, and goes farther than Balme on a matter that interests us: 'Following Balme, I take it that the first necessity mentioned is the necessity of what follows directly from the human essence, as included in it' (254). This is a tricky issue, and Cooper has acknowledged that Aristotle is less clear on it than one would like. In favor of including in the essence features explained in this mode is the opening section of *GA* V.1, cited (among other useful passages) by Cooper, and perhaps several of the ten passages analyzed below. But the issue is, as I say, complicated. First, the relevance of Cooper's *GA* V.1 passage is uncertain: cf. my n. 33 below. Then, the passages explored below show clearly that Aristotle is prepared to include some animal parts in the *ousia* of the animals that possess them. In none of these cases, however, are the parts explained in the first mode, or explained at all: in each case they do the explaining. And

Assuming that the second half of the first sentence names a mode of explanation distinct from the two that follow, we have three modes of explanation, each of which refers some part of an animal to some part of the essence of that animal.[5] On the further assumption that *PA* II–IV employs that proposed method of exposition, involving these three modes,[6] we can turn the process around and identify what Aristotle takes to be parts of the essences of his animals by seeing which features turn up in the *explanantia* of these explanations.[7]

The project, then, is to examine each explanation offered in *PA* II–IV (and *IA*) and to identify, in each case, which one of these three modes of explanation is at work, which feature of the kind of animal in question is being explained, and which feature(s) this feature is being explained by reference to. Since some of these latter features are themselves likewise explained in terms of yet other features, we will look for those features which, though not explicitly said to be part of the *ousia* or essence of the animal (and thus already collected in our project's first part), are not themselves explained in any of these ways; these features are then prime candidates for being part of the essence of the animal in question.[8] Such is the second part of our project. The two parts together promise an interesting view of the conceptions of substance and essence at work in the biology. As explained above, I restrict myself here to the project's first part—a study of the passages in which Aristotle states explicitly that some features of an animal are part of the *ousia* or *ti ēn einai* of that animal.

where a feature seems to be explained in the first mode, that feature seems *distinguished* from other features thought to constitute the *ousia* of the animal in question.

The issue is further complicated by the fact that all our passages but one speak of the *ousia*, and not the essence, of the animal in question. Cooper, by implication, appears to make no distinction between essence and *ousia*. Balme, who has long been insistent that animal *ousia* include animal parts, makes a firm distinction between essence and *ousia* (or *logos ousias*): see Balme 1980, 4ff. (= 1987d, 294ff.). Where a passage below appears to bear on this issue, I will note that in a comment.

[5] 'An animal', 'that animal': received opinion would perhaps expect that all such explanations would be at the level of 'species', traditionally understood. That this is *not* the case is one important result of this study, and will receive attention in the concluding section.

[6] There is no question but that they do, as even a casual reading of *PA* II–IV shows. Cf. Balme 1972, *ad* 640a33, the many passages cited by Cooper 1987, and Gotthelf 1987a (= chapter 7, above).

[7] On Cooper's reading of the first mode (above, n. 4), the *explanantia* of explanations in that mode will also be part of the essence.

[8] I say 'are prime candidates for being' and not 'must be' for perhaps obvious reasons. That a feature does not get an explanation in *PA* II–IV does not guarantee that Aristotle would not have been prepared to give such an explanation; but it does render it likely. One needs to consider the causal role of the feature in the light of one's understanding of Aristotle's overall view of the functions and interrelations of the organic parts of the animal in question, as well as occasionally his reports on its lifestyle, and to treat the matter, as all things animal should be treated, with a gentle hand.

The explicit passages

I list the passages in the order in which they occur in the text. I follow Peck's *PA* and Forster's *IA* texts (both in Peck and Forster 1961) with one exception where I (uncertainly) follow a suggestion of Ogle's 1882; and (although it is unlikely to be of any use in our discussion) I include any editors' notes indicating divergence from Bekker's text. All translations are my own unless otherwise indicated. My remarks on the passages are numbered. Since David Balme has discussed, or at least cited, each of these passages in one or another of his studies, my remarks consist largely of commentary on his discussions.

1. PA *II.8 653b19-27*

We must now examine the other homoiomerous parts, and first flesh in those that have flesh and its analogue in the others; for this [*sc.* flesh-or-its analogue] is of animals an origin and a body in itself. This is clear also from **the *logos*, for we define an animal by its having sensation**, and first the primary sense, which is touch, and such a part is the sense organ of this—either its primary organ, as the pupil is of sight, or the organ taken together with that-through-which, as if someone took the whole of the transparent together with the pupil.[9]

(1) The bold-faced portion is of course no surprise: cf. such passages as *An.* II.2 413b1-7. *An.* I.1 402b5-6 says that there is a single *logos* of animal and the particular one given in this *PA* passage is what we would expect. Whether 'living thing having sensation' is the *whole logos*, is another matter; it is in fact very unlikely, as *prōton de tēn prōtēn* etc. suggests. (I might well have boldfaced these words too in the translation above.) A definition of animal would have to capture the fact that animal kinds all have touch but otherwise differ in the senses they will possess, thus permitting division down by 'successive differentiation', which specifies which of the disjunctive possibilities is realized in its case (*PA* I. 643b10-26 and Balme 1972, 102).[10] The present definition is thus only a partial one (and may be so in other ways as well), though it is fully sufficient for its argumentative purpose here. Aristotle's willingness to treat disjunctions of contrary features (broadly understood) as units of thought is clear from *touto* in line 653b21, which clearly has as its

[9] *περὶ δὲ τῶν ἄλλων μορίων τῶν ὁμοιομερῶν σκεπτέον, καὶ πρῶτον περὶ σαρκὸς ἐν τοῖς ἔχουσι σάρκας, ἐν δὲ τοῖς ἄλλοις τὸ ἀνάλογον· τοῦτο γὰρ ἀρχὴ καὶ σῶμα καθ' αὑτὸ τῶν ζῴων ἐστίν. δῆλον δὲ καὶ* ***κατὰ τὸν λόγον· τὸ γὰρ ζῷον ὁριζόμεθα τῷ ἔχειν αἴσθησιν****, πρῶτον δὲ τὴν πρώτην· αὕτη δ' ἐστὶν ἁφή, ταύτης δ' αἰσθητήριον τὸ τοιοῦτον μόριόν ἐστιν, ἤτοι τὸ πρῶτον, ὥσπερ ἡ κόρη τῆς ὄψεως, ἢ τὸ δι' οὗ συνειλημμένον, ὥσπερ ἂν εἴ τις προσλάβοι τῇ κόρῃ τὸ διαφανὲς πᾶν.* Cf. III.4 666a34: *τὸ μὲν γὰρ ζῷον αἰσθήσει ὥρισται* (See now also Balme 'for an animal is defined by perception') and II.10 655b29-656a6.

[10] Balme's clearest statement of this view, applied at the level of kinds such as Bird and Fish, appears perhaps in the excerpt from his letter to Marjorie Grene, published in Grene 1974, 125-26 n. 10. See now also Balme 1987b and 1987d.

antecedent 'flesh or its analogue', and perhaps from *APo.* I. 73a37-40, b19-21, 78a16-21, 84a20.

(2) In regard to the possible difference, remarked in note 5 above, between John Cooper and Professor Balme as to the first mode of explanation in the *PA* passage, we can notice that though Aristotle could have said here that the *organ* of touch is part of the *logos*, he says only that the *sense* (called a *dunamis* in *An.*) is. He does, however, call the flesh (and its analogue) an *archē*, and that is probably because, in the broader sense indicated by the analogy, it is the organ of touch—touch being the primary (and essential) sense for any animal. (The broader sense is 'organ + medium', since flesh is strictly the medium rather than the organ, the heart or its analogue being that.) James Lennox has suggested that we should take the *kath' hauto* technically: animals have flesh-or-its-analogue in virtue of being animals. Even if accepted, this need not decide the question though, as the feature could be *kath' hauto sumbebēkos*.

(3) As a *dunamis* or *ergon* of animals, *aisthēsis* is just the sort of feature we should expect to find in a definition. The object of definition is at a rather high level of generality, but animal is not here said to have an *ousia*, only a *logos*, and certainly kinds at all levels can have *logoi*. On the other hand, the verb used (*horizometha*) suggests that we do have here a definition in a strict sense, and I will assume this in what follows.

2. PA *III.6 669b8-12*

Generally, then, the lung is for the sake of respiration, though it will be bloodless and such [as has just been described] for the sake of a certain animal kind [sc. the egg-bearing]. But there is no common name in the case of those <which have a lung>, not like 'bird' is applied in the case of a certain kind. On account of which, just as the being for a bird is out of something, also in these cases **having a lung is present in the *ousia*.**[11]

(1) 'There is no common name': alternatively, 'that which is common is nameless'. Not the feature which is in common—that of course has a name (viz., 'lung')—but the class of possessors of that feature. On why 'lung-possessors' does not count as a name, see Balme 1962a, 90-91: 'the reason is that an adjective is not an *onoma* even when used "substantively", unless it has really become a noun'. (This must also apply to the relevant sense of *pezon* discussed in comment (3) below.)

(2) The lung-possessors, according to Aristotle, are man, the live-bearing and egg-bearing quadrupeds, the snakes, the birds, and the cetaceans—in short, all and only the blooded animals except the fishes. This is a grouping above the level of the *megista genē* in *HA* I. 490b7-14 (quoted in comment (1) to passage 3), but below

[11] *ὅλως μὲν οὖν ὁ πλεύμων ἐστὶν ἀναπνοῆς χάριν, ἄναιμος δὲ καὶ τοιοῦτος γένους τινὸς ἕνεκεν ζῴων· ἀλλ' ἀνώνυμον τὸ κοινὸν ἐπ' αὐτῶν, καὶ οὐχ ὥσπερ ὁ ὄρνις ὠνόμασται ἐπί τινος γένους. διὸ ὥσπερ τὸ ὄρνιθι εἶναι ἔκ τινός ἐστι, καὶ ἐκείνων* **ἐν τῇ οὐσίᾳ ὑπάρχει τὸ πλεύμονα ἔχειν**.

the groupings of *megista genē* into blooded and bloodless kinds. In comparing the unnamed group of lung-possessors to Bird, one of the *megista genē*, and ascribing to the group an *ousia*, Aristotle would seem to be installing it as a true *genos*, in the strong sense, appropriate to division and definition, of *PA* I.2-4. As David Balme pointed out long ago, in his celebrated paper for the Second Symposium Aristotelicum (1960, publ. 1961, cf. Balme 1975), not every 'differentiae class' (= class of objects possessing some feature in common) is a true *genos* for Aristotle. *PA* I.4 suggests as a requirement of this that the class in question 'possess both a single nature in common and forms not far separated in them' (644b3-4, tr. Balme 1972, but with 'forms' for 'species'), and allows unnamed kinds as *genē* (b4-5). It would seem relevant to ask, if we are to know how seriously to take the ascription to it of an *ousia*, whether in fact lung-possessors constitute such a group: if Aristotle were in the habit, in the biology, of ascribing an *ousia* to any class, e.g., to pale-skinned humans, we should draw different conclusions than if he were as circumspect as he suggests in *Metaph.* *Z* 4 one ought to be.[12]

(3) In his Second Symposium paper Professor Balme denies *genos*-status to lung-possessors:

> [W]hen he states the need of a new group-name which overlaps or splits genera, it is clear that he is naming a differentia and not a genus. This is the case when he suggests that *lung-possessors* would be recognized as *a kind of animal* (*PA* III, 669b9); he often makes use of this category but he himself provides the reasons why it could not be a systematic division (*HA* VIII [= VII], 589a10ff.). (Balme 1975, 189)

HA VII 589a10ff. argues that there are various ways in which an animal can be said to be a land (*pezon*) or water (*enhudron*) animal, and that many animals which are land animals in one way need not be land animals in another. One of these ways is 'taking in air to cool its internal heat'. This sense of *pezon* applies to all and only the lung-possessors, as the lung is the organ for that purpose. (Other ways of being *pezon* include spending one's time on land, feeding on land, breeding and/or giving birth on land.) Is Professor Balme's point that this passage shows the many fundamental ways in which the lung-possessors differ from each other—too many ways for them all to constitute a single *genos*, the sub-kinds being 'far separated' or perhaps not even having a 'single nature'?

A further reason for denying *genos*-status to lung-possessors is suggested in Balme 1972, quoted below in comment (1) on passage 3. Not only must a true *genos* have many features in common but many of these features must be the same in *eidos*, differing only in the more and the less, and not merely the same by analogy (*PA* I.4

[12] That Aristotle *is* circumspect in his use of *ousia* in *PA* is made clear at 649b20-35, a passage well worth study in its own right. *ho leukos anthrōpos* even makes an illustrative appearance at b26. There are surprisingly few occurrences of *ti ēn einai* in the innards; our passage 10 below is a notable exception.

644a16-24 [cf. 5 645b3-10], *HA* I. 486a14-25 [cf. II. 497b6-13]). It might be argued that apart from lungs, there is little that all lung-possessors have in common except by analogy, and that their forms are again just too 'far separated'.

(4) On the other side are the facts, first, that Aristotle nowhere else in *PA*, or in the biology generally, ascribes an *ousia* to anything other than clear-cut *genē*;[13] and, second, that he here compares lung-possessors with birds, an unquestionable *genos*, in respect of possession of an *ousia*. Furthermore, while Aristotle's purpose in the *HA* VII passage Balme cites is not entirely clear to me, the variations among lung-possessors it points to seem compatible with our finding *much* in common among such animals. After all, the claim that lung-possessors possess little else in common is certainly overstated: all blooded animals have viscera, and surely these—and other internal parts as well—will differ only by the more and the less, and Aristotle has much to say in *HA* II–III about these similarities. (See also *Sens.* 444b2-4.)

Still, overall similarities among lung-possessors are more general than among, say, birds, and many of their likenesses *are* only by analogy. The question we thus have to ask is whether the unity possessed by organisms in virtue of their sharing *analogous* organs, if there are enough of them, is not itself sufficient to constitute the class of such organisms as a true *genos* for divisional and definitional purposes. After all, Aristotle does sometimes divide down from 'animal', and one might expect, as intimated in the discussion of passage 1 above, that a full definition of animal would express disjunctively at least the vital functions shared by all animals—sensation of types, touch, or touch and XY, or touch and XYZW, and so on (cf. *HA* IV.8); locomotion of type X, or Y, or Z, or XZ, or none, and so on. *Pol.* 1290b25ff. perhaps points to such an enterprise, if one reads it to permit such disjunctive terms, their contents being arrived at inductively.

(5) All these issues are very complex, and their resolution depends on answering yet other questions about the precise nature of an Aristotelian *genos* and about the role of definition in biological explanation, questions which I cannot attempt to deal with here.[14] What warrants universal agreement, I think, is that the ascription of an *ousia* to lung-possessors is to be taken quite seriously, as revealing something of the substance or essence of the animals to which it is ascribed. Professor Balme himself wrote that 'Aristotle indicates sometimes that he would like to recognize larger genera such as *lung-possessors* (*P.A.* III. 669b9)' (1972, 120: see comment (1) to passage 3 below), and in his paper to the 1976 Princeton Colloquium on Aristotle's Philosophy of Biology (1976, 8-9 and n. 26), he cited

[13] Never? Well, hardly ever, if passage 3 below is to be viewed equally skeptically: the same question perhaps arises there for the class of blooded animals and the class of bloodless animals (unless I am right that the real subject of the *ousia*-ascription there is the individual *megista genē*; see comment (3) on that passage).

[14] These issues are treated by Balme 1987d, Pellegrin 1987, and Lennox 1987b. I will say something on the second issue in my concluding section below.

this passage as an example of Aristotle's making 'logos ousias . . . include bodily parts which are *for* functions'.[15]

(6) Finally, in our concern over the nature of the true subject of this ascription, we must not fail to note the nature of what is ascribed: here, clearly, an organic *part* is said to be part of an *ousia*. But the surprises, if this was one, are not over.

3. PA *IV.5 678a26-34*

The commonly called cephalopods and crustaceans [Peck: 'softies' and 'soft-shells'] are very different from those [sc. the blooded animals], since they have nothing at all of the nature of viscera. Nor, likewise, do *any* of the bloodless animals. (There are two kinds of the bloodless remaining, the testacea [Peck: 'potsherd-skins'] and the insect-kind.) For, none of them has that out of which viscera nature is composed—blood—because **some such condition belongs to their *ousia*; for that some animals are blooded, while others are bloodless, will be present in the *logos* defining their *ousia*.**[16]

(1) '*megista genē* of the animals, into which some animals have been divided, are these: one of birds, one of fishes, and another of cetacean—and these are all blooded. Another *genos* is that of the testacea ['potsherd-skins'], which are called oysters; another, that of the crustacean ['soft-shells'], nameless by one name, . . . ; another, that of the cephalopods ['softies'], . . . ; and yet another, that of the insects. These are all bloodless . . . ' (*HA* I. 490b7-14)

Aristotle's broadest distinction among animals is between the blooded and the bloodless (though, as noted previously, he also often works with a revised version of the Platonic land/water or land/water/flying divisions). Do the blooded (or the bloodless) animals constitute a single, overarching *genos*? Again, David Balme suggests not:

These [major] genera cannot be grouped in overarching genera, presumably because their attributes are not basically the same but are merely analogous, as feathers are to scales. Analogous attributes cannot be shown as differentiae of a single attribute. It is not enough that *some* attributes are common to different genera (e.g. breathing). Aristotle indicates sometimes that he would like to recognize larger genera such as *lung-possessors* (*P.A.* III. 669b9), and *G.A.* II. 732b27f. shows him groping towards a basic differentiation by bodily

[15] A full treatment of the issues surrounding passage 2, it is worth noting, should take account of the interesting discussions of lung-possessors in *Resp.*, esp. ch.13, in which at 477a23-24 the lung is said, strikingly, to be *aition* of the *ousia* of man and all other animals that possess it. (Ross [1955, ad loc.] not implausibly, takes 'is *αἴτιον* of' to mean 'contributes to'.)

[16] *τὰ δὲ καλούμενα μαλάκια καὶ μαλακόστρακα πολλὴν ἔχει πρὸς ταῦτα διαφοράν· εὐθὺς γὰρ τὴν τῶν σπλάγχνων ἅπασαν οὐκ ἔχει φύσιν. ὁμοίως δ' οὐδὲ τῶν ἄλλων ἀναίμων οὐδέν. ἔστι δὲ δύο γένη λοιπὰ τῶν ἀναίμων, τά τε ὀστρακόδερμα καὶ τὸ τῶν ἐντόμων γένος. ἐξ οὗ γὰρ συνέστηκεν ἡ τῶν σπλάγχνων φύσις, οὐδὲν τούτων ἔχει αἷμα* **διὰ τὸ τῆς οὐσίας αὐτῶν εἶναί τι τοιοῦτον πάθος [αὐτῆς]*· ὅτι γάρ ἐστι τὰ μὲν ἔναιμα τὰ δ' ἄναιμα, ἐν τῷ λόγῳ ἐνυπάρξει τῷ ὁρίζοντι τὴν οὐσίαν αὐτῶν.**

* *αὑτῆς sed.* Peck

heat which he regards as the underlying cause of difference in reproduction, respiration, posture, locomotions, and psychology (cf. A. L. Peck, Loeb edn. of *H.A.*, Introd. p. xv). The *blooded* and *bloodless* are not higher genera, but groupings of genera; blood is a symptom of heat. (Balme 1972, 120–21)[17]

(2) Though the force of the last clause of Professor Balme's remarks is not clear to me, surely neither of the reasons he gives for Aristotle's not seeing the blooded and bloodless (roughly, our vertebrates and invertebrates) as true genera prevents *blooded* or *bloodless* from appearing in the definition of animal kinds, and Professor Balme himself does not mean to imply that they do (see, e.g., Balme 1980, 7 n. 32 [= Balme 1987d, 297 n. 34]). 'Blooded' is said to be in the *ousia* of birds in passage 8 below, of fishes in passage 9, and of snakes in passage 10—so Aristotle means it.

(3) There is, or at least appears to be, one interesting difference between the present passage and the previous one. In passage 2, it seemed that Aristotle was asserting the feature in question (viz., the having of lungs) to be part of the *ousia* of lung-possessors *qua* lung-possessors (though there was question about that). In the present passage, however, the subjects whose *ousiai* are said to include being blooded (or bloodless) would certainly appear to be the individual *megista genē* themselves.

(4) Passage 2 asserted an *organic* part to be part of the *ousia* of an animal kind. The present passage is even more striking: here we find a *homoiomerous* part (blood or, in effect, its analogue) asserted to be part of an animal's *ousia.*[18]

[17] Balme 1972, 120–21. On the other hand, *PA* I.2 642b10–16 (esp. 15–16) might suggest that Aristotle does view *ta enaima* as a nameless but otherwise true *genos*. In a recent study of Aristotle's use of the notions of *genos, eidos, diaphora,* and *analogia* in Aristotle's biology, Pierre Pellegrin (1986, ch. 2; cf. Pellegrin 1985, 1987) has shown that although there are fairly strict logical relationships between these concepts, these relationships can appear at virtually all levels of generality.

If Pellegrin is right in his claims that (i) *eidos*, whenever it stands for a group of animals, always designates a subgroup of some wider *genos*; (ii) features are said to be the same *kat' analogian* when the groups said to possess them are viewed as *γένη* relative to explicit or understood subgroups (= *eidē*): and (iii) even the *megista genē* of *HA* I are capable of designation as *eidē* relative to some wider *genos* (as in fact they are in that very passage—cf. 490b17, which Pellegrin does not mention in his book); then it is not clear to me if Balme's argument against viewing the blooded, or bloodless, animals as true *genē* still has force. Deciding that seems to require more reflection than I can give it right now, but Pellegrin's work, and other studies in Gotthelf and Lennox 1987, seem to call for an entire re-examination of the nest of issues, referred to also in n. 13, which surround Aristotle's key metaphysical notions. For the purposes of the present chapter, though these issues had best be kept warm, a back burner will do.

[18] On the *homoiomerous* character of blood, see *PA* II.2–4. One might worry, in the case of the bloodless animals, about the assignment of a privation as part of the *ousia* of something, in view of the apparent strictures against this in *PA* I.3. It should be noted first of all that these strictures appear in a critique of dichotomous division, where further differentiation of that very privative feature will be required. If one works, as Aristotle insists one must, with multiple differentiae, and

4. PA *IV.6 682a35-b32*

Now, then, the insects . . . are all many-footed, in order to make their motion quicker and to counteract their natural slowness and coldness, and those that are especially cold on account of their length are very many-footed, such as the centipede-kind. Furthermore, because they have many origins [of life-activity], they have their 'insections' and are correspondingly many-footed . . .

They are insected both on account of the aforementioned causes and in order that they may protect themselves by curling up to avoid injury. For those of them that are long coil themselves up, and this could not happen if they were not insected. Those that do not coil up make themselves harder by pulling their sections together . . . And it is necessary for them to be insected; for **having many origins is present in their *ousia*** and in this they are similar to plants. For, just as plants, they also can live when cut up, though the latter live only a while, whereas the former become complete in nature and two or more in number from one.[19]

(1) The connection between having insections and having many origins is stated twice, first in the process of offering a second explanation of why insects are many-footed and then in explanation of the insections themselves.

(2) The passage would seem to suggest that all insects can live if divided, at least if both explanations of insections are to hold in all cases. However, at *HA* IV 531b30ff. Aristotle says:

divides them down simultaneously, one might well reach a privative feature—Aristotle himself, for instance, treats footlessness as central to snakes. If further division is required, the other, non-privative, differentiae are available to be divided. Aristotle is quite explicit about this at *PA* I.3. 643b10-26.

It is only where the privative designation in some way hides variation among the subdivisions that it will not be proper to designate the generic feature privatively. Such may well be the case in the present passage, where 'bloodless' is probably a stand-in for the unnamed fluid which is the analogue of blood in the bloodless animals. For, just as Aristotle believes that the character of blood differs importantly from blooded kind to blooded kind (*PA* I. 643a4-5; II. 647b29-35, 648a9-11, 650b14-651a19 [a fascinating passage quite apart from this issue]), so he believes that the unnamed blood-analogue differs across the bloodless kinds: cf. *PA* II. 647b35-648a7, 650b18-33. (On the fluid's namelessness, see Peck's note to *HA* I. 489a23.)

[19] *τὰ μὲν οὖν ἔντομα . . . πολύποδα μὲν . . . ἐστι πάντα διὰ τὸ πρὸς τὴν βραδυτῆτα καὶ κατάψυξιν τῆς φύσεως τὴν πολυποδίαν ἀνυτικωτέραν αὐτοῖς ποιεῖν τὴν κίνησιν· καὶ μάλιστα πολύποδα τὰ μάλιστα κατεψυγμένα διὰ τὸ μῆκος, οἷον τὸ τῶν ἰούλων γένος. ἔτι δὲ διὰ τὸ ἀρχὰς ἔχειν πλείονας αἵ τ' ἐντομαί εἰσι καὶ πολύποδα κατὰ ταῦτας★ ἐστίν . . . ἔντομα δ' ἐστὶ διά τε τὰς εἰρημένας αἰτίας, καὶ ὅπως σῴζηται δι' ἀπάθειαν συγκαμπτόμενα· συνελίττεται γὰρ τὰ μῆκος ἔχοντ' αὐτῶν, τοῦτο δ' οὐκ ἂν ἐγίνετ' αὐτοῖς μὴ οὖσιν ἐντόμοις. τὰ δὲ μὴ ἑλικτὰ αὐτῶν σκληρύνεται μᾶλλον συνιόντα ἐς τὰς τομάς . . . ἀναγκαῖον δ' ἐντόμοις αὐτοῖς εἶναι· τοῦτο γὰρ* ***ἐν τῇ οὐσίᾳ αὐτῶν ὑπάρχει τὸ πολλὰς ἔχειν ἀρχάς***, *καὶ ταύτῃ προσέοικε τοῖς φυτοῖς. ὥσπερ γὰρ τὰ φυτά, καὶ ταῦτα διαιρούμενα δύναται ζῆν, πλὴν ταῦτα μὲν μέχρι τινός, ἐκεῖνα δὲ καὶ τέλεια γίνεται τὴν φύσιν καὶ δύο ἐξ ἑνὸς καὶ πλείω τὸν ἀριθμόν.*

★*ταύτας* Peck; *ταύτ'* Bekker

All insects when cut in two retain life, except those which are excessively cold to begin with, or which on account of their smallness are quickly chilled: even wasps continue to live after being cut in two. So long as it has the middle piece, the head or the stomach can live; but without it the head cannot do so. The long insects with numerous feet live a long time after division, and the severed portion can move in either direction . . . (tr. Peck 1970)

Several other mentions of the survivability after division of ('some') insects confirm that this is Aristotle's consistent position.[20]

That raises a problem however: Are we to say that all insects have having-many-origins in their *ousia*, but the very cold ones are deformities and have this in a very limited way or not at all? Or that it is only the class of insects-which-can-live-if-divided which has having-many-origins in its (their?) *ousia* (thereby giving an *ousia* to a not-obviously-natural cleavage of insects)? Or should we say that having-many-origins is necessary but not sufficient for being able to live if divided? Without attempting to resolve this, I will proceed as if the second of these three alternatives is not true, and assume that Aristotle is ascribing an *ousia* to insects as such.

(3) Having 'insections' is *not* said to be in the *ousia* of insects, but what insections somehow make possible—viz., having many origins—is. The necessity which flows from having many origins to being insected appears to be a case of the *first* of the three modes of explanation indicated in *PA* I. 640a33ff. (quoted above) and not a case of the second mode, standard 'hypothetical' necessity. The other explanation Aristotle gives for insections, the defensive tactics they permit, is on the other hand a clear-cut case of hypothetical necessity.[21]

(4) One might think that when Aristotle says that these insects have many origins he means that they have many *souls*, and *to tēs toiautēs archēs morion* (682a1-2) suggests that. On the other hand, the relevant passage as a whole (681b31-682a8) seems to allow for his meaning by 'multiple origins' multiple *organs*, viz., the heart-analogues he mentions at 681b14-16; and he does regularly refer to the heart as an origin. *Juv.* 468a27-29 says clearly that the nutritive organ (sc. heart-analogue) is actually one but potentially many, as is the nutritive soul (b2-4). (However, two MSS are missing the words that make the former claim.) *archē* is used for the organ (*hē tēs threptikēs archē psuchēs*) at 468a21. But *Resp.* 478a31-479a7, with its peculiar occurrence of *ousia* (b3, cf. n. 15 above), *appears* to use *archē* for the soul three times (b33, a2, a7), although each occurrence is perhaps ambiguous. I do not see how we can

[20] In his *Index*, Bonitz cites: *An.* 411b20, 413b20; *Long.* 467a19; *Juv.* 468a25, b2; *Resp.* 471b21, 479a3; *PA* 682a5; *IA* 707a28, b2.

[21] 'When stepped on, a worm doubles up. That is clever. In that way he lessens the probability of being stepped on again. In the language of morality: *humility*' (F. Nietzsche, *Twilight of the Idols*, trans. Kaufmann 1968, 471).

decide this, and thus whether to say that we have here another case of an organ being said to be part of an animal's *ousia.*

5. PA *IV.8 684a32-b1*

In lobsters only, whether male or female, it is a matter of chance which claw is bigger. The cause of their having claws is that **they are in the kind that has claws**; but they have them irregularly because they are deformed [Balme: 'imperfect'] and do not employ them for their natural use but for walking.[22]

(1) There is no mention of *ousia* or essence in this passage, but since lobsters comprise a true *genos* the suggestion might be that having claws belongs in the definition of lobsters. Balme 1980, 7 n. 32 [= 1987d, 297 n. 34] concurs (though, in the Princeton Colloquium paper [1976] from which this later one was derived, he wrote, citing the present passage, 'it is rare for him to account for a feature on the grounds that the parents of the species are formed so, and my impression is that he only falls back on this when at a loss for a functional explanation'). I want—in a rather roundabout way, since the matter is complicated and several interesting side issues emerge—to explore this suggestion.

(2) The *genos* of lobsters is a subdivision of the *megiston genos* crustacea (*ta malakostraka*). All crustaceans are bloodless and have no viscera (passage 3 above). They are distinguished from other bloodless genera—the cephalopods, testacea, and insects—most notably in the degree of hardness of their earthy, supporting part, and its presence as a protective outer covering rather than as an internal, only structural support. The cephalopods—e.g., squids and octopods—are the reverse, though mostly soft and fleshy; testacea—e.g., oysters and clams—are the same as crustaceans in this regard, but their outer covering is much harder, and 'can be broken with a clean crack' while a crustacean shell 'must be crushed' [*HA* IV.1 523b5-11]; insects, finally, are hard all the way through, though it is a fleshlike hardness. (Texts: *PA* II.8 653b36-654a31; IV.5 679b31-34.) Crustacea differ from members of the other genera in numerous other morphological and functional ways which Aristotle happily catalogues in *PA* IV.5-9; *HA* IV.1-7, V.15-31, and elsewhere.

In particular, in contrast to testacea, 'all the crustacea can walk on land as well as swim, and hence they all have numerous feet' (683b25-26). The *genos* of lobsters is one of four main subdivisions of crustacea. Two of the other three main groups have claws, 'not for the sake of walking but for catching and holding, in place of hands; that is why they bend in an opposite direction to the feet' (683b26-684a1). In these two main groups, all the animals have the right claw bigger and stronger than the left, since animals, Aristotle believes, are generally structured to get more

[22] *οἱ δ' ἀστακοὶ μόνοι, ὁποτέραν ἂν τύχωσιν ἔχουσι μείζω τῶν χηλῶν, καὶ αἱ θήλειαι καὶ οἱ ἄρρενες. αἴτιον δὲ τοῦ μὲν ἔχειν χηλὰς ὅτι* **ἐν τῷ γένει εἰσὶ τῷ ἔχοντι χηλάς**· *τοῦτο δ' ἀτάκτως ἔχουσιν ὅτι πεπήρωνται καὶ οὐ χρῶνται ἐφ' ὃ πεφύκασιν, ἀλλὰ πορείας χάριν.*

use out of their right side and can thus make use of a stronger limb on that side (684a25-31). At this point the text quoted above (passage 5) appears.

(3) Lobsters observably do not uniformly have the right claw larger; in fact, the distribution in their case seems entirely random. (Note that Aristotle has made sure that the variation does not correspond to sex, which is a common source of variation within kinds.) Furthermore, we observe that lobsters do not use their claws for the activity for which they are used by all other claw-possessors and for which they are naturally suited—once you have seen a claw used for catching and holding, that use will surely seem more natural or appropriate to the claw than its use for walking. Lobsters instead use their claws for walking, along with their ten or so (other) feet. (Cf. *HA* IV. 526a11-30, where the claws are in fact called 'feet'.) How are we to explain these differences from the other crustacea, which lobsters otherwise resemble in so many ways? Aristotle's answer is that lobsters are 'deformed' or 'defective' (*pepērōntai*). What does that mean—and how can a whole Aristotelian kind be deformed? Relative to what?[23]

(4) LSJ gives 'maimed', 'mutilated', 'defective' for the term, and cites several Aristotelian passages. The seal's puny feet make it, *qua quadruped, hōsper anapērian* (*HA* II. 498a32)—'a sort of stunted quadruped' (Peck 1965). Dwarfs, and other defective individuals of various species, are said to be deformed in the uterus, and the concept plays a role in the discussion of hybrids (*GA* II.8 748b34-749a6). Moles, whose blind eyes got 'stunted in their *genesis*' (*HA* I. 491b34; IV. 533a11-13; tr. after Peck), are themselves said to be 'stunted' (IV 533a2; *De An.* 435a10-11). This reference to generation—and an insistence that a type of animal will be 'deformed' as a whole *only* relative to some wider type to which it belongs—will be the clues to understanding the meaning of the term.

(5) *Deformity is relative to a wider kind.* Dwarfs and so on (including, alas, females, whose femaleness we should view *hōsper anapērian* [*GA* IV.6 775a15-16]) are defined of course relative to the normal (male) members of their kinds. The seal is a (sort of) deformed *quadruped*. Interestingly, in speaking of kinds that seem to defy classification into wider groups because they have several features, but not all, of each of two such groups (*PA* IV 13—our classic 'borderline cases'), Aristotle explains that seals 'dualize' (*epamphoterizein*: Professor Balme has suggested 'tend in both directions') with respect to being land and water animals, and that their hind

[23] This view is sometimes cited as an illustration of the influence of pre-philosophical popular belief on Aristotle's biological thought. G. E. R. Lloyd (1983, 40-43), in a discussion sympathetic to such claims, finds limited evidence in this case, though he is reluctant to abandon the anthropological interpretation completely. In light of the extent to which Aristotle's view, as I will show, is integrated into his general biological theory, I see no basis whatever for the claim of a *distinct influence* from popular belief, whatever the 'parallels' might be. The same is true, I would maintain, for Aristotle's 'dualizers', *contra* Lloyd's more confident assertions of influence here (48 and 50). (For trenchant critiques of Lloyd 1983 on these matters, see Lennox 1985c and Parker 1984.)

feet can be considered completely fishlike (697b1-7)—so they are not really quadrupeds. Yet, if we consider them *hōsper* quadrupeds, as we need not but can, they do not do a very good job of it. The mole, finally, is defective relative to *other* eye-possessors (*HA* I. 491b26-34; IV. 532b30-533a15: cf. the explicit *kata to genos* at *Metaph*. *Δ*. 1022b26-27).

Pace anthropologizing interpretations, then, the biological point is that we have organs that cannot perform (as well) the functions they are most suited for and have developed for (as I will explain in a moment), by the standard of the performance of the version of that organ in the other members of that wider kind. Where the organs perform another function sufficiently well, as the fin-like hind feet do in seals (697b4-6), Aristotle will say, it seems that they are only *hōsper* deformed. The female, who has a natural and necessary function, which she performs perfectly well, has her 'deformity' doubly qualified (and perhaps triply, if *anapēria* is weaker than *pērōma*—but LSJ is no help towards deciding this).

(6) *Deformity occurs in generation*. Let us remind ourselves, though only sketchily for considerations of space, of two of Aristotle's theses about generation: (i) The embryo gets progressively differentiated from more generic to more particular features (*GA* II.3 736a32-b5, and II.3-6 and IV. 3-4 *passim*); and (ii) There is a limited amount of material that can be used to form the embryo—the mother only supplies so much and the organism can only absorb so much—thus if, for instance, it needs solid (rather than cloven) hooves, it may only have material enough for one horn (*PA* III.2 663a31-34, and II–IV *passim* for many other examples). In the cases of so-called 'deformity', as a member of the wider kind against the background of which it is viewed as deformed, the animal in question will go through embryonic stages in which it is indistinguishable from members of other subdivisions of that kind. Insofar as seals are like quadrupeds, for instance, they *could have* gotten fully developed legs, which would have made them nifty land-movers and not the clods (*pērōmata*) they are; but that would have made them awfully inept at feeding in water which, by the 'blend of their elements' and probably their constitution generally, they must do, so their nature does not develop that way.[24] What, then, about lobsters?

[24] It may sound peculiar to attribute to Aristotle the view that the blend of material elements makes it causally necessary that an animal feed in a certain place: one would have expected the causation to go, as it so usually does in Aristotle, *from* the *bios to* the materials necessary for such a life. But the former is clearly Aristotle's view: cf. *HA* VII. 589a5-8, *Resp*. 14; *PA* III. 669a9-11; *GC* II. 335a10. At *HA* VII. 589b29-590a12, Aristotle seems actually to claim that the live-bearing animals with a watery blend, which blend requires them to feed in water, are formed with organs appropriate to that environment by a 'deformity' in the generative process. These passages belong more properly in the second part of the research program I have described, as sources for the animal features which Aristotle takes to be explanatorily fundamental and thus parts of the *ousia* of those animals that possess them.

(7) Being in the *genos* of claw-possessors is supposed to explain lobsters' possession of claws. Why does Aristotle not say that their use in walking explains their presence? Well, for one thing, then they would not be *claws*, they would be *feet* (which, as we saw, Aristotle is tempted to call them in *HA*): an organ is defined and ultimately named (in part at least) by its function. But they *look* so much like claws, and—much more importantly—they do so in the context of the lobster-kind's being in many ways much more like the crustacea-kinds that possess claws than like the kinds that do not. That is: I want to suggest that the overriding similarities to other claw-possessors puts lobsters in a group with them; accordingly, we should see their generation as out of roughly the same material and following sufficiently the same plan so that they pass through the same early stages. Relative, then, to the development of the claw in that larger group, we can say that the lobster claw does not develop all the way towards the maximal possible use of it, perhaps because the rest of the animal develops in such a way that it cannot make use of a catcher and holder of that sort (as the mole develops from an embryo much the same as that of the other smallish live-bearing quadrupeds into a burrower which does not particularly need eyes).[25]

(8) To return finally to the main text under review (passage 5 above), the preceding discussion suggests that having a claw will *not*, after all, be in the definition of lobsters, not even in its generic portion. What may appear in its place would be the causally fundamental features, whatever they are, shared by all claw-possessors that make them so alike relative to the non-claw-possessing crustacea, and which in the non-lobster cases hypothetically necessitate claws. The claw in the lobster will then be explained in terms of the lobster's variations on those causally fundamental features such that its mode of life (*bios*) will not require catchings and holdings. Subject to correction, then, I will *not* include this passage among those giving us features that are part of the *ousia* or essence of the animal in question.

6. PA *IV.9 685b12-16*

Now, the others [that have suckers] have two rows of suckers, but one kind of octopus has one row of suckers. The cause of this is **the length and slimness of their nature**, for a narrow one will necessarily have only one row. Thus it is this way not as being the best but as necessary because of **the particular *logos* of its *ousia*.**[26]

[25] Notwithstanding the texts that I have referred to, I view this as a largely speculative account of the matter. Skeptics have my sympathy, but only if they view the current account as a stimulus to further reflection on the issues.

[26] *τὰ μὲν οὖν ἄλλα δικότυλά ἐστι, γένος δέ τι πολυπόδων μονοκότυλον. αἴτιον δὲ* **τὸ μῆκος καὶ ἡ λεπτότης τῆς φύσεως αὐτῶν·** *μονοκότυλον γὰρ ἀναγκαῖον εἶναι τὸ στενόν. οὐκ οὖν ὡς βέλτιστον ἔχουσιν, ἀλλ' ὡς ἀναγκαῖον διὰ* **τὸν ἴδιον λόγον τῆς οὐσίας**.

(1) It seems clear that the length and slimness (or perhaps just the slimness) is here said to be in the particular *logos* of its *ousia* ('particular', to connote 'versus that of other kinds of octopus'). If so, this is our first case of a feature in the *ousia* of an animal which is neither a function nor a part: the feature in question is, roughly, shape—or shape relative to size. This occurs again in passage 10 below, and in several passages more appropriate to the second part of our project.[27] Using a term suggested by Montgomery Furth, I shall call these 'dimensional features'. James Lennox reminds me that these are just the sorts of features that Aristotle says in *PA* I.4 and elsewhere will vary in 'the more and the less' (see Lennox 1980).

(2) Aristotle does not say that having one row of suckers is part of the *ousia* of this kind of octopus. Rather, he explains it as a necessary consequence of the dimensional feature which he does say is part of this octopus's *ousia*. What is most striking here is that the necessity in question is *not* the hypothetical necessity of the second mode of explanation indicated in *PA* I 640a33ff. (quoted above). It is rather a sort of geometrical necessity: given that a sucker must be at least so large, and that this octopus's arms can only be so large, no more than one row could fit. This may—interestingly—be a case of the *first* mode of explanation. Notice too that the second mode is not mentioned at all here by Aristotle, who only rules out the third mode (unless 'best' is an ellipsis for 'necessary or best for some end', which it may well be).

7. PA *IV.10 686a25-31*

Now, man, instead of forelegs and forefeet, has, as we call them, arms and hands. For he alone of the animals stands upright, on account of **his nature and *ousia* being divine**, and the function of that which is most divine is to think and reason; and this would not be easy if there were a great deal of the body at the top weighing it down, for weight hampers the motion of the intellect and the common sense.[28]

(1) I take it that man's ability to reason is part of his *ousia*, and that Aristotle is saying, not that *being divine* is part of man's *ousia*, but that those rational capacities

[27] The elephant's size, for example, appears to be causally primary in the explanation of the unique (*idiōtaton*) size and character of its 'nose' (*PA* II. 658b33-659a8, discussed in chapter 8 above). Montgomery Furth has brought my attention to *GA* II.6 745a5-6: *esti gar ti pasi tois zōios peras tou megethous*, which he translates, 'for all animals there's a particular limit to their size'. While this does not show that size is *causally primary* (and thus part of the *ousia* or essence) in every case, it does suggest that it may well be so more often than we (and especially those amongst us influenced by strongly functionalist interpretations of Aristotelian form) might have thought. (This passage is discussed briefly in chapter 4 above, pp. 113-15.)

[28] *ὁ μὲν οὖν ἄνθρωπος ἀντὶ σκελῶν καὶ ποδῶν τῶν προσθίων βραχίονας καὶ τὰς καλουμένας ἔχει χεῖρας. ὀρθὸν μὲν γάρ ἐστι μόνον τῶν ζῴων διὰ τὸ* ***τὴν φύσιν αὐτοῦ καὶ τὴν οὐσίαν εἶναι θείαν****· ἔργον δὲ τοῦ θειοτάτου τὸ νοεῖν καὶ φρονεῖν· τοῦτο δ' οὐ ῥᾴδιον πολλοῦ τοῦ ἄνωθεν ἐπικειμένου σώματος· τὸ γὰρ βάρος δυσκίνητον ποιεῖ τὴν διάνοιαν καὶ τὴν κοινὴν αἴσθησιν.* Cf. II.10 655b29-656a13, esp. 656a4-8, quoted in comment (2) below.

which *are* part of man's *ousia* are divine or most divine among animals. (Has Aristotle here used two verbs, and the particular ones he did, in order to capture both the theoretical [*noein*] and the practical [*phronein*] sides of man's intellect?)

(2) 656a4-8: 'The greatest variety [of structure and function among organisms possessing sensation, i.e., animals, is] found among those whose nature partakes not only of living but of living well. Such is the *genos* of men, for he alone of the animals known to us has a share of the divine, or most of all.'

(3) Cf. *GA* II.3 736b30f.: 'Now the *dunamis* of all soul seems to be associated with a body different from and diviner than the so-called elements'; and Balme (1972, 163 *ad loc.*): 'The very use of the comparative "divin*er*" excludes a definite boundary between divine and non-divine. He need only mean here what he meant at 732a2 and often: less grossly material, purer, superior.'

8. PA *IV.12 693b2-13*

Two legs like man, bent inward like quadrupeds and not outwards like man; and the wings, as the fore legs of quadrupeds, with the convex side outward. They are biped of necessity, for **the *ousia* of a bird is among the blooded,** and at the same time also winged, and the blooded do not move on more than four points. So birds have four attached parts, just as the other [blooded] ones, i.e., the ones that live and move on the ground; but whereas these have arms and legs (quadrupeds having forelegs), birds instead of forelegs or arms have—uniquely—wings, for they can be stretched out with these, and **being a flier is in the *ousia* of a bird.**[29]

(1) 'the *ousia* of a bird is among the blooded': an unusual locution, but I assume that its force is 'a bird, in respect of its *ousia*, is among the blooded', in which case we have one more instance in which the material feature of being blooded is said to be part of the *ousia* of an animal kind.

(2) 'being a flier' (or: 'being able to fly'): a functional feature, in particular, a mode of locomotion. The mode of locomotion is given as part of the *logos tēs ousias* of fishes in the next text (passage 9); and in his logical and metaphysical works, of course, Aristotle frequently uses 'two-footed animal' as the *logos* of man (though one never knows how seriously to take this). Is *two-footed* a morphological or a

[29] *σκέλη δὲ καθάπερ ἄνθρωπος δύο, κεκαμμένα καθάπερ τὰ τετράποδα εἴσω, καὶ οὐχ ὥσπερ ἄνθρωπος ἔξω· τὰς δὲ πτέρυγας, ὡς τὰ πρόσθια σκέλη τῶν τετραπόδων, ἐπὶ τὸ περιφερές. δίπουν δ' ἐξ ἀνάγκης ἐστίν·* **τῶν γὰρ ἐναίμων ἡ τοῦ ὄρνιθος οὐσία,** *ἅμα δὲ καὶ πτερυγωτός. τὰ δ' ἔναιμα οὐ κινεῖται πλείοσιν ἢ τέτταρσι σημείοις. τὰ μὲν οὖν ἀπηρτημένα μόρια, τέτταρα, ὥσπερ τοῖς ἄλλοις τοῖς πεζοῖς καὶ τοῖς πορευτικοῖς, ἔστι καὶ τοῖς ὄρνισιν, ἀλλὰ τοῖς μὲν βραχίονες καὶ σκέλη, τοίς δέ τετράποσι★ σκέλη τέτταρα ὑπάρχει, τοῖς δ' ὄρνισιν ἀντὶ τῶν προσθίων σκελῶν ἢ βραχιόνων πτερύγιον κοινόν ἐστιν κατὰ ταύτας γὰρ τονικοί★★ εἰσι,* **τῷ δ' ὄρνιθι ἐν τῇ οὐσίᾳ τὸ πτητικόν ἐστιν.**

★*sic* PY*b*, Ogle; *σκέλη, τοίς δέ τετρ. om. vulg.*)

★★*πτητικοί* conieci; idem Th. (*volatiles* Gaza)

functional differentia (in the language of *HA*, is it a *morion* or a *praxis*), or is it in some sense both? (Aristotle does not call fishes 'finned animals', yet live-bearing *four-footed* animal is a *megiston genos*, and the words usually translated 'flying' sometimes seem to mean 'winged'.)

9. PA *IV.13 695b17-25*

Fishes do not have attached limbs because **their nature according to the *logos* of their *ousia* is being a swimmer,** since nature makes nothing either superfluous or pointless. And since **they are according to their *ousia* blooded,** because they are swimmers they have fins, while because they do not walk they do not have feet, for the addition of feet is useful for moving on the ground. At the same time it is not possible for them to have four fins and feet as well, or any other such limbs since they are blooded.[30]

(1) As the *ousia* of birds includes being blooded and being a flier, so the *ousia* of the coordinate *megiston genos* of fishes includes being blooded and being a swimmer.

(2) That all blooded animals must move on more than four points is argued in *IA* 7.

10. IA *8 708a9-20*

The cause of the footlessness of snakes is *both* that nature makes nothing pointless, but always with a view towards the best for each thing from among the things which are possible to it, preserving **the particular *ousia* and essence of each,** *and* what we have said earlier, that it is not possible for blooded animals to move with more than four points. For, it is evident from these things that **all blooded animals which are disproportionate in length relative to the rest of the bodily nature, as are the snakes,** cannot possibly be footed. For it is not possible for them to have more than four feet (since then they would be bloodless), and if they had two or four feet they would be pretty much unable to move at all; this way their movement would necessarily be slow and useless.[31]

(1) This passage contains features of both kinds of texts with which our overall project is concerned. On the one hand, it enunciates a principle which underlies at

[30] οὐκ ἔχουσι δὲ ἀπηρτημένα κῶλα οἱ ἰχθύες, **διὰ τὸ νευστικὴν εἶναι τὴν φύσιν αὐτῶν κατὰ τὸν τῆς οὐσίας λόγον,** ἐπεὶ οὔτε περίεργον οὐδὲν οὔτε μάτην ἡ φύσις ποιεῖ. ἐπεὶ δ' **ἔναιμά ἐστι κατὰ τὴν οὐσίαν**, διὰ μὲν τὸ νευστικὰ εἶναι πτερύγια ἔχει, διὰ δὲ τὸ μὴ πεζεύειν οὐκ ἔχει πόδας· ἡ γὰρ τῶν ποδῶν πρόσθεσις πρὸς τὴν ἐπὶ τῷ πεδίῳ κίνησιν χρήσιμός ἐστιν. ἅμα δὲ πτερύγια τέτταρα καὶ πόδας οὐχ οἷόν τε ἔχειν, οὐδ' ἄλλο κῶλον τοιοῦτον οὐδέν· ἔναιμα γάρ.

[31] τοῖς δ' ὄφεσιν αἴτιον τῆς ἀποδίας τό τε τὴν φύσιν μηθὲν ποιεῖν μάτην, ἀλλὰ πάντα πρὸς τὸ ἄριστον ἀποβλέπουσαν ἑκάστῳ <ἐκ> τῶν ἐνδεχομένων, διασώζουσαν **ἑκάστου τὴν ἰδίαν οὐσίαν καὶ τὸ τί ἦν αὐτῷ εἶναι**· ἔτι δὲ καὶ τὸ πρότερον ἡμῖν εἰρημένον, τὸ τῶν ἐναίμων μηθὲν οἷόν τ' εἶναι πλείοσι κινεῖσθαι σημείοις ἢ τέτταρσιν. ἐκ τούτων γὰρ φανερὸν ὅτι **τῶν ἐναίμων ὅσα κατὰ τὸ μῆκος ἀσύμμετρά ἐστι πρὸς τὴν ἄλλην τοῦ σώματος φύσιν, καθάπερ οἱ ὄφεις,** οὐθὲν αὐτῶν οἷόν θ' ὑπόπουν εἶναι. πλείους μὲν γὰρ τεττάρων οὐχ οἷόν τε αὐτὰ πόδας ἔχειν (ἄναιμα γὰρ ἂν ἦν), ἔχοντα δὲ δύο πόδας ἢ τέτταρας σχεδὸν ἦν ἂν ἀκίνητα πάμπαν· οὕτω βραδεῖαν ἀναγκαῖον εἶναι καὶ ἀνωφελῆ τὴν κίνησιν.

least the third mode of explanation identified at *PA* I. 640a33ff. and then proceeds to explain *via* that principle (and in that mode) a feature (viz., being footless) in terms of some other feature (being unusually long relative to the rest of its body—unusually, I suppose, by comparison with other blooded animals). This latter feature is not itself explained anywhere in terms of yet other features and thus, by the considerations of this second part of the program, is a candidate for being viewed as part of the essence of that animal.

On the other hand, the reference to the 'particular *ousia* and essence' in the context of a remark about the length-relative-to-the-rest-of-the-body (apparently) as compared with other blooded animals, leads one to take this as an explicit statement that such a relationship of length relative to thickness *is* in the *ousia* of snakes.[32]

(2) If so, we have a second case—along with the length and slimness of the type of octopus mentioned in passage 6—of a 'dimensional' feature pertaining to a size–shape relationship, which is in the *ousia* of an animal kind.

Interim conclusions

A fuller picture of the conceptions of substance and essence at work in the biology must await the results of the second part of this project, the examination of the explanations in *PA* II–IV for the causally primary features. But the results of our examination of these nine partial definitions can be summarized and some preliminary reflections ventured.

Each definition (as I will call what are in fact only partial definitions) asserts some feature of an animal to stand in some, shall we say, *definitional relation* to that animal ('is present in the *ousia*', 'have their nature according to the *logos* of their *ousia*', and so on). We may call the feature(s) asserted to stand in that relationship the *content* of the definition, and the animal kind of which it is asserted the *object* of the definition. It will be interesting to explore the sorts of things that turn up in each of those roles (relations, features, animal kinds) in our definitions, and we may do so by means of the following chart:

[32] There is of course no conflict between (a) claiming to find an explicit occurrence of a feature being said to be part of some animal's *ousia* and (b) finding that feature's presence quite independently to be the basis for explaining some other feature's presence in that type of animal. Quite the contrary, for we can see from passages 1-9 that in every case in which a feature is identified as part of the *ousia* of some animal kind, it is so identified as part of the attempt to explain some other feature's presence. This should come as no surprise, since there is surely some reason to think that the very basis for identifying some feature as part of the *ousia* is precisely its being causally primary relative to other features of the animal. On this, see Bolton 1987 and chapters 3 and 7 above.

Relation	*Content*	*Object*
1. *logos*,★ we define . . . by . . .	having sensation (and first the primary sense)	an animal (*to zōion*)
2. is present in the *ousia*	having a lung	nameless (the lung-possessors)
3. belongs to the *ousia*	(a) not having blood, being bloodless★★	(a) bloodless animals★★★
is present in the *logos* by which we define the *ousia*	(b) being blooded	(b) blooded animals★★★
4. is present in the *ousia*	having many origins	the insects (*ta entoma*)
5. [not a definition]		
6. the particular *logos* of the *ousia*★	the length and slimness of their nature	a certain *genos* of octopuses
7. the *ousia*★	thinking and reasoning	man
8. (a) the *ousia* of . . . is of [= made of, among] . . .	(a) blooded	a bird (*ho ornis*)
(b) is in the *ousia*	(b) being a flier	
9. (a) the nature according to the *logos* of the *ousia* is	(a) being a swimmer	the fishes
(b) are according to their *ousia*	(b) blooded	
10. the peculiar *ousia* and *to ti ēn einai*★	unusually long relative to the rest of the body	snakes★★★★

★ In these cases Aristotle does not explicitly identify the relation (*in* the particular *logos tēs ousias, is part of* the *ousia*, etc.), though one will infer some such relation. These occurrences are indicated here by the absence of a verb in the Relation column.

★★ As discussed above, probably equivalent to 'having the nameless fluid analogous to blood'.

★★★ As above, probably not under that description, but *qua* the various bloodless or blooded *megista genē*. Or perhaps Aristotle is speaking of *all* bloodless or *all* blooded animals, whether or not they fall under any *megiston genos*. Does that imply, then, 'however identified'?

★★★★ Equally, '*all* blooded animals which are disproportionately long relative to the rest of their body', including, e.g., eels as well as snakes (though eels are also footless because they are swimmers, as in passage 9). As with the case marked '★★★', the proper level of identification, if there is one, is not clearly indicated. If the explanation included in passage 10 is taken to apply only to land animals, then the only kind involved would seem to be the snake. But there are water as well as land snakes (*ophis* even includes 'sea serpents'—cf. *HA* II.14), so this is not clear.

1. *The definitional relation*. The most striking feature of this column is the variability of expression, the lack of any canonical form in which the relation is stated (as there is for the relations studied in the *Prior Analytics*, for instance). Verbs used include: *esti, huparchei, enhuparchei*. The definitional character is indicated by such varied expressions as: *ousia, logos tēs ousias*, the peculiar *ousia* and *ti ēn einai*. Connectives include: *en* + dative, various genitives, and so on. At times the precise relationship is not explicitly stated (such cases are marked '★' in the chart).

The same core notion of *ousia/logos tēs ousias* seems involved in each case (including the first, where the term *ousia* is not used), though the variability of exact terminology should warn us against over-systematization (as David Balme would be the first to stress). As observed above (in n. 32), each feature which is said to be part of the *ousia* of some *genos* of animals serves to explain some other feature(s) of the animals of that *genos*, and it would seem as if each such feature has its *ousia*-status precisely in virtue of playing that role.[33]

It is worth noting, as observed earlier, the paucity of occurrences in these passages of *to ti ēn einai*. In its only appearance, in passage 10, it is unclear whether it is offered as equivalent to *ousia* (epexegetical *kai*) or as a distinct notion, Aristotle intending his generalization to cover both *ousiai* and essences. (See above n. 4.)

2. *The definitional content.* Here the results are clearly significant. In addition to

(i) the standard soul-functions (sensation of various types, thinking and reasoning, locomotion of various types), we find:
(ii) animal parts, both
 (α) non-uniform parts (lung, possibly heart-analogue in insects) *and*
 (β) uniform ones (blood, the blood-analogue in bloodless animals); and
(iii) dimensional features—size-relative-to-shape (length-and-slimness in one type of octopus, length-relative-to-width in snakes), with hints (above nn. 24, 27) that size itself may be definitional as well; and very possibly
(iv) the blend of material elements that constitute the animal.

Categories (ii) and (iii) and of course (iv) (if it pans out) constitute difficulties for strongly functionalist accounts of Aristotle (see Nussbaum 1978, 59-106, 143-64), which latter wish to see only functions, or parts functionally described, as belonging to the *logos ousias* of animals. The parts in (ii) are precisely not defined (only) functionally: Aristotle's distinction between a part and its analogue is precisely one

[33] The issue is complicated by the fact that the possession of a lung, stated to be present in the *ousia* of all animals that have one, is itself explained by reference to the lung's respiratory (i.e., cooling) function, a function which is needed by virtue of these animals' unusually hot nature. The degree of heat is what is explanatorily fundamental—this heat not being explained in terms of anything more basic about these animals—rather than the possession of a lung. (Cf. *PA* 669b8-9, 668b33-669a13; *Resp*. 8-13 [esp. 13]; *GA* 732b26-34.) As the organ which performs this cooling function (see 669a13-14), the lung in such an explanation would have a necessity of the sort given by the first mode of explanation in *PA* I. 640a33ff. and correctly denied to be a 'hypothetical' necessity by John Cooper in Cooper 1987. Cooper himself, as we have noted, treats necessities of the first mode as 'the necessity of what follows directly from the . . . essence, *as included in it*' (Cooper 1987, 254). An important part of Cooper's evidence for this is *GA* V.1 778a34-35, b16-17, where Aristotle *appears* to say just that of the eye (again, allowing the equation of essence to *ousia*). However, the passages Cooper cites are less clear than one would like. For instance, 778a34-35 commits Aristotle only to the claim that the eye *sunteinei pros logon ton tēs ousias* ('reaches to the definition of the substantial being') and this is compatible with the eye being hypothetically necessary and not *included in* the *logos ousias*; *sunteinei pros* is a weaker relation than, for instance, *huparchei en*. In fact the entire opening argument of *GA* V.1, bearing as it does on Aristotle's practice in the explanatory treatises, needs further attention, especially at the peculiar 778b10-13. (See now chapter 5 above.)

between two parts with (essentially) the same function but which differ in material constitution and structure, and do so by more than 'the more and the less'.[34] The material constitution and structure of such parts thus enter into their nature for Aristotle, and this seems ignored by functionalist interpretations. Likewise for dimensional features. The frequent conjunction of *eidos* and *morphē* reminds one of the presence of shape in the *ousia* of certain inanimate objects (insofar as they have an *ousia*), but Aristotle's frequent insistence that shape is not sufficient (e.g., *PA* I. 640b30-641a7, 'a hand in name only') could easily be taken to imply an equation of *ousia* (*via* form) with function, especially in light of Aristotle's standard equation of soul with form. At least in the biology proper, this is not the whole story. And though it is outside our purview in this chapter to attempt to extract interpretative insight into the *Metaphysics*' theory of substance from the conception at work in the biology, one cannot help but suspect that some of our ten passages might throw strong, useful light on the strictures in *Metaph.* Z.10 and (especially) 11 against excluding all reference to matter from definitions of animals and animal forms.

3. *The Object of Definition*. Here the striking feature is the existence of definitions at such widely varying levels of generality, with the absence of any suggestion that there is a privileged lower level (the '*infima species*') which gets special status or attention. More often than not, in fact, the *ousiai* are at quite high levels of generality: the birds, fishes, and insects are *megista genē*, and the lung-possessors, and the blooded and bloodless animals (if they are to be viewed as units), are even broader groups, as of course are the animals as such.

Now, one may well wish to maintain that for Aristotle in the biology *ousia* is *really* given at the species level, and that these higher-level *ousiai* are components of the definition of some genus under which the species falls (perhaps as in *Metaph.* Z.12), which become parts of an *ousia* only when brought down into the ultimate definition of the species. But there is nothing here to suggest that. If any 'bringing down' is to be done, it may just as well be all the way to the individual level as to any level before that. But that would be speculative too, at this point. An easier but important suggestion can be made first about these *ousia*-ascriptions, one which takes off from the explanatory primacy of the features so ascribed. If features are included in the *ousia* in virtue of their role in explanations, and such features occur at varying levels, it must be because *explanations* occur at varying levels. And of course they do. That explanation should be at the 'commensurately universal' level, whichever it is, is standard issue in the *Posterior Analytics* (though perhaps more central to the whole of that treatise's doctrine than has usually been realized); but that this concern pervades the practice in the biological works has only recently

[34] *HA* I. 486a14-487a1 (esp. 486b17-21), II. 497b5-13; *PA* I. 644a16-23: Bonitz, *Ind. Arist.*, 47b53-57, 48a18, 31-47.

been argued with any force, most fully by James Lennox in 'Divide and Explain: The *Posterior Analytics* in Practice' (Lennox 1987a).

Taking the lead from Lennox's study, one might observe that if there is to be explanation at various levels, there will be causally primary features and thus *ousia* at various levels. But just as it is the possession of features by *individuals* that is being explained under commensurately universal identifications of those individuals, so it is perhaps individuals that are primary possessors of *ousia*.

Such a thesis would give no difficulty to species-form-oriented interpretations, so long as the individuals in question are individuals-of-a-kind, with the 'kind' given by their *infimae species*. It would become a controversial, profoundly interesting thesis only if one were unable to find something like *infimae species* in the biology, or if one found them only as constructs out of, or derivatives from, individuals with more determinate natures. Such claims *have* in fact been made recently from varying directions.[35] While this is not the place to evaluate them, it does seem clearer than ever that any serious evaluation of this and other issues of Aristotelian metaphysics will need, as David Balme has long gently been reminding us, to make full and serious use of the biological works.[36]

[35] See Balme 1987d, Pellegrin 1987 (although I would take exception to his claim that Aristotle has no interest in definitions of animals in his biology: what we have been looking at are partial *definitions*, and if Aristotle offers no full definitions in *PA* [or elsewhere] that is because his purpose there does not call for them—but it does call for him (i) to believe that they have definitions, and (ii) to know at least many of the components of these definitions), and esp. chapter 7 above.

[36] A preliminary version of this chapter was prepared for an NEH Research Conference on Aristotle's Epistemology and Metaphysics at Florida State University, Tallahassee, January 5-9, 1983. I am indebted to John Cooper for some helpful remarks thereafter and, especially, to James Lennox for valuable written comments, including the suggestion to include the chart that appears above, and for subsequent discussions. The leading ideas of the chapter were first developed, and the ten passages were located and explored, in connection with an undergraduate honors seminar 'Philosophical Issues in Aristotle's Biology' at Trenton State College, Fall 1982. Two stalwart souls—Dana Ramonas Matioska and Jerry Moonan—persisted, and deserve thanks for lively and productive discussions.

11

A Biological Provenance

Reflections on Montgomery Furth's *Substance, Form and Psyche: An Aristotelean Metaphysics*

I begin with a quotation:[1]

> There is a second major matter of method in this study requiring mention here as not altogether standard, having to do with my use of the biological writings. It springs from a hypothesis, stated more fully in §9 where it first comes into operation, that Aristotle's metaphysical inquiries about essence, substance, individual, and so on were (or came to be) motivated in an extremely concrete and specific way by his theoretical preoccupations in biology. It is of course well known that these biological researches of his were of unexampled depth and extent; but my belief—and one of the interpretive hypotheses in this work—goes farther: it is my view that Aristotle asked himself a series of very profound questions about the animal kingdom, to which the metaphysics of substance was part of a response, so that the latter was to a sizable extent motivated, and thus should be read, as a deep theoretical foundation, a 'methodology' in Tarski's sense, for the biological sciences—much as the work of Frege and his successors was to stand to the 'deductive sciences' two millennia latter.... The idea of a biological provenance for his 'more strictly philosophical work' (in D'Arcy Thompson's phrase) is not itself new and has indeed been pursued in the present century by several distinguished interpreters; it is pursued here on a more specific basis and from a more systematic and theoretically-oriented standpoint than other treatments to date.

So Montgomery Furth in the introduction to *Substance, Form and Psyche: An Aristotelean Metaphysics*, an introduction labeled, as is so characteristic of Furth's literary style, not §1 but §0—'a short discourse on method', as he called it. My aim in this presentation is to examine some of the evidence and argument Furth offered for this hypothesis that (to put it crudely) Aristotle's metaphysics is motivated by his

This chapter was presented as a paper to an APA Pacific Division memorial symposium for Montgomery Furth on the tenth anniversary of the publication of his *Substance, Form and Psyche: An Aristotelean Metaphysics*. The other symposiasts were Alan Code and Aryeh Kosman; the session was chaired by its organizer, Georgios Anagnostopoulos. The three papers were published in *Philosophical Studies* 10 (1989). This has been my one occasion to date to present in print any part of my interpretation of the central books of Aristotle's *Metaphysics*.

[1] Furth 1988, 5-6.

biology, and to offer some reflections both on his argument and on the hypothesis (and surrounding issues) itself. Let me start with the revolutionary character of Furth's hypothesis. This is clearer to those of us who have been around at least since the early to mid-1970s. Furth's book was not published until 1988, by which time the value of the biology for understanding central aspects of Aristotle's 'more strictly philosophical work' was already well on its way towards being firmly established.[2] Because of this, Furth's book arguably made less of a splash than it otherwise might have. Still, as many of us knew, much of the book had been written years earlier: the work for it was begun around 1970—and the central biological portions, §§9-15(ii) in the table of contents (see this chapter's appendix, pp. 255-57)—had almost all been drafted, in some form, by 1976. And so, in that year, when David Balme—the pioneer in the modern study of the philosophical implications of Aristotle's biology, who was already publishing in this area at the beginning of the 1960s—was a visiting member of the Institute for Advanced Study in Princeton, and David Furley sought to build that year's annual Princeton University Classical Philosophy Colloquium around Balme, he chose as co-speakers Marjorie Grene, whose 1963 introductory text, *A Portrait of Aristotle*,[3] had stressed, very programmatically, Aristotle's biological orientation—and Montgomery Furth.[4] So Furth was one of the first. But because the book was not published until 1988, it did not have the impact, or the reception, it would have had, had it been published, say, in 1980.

But Furth's hypothesis was not just revolutionary relative to Aristotelian studies in general; it was also, even for partisans of the biology, quite delightfully radical in content. D'Arcy Thompson, in a prefatory note to his 1910 translation of *Historia Animalium*,[5] and with him John Herman Randall, Jr. and other Columbia University students of Frederick Woodbridge in the first part of the twentieth century (building on the dating arguments of Thompson and later Desmond

[2] A year earlier, a book called *Philosophical Issues in Aristotle's Biology* had appeared on the scene, to quite good reviews and—if its senior editor can be a judge—it consolidated something of a revolution that was already well under way. Indeed, it came to be seen as a sort of flagship of this movement, and Furth's book as only one of several works appearing in its wake—notwithstanding that Furth had been in on this revolution, or movement, from the beginning. In fact, *Philosophical Issues* (Gotthelf and Lennox 1987) contained an important excerpt from Furth's then-forthcoming book, and Jim Lennox's and my introduction placed Furth as one of only a few scholars who had by 1987 already spent over a decade with Aristotle's biology and its philosophical significance.

[3] Grene 1963.

[4] Martha Nussbaum, who had finished her PhD a couple of years earlier with her work on *De Motu Animalium* (Nussbaum 1978), was Furth's commentator. Jim Lennox, who was finishing his PhD, and I, who had finished a year or so before, were in the audience. (I don't think any of us, except perhaps Balme, knew then of the work of Wolfgang Kullmann, whose book, *Wissenschaft und Methode*, had just been published in Germany in 1974.)

[5] W. D., Ross, ed. (1908-1952), *The Works of Aristotle Translated into English*, 12 vols. (vols. 4, 5, and 8 under editorship of J. A. Smith and W. D. Ross) (Oxford) (= Thompson 1910).

Lee[6]) had argued that Aristotle's key philosophical conceptions arose out of his detailed study of animals. Randall, in his 1960 *Aristotle*,[7] summarized approvingly the thesis of a 1939 Columbia dissertation of Harold Hantz:[8] 'Hantz points out that the characteristic Aristotelian distinctions emerge naturally from the inquiry into biological problems of function and generation, and do not appeal to other inquiries or writings.' Randall also quotes Hantz as saying: 'In short, in the biological works, the reader witnesses the emergence of "form", "matter", "end", "mover", "potentiality", and "actuality", the terms which carry Aristotle's system of thought....'[9] But Randall's own treatment of the *Metaphysics*[10] does not quite reflect this picture, since he sees the questions and theses of that treatise as arising in part out of Aristotle's reflection on his predecessors in Plato's Academy, and in part out of a general reflection on what Randall calls 'process', *kinēsis*, understood as the earlier twentieth-century process philosophers did. And Marjorie Grene, although she spoke intriguingly of the influence of Aristotle's biology on his concept of essence, *to ti ēn einai*, presented in her chapter on the *Metaphysics*[11] Joseph Owens's interpretation,[12] which does not in any way rely on or refer to the biological treatises.

Still, Randall and Grene had asserted the thesis of the biological roots of the metaphysics, and it was against this view that David Balme, in his 1965 inaugural lecture and elsewhere, had argued rather strongly that all the philosophical concepts, and the key metaphysical doctrines (e.g. of species-form, and of the teleology as directed to species-form), were in place when Aristotle wrote his biological treatises; indeed, Balme argued, there were important conflicts between the biology and the metaphysics, e.g. on the nature, and importance, of species, on the very notions of *genos* and *eidos* in general, on the teleology, and even on spontaneous generation.[13] It was against this background that Furth began working in the early 1970s, and against this background that his thesis was both radical, and in an exciting way 'reactionary': he was returning to the earlier intimations of a biological provenance for the metaphysics, now as I quoted before, 'on a more specific basis and from a more systematic and theoretically-oriented standpoint than other treatments to date'.[14] During that same period, Balme's views were changing, and as the result of a reinterpretation of the metaphysics in many ways extremely different from Furth's, he came independently to see an essential harmony between the central concepts and theses of the metaphysics, now re-interpreted, and those of the biology. It would be instructive to contrast and compare their respective views of this relationship, but I want now to focus on Furth by noting that Balme has never gone as far as Furth in stating the very strong claim that the metaphysical theory of material substance was 'motivated as a deep

[6] Lee 1948; see also Lee 1985. [7] Randall 1960.
[8] Hantz 1939. [9] Randall 1960, 224-25 n. 7. [10] Randall 1960, ch. VI.
[11] Grene 1963, ch. VI. [12] Owens 1951.
[13] Balme 1965; see also Balme 1962a and 1962b. [14] Furth 1988, 6.

theoretical foundation...for the biological works'. For there is an ineluctable chronological claim here, that much of the work reported in the biological treatises—and the departures from the *Organon* that it reflects (e.g., on differentiae[15])—was essentially completed *before* the main theses and arguments of the central books of the *Metaphysics* were laid down.

It is important to stress this, because there are, in Furth's programmatic statements, and generally in the statements of those who have stressed the importance of the biology for the philosophy, a *weaker* and a *stronger* thesis in regard to the metaphysics. The weaker thesis has been that the study of the biology can illuminate our understanding of the metaphysics (and the *Metaphysics*), if only because living things, intuitively understood, were for Aristotle the paradigmatic objects, and because Aristotle's intuitive vision of those objects was shaped, or at least reshaped, by his biological study. But that view is consistent with the thesis that major problems and central lines of argument of the central books of the *Metaphysics* are shaped by concerns arising, even wholly, from elsewhere than the biology. Furth's thesis that the metaphysics is a 'deep theoretical foundation...for the biology' is a stronger claim about these major problems and central lines of argument. And its support requires showing, not just that there are the heuristic benefits to us readers of the *Metaphysics* of having read the biological treatises, but that central stretches of the *Metaphysics* embody—and thus make sense only against the background of—conceptions and theses unique to the biological treatises. My focus will be on Furth's argument for that *stronger* thesis.

Now, it is striking that Furth insists from the very first that the argument for his (stronger) thesis—for the hypothesis he advocates—must be *global*. For, in this case we've got to argue, not 'from [the philosopher's] text to his meaning', but by:

> recreat[ing] in imagination the world the philosopher saw, the problems it presented to him, and the conceptual instrumentalities with which he sought to reduce what was problematic to intelligibility and order—to do this so completely and so vividly that in the light of this understanding arrived at in our own terms, it becomes possible to '*deduce* the text', to validate a reaction that goes: 'But someone who assumed *that*, or looked at the world in *that* way, or noticed and was impressed by very general facts of *that* sort, would quite naturally say things like *this*'—where *this* is some characteristic verbal behavior with well-established standing in the philosopher's text, which has discouraged, if not outright baffled, standardly cautious essays at picking out a plausible meaning directly *ex ipsissimis verbis*.[16]

And so, what Furth does, as you can see from his table of contents in my appendix, is first to set the problems of understanding the nature of a material substance, and the basis of its synchronic and diachronic individuation, as these arise from the 'miniature or cut-down' world of the *Categories*; and then to describe in rich detail, and in my judgment with great illumination (except on two or three

[15] Cf. Balme 1987b and Furth 1988, 95ff. [16] Furth 1988, 1.

key matters to be mentioned later), Aristotle's rich world of biological objects. He does this with special attention to the very general facts about these objects from which Aristotle started, and to the concepts of matter and form he forged to explain those general facts, both structurally and developmentally, drawing morals on the way for the full understanding of the nature and individuation of a material substance. Then in §17, he sums up three main lessons of the biology: (i) the size and depth of the world—i.e., the way these material substances sit in, and are constructed out of, a sea of Empedoclean matter satisfying a mass-logic, and themselves have an inner material structure and a formal dimension; (ii) the constructional interpretation or vertical dimension of differentiae—i.e., the fact that in the biology the classificatory notion of a difference *from* something else has given way to the constructive notion of a difference *of* something, a specifying down from a generic potential (say, being feathered, or footed) to a specific differentiation of, and out of, that (e.g. long-feathered, or two-footed); and (iii) the connection between being *formed* and being *one*—i.e., the way in which, both in the structural differentiation just described within a mature organism, and in the generative process (in which an animal fetus, through successive differentiation, articulates out from being merely alive ['plantlike'] to being animal to being human), the advent of substantial form and the advent of individuality are one and the same process.

Having 'recreated in imagination', indeed with extraordinary vividness, Aristotle's biological universe, Furth then returns, in the remaining sections, to the *Metaphysics*, to show how much sense its recalcitrant doctrines and texts now make, when looked at in the light of that recreation. He discusses most of the chestnuts of *Metaphysics* interpretation, cracked open now to reveal their biological matter and form—proceeding for some 100 pages and eight sections, through §24 'Substantial being: the final analysis' (which in Aristotle's text is largely a static analysis), to a final exploratory chapter in which Furth attempts to extend the analysis into the transtemporal dimension. The book ends abruptly, indeed on a parenthetical remark. There is no concluding chapter and no sense of closure or completion.

How convincing is the global argument? Let me answer via two more points of history. I was one of the two readers of Furth's book for Cambridge University Press, and in a long referee's letter, I raved about the book, urging publication. Even though I disagreed significantly with several of the main lines of Furth's interpretation of the metaphysics, and thought that because of his commitment to those lines he had read parts of *GA* wrongly, I thought the treatment of Aristotle's biological universe, both the seven general facts, and the detailed analysis of biological matter and form in §§9-13, were alone worth the price of the book. I found Furth's discussion marvelously illuminating, helpful to me as both a scholar and a teacher. It seemed to me, upon reading the book, as intuitively clear as it seemed to Furth that this rich hierarchical picture of matter, the identification of where form entered, the wonderful account of what was non-uniform in the non-uniform bodies, and so on, lay behind the theory of the unity of form and matter in natural substances developed in the metaphysics, even if as I say we differed about what exactly that theory was. But I was aware that others, not already predisposed to a thesis of biological influence, would hold out. After all, Furth's very need to make that global claim suggested that individual metaphysical texts did not scream out their biological provenance, and great strides had already been made by Aristotle scholars in understanding the metaphysics since, say, Furth was a student, without much use of the biology. So I knew it would be an uphill battle for Furth, and in my letter to the Press I urged that he write a concluding chapter, reviewing the case, identifying specifically those *Metaphysics* texts which we now see—having read the book and followed Furth's analyses thereof—can only, or at least best, be read as reflecting powerful biological influence. Furth didn't take my advice, and the book ends as I just said it does. I think Furth paid the price for that at Oxford in 1987. He had returned after many years away, an event of some personal importance to him, and he presented an overview of his approach. The paper was not well received. I think it was Michael Ayers who asked the inevitable question, to the effect of 'Well, this is an interesting story you're telling about a biological background to the metaphysics, but what's the basis in the text for thinking the story's right?' Furth said something about the need to see it globally, and referred to the forthcoming book. Having just read the book for the Press (unbeknownst then to Furth), I suggested that he could surely do more in support of his case, and mentioned a couple of texts I knew he thought could only, or best, be read according to his story, and invited him to say how that was the case about these texts, and others. He declined, and oddly just reiterated the need to read his whole case globally. No one in the Oxford audience appeared convinced. The lesson these stories invite us to draw is that the intuitive plausibility of the global picture is not a substitute for a focused review of particular passages, even if they must be seen in light of the whole.

So, in preparation for our symposium, I decided to review Furth's overall case, looking for those texts and aspects of the central books of the *Metaphysics* which can only, or best, be understood in light of the hypothesis that the theory of substance is a theory specifically of objects with the complexity and richness spelled out in §§9-17. Let me stop for a moment to clarify this. Furth was not the first, nor was he the last, to suggest that the problems of the *Metaphysics* arise out of the introduction of the notion of matter into the world of the *Categories*. That's been the standard view, I think, for a long time. But the assumption most frequently has been that it is the concerns of *Physics* I, and of *Generation and Corruption*—that is, the general analysis of change, and the recognition of a distinction between substantial and non-substantial change—that were at work. On that view, living organisms, understood intuitively, would be examples of form–matter compounds—like bronze spheres but more interesting than bronze spheres, because, even intuitively, the form and matter of a living thing seems more intimately connected than do the brazenness and sphericity of the simpler object. What Furth needs to show is not merely that biological objects fit the general account of material substance in the metaphysics, but that they do so in virtue of facts about them that were uncovered in the course of sustained biological research and theorizing. Does Furth show this? Here's what I found. You can follow along in his table of contents (in the appendix below).

§§1-8 are preliminary and don't adumbrate anything that can't be found in the later sections. So, let's look first at §§9-17, then at §§18-25.

In §§9-17 passages in the *Metaphysics* are occasionally cited as compatible with the theoretical account of biological matter and form, and their unity in the biological substance, but it isn't argued that these passages would fail to make sense on the weaker model I just sketched, in which living organisms are just good exemplifications of theses that apply more widely than to living things. The only substantive claim about the dependence of metaphysical texts upon biological theory comes in §13, after Furth's exposition of the move in the biology from the 'horizontal', classificatory notion of a difference-*from*-something to the 'vertical', constructive notion of a difference-*of*-something. At that point[17] Furth introduces a quotation from *H*.2 with the words 'Looking ahead momentarily to the advanced stage of development of the theory represented by *Metaphysics H*.2, the constructional or organizational moment in differentia is quite marked.' *H*.2 is discussed much more fully in §24, and I'll return to its connection with the biological theory in a moment. §17, as I've mentioned, identifies the three lessons of the biology, but points ahead to the application of these lessons in §§18-25, to which we need to turn in our search for argument in support of the stronger thesis.

Because of time limitations, I need to be rather abrupt and dogmatic about these sections. I think that with one exception there is actually very little case made that the biology is required to make sense of metaphysical texts, and the great bulk of Furth's often subtle and interesting, and sometimes very important, discussion in these sections makes no significant mention of earlier biological doctrine. The big exception is the discussion of *H*.2, which depends explicitly on a theory of differentiae of the sort Furth has extracted from the biology. Because Furth is apparently more interested in showing that the same theory of differentiae is to be found *both* in the biology *and* in *H*.2, than he is in arguing that *H*.2 makes sense only on the assumption that the full biological work has already been done, he doesn't try out an alternative hypothesis I want to mention, and thus doesn't really argue for the stronger thesis. The key notion for this new theory of differentiae, not found anywhere in the *Organon* (including *APo*. II.13), is the vertical, constructive view of differentiae, which is deeply embedded in biological theory, which surfaces

[17] Furth 1988, 102-103.

in *Z*.12, and which is deployed again and more metaphysically in *H*.2—the notion of a differentia as a determination of a generic potentiality which could be determined in several different ways.

Now, here I confess to being torn. On the one hand I want to say that it seems to me, as it seemed to Furth, extremely plausible that the rich notion of differentiation deployed in *H*.2 arose out of the work reported in the biology. On the other, I don't know what to say to someone who would argue, recalcitrantly perhaps, that we don't need heavy biology to observe that two-footed is a differentiation of footed—or breakfast a differentiation, via time, of meal-eating, or a threshold a differentiation, via place, of supporting-beam or ice a differentiation, via solidity, of water. This rich notion of differentia as *differentiation* may well have derived from the biology—but then again, the argument goes, it may have been developed in a general way from everyday examples, including plants and animals commonsensically understood, and then applied with unexpected richness when Aristotle got around to doing his own biological theorizing. The problem is that we need to infer to the best explanation of the *H*.2 account, and it's not clear what the rules are for making such an inference on this issue. My hunch is that Furth is right: some of the key examples are biological, and the notion of differentia-of-a-differentia was arguably adopted around the same time that the biological work was being done, as Balme has importantly argued.[18] Although I by no means find this argument conclusive, I'm willing to give this round to Furth and say that it's at least as reasonable that the new theory of differentiae was first developed within the biology and then later applied to the metaphysical theory, as the other way around—and maybe more reasonable.

But elsewhere through §§18-25 I don't find any case where a passage in the *Metaphysics* must be read as reflecting rich biological content. What I find, though, is something else, which I do want to mention. What Furth does, and does with extraordinary impressiveness, is to provide what might be called a metaphysics of biological objects. That is, he shows time and again *the way in which biological objects happen to instantiate or embody, in rich and unexpected ways, the general metaphysical theses and theories he ascribes to Aristotle.* A rich understanding of the biology, Furth has unequivocally established, shows us how biological objects exhibit for Aristotle metaphysical truths—and that may have been what impressed Furth most of all. But I don't see that this *by itself* establishes, even as only the most likely hypothesis, that the metaphysical theory was *aimed* at providing a theoretical foundation *for* just these objects, as such, based on specialized knowledge about them. That's the stronger thesis. Insofar as biological objects instantiate or embody, in remarkably rich ways, the metaphysical theory, they are going to be great teachers of that theory, and the biological works a rich heuristic for that theory. But that's the

[18] Balme 1987a and 1987b.

weaker thesis. This is not an insignificant claim, as I've spent a bit of my career arguing,[19] but not yet the stronger thesis.

Is there more to be said in favor of the stronger thesis, though—some premises or texts that we might supply which, together with Furth's marvelous case for the weaker thesis, might lend more support to the stronger one? I don't know. But I do think there's one direction we might pursue, on which I think the book is somewhat inadequate, so let me take a few minutes to do so. Although I think the book's discussion of matter is wonderful, the discussion of form from a biological perspective is rather limited. Furth stresses the need for the introduction of form, to explain the 'knots' in the sea of Empedoclean matter that living things are, and insists that form is a real cause, not to be reduced away. He speaks of form as providing *structure*, and as connected with the future-looking activities, and indeed with the soul. But he does not go beyond that, and does not make sufficient use of a coexistent body of work which even by the early 1980s had made good use of the biology—viz., the work on natural teleology and its connection to formal natures, and the related work on hypothetical or conditional necessity. Now that's a paper in itself, and indeed it's going to suggest yet another limitation in Furth's discussion of form—this time in his discussion of the unity of form—which will have implications for the extent to which the central books of the *Metaphysics* indeed reflect the full biological theory, so I'll have to be very brief about it all. But let me sketch out these issues, in hopes of stimulating further reflection. I have in mind, to start, a thread of connection between Furth's favorite *Z*.16 passage and the work of John Cooper and myself and others, including David Charles, on Aristotle's natural teleology. Just a thread, but perhaps a provocative one. In the book Furth quotes only the first part of the relevant passage, 1040b5-9:

> It's evident that even of the things supposed to be substances, most are [really] potencies, both the parts of animals (for none of them exists separately; and when they are separated, then too they 'are', all [only] in the sense of matter) and earth and fire and air; for none of *them* is 'one', but they're like a heap, until they're worked up and some kind of unity comes-to-be out of them.[20]

But the sentence I'd like to focus on occurs a few lines later, at 1040b14-15, where Aristotle, almost as an aside, identifies *the kind of unity* required for substance, which that working-up produces. Here I quote from Furth's famous 'Eek' translation:[21]

> But all the same, the parts all will 'be' [only] potentially, when they are [fused] into *one* organism and are continuous by nature—but not by force, or by [an abnormal] growing-together; for that sort of thing is a deformity.

It's the notion of being *continuous by nature* that I want to focus on. Nothing more is said about this in *Z*.16. *Z*.17 opens with the suggestion that the key to what makes

[19] See, e.g., chapters 7 and 10 in this volume.
[20] Furth 1988, 125.
[21] Furth 1985.

something a substance will be found in what *causes* it to be one, but little is said about *what it is* to be one and not much is said about what a form *is*. It might thus catch a reader by surprise that Aristotle, after stating his conclusion at 1041b27—'And this [sc. form] is the substance of each thing, the primary cause of its being'—writes:

> But while some things aren't substance, still such substances as there are, are composited[22] in accordance with a nature and by way of a nature of their own, and that very nature would appear to be [their] substance . . .

The surprise is that only a nature will count as a substantial form. Nothing in *Z*.17 prepares us for this, but I want to suggest that we have indeed been prepared for it in *Z*.16 in the passage we just noted in which we were told that the sort of unity required for a substance is *being continuous by nature*. How this works we can see, I think, if we ask what it is to be 'continuous by nature'. In *I*.1, Aristotle begins a review of the various ways in which something may be said to be one. The first item on his list is this, at 1052a19, also in Furth's 'Eek' translation:

> That [is one] which is continuous, either simpliciter, or especially that which [is continuous] by nature, and not by contact or by tying (and of these, that is more 'one' and prior, whose movement is *more* indivisible and more simple.) Even more of this sort [= continuous?] is that which is a whole, and has a certain shape and form; and especially if something is of this sort [= is a whole having a certain shape and form?] by nature, and not by force, as are such things as are [wholes, etc.] by gluing or nailing or tying together, but has in itself the cause of itself being continuous. It is of this sort [= as described] because *its movement is one and indivisible in place and time*.[23]

To be continuous *by nature*, then, is to have one movement *by nature*. That is to say, it is to have a nature as form, a formal nature in the precise sense of *Physics* II. It is, then, to have a mode of action *as a whole*, which is not reducible to the natures and potentials of its material constituents—in whatever sense of irreducibility you think Aristotle's texts support. What I want to suggest is that the very same irreducibility that underwrites Aristotle's natural teleology underwrites his theory of substance. This irreducibility also underwrites the *De Anima* theory of soul as form and first actuality—since soul is essentially a causal or explanatory notion in Aristotle.[24] Roughly and simply, it's having a single mode of life, a unified set of soul-functions manifested in a distinctive environment; it's because *life* is a 'movement one and indivisible in place and time', that a living thing is a substance; and

22 *sunestēkasi*, a word which has, as does *pephthēi* ('worked up', literally 'concocted') in the first *Z*.16 passage quoted just above, a distinctly *biological* provenance. Furth might have made a good deal more of the appearance of these terms in these pivotal places in the *Metaphysics* than he did.

23 1052a19-26; my emphasis in last clause. Bracketed interpolations appear in the original translation.

24 Cf. Irwin 1988, ch. 13 and *passim* and now Frey unpublished.

this movement is one because it is the expression, somehow, of one potentiality for one end. The various parts of the organism are unified by all being necessary for that life—or best for that life—or byproducts of that life. Understanding hypothetical necessity, then, helps us to understand the unity of a material substance, and the work already done on natural teleology and hypothetical necessity should help us with the metaphysics—including *Z*.11 where the argument against the younger Socrates makes covert reference to hypothetical necessity (or something like it) as relevant to determining what belongs 'in the form' or 'in the definition' of a natural substance.[25] There's a rich mine of connection there, which might support Furth's stronger thesis at least to the extent that the teleology itself depends on biological theory.[26] But Furth doesn't pursue these connections in his book and, as a result, I think, his treatment of form in his central biological sections is much weaker than his treatment of matter.

But once one gets that far, one can see, I think, another weakness in Furth's treatment of form, this time in his treatment of the unity of form. It is natural to think that for Aristotle, form is soul and soul is something unitary, so form is itself unitary, as Aristotle's theory anyway requires; and Furth makes and explains these points in rich ways. And there seems nothing in either the *Metaphysics* or the *De Anima* to contravene this simple picture. But there is one clue in *PA* I.1, and a bunch of clues across the explanations in *PA* II–IV, that suggest matters are not that simple. At *PA* I.1 641a18, speaking of the form of a living thing, Aristotle is oddly hedging:

> Now if this is soul, or part of soul, *or not without soul*, . . . it will be for the natural philosopher to speak and know about soul . . .

So form involves soul, but it may involve more, or so the third option suggests. Why would Aristotle bother to say that? Well, recall the important claim in *PA* I.3, which Furth rightly treats as very significant, that living organisms must be defined by multiple differentiae all at once. David Charles and I have shown that the explanatory practice in *PA* II–IV mirrors this theoretical claim: explanations proceed not only from essential soul-functions but also from other features of animals that are treated as equally basic and essential (e.g., the large size of the elephant, the long-and-slender shape of the snakes, the bloodedness of blooded animals, etc.).[27] Furth notes the threat to the unity-of-form doctrine posed by the

[25] 1036b24-30.

[26] The extent to which this is the case, though, is a tricky issue. My own view is that the teleology is defended in roughly two stages: first, from *phainomena* available to all observers of the general scene, as we see it defended in *Phys*. II; but then via the work of the biology, which is brought in to support, and refine, those general claims and arguments. I don't think it's conclusive that *Metaph*.'s dependence on the teleology requires the biological work, although I often doubt that Aristotle would be as confident about irreducibility as he was, without the biology. See also below, in the Coda, p. 392 n. 38.

[27] See, in the present volume, chapters 7, 8, and 10, and also Charles 1997 and 2000, ch. 12.

PA I.3 statement, and proposes in §21(iv) that the theory of vertical differentiation—the differentia-of-the-differentia thesis about lower to higher differentiae—solves this, referring us to *Z*.12 and *H*.6—and to *PA* I.3, where that view of differentiae is also endorsed. But I think it important to see that while the *Z*.12/*PA* I.3 notion of differentia-of-the-differentia does indeed solve one unity problem, it doesn't solve the multiple-differentiae one. Think of it this way. A definition of an animal form (or 'species') by multiple differentiae could be represented like this (S = 'species', G = 'genus', D = 'differentia'):

$$\begin{aligned} S = G + D_1 &+ D_2 + D_3 + \ldots \\ + D'_1 & \\ + D''_1 & \end{aligned}$$

The view of lower differentiae as specifications of higher ones—of D'_1 as a specification of D_1, and of D''_1 as a specification of D'_1 (from the top down, footed—two footed—bent-over-two-footedness, as in birds), etc.—solves what we might call the vertical unity problem, but not the horizontal one: the unity of D_1, D_2, D_3, etc. (e.g., two-footed and winged and beaked, etc.), is nowhere addressed in the *Metaphysics* (unless it is alluded to in the cryptic remark at the close of *Z*.12, that this is only a first shot at definition by division).

So, although *Z* certainly does reflect, it seems to me, a well-worked-out doctrine of natural teleology and hypothetical necessity, and although *H*.2 probably does, as Furth insists, reflect an awareness of a biologically oriented theory of differentiae,[28] there are also some issues that the biological works raise—such as the need to define living things by multiple differentiae—which *ZH* pointedly seem *not* to reflect. That there was interaction between work reflected in the biology and the work reflected in the metaphysics seems to me certain; but sorting out how this interaction occurred, and in what temporal and conceptual relationships, is clearly more complicated than Furth suggested, and there is much more work to be done.

As I've intimated, I think as well that there are weaknesses in Furth's use of *GA* to support his account of the metaphysics. I think, for all the reasons that David Balme,[29] and after him John Cooper,[30] have given—which Furth does not in my view come near to successfully countering in his discussion in §§15(iii)—that the theory in *GA* IV.3 is completely compatible with, and indeed clearly foreshadowed in, *GA* I–II. That the IV.3 theory countenances subspecific forms must matter, then, for our understanding of *GA's* overall view of the unity and individuation of material substances. Indeed, I think Furth misreads a passage in

[28] And there is some terminology in *Z* which has a distinctively biological provenance (above n. 22).

[29] Balme 1987d, esp. n. 14.

[30] Cooper 1990.

GA II of which he makes much, in ch. 3 at 736b2-15. He offers the passage in support of his 'suggestion that according to Aristotle's view of these events, the developmental process that takes matter into specific form and the process that takes catamenial mass into discrete individual are one and the same process'.[31] As he explains, 'The only reasonable choice on this theory for the actual coming to be of *Socrates*, i.e. *the individual man*, seems to be: the point at which differentiation becomes complete.' The passage reads:

For it's not simultaneously that *an animal* and *a man* come-to-be, nor *an animal* and *a horse*, and likewise in the case of the other animals (= types of animal); no: The end-product or completion (*telos*) comes-to-be *last* of all, and the most distinctive character (*idion*) of each thing is (comes at) the completion of its genesis.[32]

Furth takes Aristotle's example of the *relatively later* articulation-out of features that establish it as actually a man or horse, as a statement of what is *last* articulated, and as support then for specific form as the last formal articulation. But the text doesn't say that, and is perfectly compatible with the alternative reading suggested by *GA* IV.3, in which what is *most idion* is something peculiar to the individual qua horse or qua human being. Thus at 767b23-768a2, in IV.3, Aristotle expresses very similar sentiments to our book II passage but clearly sees the end-product or completion—and the *idion*—as something subspecific. In Furth's translation:

The progenitor isn't just *male*, but he's such-and-such a male, as Coriscus or Socrates; and he's not just Coriscus, but also *man*. And so in this way some of the things belonging to the progenitor are nearer, some remoter—according as he has the generative power, not as what he may just 'happen' to be, like the male progenitor being grammatical or someone's neighbor. And what's *peculiar* (*idion*)[33] and *particular* (*kath' hekaston*) exert the stronger influence on the genesis. For Coriscus is both man and animal, but man is nearer to what's peculiar (*idion*) than is animal. Both the particular and the kind (*genos*) generate, but more the particular; for that's the substance. As a matter of fact, the product indeed comes-to-be of-a-particular-sort (*poion ti*)—but simultaneously also a particular this (*tode ti*)—and that's the substance. For which reason it's from the *dunameis* of *all* these sorts of things that the motions in the semen are constituted—potentially indeed from (the *dunameis* of) the earlier ancestors, but more especially from those of the one which in each case is closer to some particular—I mean by particular, Coriscus, or Socrates.[34]

Some may consider this unfair, but I can only think that Furth's penchant for interpreting both the *GA* IV.3 and the II.3 passages as he does is strongly influenced by his prior view of the centrality of species-form, rather than providing actual independent evidence of it. In any case, we can see why Furth viewed *GA* IV.3 as

[31] Furth 1988, 123. [32] Furth 1988, 112.

[33] Note the change in Furth's translation of *idion* from 'distinctive character' in the II.3 passage just above to the less theoretically significant 'peculiar' here.

[34] Furth 1988, 129-30.

so important. For if indeed the formal determination that establishes individual substance in *GA* was itself individual, or in any case subspecific, and the metaphysics was a deep theoretical foundation for the work already done in the biology, then the Furthian metaphysics of specific form begins to unravel as an interpretation of Aristotle—as it did for David Balme in the 1970s, when he came to see (i) the possibility of reading the *Metaphysics* in terms of individual forms, and (ii) the likelihood that the theories of the *Metaphysics* and the *Generation of Animals* arose, consistently, at more or less the same time.[35]

So, in sum, I have some reservations about Furth's readings of some biological texts, and of what their lessons are for the metaphysics, and about whether he has established his stronger thesis. And others will have other reservations. But, I want to stress: I come not to criticize Furth but to praise him. As I found in preparing for this session, the book, whatever its limitations, is not in the least dated. Its middle part on matter and form, §§9-13, and much of the moral-drawing in §§14-16, are still thrilling and invaluable, establishing (as I think beyond question they do) the weaker thesis and provoking questions about the appropriateness of some sort of stronger thesis. The 'size and depth' of Furth's own philosophical world—the questions he asks, the issues he explores, and the subtlety and challenge of his interpretative accounts—can be a powerful stimulus in our own work, and will deserve attention from Aristotelian scholars for a long time to come.[36]

[35] This view was first presented at the 1976 Princeton Colloquium referred to earlier, and then published in 'Aristotle's Biology was not Essentialist', which appeared first in *Archiv für Geschichte der Philosophie* 62 (Balme 1980). It was slightly revised (and n. 14 was added for publication) as Balme 1987d.

[36] It was an honor and a pleasure to participate in this program. I would like to thank George Anagnostopoulous for organizing, introducing and chairing the session, and my fellow symposiasts and all those who participated in discussion for making it a very lively event of the sort I'd like to think Furth would himself have enjoyed.

Appendix. The original Table of Contents of Montgomery Furth's *Substance, Form and Psyche: An Aristotelean Metaphysics* (Cambridge 1988).

Appendix. Continued.

Appendix. Continued.

PART IV

Starting a Science: Theoretical Aims of the *History of Animals* (*HA*)

12

Data-Organization, Classification, and Kinds

The Place of the *History of Animals* in Aristotle's Biological Enterprise

Introduction

I have long believed that scholars of ancient philosophy, and especially of Aristotle, have much to learn about his philosophy from a close study of his biological writings. These writings comprise over 25 percent of his surviving corpus, and they must have taken quite a few years of Aristotle's relatively short time on earth. Yet, through the ages, they have not received an amount of study commensurate with Aristotle's own efforts. Although these treatises were an object of careful study by Theophrastus, they very soon thereafter fell out of favor among philosophers in the ancient world, a sign of which is that there is no record of any ancient commentary on any of them.[1] They were translated into Arabic in the ninth century AD, under the collective title which translates into Latin as *De naturis animalium* and soon became known as *De animalibus*, and there are commentaries on some of these treatises by Avicenna, Averroes, and Avempace (Ibn Bâjja), among others. In the Byzantine world, writing in Greek, Michael of Ephesus, apparently at Anna Comnena's urging, attempted in the eleventh century to fill the gap left by the ancient commentators, writing commentaries on *PA, GA, IA, MA*, and *PN* (though not *HA*). The first Latin translation was produced early in the thirteenth century, from the Arabic, by Michael Scot; this translation was the basis

This chapter is a slightly updated version of a two-part seminar given at Tokyo Metropolitan University in 1994. I have retained the informal character of the public presentation, omitting only the opening expression of gratitude to my hosts and the description of the three two-part seminars that followed this one. I thank David Charles and Jim Lennox for valuable comments on the penultimate version of the chapter. I have also added as an appendix an excerpt from a paper (not otherwise included in the present volume, except as a reference) which I co-authored with Pieter Beullens, 'Theodore Gaza's Translation of Aristotle's *De Animalibus*: Content, Influence, and Date' (Beullens and Gotthelf 2007).

[1] See Lennox 1995 and (on the early post-Aristotelian period, which must be a crucial part of the story, and indeed requires yet a good deal more work) Sharples 1994, esp. 32–41.

for the magisterial paraphrase and commentary by Albertus Magnus, to which he added much new material based on his own study of animals in northern Europe. Soon after, in the mid-thirteenth century, William of Moerbeke produced Latin translations from the Greek.[2]

Retranslations in the mid-fifteenth century by George Trebizond and especially Theodorus Gaza, and the burst of Aristotelian scholarship in Italy in the late fifteenth and throughout the sixteenth century, sparked a renewed attention to these treatises by philosophers interested in questions of scientific method, natural philosophy, and metaphysics;[3] but this interest soon died out, as what we know as early modern philosophy replaced the more Aristotelian concerns of the Northern Italian Renaissance. Aristotle's biological corpus was indeed the foundation of early modern biological science, and was a standard reference work and starting point for biological research well into the eighteenth, and even the nineteenth century—see for example the recent edition, published by the University of Chicago Press less than twenty years ago, of 'The Hunterian Lectures of Richard Owen', given in 1837, in which Aristotle is still treated as an earlier colleague, on whose work one must build, rather than as the antiquarian figure he certainly became after Darwin.[4]

[2] **The 9th-c. Arabic translations:** *GA*: Brugman and Drossaart Lulofs 1971; *PA*: Kruk 1979; *HA*: Filius et al. (in preparation) and Filius 2007, 2008. **Avicenna:** Madkour 1970; **Avempace:** Al-'Imâratî 2002 and Kruk 1996 (and in general for the Arabic tradition of the use of Aristotle's biological writings: in addition to the introductions to the aforementioned works, Kruk 1990 and 2003 [see also Kruk's list of publications through 2006 in her Festschrift, Vrolijk and Hogendijk 2007, xiii f.]). The Arabic translations gave rise to a Hebrew translation and commentary tradition currently under study: see Zonta 1999. **Michael of Ephesus:** *PA*: Hayduck 1904; *GA*: Hayduck 1903; *MA* and *IA*: Hayduck 1904, Preus 1981; *PN*: Wendland 1903; there is no ancient Greek commentary on *HA*. **Michael Scot:** Latin translations from the Arabic—of Aristotle: Van Oppenraaij 1992- ; of Avicenna: Van Oppenraay (in preparation). **Albertus:** Stadler 1916-20, Kitchell and Resnick 1999, Resnick and Kitchell 2008. **Moerbeke:** *PA*: Rossi forthcoming; *GA*: Drossaart Lulofs 1966; *HA*: Beullens and Bossier 2000- . *IA* and *MA*: De Leemans 2011. There is a thirteenth-century Spanish translation of *HA*, made from Arabic and Latin into Castilian: Dehmer 2007. The earliest surviving Latin commentary on *HA* is included in Peter of Spain's *Quaestiones super libros de animalibus*, on which see esp. the writings by Miguel de Asúa cited in Asúa 1999 and, most recently, the text, translation, and study in Navarro Sánchez 2009; see also Perfetti 2000, 2 n. 4 for mention of commentaries that have not survived, e.g., by Roger Bacon and Boethius of Dacia (though '1520' should read '1250'!).

[3] Beullens and Gotthelf 2007; and Perfetti 2000 (each with additional references).

[4] Sloan 1992. Darwin himself cites Aristotle on occasion in various of his books, as a source of a variety of observational data, always quoted from other writers, but always accurately, and there is, of course, his famous remark that 'Linnaeus and Cuvier have been my two gods, though in very different ways, but they were mere schoolboys to old Aristotle', a remark that was preceded by the comment that 'from quotations which I had seen, I had a high notion of Aristotle's merits . . . '. This latter remark was preceded in time by the following one, to be found in an as yet unpublished letter in the Darwin archives at Cambridge University Library, written a few years earlier to a young man who had asked him about Aristotle: 'From extracts which I have seen, I have an unbounded respect for him, as one of the greatest, if not the greatest observers, that ever

But this general lack of interest in the *philosophical* significance of Aristotle's biological corpus after the Renaissance continued well into the twentieth century, with a few partial exceptions, among French writers,[5] until David Balme in the late 1930s, on a suggestion by his Cambridge supervisor, the famous scholar F. M. Cornford, began a lifetime's study, which issued in many seminal philosophically oriented papers, a Clarendon Aristotle volume, and (posthumously) a new text, with commentary, of the *History of Animals*, issued first in part, in the Loeb Classical Library, and then in whole, in the Cambridge Classical Texts and Commentaries series.[6] Balme showed us the many ways in which Aristotle's biological writings helped shed light on aspects of his philosophy, and even were meant to contribute to his philosophical work, and some of us are carrying forward Balme's legacy.[7]

Balme's rejection of the traditional pictures of *HA*

Our subject, in this chapter, is the organization of the data about animals in the *HA*, and the place within that organization of the grasp and identification of kinds of animals. Its textual center is the question, somewhat controversial, of the aims of *HA*, and the place within those aims of animal *classification*. Specifically, our issues are:

(i) the aims of *HA*; and
(ii) in particular, the extent to which a systematic classification, or taxonomy, of animals is one of those aims; and
(iii) more generally, the place of kinds in those aims, whether those kinds are embedded in a systematic classification or not; and finally
(iv) the method or methods by which kinds are recognized, or established, as such, so far as they are part of those aims; and
(v) the connection of all this to the *Analytics* theory of science.

lived.' On all this, and on the relationship of Darwin's teleology to Aristotle's, see further now chapter 15 below, 'Darwin on Aristotle'.

[5] See the writings by Le Blond (1939, 1945), Robin (1944), Bourgey (1955), and Louis (1955, 1975), among others, listed in the bibliography of Pellegrin 1986. The French tradition of the study of Aristotle's biological work goes back at least to Camus 1783 and the series of translations (with notes) by J. Barthélemy Saint-Hilaire (*HA* 3 vols.,1883-87; *PA* and *IA*, 1885; *GA* 2 vols., 1887). Manquat 1932 was frequently cited in its day. For work in the German tradition, already underway in the early nineteenth century, see [Anon.] 1864 and Kullmann 1974 and the bibliography therein; and for the recent Italian tradition, the bibliography in Lanza and Vegetti 1971 (reviewed in Balme 1974).

[6] Balme 1991, 2002, and forthcoming. For a full bibliography of Balme's writings on ancient philosophy and science, including especially on Aristotle's biology, see together Gotthelf 1985 a, 'Introduction' and Gotthelf 1992. For work of some of the Balme-influenced scholars, see (e.g.) the essays in Gotthelf and Lennox 1987.

[7] For an overview of Balme's work, beyond what is described later in this chapter (263-75 *et passim*), see the introduction to Gotthelf 1985; chapter 11 above, p. 243; and chapter 14 below, pp. 310-12.

We start with *HA* proper. The key year is 1961, which saw the publication of David Balme's contribution to the Second Symposium Aristotelicum, held in Louvain in 1960; the title of his paper was 'Aristotle's Use of Differentiae in Zoology'.[8] In 1961 also, Balme submitted to *Classical Quarterly* a related paper entitled '*GENOS* and *EIDOS* in Aristotle's Biology'[9] and G. E. R. Lloyd's paper, 'The Development of Aristotle's Theory of the Classification of Animals', written without knowledge of Balme's paper, was published in *Phronesis*.[10]

Prior to Balme, reaching back to the sixteenth century, it had often been held (a) that, given certain things Aristotle says in the *Topics*, and given the customary practice in biological science from the Renaissance onwards,[11] Aristotelian biology *ought* to contain a systematic classification, or taxonomy, of animals; and (b) that Aristotle's biological treatises, especially *HA*, either *did* contain a systematic classification of animals (although no two proponents seem to agree on what that classification system was), or was aimed at one but failed (perhaps failed miserably) to achieve it.[12]

By a 'systematic classification (or taxonomy) of animals' I mean: a grouping of animal forms (*atoma eidē*) under a hierarchical series of larger kinds (*genē*) which aims to be (whether or not it actually succeeds at being):

(i) mutually *exhaustive* at any upper level, and especially at the highest level, of the lower kinds and forms;
(ii) mutually *exclusive* at any upper level of the lower kinds and forms; and
(iii) a grouping of animals by *essential nature*, so that
(a) each is an authentic *kind*, in the sense that the kind name is always in the first category of being, identifying *what something is*;[13] from which it follows that
(b) there is only one such grouping of the animals there are, which entails that the systematic classification is
(iv) unique.[14]

[8] Balme 1961, reprinted in somewhat revised form as Balme 1975, and again, in 1987, with a substantially new beginning section on Aristotle's developing conception of the nature and use of division in *HA* and elsewhere, and with the new title, 'Aristotle's Use of Division and Differentiae' (Balme 1987b).

[9] Balme 1962a.

[10] Lloyd 1961, reprinted with a new introductory comment in Lloyd 1991.

[11] e.g., Ogilvie 2006, ch. 5.

[12] Balme (1987b, 80–81) cites Scaliger 1619 at one extreme, Whewell 1837 at the other (n. 16 below), and Meyer 1855 in between. Pellegrin cites Cuvier 1841, Le Blond 1939, Robin 1944, Louis 1955 and 1975, Lloyd 1961, and Petit and Théodoridès 1962 (see Pellegrin 1986, 1–12, 16–20).

[13] Versus, e.g., 'the water animals', which includes only some birds but not others, or 'the animals with bladders', which groups together animals that do not share a common essence, since this group includes all of the live-bearing animals *and* the turtle.

[14] (iii.b) and its restatement, (iv), follow from, and are not independent of, (i)–(iii.a), but are worth stating separately for emphasis.

Since *HA* obviously contained an immense amount of information about animals, and since it was clearly *not* organized first and foremost by animal kind, it was held by some that the classification scheme must be more subtly embedded in the text; in particular, many scholars held that within the discussion of particular features, e.g. teeth, the information would be collected under animal headings. This explains both the marginalia in some manuscripts and early printed editions identifying the animal kind being discussed at that point, and some of the weird chapter divisions—which for the biology was supplied only in the late fifteenth century by two obscure editors for two different Venetian presses, one building on the work of the other.[15]

The result of this apparent chaos was first the denial, e.g. by William Whewell in his *History of Inductive Sciences*, that Aristotle had any systematic classification scheme at all,[16] then the denial by others that there was any organization of the data whatsoever. Thus George Henry Lewes, author of the only complete survey of Aristotle's biology I know of in the nineteenth century, who is generally sympathetic to Aristotle, threw up his hands in exasperation over *HA* and complained that it was

> an ill-digested, ill-compiled mass of details, mostly of small value, with the occasional gleam of something better. There is, strictly speaking, no *science* in it at all. There is not even a system which might look like science. There is not one good description. It is not an anatomical treatise; it is not a descriptive zoology; it is not a philosophy of zoology; it is a collection of remarks about animals, their structure, resemblances, differences, and habits. As a collection it is immense. But it is at the best only a collection of details, without a trace of organization; and the details themselves are rarely valuable, often inaccurate.[17]

Lewes was certainly right to reject one traditional view that *HA* was a descriptive zoology or 'natural history' as it is often called, of the sort one finds from Gesner in the sixteenth century through the myriad volumes in bookshops today with titles

[15] See now Beullens and Gotthelf 2007, 477-83. (As explained there, the Venetian editors were Ludovico Podocatharus, for the 1476 *editio princeps* of Gaza's translation, printed by Ioannes de Colonia and Ioannes Manthen, and Sebastian Manilius Romanus for the 1492 edition, printed by Ioannes and Gregorius de Gregoriis.)

The most absurd example of chapter divisions reflecting the assumption that Aristotle is preoccupied with animal kinds is perhaps *HA* II.2-5, in which the discussions of teeth in dog, horse, man, and elephant are each given a separate chapter, although they are only several lines long, and although this scheme clearly does not capture what Aristotle had in mind, since the end of ch. 1 had already discussed the shedding of teeth in men, and immediately following the discussion of teeth in elephants, Aristotle adds a sentence about the elephant's tongue, which our editor, in desperation, labels 'chapter 6' (!), followed by a discussion of the mouth generally, which is labeled ch. 7, although it too only runs a few lines.

[16] Whewell 1858–59, vol. 2, 412-17 (1st edn. 1837).

[17] Lewes 1864, 271; from ch. 15, 'The History of Animals'.

of the form 'The Natural History of...'.[18] Lewes, like others of his time, appears to treat *HA* as a kind of *notebook* to which data were regularly being added.[19] He did not rule out the existence of a systematic classification behind the details, claiming only that the data were not *organized according to* any scheme of animal classification.

As to the dispute regarding whether there was a classification scheme in the background of *HA* and the other biological treatises, some people in the twentieth century—e.g., Ingemar Düring in his 1943 study of *De Partibus Animalium*—observed that insofar as Aristotle does classify animals into larger groups, he does so in various ways in various places, and his accounts of division and definition in various places in his corpus would suggest other, yet different, schemes.[20] Düring's solution followed the lead which Werner Jaeger had set for Aristotelian scholarship, namely to resolve apparent contradictions by appeal to 'development', to

[18] Gesner 1551-87. For a beautiful copy of the first two volumes of Conrad Gesner's *Historiae Animalium* (Zürich, vols. I–IV: 1551-58, vol. V: 1587—the copy is the one belonging to the U.S. National Library of Medicine), see <http://archive.nlm.nih.gov/proj/ttp/flash/gesner/gesner.html>. As two random examples in a flood of recent material, see *The Natural History of Deer* (Putnam 1989) and the well over 20 volumes in the Australian Natural History Series of CSIRO Publishing (1996-2010 and forthcoming).

The grandfather of this genre is of course Pliny the Elder (10 vols. in the Loeb Classical Library; see Healey 1991 and Beagon 1992), but its master is Buffon, whose *Histoire naturelle, générale et particulière* (*HN*), published across the last forty years of his life (1749-89), ran 35 volumes; 9 additional volumes, with contributions by collaborators, were published posthumously between 1789 and 1804. (For images of these volumes, and a wealth of information pertaining to Buffon, see the website of the international Buffon project at CNRS: <http://www.buffon.cnrs.fr>, created and edited by P. Corsi and T. Hoquet.) On Buffon as natural historian, see the fascinating study by Jacques Roger (Roger 1997). The title Roger gives to his ch. XVII, 'To Identify, Name, and Describe', well captures Buffon's own aims in his majestic study (see Roger's citations from Buffon therein). Buffon certainly had a strong theoretical bent (and his reflections thereupon influenced later thinkers, such as Lamarck and Cuvier), but the descriptive portions of *HN* were not tailored specifically to facilitate the reaching of these theoretical conclusions nor to reaching a systematic explanatory science generally, as was Aristotle's *HA*, and this is the difference that Balme (see below) has focused on. A close study of the various relationships, historical and theoretical, between Buffon's *HN* and Aristotle's *HA* would be of great interest.

Interestingly, as far back as 1590, the Jesuit José de Acosta, who spent much of his life in the New World, in Peru, showed an awareness of the distinction between Aristotle's project and that of the natural history tradition from Pliny downwards. In his *Historia Natural y Moral de las Indias*, Acosta, while acknowledging that 'many have written sundry books and discourses of the New World and the West Indies, wherein they describe new and strange things discovered in those parts', introduces his own approach (which is based in his Aristotelian education at Alcalá de Henares) by saying that 'hitherto I have not seen any other Author which treats of the causes and reasons of these novelties and wonders of nature'. (Quoted, from a 1604 English translation, in Asúa and French 2005, 78; cf. 76-85 and, with it, 62-74.)

[19] This corresponds to what I call, in the Coda to this volume (in which I identify three stages in Aristotelian science: Collection, Organization, Explanation), the 'Collection' stage (below, pp. 383-87).

[20] Düring 1943, 109ff.; cited in Lloyd 1991, ch. 1, nn. 7, 9.

chronology: Aristotle had different classification systems at different times. Lloyd in his 1961 paper followed Düring on this, but offered a different and more interesting solution. What is interesting about this paper for our present purposes are two assumptions he makes:

(1) that Aristotle's aim in discussing the proper methods of division in *PA* I.2-3 is *classification* rather than definition; and
(2) that Aristotle's aim when he groups animals by their mode of reproduction in *GA* II.1 (which Lloyd views as developmentally last) is to produce a systematic classification of *animals* by essential nature, rather than a systematic grouping of the attributes under study, namely *mode of reproduction*.[21]

What is the difference in each case? Well, with regard to the first assumption, if you are interested in defining one particular animal form, then when you divide down according to your rules of division, you will ignore the side of the division in which your animal does not fall, and will not divide that side any further. If Aristotelian division is a method of reaching *definitions*, then a concern with division, which Aristotle clearly has, does not as such evidence a concern with systematic classification. But as Balme and, later, Pellegrin showed, the aim of division for Aristotle—as for Plato, in the *Sophist* and *Statesman* at least—is indeed definition and not classification.[22] This is clear from the opening sentence of *PA* I.2, which announces the aim of the ensuing discussion as 'grasping the particular [= *eidos*]', and from the discussion of division in *APo*. II.13, which is introduced as a technique for 'hunting down' (*thēreuein*) 'the definition of what something is' (96a22), in which Aristotle even criticizes the idea of Speusippus that one must know both sides of the division in detail if one is to define anything (97a6 ff.).[23]

With regard to the second assumption, any classification of the different types of some animal *feature*, e.g. mode of reproduction, does indeed generate a grouping of animal kinds, namely for any particular mode of reproduction, the animals that reproduce in that mode; but another feature, such as mode of locomotion, may generate a *different* grouping of animals—and in fact does, as Aristotle himself says in that very passage in *GA* II.1![24] There is no justification in Aristotle's text for singling out that one feature, mode of reproduction, as the basis of a systematic classification by essential nature; in fact, Aristotle insists in *PA* I.3 that any single-differentia definition of animal forms must fail, and explains that animals must be defined by multiple differentiae.[25] Once you come to see that Aristotle's focus in *HA* and elsewhere is on animal *features*, the temptation to see animal groupings

[21] Lloyd 1961, 75-79 (repr. 1991, 21-24).
[22] Balme 1987b (orig. 1961), Pellegrin 1986, 36-49 *et passim*.
[23] See previous note. 'Grasping the particular': *PA* I.2 642b5; 'hunting down the definition of what something is': *APo*. II.13 96a22-23. For Speusippus' views and Aristotle's criticism of them, see Tarán 1981.
[24] 732b15-29. [25] Cf. 643b9-644a11.

based on those features as necessarily a *classification scheme* disappears. And that was David Balme's great advance.

Balme had picked up on those two points—that in his discussions of division Aristotle is concerned with definition and not classification, and that a grouping of animals may reflect a classification of animal *features* and not of animal kinds—and others as well in rejecting the two-part traditional view (a) that Aristotelian science requires a systematic classification, or taxonomy, of animals and (b) that *HA* embodied such a classificatory scheme; but the great advance was in his positive account of the actual aims of *HA*, and the much richer explanation that account gave of the structure and organization of data in *HA*, which no one had seen before. Let me first list the other objections he made against the traditional view, and then we'll go on to his own positive thesis.

I've already mentioned his first two objections: (i) that division was aimed at definition and not classification, and (ii) that what has looked to other scholars like a classification of animals was really a classification of animal features (e.g., modes of reproduction), and that different features generated different animal groupings, none of which was intended as a systematic classification (and thus there was no contradiction that required a developmental story).

Balme also argued (iii) that there is no terminology for fixed classificatory levels in Aristotle: the only classificatory terms are *genos* and *eidos*, and they are both used at all levels of generality. Balme originally thought that the terms were used by Aristotle interchangeably, as they are by Plato, but later accepted Pellegrin's thesis that *genos* is the general term for a group, while *eidos*, when it is used for a group, almost always refers to a *subdivision* of some larger group, some *genos*, which is either named, or is implicit, in the context. Also, Balme showed that the term *genos* is used for groupings that are not *natural kinds*, as in 'the *genos* of animals with bladders', so that there is not much about Aristotle's terminology that would suggest a concern with producing a systematic classification, or taxonomy. This last point is not decisive, since one can mark out natural kinds without having a single word that exclusively means 'natural kind'—after all 'natural kind' is *two words* in English, and 'kind' by itself has, if not as much flexibility as *genos* in Greek, still enough to require the adjective 'natural' when we want to refer exclusively to a grouping according to essential nature; but, although not decisive, the absence of a fixed vocabulary for natural kinds does create a burden of proof for any interpretation that sees systematic classification as an aim of Aristotelian biology.

Balme acknowledged that Aristotle did indeed recognize some nine *megista genē* ('very large kinds') of animals as such, at two places (*HA* I.6 490b6, II.15 505b26), seven of which Aristotle implies were correctly recognized in common usage, even if not all had single-word names: birds, fishes, cetaceans (i.e. water mammals), shell-skinned animals (e.g. oysters, clams), soft-shelled animals (e.g. lobsters, crabs, shrimp), soft-bodied animals (e.g. octopus, squids), and insected animals, to which he adds two more not previously recognized as unified kinds: the four-footed live-bearing animals (our mammals excluding man), and the four-footed egg-bearing animals. We will have a closer look at this list ourselves later; but Balme's fourth objection against the view that *HA* was concerned with systematic classification was (iv) that for two reasons we cannot count this list of 'very large kinds' as evidence of such a concern: first, the list is not exhaustive (e.g. man and snakes, and others, are not included), and second (and most important to Balme), one cannot find in the biological corpus the sort of *intermediate kinds* that a taxonomical concern would entail. Virtually all the candidates for such which previous scholars had proposed, Balme showed to be what he called 'differentiae-classes'—classes named descriptively such as 'the crook-taloned birds', 'the horned animals', the 'web-footed birds'. In virtually every case, Balme showed, Aristotle recognized that such groupings failed to constitute a natural kind.[26] They failed because none of them met what Balme took to be Aristotle's condition for there to be a large natural kind, namely that there must be a set of significant features, say D1, . . . , Dn, possessed in common by all the subgroups of the supposed kind. They failed to meet this condition because in virtually every case the situation was something like this (assuming for the sake of argument that n = 4):

The sub-kinds (*eidē*) F and G each possess	D_1, D_2, D_3, D_4
but sub-kind H and I each possess	D_1, D_2, D_3 (but not D_4)
while sub-kind J possesses only	D_1, D_2, D_4
and sub-kind L possesses only	D_1, D_3, D_4

Those familiar with the later Wittgenstein on meaning and reference, especially as discussed by Kripke in Lecture I of *Naming and Necessity*,[27] might be tempted to find a cluster theory in Aristotle, but that wouldn't be right, since in the case of the *megista genē*, Balme apparently held, Aristotle *does* believe that there are features D1, . . . , Dn that *all* sub-kinds of the *genos* share in common. That is the criterion (roughly speaking) for having a natural kind, and virtually all candidates for intermediate kinds fail to meet it. This suggested to Balme that the grouping into *megista genē* had some other purpose than classification, and that brings us to Balme's positive account of the aims of *HA* including the aims of identifying such natural kinds as Aristotle does think can be identified.

HA I.1-6 and Balme's positive picture of the aims of *HA*

Balme's positive picture of the aims of *HA* may be said to have three parts:

(1) 'laying out the differentiae'
(2) 'significant groupings of differentiae'

[26] Balme 1987b, 82-83. [27] Kripke 1980.

(3) ultimate aim(s):
 (i) causal explanation
 (ii) definition of animal kinds

Let's look at each one.

1. 'Laying out the differentiae'

If we are to understand what Aristotle's aims are in *HA*, Balme suggested, why not look to see how Aristotle starts off *HA*, and what he *says*, in *HA*, they are. Aristotle begins the treatise by distinguishing two broad categories of parts: incomposite and homogeneous, and composite and non-homogeneous. He then identifies at some length various ways in which parts can be the same or different.[28] At 487a10 he notes that the differences that animals possess relate to their modes of life and their activities and their 'characters' as well as their parts. The term for 'differences', *diaphorai*, Balme insisted, is used in a quite general way, for any feature in which one animal may differ from another, and not only for defining features, since ahead of causal explanation, one cannot know which features are defining and which aren't.[29] The balance of *HA* I.1-5, Balme maintained, gives important *examples* of the ways animals can differ in these respects. Translating from 487a10:

> The differences that animals possess relate to their modes of life and their activities and their characters and their parts, concerning which let us speak first in outline, and then later we will discuss them in detail, speaking about each kind. In respect of their modes of life and their characters and their activities there are differences such as these: . . .

After a long list of differences, for most of which he mentions at least one sort of animal that has that difference, Aristotle continues, at 487b32:

> There are also, in respect of their modes of life and their activities, such differences as these: . . .

And another list follows, many of which are also exemplified. Finally, at 488b11, we are told that animals

> differ also with respect to character by way of differences such as these . . .

and each of a long list of 'character traits' applicable to animals is illustrated with one form of animal that possesses it.

That ends what our editor calls chapter 1, but by means of a *men/de* construction ('on the one hand'/'on the other hand') that bridges the chapter division, Aristotle continues right on, in what our editor calls chapters 2-5, to do the same for parts,

[28] Note his conclusion at 486b22-23: 'So then the parts, as many as each of the animals have, are in this way different and also the same . . .'.

[29] Cf. Balme 1987a, 19.

starting from the parts that all animals have (488b29-32) and moving through parts that most animals have (488b32-489a8), on to the broad categories of parts that correspond to soul-functions such as reproduction and locomotion (489a8-490b7), with an outline of the sorts of variations, or differences, that exist under each heading, again supporting each claim that a difference exists with an example of an animal kind that possesses that difference.

In chapter 6, Aristotle turns to a discussion of the very large kinds of animals that can be found—the *megista genē* that I mentioned above and which we will examine in more detail shortly. Then, at 491a7 he provides his own statement of his aims in this treatise:

> We have stated these things in this way now—in outline form—to provide a taste of the range and sorts of things we must study; later we will have to speak in greater detail, in order that we may grasp the differences and <in general> the features that belong to all. After this we must try to discover the causes. For that is the natural way to pursue such an inquiry, once one has completed an investigation (*historia*) concerning each of these: for it becomes apparent from this investigation both about which things the demonstration must be, and from which things it must proceed.[30]

The key statement of the immediate aim is at a8-10:

> later we will have to speak in great detail, in order that we may grasp the differences and <in general> the features that belong to all.[31]

That is, we are to do in greater detail, or with greater precision, what we have done in outline so far, namely to present in organized fashion all the differences there are, with examples of animals that possess each one, and to do so in order that we may grasp the differences that are present and the features belonging to all. My translation above, 'in order that we may grasp the differences and <in general> the features that belong to all', follows Balme's interpretation on two counts. First, it takes Aristotle's declared aim to be the identification of all the differences that are to be found in the animal kingdom, not the identification for each animal of each of its differences. Second, it takes 'differences' here to have the same non-technical meaning it has in its occurrences before and after this passage, signifying any way in which an animal kind differs from another. In this case, 'differences' (*diaphorai*) and 'attributes' or 'features' (*sumbebēkota*) have the same extension, so that the sense is

[30] Translation of the text in Balme 2002. The Greek reads: *ταῦτα μὲν οὖν τοῦτον τὸν τρόπον εἴρηται νῦν ὡς τύπῳ, γεύματος χάριν περὶ ὅσων καὶ ὅσα θεωρητέον· δι' ἀκριβείας δ' ὕστερον ἐροῦμεν, ἵνα πρῶτον τὰς ὑπαρχούσας διαφορὰς καὶ τὰ συμβεβηκότα πᾶσι λάβωμεν. μετὰ δὲ τοῦτο τὰς αἰτίας τούτων πειρατέον εὑρεῖν. οὕτω γὰρ κατὰ φύσιν ἐστὶ ποιεῖσθαι τὴν μέθοδον, ὑπαρχούσης τῆς ἱστορίας τῆς περὶ ἕκαστον· περὶ ὧν τε γὰρ καὶ ἐξ ὧν εἶναι δεῖ τὴν ἀπόδειξιν, ἐκ τούτων γίνεται φανερόν* (491a7-14).

[31] *δι' ἀκριβείας δ' ὕστερον ἐροῦμεν, ἵνα πρῶτον τὰς ὑπαρχούσας διαφορὰς καὶ τὰ συμβεβηκότα πᾶσι λάβωμεν.*

'the differences *and in general* the attributes', rather than taking 'differences' to mean causally primary, defining features, and 'attributes' to mean necessary features of the kind which follow from the 'differences'; the latter distinction presupposes that one has *already done* the causal investigation that the *historia* is said to be a means to.[32]

The first aim of *HA*, then, according to Balme, was 'to lay out the differentiae', i.e., to present all of the differences that are to be found in the animal kingdom—in parts, modes of life, activities, and character—and to do so systematically, characterizing each difference precisely. And that is just what Aristotle went on to do, Balme argued. Books I–IV, he observed, are concerned with parts; V–VII with 'lives and activities' (V–VI are about generation, VII about nutrition and means of coping with heat and cold such as migration, the sloughing of skin, and so on); VIII with characters; and IX starts the study of these differences in human beings, but only gets generation off the ground.[33] Within I–IV, for example, the organization is: parts of blooded animals/parts of bloodless animals; then, within each, external/internal; then, within each of those, by *megiston genos* with remaining kinds treated separately (e.g., man, snakes). So much, then, for laying out the differentiae.

2. 'Significant groupings of differentiae'

Balme discussed this aim rather informally, as a concern for which differentiae 'can be combined' or 'are essentially associated or divergent'.[34] He correctly offered *APo.* II.14 in support of the view that by establishing such associations we will be better positioned to establish the causes. James Lennox and I took this idea much further,[35] but because we were essentially just building on Balme's idea, I will, in the course of this exposition, treat our view as part of the Balme position on *HA*. Most of the theoretical work was Lennox's: he explored the *Analytics*' insistence that explanations be at the right level of generality, i.e., that demonstrative conclusions state *kath' hauto* (*per se*)—and even primitively universal—connections, analyzing closely not only *APo.* II.14 but also I.4-5, II.15-18, and *APr.* I.27-30.[36] Along with the methodological passage I quoted above from *HA* I.6, in which Aristotle claimed that a *historia* would facilitate the discovery of causes, they all supported the view that securing propositions of the form 'all and only those

32 Cf. n. 30 above for the Greek of the surrounding context.

33 In citing books of *HA* I adopt Balme's restoration of the MS ordering, returning what modern editions call book VII back to its original place as book IX; the fifteenth-century translator Gaza had placed it seventh and most editors before Balme followed him. (For a fuller discussion of the matter, see now the appendix to this chapter.)

34 Balme 1987b, 86.

35 See chapter 14 below (the first part of which was written for, and delivered as a guest lecture in, A. C. Crombie's seminar on History of Science and Medicine at Oxford in Hilary term 1984) and Lennox 1987a.

36 Cf. Lennox 1987a, n. 18.

animals which have feature F, also have feature G' is central to the *hoti* ('that') stage of scientific inquiry identified in *APo.* II.1, and is, as II.1 claims of the *hoti*-stage, the *route* to the discovery of the *dihoti* ('why') and thus the route to demonstration, *apodeixis*.

A detailed study of *HA* by both of us, done in part together and in part separately, established that the discovery of such propositions, or propositions approximating these, was a pervasive feature of *HA*, of equal importance to that of laying out the differentiae as such. We called such propositions 'widest-class generalizations'. I noticed that they often were presented in a certain type of doubly-quantified proposition of the form:

hosa (echei) X, *panta echei* Y. ('As many as are/have X, all have Y.')[37]

Lennox counted 31 such propositions, 32 more without the *panta* ('all') but carrying the same meaning, and 16 additional with different quantifiers: 'as many as are/have X, *most/some/few/none* have Y'.[38]

We noticed also that in these 'widest class generalizations' the subject place need not be held by a single large kind, K, but may be held by:

	$K_1 + K_2 + \ldots + K_n$
or	$K_1 + K_2 + \ldots + F_1$
or (as often)	$P_1 + P_2 + \ldots + P_n$

(where F is a form of some coordinate kind K, and P a feature or predicate). For instance, in the discussion of external parts in birds, *HA* II.12, Aristotle begins by listing features that birds have in common with the very large kinds that have already been discussed, then moves to features unique (*idia*) to birds, then to features possessed by various sub-groups of birds (e.g., 504a8-9 just mentioned); where he notes an oddity within a certain sub-kind, such as the rare distribution of toes in the wryneck (*iunx*, 504a11-12) and a few other birds, he often also mentions, as he does here for the wryneck, other features that are also unique to that sub-kind.

So, finding the 'significant groupings of differentiae' is a significant aim of *HA*. But what ultimate aim or aims are served by the laying out of, and finding of significant groupings of, differentiae? That takes us to the third component of Balme's positive view of the aims of *HA*.

[37] See, for example, 504a8-9: *εἰσὶ δ' ὅσοι αὐτῶν μετεωρίζονται πάντες τετραδάκτυλοι* ('As many of [the birds] as raise themselves to a height, all are four-toed') and 505b32-33: *ὅσα μὲν οὖν ἐστι τετράποδα καὶ ζωοτόκα, στόμαχον μὲν καὶ ἀρτηρίαν πάντ' ἔχει* ('Now, as many as are four-footed and live-bearing, all have an oesophagus and windpipe').

[38] Lennox 1990, 178; for references, see also Gotthelf 1988, 110-11 (= chapter 14 below, pp. 296-97); Lennox 1991, 285; Charles 1990, 158-59.

3. Ultimate aim(s)

(i) Balme's first answer,[39] and the one he never abandoned, was the aim Aristotle himself states at 491a10–11: *causal explanation*.[40] As we have already seen, the laying out of differentiae, and the finding of significant groupings which that makes possible, facilitates the discovery of their causes. As I illustrate in the Coda below, if we want to know why those animals which have four stomachs have four stomachs, we need to discover what other features, or combination of features, all four-stomached animals, and only four-stomached animals, have in common; for the cause will surely be found among these.[41]

(ii) However, in the course of preparing his Clarendon Aristotle volume, on *PA* I and *GA* I, Balme did a close study of *PA* I.2-4 and noticed the great similarity between the account in the latter chapters as to how one should proceed if one wishes to grasp (*lambanein*) or define an *eidos*, a lower-level animal kind, and the very opening lines of *HA*.[42] The former identifies the need to divide down from a generic definition that contains multiple differentiae, and the latter discusses how differentiae may be the same or different, and what sort of sameness and difference we can expect to find at various levels of generality of animal kind. This suggested to Balme that one ultimate aim of *HA* is the reaching of such definitions. The connection with the stated aim, causal explanation, Balme thought, is clear from *APo*. II.8-10 (*et passim*), in which Aristotle insists that the defining differentiae of a kind will be those that are causally responsible for the non-defining necessary features of that kind.[43]

Noticing that no such definitions were to be found in Aristotle, and that the aim of reaching an ultimate definition is 'a somewhat naive one', since 'there is no end to the recognition of fresh significant attributes', Balme hypothesized that Aristotle may have given up this aim as his scientific work progressed.[44] I am not convinced by that, and will offer a different explanation for the absence of such definitions

[39] Balme 1961, 208-12 (repeated in 1975, 189-98 and 1987b, 85-89).

[40] *meta de touto tas aitias toutōn peirateon eurein . . . eurein* ('After this we must try to find the causes'). The *de* here answers to the *men* at 491a7, and so the causes to be found are the causes of *'the differences and <and in general> the features'* which have been laid out in the first stage of this investigation (*methodos*, 491a12; cf. 11-14).

[41] And so I sometimes wonder if (at least some of) 'Mill's methods' shouldn't be called 'Aristotle's methods'. As Jim Lennox points out to me, they should at least be called 'Herschel's methods', since Mill credits Herschel directly. (Herschel says he is just formalizing 'rules' he found in Bacon; so we are already part-way to Aristotle.)

[42] *HA* I. 486a5-487a10. The results of this study of *PA* I.2-4 were first published in Balme 1970 and then as part of Balme 1972.

[43] The revised picture of *HA*'s ultimate aim was first presented in Balme 1987a, 11.

[44] Balme 1987b, 80. I suppose Balme thought either that the body of such definitions would never be complete, or (more likely) that any attempt at causal definition of an attribute would very likely be revisable, but he doesn't say.

below;[45] in any case, Balme took it as an initial aim of *HA*, and so it belongs in our characterization of his view of that treatise's overall aims.

So, what we have is biological science without a systematic classification of animals, as we defined that notion above. This was a very interesting claim from the standpoint of the history of science. As Balme commented:

> The belief that there must be a classification in the background rests on the assumption that Aristotle, like every good pre-evolutionary zoologist, put systematics first in zoology and morphology first in systematics. But better sense can be got out of the evidence if this dual assumption is removed. Aristotle's aim was not taxonomy, nor did it involve one; the comparison of differentiae had other purposes.[46]

That result must stand. But we still have to ask: can the aim of *HA* really be stated without significant reference to the *megista genē* delineated in I.6? Does not Aristotle's concern with them represent a concern with classification (if not with a full systematic, hierarchical taxonomy) or at least a concern with natural kinds, which must be incorporated into our view of the overall concerns of that treatise, and of the stage of biological science preliminary to the discovery of causes?

The significance of the *megista genē* of *HA* I.6 for our understanding of the aims of *HA*

There are three questions, then, that we must face, and answer, if we are to have a full picture of the aims of *HA*:

1. How are the *megista genē* of *HA* I.6 established as such?
2. Where are they so established?
3. What role(s) do they have in *HA*?

I want to discuss views that have been taken on these questions by Balme, David Charles, and James Lennox.[47]

[45] p. 281. [46] Balme 1987b, 84.

[47] In the case of Charles and Lennox, I will take off from their exchange of views in Devereux and Pellegrin 1990: Charles's 'Aristotle on Meaning, Natural Kinds and Natural History' (Charles 1990) and Lennox's 'Notes on David Charles on *HA*' (Lennox 1990). Where appropriate I will supplement my account of their views from more recent writings: Charles's *Aristotle on Meaning and Essence* (Charles 2000), ch. 12 ('Biology, Classification, and Essence'), and Lennox's 'Getting a Science Going: Aristotle on Entry Level Kinds' (Lennox 2004). (The three of us also had lively discussions of these matters during our joint postgraduate seminar on 'Philosophical Issues in Aristotle's Biology' at Oxford in Trinity term 1994, and I included recollections of those discussions in the first draft of the present chapter, written within a few weeks afterwards for my Tokyo seminar. Where useful, I will draw on that record of those discussions, occasionally and with care, in what follows.) While I shall discuss Balme's, Charles's, and Lennox's views and suggestions on these topics, my main goal is to review the relevant textual evidence and, on that basis, to develop a view of my own.

Common ground for everyone are two propositions:

(1) When Aristotle introduces his initial list of seven of the nine *megista genē* at the opening of I.6 (490b7) with the words:

> Very large kinds of animals, into which some animals have been divided, are the following: one kind, of birds; one, of fishes; another of the cetacean; . . .

the force of the perfect tense, *dieirētai* ('have been divided'), is to suggest that these seven groups have already been marked off by people, and that they have been marked off *correctly*.

There are several reason for thinking this, of which the primary one constitutes the second proposition that is common ground:

(2) The activity of introducing *megista genē* in *HA* I.6 reflects the application of a methodology regarding the recognition of *genē* that is laid down in *PA* I.4. There, at 644b1-7, Aristotle says that:

> Perhaps then the right course is to speak of some affections in common by kinds, wherever the kinds have been correctly (*kalōs*) marked off by popular usage and (*kai*) possess both a single nature in common (*mian phusin koinēn*) and forms not far separated in them—bird and fish and any other that is unnamed but like the kind (*anōnumon men, tōi genei de homoiōs*) embraces the forms that are in it; but wherever they are not like this, to speak of particulars, for example about man or any other such.[48]

[48] Tr. Balme (1992), with 'kinds' for 'genera', 'forms' for 'species', and 'correctly' for 'satisfactorily'. For the Greek of the entire passage, see chapter 13, p. 297 n. 9. Lennox's translation is not essentially different; he renders *mian phusin koinēn* as 'a single common nature': 'Perhaps then the right course is this. In some cases—whenever kinds are spoken of by people in a clearly defined manner and have both a single common nature and forms in them not too distant—we should speak in common according to kinds, like bird and fish and any other there may be that, though it is unnamed, embraces, like a kind, the forms within it. But whenever they are *not* such as this, we should speak one by one, e.g. about mankind and any other such kind' (Lennox 2004, 95, quoting his 2001a). Charles (2000) renders *mian phusin koinēn* as 'one common nature apiece'. His translation, though in general more expansive, differs significantly from Balme and Lennox only at the point at which Aristotle speaks of the unnamed cases, apparently because Charles follows Peck who (against all the manuscripts, it seems) reads *homoia* instead of *homoiōs*: 'So, perhaps the right procedure is this: [a] so far as concerns the attributes of those groups (genera) which have been correctly marked off by common usage—groups which possess one common nature apiece and contain in themselves species not far removed from one another (I mean birds, fishes, and any other group which though it lacks a name yet contains species generically similar)—to describe the common attributes of the group all together; and [b] with regard to those animals which are not covered by this, to describe the attributes of each by themselves, e.g. those of man, and of any other such species' (Charles 2000, 312). Peck, who writes 'correxi' in his apparatus note to *homoia*, seems to miss the way in which a grouping without a name might be thought by Aristotle to be only 'like a *genos*'. See my discussion of this in the next chapter, pp. 299-301, which I draw on in what I go on to say here.

Not every grouping marked off by popular usage will be accepted but only those which are *correctly* (or *well-, kalōs*) marked off—and I take it that what follows these words provides the conditions for a grouping's being correct (the *kai* ['and'] being more or less epexegetical):

(i) there is a common nature;
(ii) the forms embraced by the kind are not too far apart (so that the common nature must not be too abstract or generalized)

for simplicity, we may say that condition (i) is met (and there is a 'common nature') *only* when condition (ii) is also met, although on occasion we will refer to these conditions separately. And we should note right at the outset that Aristotle is not especially clear as to *what constitutes* having a common nature.[49]

Such is common ground. Based on it, we may restate our three questions as follows:

1. In establishing these very large kinds, what does Aristotle take as evidence of a common nature?
2. Where—i.e., at what place(s) in *HA*, or in the work leading to the writing of *HA*—does Aristotle establish these as actual kinds?
3. For what purpose(s) are these kinds established, however and wherever they are established?

Balme does not comment on questions 1 and 2, so let us focus first on question 3 and start with Balme's view. As we have noted, he sees the primary aim of *HA* as the laying out of differentiae. Balme actually said very little as to the role of the *megista genē* in facilitating that aim. At one place he writes:

> The *HA* is a collection and preliminary analysis of the *differences* between animals. The animals are called in as witnesses to differentiae, not in order to be described as animals.[50]

Although very large kinds may serve as examples of differentia-possessors, that role would not explain their importance, which is evidenced by their inclusion in the introductory 'outline' in *HA* I.1-6. However, Balme's secondary aim for *HA*,

[49] I also observe, though only in a footnote, because Charles's translation (as against Balme's, Lennox's, and my own) does not commit him to this, that Aristotle allows that we may treat as a very large kind a grouping which meets these conditions even though the grouping is 'unnamed', which I take to encompass two cases: (a) where a kind has been recognized but is designated only by a descriptive phrase, typically a neuter plural article plus adjective, as in *HA* I.6 490b11: 'another, the one of the soft-shelled animals, *unnamed by one name*' (my emphasis); and (b) where a group has not yet been popularly recognized but meets the conditions nonetheless, as I believe is the case for the two groups that Aristotle himself (i) *adds* to the seven in the subsequent discussion in I.6, (ii) includes (if the text is sound) in his list of *megista genē* at the opening of II.15, and (iii) employs as very large kinds throughout: *the live-bearing four-footed animals* and *the egg-bearing four-footed animals*.

[50] Balme 1987b, 88.

identifying 'significant groupings of differentiae', does make important use of the list of very large kinds, as Lennox and I have argued in the course of giving a greater prominence to this aim than Balme did. For, it seemed to us that a very large kind was a very powerful tool in establishing what we called 'widest class generalizations'. After all, whatever else is true of a very large kind, it is indeed true that the forms in such a kind will have many features in common, and many of those features will be unique to the kind: e.g., all and only birds have feathered wings, beaks, perhaps light bones, etc. If one's goal is to get propositions of the form 'All and only animals that have F have G', a proposition of the form 'All and only Ks have G' is a very powerful tool, since it brings along at once a whole set of propositions of the form 'All and only animals that have F have G'. A study of differentiae by very large kinds greatly facilitates this second aim, Lennox and I thought, and thus deserves the prominence it gets in *HA* I.6, and even the concern with having groups that are *correctly* marked off, since the members of true kinds will have the most features, and indeed the most causally significant features, in common and thus be the most powerful tools.[51]

Charles, while not necessarily denying any of this, insists that *HA* has *another* aim, one at least as important as (and arguably more important than) the two-part one we have just found Balme, Gotthelf, and Lennox focusing on—an aim in which the I.6 establishment of *megista genē* has a central role. We can get at that aim, says Charles, by returning to the *APo.* II.1-2 statement of what scientific inquiry itself seeks. Balme, Gotthelf, and Lennox have focused on the two-stage progression in scientific investigation that first reaches a knowledge *that* (*hoti*) an attribute is possessed by a subject kind and then moves from there to knowing *why* (*dihoti*) it does. But, says Charles, *APo.* II.1-2 also speaks of the progression that first reaches a knowledge *that a kind exists* and then moves from there to knowing *what* the kind is, i.e., to knowledge of its *essence*.

As support for his view of the aims of *HA*, Charles claims, among other things, that (i) *HA* I.1-5 marks off the very large kinds according to their possession each of a common set of soul-functions, evidencing Aristotle's concern with establishing very large kinds as such;[52] (ii) *HA* I.6, after listing the *megista genē* that have been recognized, considers various other candidates, rejects at least one and accepts two more, further evidencing his concern within *HA* for establishing kinds;[53] (iii) throughout *HA*, in the course of describing which animal kinds possess which differentiae, Aristotle repeatedly confirms these very large kinds and establishes their sub-kinds, on the basis of shared soul-functions (and sometimes other criteria as well);[54] and (iv) even the enterprise of establishing 'widest class general-

[51] Cf. chapter 7 above, p. 148 and chapter 14 below, p. 294-96.

[52] Charles 1990, 157; 2000, 317-18.

[53] Charles 1990, 157-58; 2000, 318-19.

[54] Charles 1990, 159-65; 2000, 320-25.

izations' depends importantly on having kinds in the subject place, at least much of the time, if the enterprise is to facilitate the discovery of causes,[55] as Aristotle himself appears to say in several passages, e.g., *PA* I. 645b1-3, where he says:

It is necessary first to divide the attributes pertaining to each kind which are present in all the animals <of that kind> in virtue of themselves, and after that to try to divide their causes.[56]

Lennox in his 1990 reply to Charles supported, in a general way, the view that I.1-5 is concerned rather to outline the enterprise of laying out, and grouping, the differentiae, than to establish kinds. Charles subsequently pointed, in the case of the fishes, to 487a15-20 and 489b23-33 as the places where they are established as a very large kind. But the first mention of fishes, at 487a18, speaks of 'many fishes', not all, and everywhere that fishes are mentioned in those chapters, they appear to be brought in, as Balme had said, 'as witnesses to the differentiae'[57] and not for their own sake: in no place does the *genos* of fishes appear to be the actual subject of discussion, until their introduction as a very large kind in I.6. Furthermore, Lennox observed, the passages Charles cites from the remainder of *HA* after I.6 as evidence of Aristotle's concern to confirm the establishment of fishes as a very large kind, and to establish its sub-kinds (cf. above, p. 278 and n. 54), are spread all over the treatise, not concentrated in any one place, and again are all concerned, not with classification or kind-grouping, but with differentiae, the kinds being brought in, again, only as witnesses to the differentiae.

As I have argued elsewhere in this volume, Aristotle's concern that the ultimate subjects of scientific explanation be true kinds is not contradicted by an explanatory science in which much of the explanation is at a high generic level, and in which the propositions explained have as their subjects expressions that do not designate kinds but designate only what Balme called 'differentiae-classes'.[58] Thus, among the four arguments I listed above, there remains only the discussion within I.6, at 490b15-491a6, of the question whether there are any very large kinds other than the seven that have been popularly and correctly divided off. Unfortunately, this is easily the most difficult passage in all of *HA*, and there is not yet a fully convincing line-by-line interpretation of its aims and contents.[59] However, I in fact have long held, as did Balme, and does Lennox, more or less the same view of this passage as Charles does. We all think that Aristotle first rejects *four-footed animals* as a very large kind, but in the process provides the basis for establishing as distinct very large kinds

[55] Charles 1990, 158-59; 2000, 319-20.

[56] For a full discussion of this sentence, see chapter 9 above, pp. 200-14, esp. 204-206. See also, in this chapter, p. 273.

[57] Above, p. 277 and n. 50.

[58] 'Elsewhere in this volume': see chapter 8 above. 'Ultimate subjects . . . true natural kinds': Cf., e.g., *PA* I.4 644a23-33. 'Differentiae-classes': above, p. 269.

[59] I have offered my best shot at a close reading of this passage in the chapter which follows the present one.

both *four-footed live-bearing animals* and *four-footed egg-bearing animals*; and then that he rejects various other possibilities, such as *animals with hair*, and *animals with scales*. And we all think that he uses to decide all this the conditions for kind-recognition laid down in *PA* I.4, conditions we have already examined.

Now, that one excursion in *HA* I.6 into the establishment of two new very large kinds could not justify our treating the establishment of kinds as *an aim* of *HA*, much less a general aim, but Lennox, influenced by Charles's surely correct observation that the *APo.* II.1-2 thesis that a scientific inquiry into any broad *genos*, such as *animals*, includes the establishment of the various kinds and sub-kinds falling under that broad *genos*, has tried to provide other evidence for the view that *HA* indeed has as one of its aims, if a lesser one, the establishment of kinds. As I understand his argument, it appeals—apart from a reliance on the claim in *APo.* II.1-2 and the hypothesis that the biology is in line with *APo.* in this respect—primarily to two sets of texts. First, we are asked to consider places where the very large kinds to be used in the study of differentiae are introduced, such as the opening of Book IV: here, for instance, discussion of differences in bloodless animals is preceded by the introduction of the very large kinds of bloodless animals. Second, we are pointed to several passages in which Aristotle is supposed to be grouping certain forms of a very large kind together on the basis of a certain similarity, notwithstanding that, as he notices, the grouping is 'unnamed'. One example offered is 531b21,[60] and Lennox also claims that in the biological example Aristotle discusses in *APo.* II.17. 'broad-leafed trees' is recognized as a distinctive *kind*.[61]

But this last seems wrong: the *APo.* example is to me a clear case of an explanation in which none of the terms appears to designate a kind, or at least in which it does not matter whether the expression does or does not (and there is no clear evidence there that it does[62]). Nor, as Lennox himself pointed out, does Aristotle, in the admittedly very rare passages in which he recognizes such unnamed similarities, ever go on either to name a kind or to treat the grouping in any significant way as a kind. So, even were we to grant that Aristotle believes these unnamed groups to be proper kinds, and that he was noting that (alleged) fact on those occasions, we would not have in those few passages sufficient evidence of an *aim* of *HA*. And, finally, although this is a somewhat subjective matter, it does not seem to me, as I read the opening of IV.1, that Aristotle is aiming either to *establish* or to *confirm* the existence of these kinds. First of all, they were already included in the list of correctly recognized very large kinds in I.6, so that it would seem rather

[60] Lennox 1990, 180.

[61] Lennox 2004, 89.

[62] In correspondence Lennox has cited 98b35-38, but I just don't see that anything more than a descriptive phrase is necessarily called for.

late to be *establishing* their existence; and instead of confirming their existence, the passage seems rather to offer a mode of clearly distinguishing one from the other, as a basis for going on to use them to facilitate the finding, and reporting, of (what Lennox and I have called) widest-class generalizations.[63]

How, then, is the practice in *HA* to be related to the *APo.* II.1-2 theory of inquiry? First of all, I certainly agree that this theory, and much else in *APo.*, occasionally in modified form, is a prominent part of the background of Aristotle's biological enterprise; but I do not think that this means that *HA* itself, as we have it, includes that part of the enterprise which is the establishing of animal kinds (except in that one passage in I.6 we have mentioned, and perhaps in one or two others). Now, I *do* think *HA* is indeed concerned to establish that kinds exist, in preparation for the later enterprise of establishing the essential definitions of those kinds, but the kinds in question are *the kinds of differentiae*, not the kinds of animals. Note that *HA* might indeed be viewed as providing (among other things) a systematic classification of *differentiae*, presented via a significant use of division; and that the explanations in the other treatises of the possession of those differentiae by whichever animals do possess them, do generate essential definitions of those differentiae. Just as the explanation of thunder, in the *APo.* II.8-10 example, enables a definition of the essential nature of thunder, so determining the function of some animal part in *PA* enables an essential definition of that part.[64]

As for *animal* kinds, I would say first that I have been convinced by Charles that *APo.* II.8-10 and *Metaphysics* Z.17, among other texts, make clear that Aristotle envisioned the possibility of essential definitions of substance-kinds, such as animal forms and kinds, as well as non-substance-kinds, such as thunder and eclipse. And, contrary to Balme, I see no reason to think that Aristotle abandoned that goal in the study of animals.[65] I do think, however, that the path from the discovery of the causes (and thus the essential definition) of individual differentiae, to the essential definition of animal forms is a long and arduous one, and that this aim, although real, may have been seen by Aristotle as a quite distant one. But to reach definitions of animal kinds we have first, of course, to establish their existence, and the question that Charles and Lennox have urged me to consider is this: where are the kinds established and how? For even if *HA* is not itself concerned with establishing kinds, the very fact that Aristotle introduces the very large kinds in I.6 as *correctly* marked off implies that he must at some point have established to his own satisfaction that these kinds *are* correctly marked off. *HA* may presuppose such an activity even if that activity is not part of its aims. Indeed, there is no reason to

[63] See also my further comment on the *HA* IV opening passage below, p. 285.

[64] See on this, chapter 7 above, pp. 164-65 n. 35 *et passim*. For Theophrastus' reservations about the possibility of doing this, in the case of plants, even for differentiae, see below, chapter 14, pp. 325, 327.

[65] Balme 1987b, 80.

think that *HA* must include within its covers *every* activity at the pre-explanatory stage of inquiry.[66]

But once one grants that, one opens oneself to another point, which has also been urged on me by Charles and (especially) Lennox: that even if *HA* is not itself concerned to establish kinds, it may provide evidence of *how* Aristotle did think kinds were established.[67] That is to say, bringing to bear now the conditions for the establishment of a kind we have found in *PA* I.4, perhaps *HA* might provide us with signs of what Aristotle himself considered evidence of a 'common nature'. And that brings us to our final issue, this very question of how Aristotle thought kinds were established (even if that process didn't take place within *HA*)—and in particular, how he went about establishing as properly-marked-off very large kinds the seven kinds he provides us with at the start of *HA* I.6 as so established.

Evidence for *how* the very large kinds may have been established

A point of terminology

By 'establishing the kinds', or 'fixing the kinds', is meant: providing sufficient reason, at this stage of inquiry, for holding that these indeed are actual kinds, sharing a single common essential nature and embracing a range of forms (*eidē*). The stage of inquiry in question is the stage described in *APo* II.1 as the stage at which one knows *that* a kind exists, even if one doesn't know *what* it is, i.e., what its essential nature is.

Theses

Charles The enterprise of laying out the differentiae and establishing correlations presupposes (for various reasons discussed above) the establishment of very large kinds. These kinds are first established in *HA* I.1-5, prior to being listed in I.6; this demarcation is further confirmed across *HA*, where the sub-kinds of these kinds are established as well. The notion of a *common nature* that Aristotle brings to this process (cf. *PA* I.4 644b3) is theoretically based in the *De Anima* theory of soul-functions. According to that theory, a common nature is (primarily) a set of distinctive soul-functions, and Aristotle thus gives primary weight, in determining whether a group of forms constitutes a large kind, to evidence of the presence of such a distinctive set of soul-functions, although he gives some weight as well to broad perceptual similarities in the shape of the animal and its parts. The functions

[66] Just to stress: I am not conceding that *HA* is itself concerned with animal kinds. Indeed, with the exceptions noted above, I think that this view is mistaken.

[67] Cf., e.g., Lennox 1990, 173-74.

which are typically of primary interest are those of: feeding, locomotion, breathing (i.e. cooling), and reproduction.[68]

Lennox The *HA* is a report of the results of a process of collecting and correlating differentiae. The kinds are established by Aristotle (i.e., declared to be actual kinds, correctly marked off) *in that process*; his criterion for determining that we indeed have a kind is the presence across a set of forms of (i) many features in common, (ii) differing in most cases only by 'the more and the less' of their perceptible features (*pathē*). At that stage of inquiry we know that a kind must have a common nature but we do not know what constitutes its common nature (or, for that matter, a common nature in general). Evidence of a common nature is thus (only) the sharing of many features in common. Once the causes have been discovered, and found to reside primarily in the soul-functions, which are the teleological causes of the other features, the scientist has then established that a common nature is (primarily) a distinctive set of soul-functions; but this view of a common nature is not available to Aristotle at the pre-explanatory stage of which *HA* is a part. Returning to *HA*, now: to say the kinds are established during the research the results of which *HA* reports is to say that they are already established by the time that *HA* is written, and not within *HA* itself. However, various passages in *HA* and elsewhere may help us to determine the criteria Aristotle used to establish the kinds when he did establish them.

Texts

I start with the texts which Lennox has offered, or might have offered, in support of his view, and then turn to those which Charles has offered, or might have offered, in support of his, commenting on some of the texts in bracketed remarks. I will then offer an 'Assessment' in which I build on Balme's (all too brief) indication of his own position to offer (what I will call) a cautious synthesis of elements from Charles's and Lennox's views, though in certain ways I will go beyond what either of them says.[69]

The passages appealed to by Lennox include:

1. *PA* I.2 642b14: A kind that should not be split, like Birds, has a *similarity* or *likeness* (*homoiotēs*). ['The blooded', which is apparently not a *genos*,[70] also has a likeness, so having a 'likeness' must be a necessary and not a sufficient condition for a kind.]

[68] Charles 1990, 157-64 and (esp.) 2000, 317-25.

[69] To bring this discussion within manageable proportions, my comments on these passages will be brief and in note style. Likewise, my assessment will provide only the sketch of my 'cautious synthesis'. My aim is to lay the foundation for a view that better integrates the full range of relevant texts (and to encourage work by others in that direction).

[70] See my discussion of this in chapter 10 above, p. 226.

2. *PA* I.3 643b10: marked off by many differentiae. [Aristotle does not here say *which* sorts of differentiae, and one must keep in mind Charles's point that to say 'multiple differentiae' is not to say 'all differentiae, weighted equally'.]
3. *PA* I.4 644a16: marked off rightly *because* the kinds yoked together into one kind differ in the more and the less. [With this passage compare the following one:]
4. *HA* I.1 486a23: 'I call a "kind", e.g., fish or bird, for each of these has a difference in kind [= many features are the same but differ in degree] and there are numerous forms of fishes and birds.' [A long list follows of more-and-less differences which appear determinable as such prior to deciding what will count as a kind.]
5. *PA* I.4 644b7: 'The kinds have been marked off mainly by the shapes of the parts and of the whole body, wherever they have a likeness (*homoiotēs*), *for* the parts of these differ not by an analogous likeness but rather by the bodily affections (*pathē*), e.g. large/small, soft/hard, . . . , and in general by the more and the less.' [With this passage compare:]
6. *HA* I.6 491a15: we should first discuss the differences among parts, 'for it is chiefly, and first, in respect of these that the wholes also differ'—either by having or lacking them, or by position, or arrangement, or according to the differences we have spoken of [namely: same in form; same in form but different in the more and the less or by degree; different in kind but the same by analogy—a reference back to *HA* I.1 486a14-487a1].
7. *HA* I.6 490b20: the decision as to whether a grouping counts as a kind or not seems to depend on the presence or absence of correlations between such features as hair and scales (which are observable features that vary in the more and less, and not soul-functions).
8. *HA* II.13 504b13: among the water animals the *genos* of fishes is *one genos* marked off from the other animals, embracing many *ideai* [forms? visually distinct appearances?]; *for* (*gar*) they all have a head... [How is this to be understood? As a recapitulation of the process by which the kind was established? Or, perhaps the missing line that explains the *gar* is: 'And so we should expect to find common and distinctive predicates, and thus widest-class generalizations, *for they all have a head, . . .*' In that case, then the fishes' status as a kind is being assumed, and the *gar* does not indicate that the kind's establishment is being reviewed.[71]]
9. *PA* IV. 692b4: Birds differ among themselves in the more and the less, but with respect to other kinds of animals they differ according to the shape of their parts.

[71] See below, item 12, for a similar reading of the characterization of the bloodless kinds at the beginning of *HA* IV, a passage of which Lennox made much in a presentation at our Oxford seminar (above n. 47).

10. *HA* VIII chs. 19 ff.: Types and sub-types (sometimes called *eidē*) of birds are marked off not by variation in mode of performing soul-functions but by *eidos*, as it is called at 616a19, which apparently means 'appearance'—specifically, by: size, color, features of the nest, voice, locale, habitat, type of food, degree of health or 'thriving'. [However, it is important in reading these passages to distinguish *identifying remarks* from statements laying out (in Balme's sense) the differentiae. It is only the latter that are relevant to the issue between Lennox and Charles, but in my view the initial descriptions in these chapters are the former—identifying remarks that are not meant to capture features that are relevant to the *kinds* the birds are put into (though I do consider the question open).]
11. *HA* IV.1 523b1 ff.: 'The discussion of bloodless kinds perhaps shows how, in the inquiry there reported on, the kinds might have been fixed'.[72] The four bloodless kinds are distinguished by whether their hard or soft parts are inside or outside, and how the hard part responds to pressure:

Kind	Soft	Hard
Soft-bodies:	outside	inside (if any)
Soft-shells:	inside	outside—crushable not crackable
Shell-skins:	inside	outside—crackable not crushable
Insects:	[uniformly hard inside and out, a hardness intermediate between ordinary flesh and bone]	

[A comment: Taking the passage this way assumes (i) that prior to collecting these differentiae, and noting across the forms the shared softness/hardness of interior/exterior, Aristotle was not prepared to endorse the groupings already made by ordinary folk or fishermen, *and* (ii) that it was primarily on the basis of the shared softness/hardness of interior/exterior that their kind-status was endorsed, although perhaps also on the basis of the other shared features Aristotle goes on to mention, all of which are parts and none of which are soul-functions. But, surely the reason only parts are mentioned in what follows, and perhaps also in the initial characterization, is that Aristotle is here in book IV only laying out the differences that are to be found *in the parts*. Books V and VI will deal with reproduction, VII with feeding (etc.), and VIII with characters. So, with respect to (ii), it is not at all clear that the kinds would be established by reference to these features, and with respect to (i), this assumes that prior to considerations of the consistency of the materials (hard/soft, exterior/interior), Aristotle could not have distinguished them as kinds, e.g., by reference to gross shape of whole and parts and gross activity (e.g., the way they move [including, where relevant, their stationariness).]

[72] Lennox, Oxford Seminar, 9 May 1994.

The passages appealed to by Charles include:

12. *HA* I.1-5 *passim*: the kinds are first introduced as possessing distinctive modes of locomotion, reproducing, breathing (i.e. cooling). [But, as observed above, the entire discussion seems concerned to lay out *differentiae*, with kinds brought in as examples, and their kind status already established or at least taken for granted.]
13. *HA* I.6 490b19-32: as discussed earlier, Aristotle appears to be claiming that: 'The wingless four-footed animals' do not constitute a kind because their reproductive mode is too different. [Note, in favor of Charles, that nothing else is mentioned, and there is no counting up of features.] And where (as it seems) 'the four-footed live-bearing animals' and the 'four-footed egg-bearing animals' *are* admitted as kinds, all that is mentioned is their distinctive combinations of locomotive and reproductive functions. [Note: it is important to say 'distinctive set of soul-functions', and not 'set of distinctive soul-functions', since it's not obvious that (e.g.) there is a common and distinctive way of being four-footed that belongs to each of these two very large kinds.]
14. *HA passim* re fishes:[73] there is a constant focus on soul-functions (e.g. swimming, feeding, reproducing), or on parts that perform soul-functions (e.g. gills, fins), in characterizing, and sub-dividing, fishes. [Lennox 1990: On Aristotle's teleological theory, *all* common features at the kind level (ignoring more-and-less differences) are soul-functions *or* activities that serve soul-functions *or* parts that contribute to them, etc. So we cannot tell from the fact that a feature is connected with a soul-function that it helps establish a kind *in virtue of* being so connected; and in some cases one may distinguish a kind or sub-kind by an *activity* whose *function* we are unaware of—e.g., to supply an example of my own, one can imagine the *HA*-stage scientist marking off certain kinds of fishes by their traveling in schools without his knowing in the least what *function* traveling in schools has, if any.]
15. How cetaceans are marked off from fishes in *HA* I and VII and *PA*: cetaceans lack gills but have a blowhole (489b2-4) and are live-bearing (489b1). That the former, a difference in parts, is based in soul-function is clear from VII where it is tied to their distinctive way of being 'water animals' (*enhudra*), viz., taking in water only for feeding (expelling it via blowhole), taking in air for cooling (via lungs not gills).
16. *HA* I.1 487a15: [I might add, on Charles's behalf, finally, a point that Balme made to me once in discussion, about the order in which the differentiae in regard to mode of life are introduced in the initial outline in *HA* I.1: Balme suggested that the lives-in-water/lives-on-land distinction is named first because it is among the most important, being explanatory of many other

[73] For specific references, see Charles 1990, 159-63; 2000, 320-24.

> distinctions. I suppose Lennox could reply either (a) that this was viewed as an important basis for division in the Academy, and Aristotle might well have put it first for that reason, and not for any theoretical reason of his own, or (b) that the *HA* is a *presentation* of information already organized, and that in that earlier research it became clear that the differentiae lives-in-water and lives-on-land correlate with many other differentiae and therefore will be of importance in the causal theory, however that theory is going to go.[74]

An additional argument of importance used by Charles is based on the claim that what counts as a more-or-less difference vs. an analogous difference is at least sometimes dependent on whether the class of animals that possesses that difference already counts as a kind or not.[75] Some passages, such as most of those cited by Lennox above, seem to use perceptual criteria of more-or-less differences; others may seem to go the other way.

The tension the previous paragraph introduces leads to a wider caveat. It is possible that in the end the evidence may underdetermine the answer to our question of how Aristotle establishes kinds. The matter is made worse by the fact that once one grants Aristotle the recognition, implicit in his teleological theory, that a distinctive set of soul-functions entails many additional features in common, and that the possession of many features in common entails a distinctive set of soul-functions, it becomes very hard to tell which side of the double entailment is the ultimate *basis* for fixing the kind, and thus very hard to decide who is more right on that score, Charles or Lennox.

This caveat notwithstanding, I do think there is more to say on the matter.

Upshot: a synthesis and a larger cognitive story

Balme, in his 1987b (77, 78-79), sketched a position on the issue of how Aristotle sees the very large kinds to be established, a position which, though not clearly drawn, may serve as our starting point. Balme claimed that the groupings are 'intuitive' (77, 78) and 'empirical' (77) at first, citing passages 1 and 5 above, but that they come to get theoretical justification as a result of the process of grouping differentiae (i.e., discovering correlations) and discovering causes. This sounds rather like Lennox's view, though there is one important difference. Lennox takes the groupings of

[74] I leave aside consideration of the far more detailed and complex discussion of different ways of being land or water animals in *HA* VII.2, since neither Lennox nor Charles draws on it. That discussion would be worth careful study, however, since it appears to be based on views as to the causal significance of the features that create the distinctions between types of land or water animals, which though not kinds per se may shed light on how kinds are marked off by Aristotle at this stage of inquiry.

[75] This is also a point that Pellegrin seems to make; cf. Pellegrin 1987, 328ff. And see below, p. 312 n. 20.

differentiae to be the basis for marking off the kinds, but Balme's choice of the term 'intuitive' would seem to suggest, and not implausibly, that the grasp of 'likeness' Aristotle speaks of in those two passages is not to be taken as a set of distinct grasps of common features. Passages 6 and 7 suggest that the initial grasp which enables someone to form first, as we might say, the concept of, say, bird, is directed 'chiefly and first' to gross features of shape, of whole and part (cf. *Phys* I.1 on the child's grasp of an unanalyzed whole); passage 1 suggests that this initial grasp has cognitive validity, and counts in some sense as fixing the kind for the purpose of further scientific investigation. Whether it counts as 'knowing that the kind exists' in the way required by *APo.* II.1-2 is an interesting question, but even without deciding that, one may claim that this initial 'intuitive' grasp is a central step in the process of *coming to know* that the kind exists, even if it is not sufficient for it.

So, I wish to agree with Charles that the process of grouping differentiae is preceded by a grasp of kinds, but with Lennox that for Aristotle there is an initial fixing of kinds, early in a person's cognitive development, which does not depend on a theory that essential nature is to be found (primarily) in a distinctive set of soul-functions. My own tentative hypothesis is that in Aristotle's favored method for arriving at biological explanations, the teleological principle is regularly appealed to in looking for the cause.[76] Likewise, although biological research may well turn up other soul-functions than those recognized in *De Anima* (such as perhaps defense), such research will regularly start from something like the *De Anima* list of basic soul-functions,[77] if it is to discover the causes which underwrite the detailed biological explanations. While it would require much further work to establish this twofold hypothesis, let me try to set out my conjecture more fully.

An Aristotelian scientist at the *HA* stage is already aware of, and has accepted, the teleological theory and the theory of soul, and thus has already accepted that a common nature will centrally include a distinctive set of soul-functions, whatever else it may include. To that extent, I agree with Charles. Now, Lennox is right that there is no explicit mention of those theories in the passages descriptive of how kinds are marked off (1-6 above), but I find it hard to believe that Aristotle makes no use of them, or doesn't give greater weight to shared distinctive soul-functions, as Charles has plausibly suggested for the one place in *HA* where it seems to me the issue of what kinds one can establish is *actually* in view, viz., I.6 490b14-31 (see 14 above). I think we need a vision of the scientific progression in the grasp of kinds as *moving gradually from* the basic grasp Balme called 'intuitive' *to* a more articulated awareness of multiple shared features *to* a grasp of the centrality for essential nature of contribution to life-functioning, all ahead of the formal enterprise we find recorded in *HA* of laying out all the

[76] See again *Resp.* 471b23-29 and Leunissen 2010b, ch. 4 (although I think Leunissen overstates the heuristic role of the natural teleology relative to its explanatory role).

[77] For a brief discussion of this, see below, p. 386, in this volume's Coda.

differentiae and exhibiting their significant groupings. Each of these stages is cognitive, representing a *grasping* (*lambanein*), however primitive, of a group of animals sharing at least an overall resemblance. The early stages, as I have suggested, may be understood as part of a process of concept-formation and development. And the progress beyond that (if done correctly) may be regarded as a continuing 'establishment' of the kind in question. It is, in my view, an open question at which stage Aristotle would hold that the investigator knows that the kind exists in way described in *APo.* II.1-2. However, even if this is only achieved at quite a late stage, there is no need to identify it with the stage at which the kind is 'fixed' or 'established', since the process is long and continuous and each later stage essentially presupposes the earlier ones. Indeed, the very notion of 'fixing' or 'establishing' the kinds itself needs further analysis.

There is a much more precise version of this larger cognitive story to be told; I mean here only to point a direction.[78] But I do want to say that it is not *HA*'s aim to tell very much of that story; its aims are elsewhere, as I have endeavored to show.

Appendix: A case for the ordering of the books of *HA* VII–IX and a question about the biological study of man that arises therefrom (*co-authored with Pieter Beullens*[79])

Excluding Book X, every full Greek manuscript of *HA* has nine books, divided at the same places give or take a phrase or sentence—except *Laurentianus* 87,4 (C^a). This manuscript divides Book II at 504b13, starting Book III there, ending it at the same place the others end Book II, and numbering each succeeding book accordingly, so that C^a alone has ten books where the others have nine. In all manuscripts of the Arabic translation, in Michael Scotus' Arabic–Latin translation, and in the Greco-Latin ones by Moerbeke, Trebizond, and Gaza, there are uniformly nine books divided more or less at the same places as the Greek manuscripts.[80]

[78] The fullest study of concept-formation in Aristotle as it bears on the progressive grasp of animal kinds is, to my knowledge, Gregory Salmieri's unpublished dissertation, *Aristotle and the Problem of Concepts* (Salmieri 2008), though there is much of value as well in Charles 2000, chs. 5 and (esp.) 6. Salmieri shares Charles's view that what Aristotle understands as the grasp of the *universal* comes somewhere well into the process I am calling the formation and development of the *concept*, and represents an important cognitive achievement beyond the initial formation of the concept. I suspect that Aristotle would not consider the grasp of the concept to count as knowledge *that the kind exists* before the 'universal' itself is grasped. However, all of these questions need further work—work which, I believe, will advance understanding on the questions of this part of the present chapter.

[79] What follows is an excerpt (pp. 474-77) from Beullens and Gotthelf 2007.

[80] There is a slight shift in Albert the Great's commentary *Liber de Animalibus*: he starts his third book at 511b1 instead of 509a26 (= p. 277 Stadler). It is unclear whether he introduced this

Every Greek manuscript and every translation prior to Gaza's orders the books the same way, placing the discussion of human generation ninth (tenth in C^a). Indeed, references in ancient authors to the content of the book on nutrition, habitat, etc., and the book on 'characters', when they refer to these with the numbers VII, VIII, or IX, almost always seem to refer to them as VII and VIII respectively: Düring cites one occasion in Athenaeus, Keaney cites two places in *P. Oxy*. 1802 (fr. 3, col. ii 49-50, 57), one uncertain,[81] and two in Harpocration. The one exception is in Aelian, in a late manuscript, and Keaney infers that 'it must be a late insertion by a scribe who was aware of the order of the *H.A.* introduced by Theodorus of Gaza' (though he does not confirm that the date of the Aelian manuscript allows for this hypothesis).[82] Keaney here refers to the fact that Gaza, as he explains in his Preface, moved the book on human generation to seventh place, and the other manuscripts which agree in that ordering all appear to be later than Gaza's translation and suggest his influence.

David Balme, in both his *editio maior* of *HA* and his *editio minor* of Books VII–X, returned to the pre-Gaza manuscript ordering.[83] He in fact thought that Gaza was probably wrong to believe that Aristotle had intended the book on human generation to follow immediately upon Book VI. The issue here is interpretative, and has to do with one's sense of what Balme called in his Loeb introduction 'The Plan of *HA*'.

Under that heading in the Introduction, after discussing at some length the philosophical context and consequent content of the first six chapters of Book I (including their identification of the primary task of the treatise as the laying out of the differentiae, *HA* 1.6, 491a7), Balme writes:[84]

> The rest of book I with II–IV deals with bodily parts, extended to include sense organs, voice, sleep, sex differences. V–IX deal with activities, lives, characters, but these are not strictly delimited. V and VI are occupied with generation and brood care, extended naturally to include sexual behaviour and nesting; all of this is stated in VII(VIII) to be a part of 'activity and life,' but inevitably some of the data are also relevant to 'characters' and are reported again in VIII(IX) where 'character' is the focus of attention. IX(VII) concerns human generation; it is placed as book IX in all manuscripts before Gaza, but he removed it to its modern position in his Latin translation on the grounds that the books on generation belong together. The introduction to V says that in regard to generation man will be

change on his own initiative or found this variant in the manuscript of the Scotus translation he used.

[81] See *Corpus dei papiri filosofici greci e latini*, I.1 (Florence, 1989), 335-36, for a recent critical edition of the papyrus.

[82] Keaney 1963, esp. 52-58.

[83] Balme had always intended the *editio maior* to follow the pre-Gaza manuscript ordering. He had originally intended the Loeb *editio minor* to follow the modern ordering, but was leaning increasingly towards harmonizing it with the then-planned *editio maior*, and Gotthelf prepared Balme's posthumous Loeb *editio minor* accordingly.

[84] Balme 1991, 18-19.

considered last because it is the largest subject: while this might imply that IX(VII) should follow VI, the introductions to VII(VIII) and VIII(IX) do not suggest that man has already been discussed; moreover IX(VII) is evidently incomplete; so that the manuscript order is probably correct, putting first the other animals' activities (not only generation) and then proceeding to man . . .

On reflection, we find ourselves in agreement with Balme that the manuscript order was probably the original order. He is certainly right that IX is incomplete (cf. *mechri gērōs*, 581a10),[85] and that, as he remarks, neither its beginning nor its end refers forward or back; and he is right that the introductions to VII and VIII give no indication that generation in man has been discussed.[86] Gaza is certainly right that IX *could* follow right upon the end of VI, since VI has been identifying differentiae pertaining to generation among the four-footed live-bearing animals. But IX has an unusually grand opening, which suggests that it is not a continuation of the discussion at the end of VI: 'With regard to man's development, both initially within the female and subsequently until old age, the attributes due to his proper nature are as follows.'[87]

One additional (though somewhat speculative) matter worth considering is this. The differentiae that are discussed in VII and VIII—nutrition, habitat, disease, etc., and 'characters' and intelligence—are almost entirely *differentiae of the other animals*, and not of man. (A check of the index to the Loeb *HA*, s.v. *man*, confirms this.) Human food-gathering and habitat, so far as they are discussed anywhere in the Aristotelian corpus, are discussed in *Politics*, Book I—and this is perhaps no surprise given the opening of *HA* VII, which points out that the more *sunesis* ('understand-

[85] As has been thought at least as far back as Gesner, cf. *Michaelis Ephesii Scholia in Aristotelis libros aliquot* (Basel [1541]) 5-6: 'Non dissimilis commissus est error in Historia animalium ubi decimus factus est liber ab imperitis ex avulsa parte quæ septimo libro continua esse debebat. Quod constat ex argumenti similitudine et quod idem septimi finis sit, qui principium decimi, qui Latine nondum habetur.' The argument was expanded by Scaliger, who found a further clue for the hypothesis in the omission of the final words of IX by Gaza, as they are repeated at the beginning of X, thus forming a token for their connection: 'Theodorus omisit προϊούσης δὴ τῆς ἡλικίας. Itaque cum proposuerit se dicturum μέχρι γήρως ὅσα συμβαίνει, neque hic explevit. Necesse est huic libro subdi eum quem decimum vocant: qui sic item incipit tamquam a tessera προϊούσης δὴ τῆς ἡλικίας. Sed quia sententia de semine muliebri contra Aristotelis opinionem est, sustulere illum atque reiecere in locum decimum' (Scaliger 1619, 850, cf. also 1186). Gesner and Scaliger apparently thought our Book X is the second half of IX, but that is certainly not possible, as Balme makes clear in his 1985 and 1991.

[86] περὶ δ' ἀνθρώπου γενέσεως τῆς τε πρώτης τῆς ἐντῷ θήλει καὶ τῆς ὕστερον μέχρι γήρως, ὅσα συμβαίνει διὰ τὴν φύσιν τὴν οἰκείαν, τόνδ' ἔχει τὸν τρόπον. The introductions are, we suppose, consistent with there having already been a discussion of human generation, but as we will go on to argue, the burden of proof is with Gaza, and so the absence of clear indication of a preceding discussion is significant.

[87] 581a1-9, tr. Balme. Note, by contrast, the smooth transition from the end of Book V through the beginning of VI. Friederike Berger, who claims that Book IX must be considered as the complement of Books V and VI, considers the opening lines as 'secondary' (Berger 2005, 10).

ing') and *mnēmē* ('memory') a kind of animal has, the more complex will be its *ēthē* as well as its *bioi* and *praxeis—including generation*, where it will have a *politikōteron* ('more social') relationship with its young (588a16-31, b26-589a2). In particular, given the full extent of man's cognitive abilities, as a result of the possession of *logos*, all humans, adults as well as children, live naturally and best in cities. So, the education which is central to the rearing of human youth is discussed in . . . *Politics* VII and VIII. The study of *man's praxeis* and *bioi* (and indeed *ēthē*), then, might well have seemed to Aristotle, when finishing *HA* VI (or earlier), to be a complex mix of theoretical and practical understanding, much of which should be set aside for special treatment (especially as the latter requires moral and political virtue).

Finally, there is the matter of where the burden of proof lies. Surely it belongs with those who would change the manuscript ordering (especially where such numbering as is used by later ancient scholars is in agreement with that ordering). Although Gaza's argument is not implausible, we do not see that it meets that burden. Thus, with Balme, we view Gaza's reordering as an intrusion.

13

History of Animals I.6 490b7-491a6: Aristotle's *megista genē*

In the previous chapter I argued that the establishment of animal kinds is no part of the fundamental aim or set of aims of the *History of Animals*, even if some such establishment is presupposed in *HA*. But I acknowledged that it *is* a concern of at least a portion of the sixth chapter of book I.[1] That stretch of text—490b7-491a6—is possibly the most difficult one in *HA* and it needs closer examination, as part of our effort to understand the place of the search for kinds in the project of *HA*.

HA I.6 as a whole contains the treatise's most sustained methodological discussion, and each of its three parts in fact deserves further attention.[2] The first part, already mentioned, and cited in this chapter's title, runs for thirty-four lines, and

This chapter was first drafted and revised, based on comments I received from David Balme, in 1984. I tinkered with it on a few occasions since, but have only figured out the point of Aristotle's discussion of the snakes (something I hadn't realized Balme had all along), and completed the paper, in 2010. With a slightly varied opening, it will also appear, in due course, in the Festschrift volume for Pierre Pellegrin (Cerami 2012).

[1] Above, pp. 281, 289.

[2] The chapter divides naturally into three parts, 490b7-491a6, 491a7-14, 491a14-26. (It should not be forgotten, however, that these parts form a continuous whole, both among themselves, and with the preceding and succeeding chapters. In fact, the chapter divisions in the biological works have no ancient authority; for *HA* they go back only to the first printed editions in the latter part of the fifteenth century, on which see Beullens and Gotthelf 2007, 477-83.) There is an excellent discussion of 491a7-14 in Lennox 1987a, 101-2 [= Lennox 2001b, 17-18]. My understanding of this passage surely benefited from the work Jim Lennox and I did together on *HA* in 1983 (below, p. 315), though I no longer remember any specifics of our discussion of the passage. I do remember a long discussion with David Balme in 1984 that sharpened much of the detail of what became this chapter; his own views, from which I borrow significant points, are presented in condensed, commentary form, in Balme forthcoming, ad loc. A subsequent close reading of this portion of the chapter with Lennox and David Charles in Oxford in 1994, while the three of us were giving a joint seminar called 'Philosophical Issues in Aristotle's Biology', helped as well. I also benefited, at the end stage, from a sustained discussion of the first half or so of the text with which this chapter is concerned at a meeting of the Euthyphrones at the University of Texas at Austin in fall 2009 (my thanks in particular to Alex Mourelatos, Lesley Dean-Jones, Jim Hankinson, and Brian Battiste), and of the whole of it at the University of Pittsburgh in our 2009-10 reading group on *HA* I.1-6.

introduces Aristotle's *genē megista tōn zōiōn*—'very large animal kinds'. Let's have a close look at this very difficult passage.[3]

I translate Balme's text[4] as follows:

490b7 Very large animal kinds into which some of the animals have been divided are: one, of birds; one, of fishes; another, of cetacean. Now, all these are blooded. Another kind is that of the hard-shelled, which are called 'oysters'; another is that of the soft-shelled, unnamed by one name, e.g., crayfish and some kinds of crabs and lobsters; another is that of the soft-bodied, e.g., the *teuthides* and *teuthoi* and squids; a different one is that of insects. Now, all these are bloodless, and as many as have feet are many-footed; and of the insects some are also winged. Of the remaining animals the *genê* are not large, for one form does not embrace many forms, but in one case it is itself simple, the form not having differentiation, viz., man, while other cases have <differentiation>, but the forms in question are unnamed. For (*gar*) all the four-footed wingless animals are blooded, but some of them are live-bearing while others are egg-bearing. Now, as many as are live-bearing all have hair,

[3] A. L. Peck found an apparent contradiction (Peck 1965, lxvi); D'Arcy Thompson wrote that 'the whole passage is very troublesome.' These words open a note to 490b16, in his Oxford translation of *HA* (Thompson 1910; virtually all of the notes to the first edition Oxford translation [Ross 1908-52], including Thompson's, were excluded from the *Revised Oxford Translation* [Barnes 1984]). Thompson's summary, in that note, of Aristotle's aims in our passage, though unargued, is on the right track. His note at b25 regarding the *most* troublesome part of the passage, b23-30 is, however, less plausible.

[4] Balme 2002. It reads as follows:

490b7 *γένη δὲ μέγιστα τῶν ζῴων, εἰς ἃ διῄρηται τἆλλα ζῷα, τάδ' ἐστίν, ἓν μὲν ὀρνίθων, ἓν δ' ἰχθύων, ἄλλο δὲ κήτους. ταῦτα μὲν οὖν πάντα ἔναιμά ἐστιν. ἄλλο δὲ γένος ἐστὶ τὸ τῶν ὀστρακοδέρμων, ὃ καλεῖται ὄστρεον· ἄλλο τὸ τῶν μαλακοστράκων, ἀνώνυμον ἑνὶ ὀνόματι, οἷον κάραβοι καὶ γένη τινὰ καρκίνων καὶ ἀστακῶν· ἄλλο τὸ τῶν μαλακίων, οἷον τευθίδες τε καὶ τεῦθοι καὶ σηπίαι· ἕτερον τὸ τῶν ἐντόμων. ταῦτα δὲ πάντα μέν ἐστιν ἄναιμα, ὅσα δὲ πόδας ἔχει, πολύποδα· τῶν δ' ἐντόμων ἔνια καὶ πτηνά ἐστιν. τῶν δὲ λοιπῶν ζῴων οὐκ ἔστι τὰ γένη μεγάλα· οὐ γὰρ περιέχει πολλὰ εἴδη ἓν εἶδος, ἀλλὰ τὸ μέν ἐστιν ἁπλοῦν αὐτὸ οὐκ ἔχον διαφορὰν τὸ εἶδος, οἷον ἄνθρωπος, τὰ δ' ἔχει μέν ἀλλ' ἀνώνυμα τὰ εἴδη. ἔστι γὰρ τὰ τετράποδα καὶ μὴ πτερωτὰ ἔναιμα μὲν πάντα, ἀλλὰ τὰ μὲν ζωοτόκα τὰ δ' ᾠοτόκα αὐτῶν. ὅσα μὲν οὖν ζωοτόκα πάντα τρίχας ἔχει, ὅσα δ' ᾠοτόκα φολίδας ἔχει· ἔστι δ' ἡ φολὶς ὅμοιον*

while those which are egg-bearing have horny scales.[5] (The horny scale is similar in place to the fish scale.) Footless by nature, blooded, air-breathers—such is the *genos* of snakes; and this <*genos*> is horny scaled. But while the other snakes are egg-bearing, the viper alone is live-bearing. For the live-bearers do not *all* have hair (for even some of the fish are live-bearing); however, as many as have hair all are live-bearing (for one must posit as a form of hair the spine-like hairs of the sort which the hedgehogs and porcupines have, for they do the job of hair, but not of feet as in the case of the sea urchins). Of the live-bearing and four-footed kind of animals there are many forms, but they are unnamed; rather they are <taken> individually, so to speak, as we said in the case of man: lion, deer, horse, dog and the others in this manner, although (*epeí*) there is one such kind in fact in the case of the so-called 'neck-maned ones', such as the horse, ass, mule, *ginnos*, *hinnos*, and those called in Syria 'half-asses'—which are called 'half-asses' on account of a similarity, not being simply the same form, for *they* mate and produce offspring with one another. That is why it is necessary to study them separately by grasping their natures individually.

490b7-14 identifies seven very large kinds; at **b15-16** Aristotle states that there are no others, and at **16-19** he explains why. The purpose of **b19-30** is not clear, though the discussion has been thought to fall into two parts, **19-23** and **23-30**. At

χώρᾳ λεπίδος. ἄπουν δὲ φύσει ἐστὶν ἔναιμον πεζὸν τὸ τῶν ὄφεων γένος· ἔστι δὲ τοῦτο φολιδωτόν. ἀλλ' οἱ μὲν ἄλλοι ᾠοτοκοῦσιν ὄφεις, ἡ δ' ἔχιδνα μόνον ζωοτοκεῖ. τὰ μὲν γὰρ ζωοτοκοῦντα οὐ πάντα τρίχας ἔχει· καὶ γὰρ τῶν ἰχθύων τινὲς ζωοτοκοῦσιν· ὅσα μέντοι ἔχει τρίχας πάντα ζωοτοκεῖ. τριχῶν γάρ τι εἶδος θετέον καὶ τὰς ἀκανθώδεις τρίχας οἵας οἱ χερσαῖοι ἔχουσιν ἐχῖνοι καὶ οἱ ὕστριχες· τριχὸς γὰρ χρείαν παρέχουσιν, ἀλλ' οὐ ποδῶν, ὥσπερ αἱ τῶν θαλαττίων. τοῦ δὲ γένους τοῦ τῶν τετραπόδων ζῴων καὶ ζωοτόκων εἴδη μέν ἐστι πολλά, ἀνώνυμα δέ· ἀλλὰ καθ' ἕκαστον ὡς εἰπεῖν, ὥσπερ ἄνθρωπος εἴρηται, λέων, ἔλαφος, ἵππος, κύων καὶ τἆλλα τοῦτον τὸν τρόπον, ἐπεί ἐστιν ἕν τι γένος καὶ ἐπὶ τοῖς λοφούροις καλουμένοις οἷον ἵππῳ καὶ ὄνῳ καὶ ὀρεῖ καὶ γίννῳ καὶ ἴννῳ καὶ ταῖς ἐν Συρίᾳ καλουμέναις ἡμιόνοις, αἳ καλοῦνται ἡμίονοι δι' ὁμοιότητα, οὐκ οὖσαι ἁπλῶς τὸ αὐτὸ εἶδος· καὶ γὰρ ὀχεύονται καὶ γεννῶνται ἐξ ἀλλήλων. διὸ καὶ χωρὶς λαμβάνοντας ἀνάγκη θεωρεῖν ἑκάστου τὴν φύσιν αὐτῶν.

[5] In b21 Balme follows two manuscripts and Michael Scot's Latin translation in not printing an *οὐ* before *πάντα*. It is clear from the second clause of that sentence that it is live-bearing *four-footed animals* that are being spoken of, and not live-bearers in general.

b31-32 live-bearing four-footed animals are described in a way which suggests that they *do* constitute a large kind, and something is said immediately thereafter about how they must be studied.[6]

The difficulties are threefold:

(1) From Peck 1965 *ad* 490b21: 'it is not clear why Aristotle does not reckon the Viviparous Quadrupeds among the "main groups" here (as he does at 505b28, together with the Oviparous Quadrupeds), since he says below (490b32) that they are a group which includes many kinds.'

(2) It is not clear what the purpose of b19-30 is, or how it fits in with the immediately preceding and succeeding lines. Because of the *gar* at b19, these lines appear to present an argument in support of something that has just come before, but it is not clear what, nor how the argument goes. And the biological information—about hair and horny scales, vipers, and live-bearing fishes—seems quite extraneous. As a result,

(3) it is not clear how b31-491a6 is connected with the foregoing. Is it the conclusion of an argument contained in b19-30? Is it an independent point of methodology? How does the passage hang together as a whole? Let's begin at the beginning, with 490b7.

The first thing to note is the perfect tense of *diēiretai*. Aristotle does not say that 'the following (seven) groups constitute the very large kinds', but: 'the following are very large kinds into which some of the animals *have been divided*.'[7] Divided by whom? It is probably: by people in general, as reflected in names in common language. While the reference *could* be to an accepted tradition of thinkers, such as the Hippocratics[8] or Academics, or to Aristotle's own prior work or practice, the first alternative is more likely, especially since the passage is reminiscent of *PA* I.4, which is critical at least of Plato and returns us, though not uncritically, to popular groupings:

[6] 'Live-bearing (animals)' is my translation of *zōotoka*, a term that, until recently. was standardly translated 'vivipara', as in the quotation from Peck in the next paragraph. (Balme follows LSJ in not adding an ι-subscript to the ω in words with root ζωοτόκ-.) In Aristotle's usage the term is still adjectival in connotation, and does not have the technical-term feel of the English 'vivipara'. Likewise for its coordinate *ōiotoka*, once standardly translated 'ovipara', and which I translate by the slightly archaic 'egg-bearing' (animals). (Lennox's 'egg-laying' is more natural English [2001a, 379], but I prefer to reflect in the English that the two Greek words have the same ending.)

[7] 'some of the animals' = *talla zōia*. *talla* ('the others') does not mean 'other than the ones that fall under the *megista genē*', which would not make any sense; the contrast, as Balme notes (cf. Balme forthcoming, ad loc.), is proleptic: 'other than' some group of animals to be mentioned shortly—in particular, other than *ta loipa zōia* at b15-16. This is a standard construction for *alla* (cf. Smyth §1273), and why I (imperfectly) translate *talla zōia* as 'some of the animals'.

[8] Cf. Burckhardt 1904 (cited by Peck 1965, li).

The question arises why there is no popular designation which includes both the water animals and the fliers under one higher name, denoting a single kind which embraces both. For they have certain affections in common, and so have all the other animals. Nevertheless they have been rightly distinguished in this way.

… Perhaps then the right course is to speak of some affections in common by kinds, wherever the kinds have been correctly (*kalōs*) marked off by popular usage and possess both a single nature in common (*mian phusin koinēn*) and forms not far separated in them—bird and fish and any other that is unnamed but like the kind (*anōnumon men, tōi genei de homoiōs*) embraces the forms that are in it; but wherever they are not like this, to speak of particulars, for example about man or any other such.[9]

That is precisely the course Aristotle follows in our *HA* passage. He first lists seven very large kinds that have already been recognized, and have been designated in the language by either names or name-like expressions (starting, as it happens, with bird and fish). When he then observes that there are no other large kinds,[10] we must supply 'into which the remaining animals have already been divided (in the same way)'.

If so, then (*pace* Peck) the exclusion of both 'the live-bearing four footed animals' and 'the egg-bearing four-footed animals' is no surprise, since these are not ordinary expressions and do not represent groups into which animals have already been

[9] 644a12-16, b1-7, tr. Balme (1992), with 'kinds' for 'genera', 'forms' for 'species', and 'correctly' for 'satisfactorily'. Bekker's text reads:

> 644a12 *ἀπορήσειε δ' ἄν τις διὰ τί οὐκ ἄνωθεν ἑνὶ ὀνόματι ἐμπεριλαβόντες ἅμα ἓν γένος ἄμφω προσηγόρευσαν οἱ ἄνθρωποι, ὃ περιέχει τά τε ἔνυδρα καὶ τὰ πτηνὰ τῶν ζῴων. ἔστι* (a15) *γὰρ ἔνια πάθη κοινὰ καὶ τούτοις καὶ τοῖς ἄλλοις ζῴοις ἅπασιν. ἀλλ' ὅμως ὀρθῶς διώρισται τοῦτον τὸν τρόπον. …* (644b1) *ως μὲν οὖν ὀρθῶς ἔχει τὰ μὲν κατὰ γένη κοινῇ λέγειν, ὅσα λέγεται καλῶς ὡρι σμένων τῶν ἀνθρώπων, καὶ ἔχει τε μίαν φύσιν κοινὴν καὶ εἴδη ἐν αὐτῷ μὴ πολὺ διεστῶτα, ὄρνις καὶ ἰχθῦς, καὶ εἴ τι* (b5) *ἄλλο ἐστὶν ἀνώνυμον μέν, τῷ γένει δ' ὁμοίως περιέχει τὰ ἐν αὐτῷ εἴδη· ὅσα δὲ μὴ τοιαῦτα, καθ' ἕκαστον, οἷον περὶ ἀνθρώπου καὶ εἴ τι τοιοῦτον ἕτερόν ἐστιν.*

(Lennox's translation, though closer to the Greek word order, does not differ in substance from Balme's. On David Charles' translation of b4-5, see n. 13 below.) Aristotle's mention, at a12, of the option of designating 'the water animals and the fliers under one higher name' is an option he may have thought Plato took up; cf. *Sophist* 220a6-b2:

VISITOR: And wouldn't it be right to speak of a double form of animal-hunting, the one of the land-dwelling kind, which is divided into many forms and names, viz., land-hunting, and the other of *swimming animal*, all an aquatic-hunting?

THEAETETUS: Sure enough.

VISITOR: Now, of *swimming <animal>*, we do see the winged race and also the water-dwelling one?

THEAETETUS: Of course.

My translation and my emphases. 'Swimming animal' renders *neustikon zōion*.

[10] I am taking *genē megala* in b16 to have an extension either the same as or wider than that of *genē megista* in b7.

divided (in that way). But that does not mean that Aristotle cannot himself treat them as very large kinds, for that is precisely, I shall maintain, what he goes on to argue at b19-23 one *can* do. The bridge to, and the basis for, his doing so appears in b16-19:

οὐ γὰρ περιέχει πολλὰ εἴδη ἓν εἶδος, ἀλλὰ τὸ μέν ἐστιν ἁπλοῦν αὐτὸ οὐκ ἔχον διαφορὰν τὸ εἶδος, οἷον ἄνθρωπος, τὰ δ' ἔχει μὲν ἀλλ' ἀνώνυμα τὰ εἴδη.

which I have translated above as:

for one form does not embrace many forms, but in one case it is itself simple, the form not having differentiation, viz., man, while other cases have <differentiation>, but the forms <in question> are unnamed.

The sentence gives two reasons why we do not find other very large kinds divided off in popular usage. First, there may be a popular term for a kind, as there is for man, but in this case we do not have a large kind (since man does not embrace sub-kinds). But, secondly, where there *are* large kinds, these large kinds are not named: there are, for example (as we'll discuss further in a moment), no popular designations for the live-bearing-and-four-footed, or for the egg-bearing-and-four-footed, animals.[11]

Two features of the wording of this second reason require attention. First, this is supposed to be one of two cases where there *is* no form embracing many forms, yet Aristotle says of this apparently non-existing form that it *has* differentiation; *it* must then exist. Furthermore, it's not immediately obvious that the forms that are unnamed are the embracing forms rather than the embraced ones.

There are several reasons, however, for insisting that it is the embracing forms which are unnamed. The primary reason, as Balme has noted (forthcoming, *ad loc.*),

[11] A fuller commentary on this passage might consider the grounds for Aristotle's choice at b17 of *eidos* rather than *genos* for the embracing form. Building on, and in important places correcting, the prior work of David Balme on '*genos* and *eidos* in Aristotle's biology' (Balme 1962a), Pierre Pellegrin, in ch. 2 of Pellegrin 1986 (as well as of its 1982 French first edition), argues that Aristotle consistently uses *eidos* to designate a *sub*-group of some explicit or implicit wider group, for which he uses *genos*, and that this usage can and does occur at all levels of generality. It is the recognition that this can take place at all levels, Pellegrin argues, which helps us to find order in the apparent chaos of Aristotle's use of these terms in the biology proper (well documented by Balme 1962a), and to show the essential unity of the account and use of these and related terms in the logical works with their use in the biology. Oddly, Pellegrin did not, in his 1982 first edition, discuss 490b17, surely grist for his mill that *eidos* can be used at all levels of generality, as the use is quite high in this case and yet arguably intended as falling under the wider *genos*, animal. But even accepting this, it would be good if we could explain Aristotle's choice to use *eidos* twice in the same phrase, each for a different level, a usage which I believe is peculiar to this passage. In his second edition, having read the original 1984 draft of the present chapter, Pellegrin agrees that this occurrence of *eidos* supports his general thesis, but prefers to take as nameless the *embraced eidē* (Pellegrin 1986, 100).

is the apparently parallel syntax of the main *men* and *de* clauses in the sentence (see both text and translation just above). The *eidē* which are unnamed are in apposition to the *ta de*, just as the *eidos* which has no differentiation is in apposition to the *to men* of the preceding clause. Further, it is not at all clear why unnamed embrac*ees* would be a problem for the existence of an embracer,[12] but an unnamed embrac*er* has, for Aristotle, a somewhat tenuous hold on existence, as in the *PA* I.4 passage just quoted: 'bird and fish and any other that is unnamed but like the kind embraces the forms that are in it' (644b5-6).[13]

The unnamed kind embraces its forms *like* a kind does—as if to say that being unnamed limits a grouping's claim to *be* a kind, leaving it only *like* a kind. Yet according to this *PA* passage, such an unnamed grouping, if sufficiently like a kind (in having one nature in common and its forms not too far apart), should be treated *as* a kind. This, I would like to suggest, is precisely what Aristotle goes on to do in our *HA* passage for the unnamed groups of live-bearing-and-four-footed, and egg-bearing-and-four-footed, animals: he argues at b19-23 that live-bearing and egg-bearing four-footed animals each have a common nature, and each should be treated as a very large kind, along with the seven other very large kinds already accepted from common language. Having done so, he will be in a position from then on to cite (as Peck puzzlingly noted he did) the live-bearing and egg-bearing four-footed animals as standard members of the list of (now nine) very large kinds.[14] The argument, alas, is vintage Aristotle: highly elliptical and hard to make out.

For (*gar*) all the four-footed /20/ wingless animals are blooded, but some of them are live-bearing /21/ while others are egg-bearing. Now, as many as are live-bearing all have hair, /22/ while those which are egg-bearing have horny scales. (The horny scale is similar /23/ in place to the fish scale.)[15]

[12] As they don't seem to be at 490b31-32 (below, p. 305).

[13] Lennox translates similarly. Charles apparently follows Peck's text in which, without any basis in the MSS (cf. Peck's 'correxi'), Peck prints *homoia*. Charles makes the best he can out of this text, but the redundancy of the translation—'birds, fishes, and any other group which though it lacks a name yet contains species generically similar' (Charles 2000, 312)—shows the weakness of Peck's suggestion. I suppose Peck must not have thought a grouping could be *like a kind* without itself *being* a kind *simpliciter*, as I go on to discuss it can in the lines that follow (and in n. 19 below).

[14] And to treat them as such in practice, even beginning with them, as he does, for instance, in his kind-by-kind survey of differences in the external non-uniform parts of non-human blooded animals in *HA* II: live-bearing four-footed animals in II.1-9; egg-bearing four-footed animals in II.10-11; birds in II.12; fishes in II.13; and snakes in II.14. (For reasons that are unclear to me, cetaceans do not get a separate discussion here; and for simplicity's sake, I have included the two chapters (8-9) on apes, monkeys, and baboons—which are said to *be* four-footed animals in some ways but *like* humans in others—under live-bearing four-footed animals.)

[15] Manuscript variants for b21 are available in Balme 2002; I follow two manuscripts as well as Michael Scot's Latin translation, Peck, and now Balme in omitting the *ou* ('whichever are [not] live-bearing'). It is clear from the second clause of that sentence that it is live-bearing *four-footed*

It will be important to determine what the antecedent conjunct of the *gar* is, but we can begin by noting that the initial subject of the argument is animals that are four-footed and not winged. One might wonder what non-wingedness adds, since when Aristotle first wrote this passage he must have held that no four-footed animals are winged.[16] The addition will make more sense, however, if one thinks of Aristotle as asking us to consider the class of four-footed animals. After all, one can imagine him asking: don't *they* form a kind too? Like the birds, fishes, and cetaceans, they too are all blooded (b20), and they too share a distinctive form of locomotion: they walk on four feet and do not fly (b19-20—in which context winglessness *is* worth stressing, since some footed animals do have wings and do fly: b14-15). The four-footed animals are thus a candidate for being a large kind.

'*But*,' Aristotle writes, suggesting that they fail to survive the candidacy, 'some of them are live-bearing while others are egg-bearing.' Now that in itself would not entail that these two kinds of four-footed creatures are 'too far apart' for us to work with *four-footed animal* as a very large kind; after all, fishes and snakes, as we are

animals that are being spoken of, and not live-bearers in general. (For the relevance of the mention of fish scales, see Balme forthcoming *ad loc.*)

[16] I say 'when Aristotle first wrote this passage' for two reasons: the dayfly and the bat. In *PA* (682a36) and *IA* (713a26) Aristotle maintains that all insects (and all bloodless animals that have feet generally) are 'many-footed' (*polupoda*), which for him means more than four, and that is what we find asserted in our passage: *hosa de podas echei, polupoda* (b14-15). However, just a few lines back, in the previous chapter (and see above, n. 2, on the deceptiveness of chapter divisions), Aristotle mentions an insect unique (*idion*) among animals, the so-called 'dayfly' (*ephēmeron*). This insect is unique not only for the (short-lived) character of its life, but for its being winged while having four legs: *ptēnon esti tetrapoun on* (490a34-b3). A fuller report is given of this creature at *HA* V. 552b17-23, where it is described as a *zōion pterōton tetrapoun*. (For more on the *ephēmeron* see Beavis 1988, 89.) The words at b19-20 in our passage, *tetrapoda kai mē pterōta*, would certainly have seemed to hearken back to *ptēnon tetrapoun on* only sixteen lines earlier, were it not for the fact that between them, and much closer to b19-20, is the insistence that *all* bloodless creatures with feet are many-footed. Given this remark at b14-15, it seems we must conclude that the non-wingedness of the four-footed animals in question at b19-20 is mentioned for some reason other than to ensure the exclusion of dayflies.

(The matter is complicated by the peculiar fact that the dayfly is presented at 490a34 *as an example of* a bloodless flier that has many feet (*pleious podas*). Furthermore, there is a mention of the dayfly at *PA* IV. 682a26-27, without reference to its number of legs, which occurs just ten lines above the previously noted assertion that all insects are many-footed. David Balme has proposed that all this is evidence of a later insertion, into an existing *HA*, of the discovery of the dayfly's four-footedness. Since both mentions in *HA* are reasonably detachable from their contexts (and Book I's appears even to contradict what is said in *its* surrounding context), this is not implausible. That would make it possible to regard *kai mē pterōta* in b19-20 as a later addition in recognition of the new information about the dayfly, though it would then be peculiar (as admittedly 490a34 is peculiar) that *polupoda* in b15 would have been left uncorrected. The interpretation of *kai mē pterōta* that I give in the text above is simpler and more plausible.)

As for the bat, while it is in *PA* and *IA* a quadruped 'in part' (697b1-13) or 'poorly' (714b13), at *HA* I. 490a5-11, less than a page earlier than our passage, it is unequivocally said to be two-footed (*dipoda*; cf. 489b19-23 for confirmation that this is meant in a way that excludes being four-footed).

about to learn, though mostly egg-bearing, have live-bearing forms too, and that doesn't exclude them from being single kinds studiable as such. (These live-bearing forms, we will discover as the study of animal features progresses, are internally egg-bearing, but this fact is not yet on the table.) In the case of the four-footed animals, however, the difference in mode of reproduction *is* suggestive, as Aristotle immediately points out, since all the live-bearing four-footed animals are hairy and all the egg-bearing ones horny-scaled. This striking, global difference in protective covering is decisive for Aristotle, in view of its correlation with numerous other differences both internal and external, all of them consequences of a basic difference in internal heat, which is also responsible for the one group being egg-bearing and the other being (in this case, internally as well as externally) live-bearing.[17]

My suggestion, then, is that b19-23 (i) has us look, among the animals included neither in the seven popularly named groups nor in the groups without differentiation (such as man), for other, not previously recognized large groups; then (ii) identifies four-footed animals and considers them as a whole; but (iii) recognizes the need to divide them into live-bearing and egg-bearing ones; and then (iv) gives reason for considering the two groups this division produces—*the live-bearing four-footed animals* and *the egg-bearing four-footed animals*—as candidates for the status of *mega* (or perhaps even *megiston*) *genos*, large (or perhaps even very large) group.

I have stuck to the looser word 'group' rather than the more technical 'kind' (*genos*), because of the strictures we found Aristotle putting on unnamed groups in *PA* I.4: as unnamed they are not yet technically kinds but can at best be *treated as* kinds. The distinction is important here because it explains the *gar* in b19: Aristotle does not in fact speak, for instance, of 'live-bearing four-footed animals', but of 'the four-footed and live-bearing animals', or 'the live-bearing and four-footed ones', and these are descriptive phrases and not names in his sense.[18] The same is true for egg-bearing four-footed animals, and since both of these groups have varying subgroups, they are prime examples of cases which 'have differentiation ... but are unnamed'; and the *gar* is present in b19 because they are offered as such.

[17] The comparison at b22 between fish scales (*lepides*) and the horny scales (*pholides*) of the egg-bearing four-footed animals confirms this reading in two ways. First, fish scales are standardly contrasted by Aristotle with feathers, and feathers with hair, as examples of analogous features—features different in kind but similar in function—that are typically found among members of different large kinds. (*HA init.*, esp. 486b20-22; *PA* I. 644a2-22, IV. 692b10-12; *GA* V. 782a16-18.) Second, the entire clause (excluding the particle), with its peculiar use of *chōra* for role or function, is an exact duplicate of a clause at *PA* IV. 691a16, where it occurs embedded in a discussion of the many shared features of egg-bearing four-footed animals.

On differences in internal heat as the basis for differences in mode of reproduction (including the difference between internally live-bearing and internally egg-bearing live-bearers), see *GA* II. 732a25-733a32, where, incidentally, being horny-scaled is twice said to be indicative of other features.

[18] For a fuller discussion of this, see Balme 1962a, 90-91.

Thus, to respond to Peck's puzzle raised above (p. 296) about b16-19: while there are no other large kinds *recognized in the language*, there are at least two groups that each share a common nature and show differentiation. Because their forms are not 'too far apart', we can treat them *as* kinds, and that is what Aristotle prepares to do.[19]

Or so it might seem, if the argument stopped here. But we still have to address the role of b23-30. The passage reads:

Footless by nature, blooded, air-breathers[20]—such is the *genos* of snakes; and this <*genos*> is horny scaled. But while the other snakes are egg-bearing, the viper alone is live-bearing. For the live-bearers do not *all* have hair (for even some of the fish are live-bearing); however, as many as have hair, all are live-bearing (for one must posit as a form of hair the spine-like

[19] It can be added that treating them as kinds (when we are justified in doing so) may actually turn them into kinds, if the descriptive phrase comes to take on the familiarity and compactness of a name, and its adjectives the force of nouns. (Cf. again Balme 1962a, 90-91.) We can see the process at work in the case of the fifth of the kinds originally listed, the *malakostraka*. They are 'unnamed by one name', but the compound descriptive adjective (equivalent to 'soft-shelleds') has become familiar enough to permit them being listed among the already named groups. For those of us reading Aristotle, I dare say the process is complete: we view the terms *malakostraka, malakia*, etc. as full-fledged names of kinds, with their descriptive meanings quite in the background (and so are startled when we come upon Peck translating *ta malakia* as 'the softies'). This way of understanding these texts has interesting implications for our understanding of the ontological status of *kinds* in Aristotle: for a set of forms to be united into a kind, the forms must have a certain kind of mind-independent 'sameness' (with differences for the most part not greater than 'in the more and the less'), but there must be a cognitive act of (at least) *recognition* of that mind-independent relationship. This provides some support for the controversial (and radically important) thesis that *universality* for Aristotle, while centrally based on shared mind-independent facts, is not 'in the world' *simpliciter*. This view is elaborated and defended by Gregory Salmieri in *Aristotle and the Problem of Concepts*, his 2008 University of Pittsburgh doctoral dissertation, and in an as yet unpublished paper, 'Aristotle's Conception of Universality'.

[20] Cf. *HA* VII. 589a12-13: 'some are called *peza* or *enhudra* because they take in air or water respectively' (tr. Balme 1991). *peza* is often difficult to translate in Aristotle, as Balme explains (Balme 1991 *ad* 589a10). Though it literally means 'walking' (or 'having feet') and is sometimes used with precisely that meaning (so that the snakes, for example, would not be *peza*, cf. VII. 589a23 and 24, and I. 487a21-23), the term sometimes connotes more specifically that the animal lives on land (*peza* here being virtually synonymous with *chersaia*—so that though land snakes are *peza*, water snakes would not be; cf. VII. 589a13-17 and ff.). But often it contrasts most directly with *enhudra* (lives in and for cooling takes in water), so that its focus is on the animal's being an air-breather, as in the passage mentioned at the start of this note—and here at 490b23. (The fact that these terms refer, in one context or another, to one or more of these different divisions, is what leads Aristotle to provide the two very careful differentiations he does of ways of being 'land-animal' and 'water-animal', the first at *HA* I. 487a15-b6 and the second—more complex and at significantly greater length—at *HA* VII. 589a10-590a18. The best indexes of *HA* (if I may say so myself) are the Greek terminological index in Balme 2002 (and especially in Bodson 2004, from which the index in Balme 2002 was drawn [cf. Balme 2002, xix-xx and 514]), and the English conceptual index to the entire Loeb edition, in Balme 1991, 547ff. In the former, cf. s.v. πεζός and in the latter, cf. esp. s.v. Footed animals; Land animals; and Walking animals.

hairs of the sort which the hedgehogs and porcupines have, for they do the job of hair, but not of feet as in the case of the sea urchins).

It is not clear what the purpose of this passage is. There are verbal echoes of the preceding lines, and one might imagine some argumentative connection—perhaps Aristotle is now considering an objection to his acceptance of live-bearing four-footed animals and egg-bearing four-footed animals as authentic kinds. Or, perhaps, as Peck's placing of this material in a separate paragraph might suggest, Aristotle is moving on to a distinct question, having to do with the status of the snake-kind (as I myself thought for some years). After all, it is not encompassed in any of the kinds recognized so far, nor is it an indivisible form such as *man*.

I want to suggest, however, that the key *is* the verbal echoes I referred to—the mention of horny scales and hair. These features were brought in at b21-22 to confirm the division of the four-footed into four-footed live-bearing animals and four-footed egg-bearing animals: the latter all have horny scales, while the former all have hair; and in each case the pair of features are significant enough to indicate a single nature and thus the existence of two more very large kinds among the blooded animals. Here, I propose, the question is whether one still has authentic large kinds if one divides without reference to four-footedness. That is, can we divide blooded animals simply by *live-bearing* and *egg-bearing* and still have in each case a single nature, and thus an authentic kind? And Aristotle's route to answering the question, I am suggesting, is to use the same test he employed just previously in considering whether four-footed live-bearing and four-footed egg-bearing qualified as authentic kinds—that is, to ask whether, in blooded animals, *live-bearing* and *egg-bearing* as such, without regard to number of feet, still correlate with a distinctive type of protective covering. The snakes, as I will argue, are brought in to answer that question, and they suggest at first a positive answer.

After all, as Aristotle immediately puts it, there *is* (sc. among blooded animals not in the already recognized very large kinds) a naturally footless kind that is also horny-scaled: the snakes.

Consider the carefulness (and significance) of the word order in this sentence. As David Balme once stressed in conversation with me, Aristotle begins it not with the name of the kind ('snakes') but with a list of those of its attributes relevant to the task at hand. Let us think of the construction of the sentence as a series of responses to a series of more specific questions:

A: We are looking for a kind wider than just the four-footed animals.
B: Well, there are certain *naturally footless* animals.
A: But, we're not talking about bloodless creatures such as worms.
B: That's fine: these animals are *blooded*.
A: But we're not talking about the fishes, since we're looking for previously *unrecognized* very large kinds.

B: Well, that's fine, too: the naturally footless blooded animals now under consideration are not fish, they are *air-breathers*.

Translating as literally as possible (and thus with subject after the predicates): 'Footless by nature, blooded, and air-breathers is the *genos* of snakes; and this <*genos*> is horny-scaled.'

We are thus to imagine as a candidate for being a very large kind the horny-scaled egg-bearers generally, regardless of number of feet. This projected kind will include the four-footed egg-bearing animals and . . . what else? The snakes? Well, not quite, and here the argument unravels: Yes, the entire *genos* of snakes is horny-scaled, '*but* (*alla*), while most snakes are egg-bearing, the viper is live-bearing.' That is to say, *egg-bearing-and-horny-scaled animals* can be thought of as a *genos* only at the price of splitting—of 'tearing apart' (*diaspan*)—the snakes. And, as Aristotle insists in his critique of an Academic type of division in *PA* I.2,[21] no division which does that grasps the nature of a kind:

[O]ne should avoid tearing each kind apart, e.g. putting some of the birds in one division and some in the other, as the written divisions have done; there, some of the birds end up divided off with the water-dwellers, some in another kind. Now this similarity has an established name, 'bird', and another has 'fish'. Other similarities are nameless, e.g. the blooded and the bloodless; there is no one established name for either of these. If then, nothing alike in kind should be torn apart, <the method of regular> division into two is worthless. For people who divide in this manner necessarily separate and tear apart . . .[22]

But that's not the only problem with this candidate division of blooded animals not in already recognized kinds, Aristotle immediately goes on at 490b26-27 to say (or so I am reading him); the grouping *live-bearing animals* doesn't capture a single nature either:

For (*gar*) the live-bearers do not *all* have hair (for even some of the fish are live-bearing) . . .

Here the problem is not that the grouping tears apart a kind, but that it puts together into one group kinds (the four-footed live-bearing animals, the viper, the selachian fishes) that are too far apart to have a single nature of the requisite sort;[23] they don't, for instance, all have hair.

Snakes are brought in, then, not as a candidate for a very large kind, but as *an argument against* treating 'egg-bearing animals' and 'live-bearing animals' as real (very large) kinds. And in pointing out that, though not every live-bearer is hairy, still every hairy animal is live-bearing, Aristotle is pointing out that it's the *hairy* live-bearers that constitute a large kind—the live-bearing *four-footed* animals. *This*

[21] The method criticized is *dichotomous* division (642b5-6, cf. b21-22). But the principle which is the basis of the criticism applies to any attempt to capture (or grasp, *lambanein*: 642b5) a kind.

[22] 642b10-18, tr. Lennox 2001a; the words in pointed brackets are my own.

[23] *PA* I.4 644b3-4; cf. above pp. 296-97 with n. 9.

division—*live-bearing four-footed* and *egg-bearing four-footed*—by contrast with the division into egg-bearing and live-bearing animals we've just entertained—does not split any authentic smaller kind into two, nor does it group together kinds that don't share a specific nature of the requisite sort. Not only, then, does the consideration of snakes *rule out* treating all the live-bearing animals as one very large kind and all the egg-bearing ones as another, it also *confirms* the claim already made that the *four-footed* live-bearers and the *four-footed* egg-bearers are actual (and previously unrecognized) very large kinds.[24]

On this reading, the next lines, 490b31-491a6, may be taken as making one more new point, in this case about the study of those four-footed live-bearers. Though this is, indeed, a very large kind, and contains many distinct forms with which we are already familiar from everyday life, it has no recognized intermediate kinds which can be studied as a whole. In looking for shared features and for differences then, we will have to consider the lowest-level forms one by one:

Of the live-bearing and four-footed kind of animals, there are many <intermediate> forms, but they are unnamed; rather, they are <to be studied> individually, so to speak, as we said in the case of man: lion, deer, horse, dog, and the others in this manner, although (*epei*) there is one such kind, in fact, in the case of the so-called 'neck-maned ones', such as the horse, ass, mule, *ginnos* [mule-mare offspring], *hinnos*, and those called in Syria 'half-asses'—which are called 'half-asses' on account of a similarity <to our true—sterile—half-asses>, not being simply the same form, for *they* mate and produce offspring with one another. That is why it is necessary to study them separately by grasping their natures individually.[25]

Aristotle's point is not that we are to write up our *historia* of the live-bearing four-footed animals one form at a time. It is rather that, in studying (say) hair, or hucklebones, in the live-bearing four-footed animals, we will not be able to look first to the shared hair or bone features of intermediate kinds, but will have to look at the low-level kinds one by one to see how they group—and they will probably group one way with regard to one feature, and another way with regard to another, with not enough significant features being common to any of these groupings to warrant its being identified as an intermediate kind.[26]

[24] This reading of 490b23-30 is obviously speculative, but I have never myself found one that makes more sense of the passage.

[25] I follow here the translation give above (p. 295), only modifying it slightly by adding clarificatory insertions in pointed brackets. On the opening sentence of this passage, see Balme 1962a, 90-91. Thompson, followed by Peck, translates *lophoura* as 'bushy-tailed animals', but the term clearly derives from the hair on the nape, or back, of the neck: see LSJ s.v. *lophos, lophoura* and *HA* II. 498b27-31. *kai innos* is probably a later gloss on *ginnos*: cf. LSJ s.v. *innos*, and on all this now Balme forthcoming, ad loc. Although singled out here as the only recognized intermediate kind under the very large kind *four-footed live-bearing animals*, the grouping is mentioned, and used, only twice in the rest of *HA*, at I. 495a4 and, II. 501a6.

[26] I say 'probably' because nothing precludes one's coming, in one's researches, to notice such intermediate kinds, and to label them with a name or 'name-like expression' (*APo*. II.10 93b30-

One can imagine Aristotle thinking that this point is worth making about the live-bearing four-footed animals here at 490b30-491a6, in part because it is not always the case that a very large kind has no recognized intermediate kinds, and in part because one might well expect that *this* very large kind has them. It is not always the case, at least because some of the large bloodless kinds have intermediate kinds: at 490b11-12, for instance, the *malakostraka* were introduced by reference to its more familiar sub-kinds, the crabs (*karkinoi*) and lobsters (*astakoi*), whose intermediate statuses seem alluded to by the *genē tina* in b12.[27] And while one might not necessarily expect there to be intermediates for the egg-bearing four-footed animals, since the number of low-level kinds is not that great, the *live-bearing* four-footed animals are many and varied, and would seem even the more so, being so familiar as farm animals and pets and objects of hunt. Alas, warns Aristotle, the latter kind does *not* have recognized intermediate kinds (with one exception), and so its lowest level kinds will have to be studied individually.

On this reading of 490b7-491a6, the passage which has been the concern of this chapter, then, Aristotle completes by the passage's end the task he had set for himself at its beginning, of laying out in outline the way he will group the animals for purposes of the study whose aims and character he goes on to sketch in the lines of *HA* that follow:

> [to] grasp the differences and attributes that belong to all <animals>. After this we must try to discover the causes.[28]

31; cf. *APr.* I.35 48a28-38). Indeed, Aristotle has just said that there *are* unnamed <intermediate> forms of the live-bearing four-footed kind. Here he is apparently describing how we must proceed, given the kinds currently recognized—that is to say, at the start of the study on which we are about to embark.

[27] At *HA* IV. 525a34-b1, Aristotle states that there are several kinds (*genē pleiō*) of crabs, and some of these kinds are listed at b4-9, prefaced (b3-4) by the remark that the *genos* of crabs is *pantodapōteron kai ouk euarithmēton* ('widely varied and not easily counted', tr. Balme, unpublished). The lobster is described at 526a11-b18 and elsewhere as if there were just one form, but 525b10-11 may contain a reference to another form. *PA* IV, on the other hand, is unequivocal: 'There are four *megista genē* [of soft-shelled animals], called crayfish, lobsters, prawns, and crabs; and of each of these kinds there are many forms, differing not only in shape but also greatly in size' (683b26-29; tr. Lennox).

The use of *megista genē* here for the four kinds of *malakostraka* is very interesting, I might add, revealing as it does that the expression does not designate a fixed taxonomic level, but simply indicates that the *genos* has a sufficiently wide and distinct variety of sub-kinds each with various forms—i.e., that it is a 'very large' kind. This usage is entirely in line with the general view of the function of the terms *genos* and *eidos* developed by Pellegrin (above n. 11).

[28] *HA* I.6 491a9-11, on which see above, n. 2.

14

Historiae I: *Plantarum* et *Animalium*

That being is manifold is clear; for perception both studies the differences and searches out the causes (or perhaps it is truer to say that it passes this on to thought, in some cases simply generating a search, in others by producing an *aporia*, through which, even if thought cannot advance, yet some light makes its appearance in our non-light as we search further).

There is no understanding (*epistasthai*), therefore, without some difference, for if things are other than one another, there is a difference; and within universals, the things that fall under the universal being several, these too must differ, whether the universals are kinds or forms. Almost all understanding (*epistēmē*) is of *idia*, for the *ousia* and the *ti ēn einai* of each thing is *idion* to it, and the attributes that things are observed to possess of themselves and not incidentally would be found to have been *ti kata tinos*. But generally it belongs to a body of understanding (*epistēmē*) to grasp what is the same in several things, whether it is said in common and universally or in a way as *idion* with respect to each, e.g. of numbers and lines, of animals and plants; a complete one has both. But while in the case of some things there is a universal end [for *episteme*] (for their cause is found in this), in others the end is in a part (*en merei*) [sc. of a universal], viz. wherever division is into the indivisibles, as in the case of things to be done and things to be made; for it is in this way that their actuality exists.

The sameness by means of which we understand (*epistametha*) is sameness either in essence or in number or in form or in kind or by analogy, or by any other divisions [of sameness] that there may be besides these; but that by analogy spans the widest interval, as though there were here the greatest separation, this [appearance of separation] being due sometimes to ourselves, sometimes to the object, sometimes to both.

Theophrastus, *Metaphysics* 8b10-9a9[1]

This chapter, published in 1988, was first drafted in 1985 for a Theophrastus conference (n. 74 below). It built on my first attempt, a year earlier, to characterize the aims of Aristotle's *HA* (n. 3 below). Although my account in chapter 12 above of the aims of *HA* in various ways supersedes the account given in the first part of this chapter, some readers may be interested in this first statement of the reading of *HA* that James Lennox and I developed together in 1983. More importantly, I believe the comparative study offered here of the *HA* with Theophrastus' *History of Plants* raises important questions regarding, and perhaps sheds additional light on, the aims of *HA* as well as of *HP*.

[1] Tr. after Ross, in Ross and Fobes 1929. Bracketed portions are not in the Greek but are added for clarification. Fobes's text reads: (8b10) τὸ δὲ ὂν ὅτι πολλαχῶς, φανερόν· ἡ γὰρ αἴσθησις καὶ τὰς

David Balme has said that 'the first difficulty in reading Aristotle's biological treatises, as often in reading Aristotle, may well be to decide exactly what their purpose is. They are so factual and so comprehensive that it is easy to mistake them for descriptive information, an animal encyclopedia, as the ancients regarded them.'[2] The same is true for Theophrastus. To understand a text such as *Historia Plantarum*, we need to be as clear as we can as to its purpose, and it is that task which I set myself in this chapter.

Theophrastus's botanical work has come down to us in two treatises, a *Historia* and a *Peri Aitiōn*. Aristotle's zoological work, though encompassing more treatises, also comes divided into a *historia* and a set of studies of causes. We may be tempted to ask if the parallels between Theophrastus' and Aristotle's biological writings run any closer. Could *Historia Plantarum*, for instance, have been modeled on *Historia Animalium*? If so, then the question of the purpose of either one of these treatises might usefully be studied in connection with that of the purpose of the other. That, at least, will be the strategy of this chapter. I will argue for a view of the aims of *HA*; and then, turning to *HP*, show that this treatise *was*, so far as its distinct subject matter permits, modeled on *HA*. Finally, I will explore *HP* in some detail in the attempt to show that its aims, both long-term and short, are much the aims of *HA*, even if Theophrastus is less sure, and less far along, in the pursuit of these aims in *HP* than Aristotle was in *HA*.

Though I will refer to each of the books of *HP* after the first, and to some of *De Causis Plantarum* (*CP*), it will not be possible in this chapter to discuss these at any length. Instead I focus on *HP* I which, like *HA* I, contains much more in the way of methodological remarks than the remainder of the treatise, and which also is significantly set off in content from the rest of the work. Similarly, we will restrict our attention to *HA* I–IV, to bring this vast text within manageable limits as well. Let us begin, then, with a careful and extended look at the question of the aim or aims of *HA*, to set the question and then formulate the competing hypotheses regarding the aim(s) of *HP*.[3]

διαφορὰς θεωρεῖ καὶ τὰς αἰτίας ζητεῖ· τάχα δ' ἀληθέστερον εἰπεῖν ὡς ὑποβάλλει τῇ διανοίᾳ τὰ μὲν ἁπλῶς ζητοῦσα, τὰ δ' ἀπορίαν ἐνεργαζομένη, δι' ἧς, κἂν μὴ δύνηται προβαίνειν, ὅμως ἐμφαίνεταί (b15) *τι φῶς ἐν τῷ μὴ φωτὶ ζητούντων ἐπὶ πλέον. τὸ ἐπίστασθαι ἄρα οὐκ ἄνευ διαφορᾶς τινος. εἴ τε γὰρ ἕτερα ἀλλήλων, διαφορά τις· ἔν τε τοῖς καθόλου πλειόνων ὄντων τῶν ὑπὸ τὰ καθόλου διαφέρειν ἀνάγκη καὶ ταῦτα, ἐάν τε* (b20) *γένη τὰ καθόλου ἐάν τ' εἴδη. σχεδὸν δὲ καὶ ἐπιστήμη πᾶσα τῶν ἰδίων· ἥ τε γὰρ οὐσία καὶ τὸ τί ἦν εἶναι καθ' ἕκαστον ἴδιον, τά τε θεωρούμενα καθ' ἑαυτὰ καὶ οὐ κατὰ συμβεβηκὸς ἦν ἂν τὶ κατὰ τινός. ὅλως δὲ τὸ ἐν πλείοσιν τὸ αὐτὸ συνιδεῖν* (b25) *ἐπιστήμης, ἤτοι κοινῇ καὶ καθόλου λεγόμενον ἢ ἰδίᾳ πως καθ' ἕκαστον, οἷον ἀριθμοῖς γραμμαῖς, ζῴοις φυτοῖς· τέλεος δ' ἡ ἐξ ἀμφοῖν. ἔστιν δ'* (9a1) *ἐνίων μὲν καθόλου τέλος (ἐν τούτῳ γὰρ τὸ αἴτιον), τῶν δὲ τὸ ἐν μέρει, καθ' ὅσα διαίρεσις εἰς τὰ ἄτομα, καθάπερ ἐν τοῖς πρακτοῖς καὶ ποιητοῖς· οὕτως γὰρ αὐτῶν ἡ ἐνέργεια. ταὐτῷ δ' ἐπιστάμεθα* (a5) *καὶ οὐσίᾳ καὶ ἀριθμῷ καὶ εἴδει καὶ γένει καὶ ἀναλογίᾳ καὶ εἰ ἄρα παρὰ ταῦτα διαιρέσεις· διὰ πλείστου δὲ τὸ κατ' ἀναλογίαν, ὡς ἂν ἀπέχοντος πλεῖστον, τὰ μὲν δι' ἡμᾶς αὐτούς, τὰ δὲ διὰ τὸ ὑποκείμενον, τὰ δὲ διὰ τἄμφω.*

[2] Balme 1987a. Cf. also Pellegrin 1986, 1: 'If anachronism is the cardinal sin for historians, perhaps historians of science are the most likely sinners.'

[3] Much of what follows on *HA* is excerpted from a fuller discussion of 'Philosophical Lessons of Aristotle's *Historia Animalium*' which was presented as a lecture to A. C. Crombie's seminar on the History of Science and Medicine at Oxford in February 1984.

Historia Animalium

HA proper consists of nine books. Parts or all of several of these have been thought at one time or another to be spurious, but David Balme has recently shown them all to be substantially authentic, and I will proceed on that basis.[4] *HA* is the largest treatise in the Aristotelian corpus, spanning 146 Bekker pages.[5] As its title indicates, the treatise is a *historia*—an investigation, or the report of one.[6] But what does an Aristotelian 'investigation' consist in?

The treatise appears, on its face, to be a massive collection of facts about animals, recording both personal observations and the reliable testimony of others. Its organization, if any, is less apparent. It has often been compared to natural histories, of the sort one still finds in the second-hand book shops: descriptive surveys, animal species by animal species, the species themselves grouped into large kinds. D'Arcy Thompson, in the preface to his Oxford translation of *HA*, even translates its title as *Natural History*, and Benedict Einarson, in the introduction to his Loeb edition, with G. K. K. Link, of the *De Causis Plantarum*, writes that 'Aristotle first classified animals and gathered information about them in the *History of Animals*.'[7] As *HA* does seem to group animals into larger kinds, and even speaks of *megista genē*, it has often been thought that it offers, or at least embodies, a taxonomy, a systematic hierarchical classification of the animal world.

Others, finding in the treatise much too little organization, have suggested that what we have are notebooks—file folders, in effect—of assorted information, perhaps aimed eventually at a taxonomic or natural-history end. Some have had less patience. G. H. Lewes, writing in the mid-nineteenth century, while admiring the advance *HA* represented over anything that had appeared before, called it:

[4] Cf. Balme 1991, Introduction. About half the manuscripts that contain all nine books also contain a tenth (though all from one family), which has long been held to be spurious. Moraux (1951, 105 n. 15) identifies it with the *Huper tou mē gennan* which appears among the biological writings in Diogenes Laertius' list of Aristotelian works. Balme has accepted this identification but argues in his introduction to Balme 1991, and more fully in a separate paper (Balme 1985) that, while not part of *HA*, it is in fact by Aristotle and is a pre-*GA* treatment of the female contribution to generation.

[5] This compares with, e.g., 58 for *PA*, 74 for *GA*, 83 for the *Physics*, 87 for the *Nicomachean Ethics*, and 113 for the *Metaphysics*.

[6] It is called a *historia* at I. 491a12. LSJ connect *historia* and the verb *historeō* with the Homeric noun *istōr* (sometimes *histōr*), 'one who knows law and right, judge', which Hesiod uses as an adjective, 'knowing, learned'. Some inscriptions have it beginning with a digamma and meaning, according to LSJ, 'witness'; they derive it from *Ϝid-tor* and connect that with **eidō* and *oida*. The first uses of *historia* they record are from Herodotus and Plato where, interestingly, the investigation is as much an inquiry into the causes of the phenomena being examined as it is a delineation of the phenomena themselves: cf. e.g., *Phaedo* 96a6–10.

[7] Einarson and Link 1976-90, vol. 1 (1976), ix. On the relation of *HA* to standard natural histories, see chapter 12 above, pp. 265-66 with n. 18.

an ill-digested, ill-compiled mass of details, mostly of small value, with an occasional gleam of something better. There is, strictly speaking, no *science* in it at all. There is not even a system which might look like science. There is not one good description. It is not an anatomical treatise; it is not a descriptive zoology; it is not a philosophy of zoology; it is a collection of remarks about animals, their structure, resemblances, differences, and habits. As a collection it is immense. But it is at the best only a collection of details, without a trace of organization; and the details themselves are rarely valuable, often inaccurate.[8]

But in a seminal and justly celebrated paper read to the Second Symposium Aristotelicum some twenty-seven years [now, in 2010, fifty years] ago, David Balme disputed this tradition.[9] He argued that *HA* does not aim to offer us a taxonomy, nor does it contain or presuppose one; that it is not a natural history, but nor is it just a large notebook. Rather, it is a theoretical treatise, with a specific theoretical purpose and a highly specific organization. Professor Balme has continued to hold that general picture, but his view of the specific theoretical purpose has developed over the years, and it will be worth taking some time to review that development, as it will provide us with several possibilities for the theoretical purpose or purposes of both of our *Historiae*.

In that paper, published in 1961, Balme argued that while Aristotle's logic and metaphysics advocated a method of systematic classification—via division, with *genē hup' allēla* ('kinds under one another'), and with *eidos* always a lower-level class, under a *genos*—the biological works (minus *PA* I) abandon this. These works, and in particular *HA*, make no use of division and offer no classification: *genos* and *eidos* are used interchangeably at all levels,[10] and—decisively—one simply cannot find in these texts the sort of intermediate genera that a taxonomic concern would entail. Virtually all the candidates for such, proposed by previous scholars, Balme showed to be what he came to call 'differentiae-classes'—classes named descriptively, such as 'the crook-taloned birds', which fail to have enough properties in common to produce a true *genos*.[11] He showed Aristotle sensitive to this failure too.[12]

Against the natural history thesis, Balme observed first that *HA* is not organized animal by animal, and that facts about one animal are spread throughout the treatise (which, of course, had no index). Second, he noticed that Aristotle's reports are

[8] Lewes 1864, 271. Lewes began his chapter on *HA* by ridiculing the following statement of Cuvier on the treatise: 'I cannot read this work without being ravished with astonishment. Indeed it is impossible to conceive how a single man was able to collect and compare the multitude of particular facts implied in the numerous general rules and aphorisms contained in this work, and of which his predecessors never had any idea' (Cuvier 1841, I. 146; cited by Lewes on p. 269).

[9] Balme 1961.

[10] This is argued more fully in Balme 1962a.

[11] *Topics* IV.2 122a3ff. and 122b15, Balme reminded us, makes a firm distinction between a *genos* and a differentia and urges us not to confuse them.

[12] See Balme 1961, 206 with n. 25 (and cf. Balme 1987b, 84).

very selective—that he fails to mention many things he surely knows about a given animal, facts that we would expect to appear in a natural history.

What, then, is *HA*'s real purpose? What is Aristotle doing with all those differentiae? Balme observed that in the explanatory treatises, *PA* and *GA*, Aristotle 'seeks the significant causal grouping of differentiae.... The method in fact is that which he briefly describes at *APo.* II.14 98a14-19: by looking for those characteristics which are regularly associated one may detect their cause'.[13] As for *HA*, Balme insists that we take seriously Aristotle's own statement of purpose at I.6 491a9: *ἵνα πρῶτον τὰς ὑπαρχούσας διαφορὰς καὶ τὰ συμβεβηκότα πᾶσι λάβωμεν.*

> The *Hist. An.* does not so much describe animals as collect and analyse the differences between them.... The result is a collection of all observable differentiae, collected *qua* differentiae. The animals are called in as witnesses, taken primarily in the order in which the differentiae are taken and then (within that order) by *megista genē*. The purpose of such a collection is no doubt adequately explained by the passage referred to above: it was to be followed by a discussion of causes: cf. 6, 491a10.[14]

Balme concludes by suggesting that the classification of animals called for by the *Organon* and by *PA* I would be somehow subsequent upon such a study of all the differentiae, but that 'however that may be, if the *Historia* is read, not as an encyclopedia nor as a collection of simple observations, but as being itself to some extent a theoretical treatise—namely a study of differentiae—then much of its apparent incoherence disappears.'[15]

By the time Balme was asked to reprint this piece in *Articles on Aristotle* for publication in 1975,[16] he had completed a close study of *PA* I.2-4 for his Clarendon Aristotle volume and had decided that the purpose of division was primarily definition, and only as a consequence classification.[17] As Balme became more and more convinced that the logical treatises and the *Metaphysics* (and *PA* I) called for definitions rather than for classification, definition came to take a more prominent place in his mind as the ultimate purpose of *HA*. Such definitions would contain multiple differentiae, and they would include both causally prior and explained features, in the spirit of the type of definition in *APo.* II.8-10, whose terms are the same as those of the demonstration but in different order. Since the aim of these definitions, which *PA* I and various passages in *APo.* II (and elsewhere) say division

13 Balme 1961, 208-209 (= 1987b, 86).

14 Balme 1961, 212 (cf. 1987b, 88). We may translate Aristotle's statement of purpose as: 'in order that we may grasp the differences and <in general> the features that belong to all.' (See the discussion of this passage in chapter 12 above, pp. 271-72.)

15 Balme 1961, 212 (cf. Balme 1987b, 88-89).

16 Barnes, Schofield, and Sorabji 1975. This study has since been reprinted once again, slightly modified (and with its first section replaced by a new paper on 'The reform of *diairesis*'), as part II of Balme 1987b.

17 Cf. Gotthelf 1992, 172-73.

helps to find, will be explanatory, they will be possible only *after* the causes have been identified. Thus *HA* is aimed at both causal explanation *and* definition.[18]

This view was briefly outlined by Balme in a paper read to the Southern Association for Ancient Philosophy (UK) in 1981, on 'Aristotle's *Historia Animalium*: Date and Authorship',[19] and developed in more recent writings.[20] There is an interesting twist in these papers, however—not anything entirely new, but a theme that moves to the foreground. Instead of speaking of *HA* as after 'significant combinations of differentiae', Balme speaks rather of its aim of 'identifying and analysing' the differentiae themselves. Whether one's aim is description, classification, explanation, or definition, the differentiae must first be identified and analyzed. I take the point to be that one cannot speak of the role of, say, being live-bearing and being egg-bearing in animal life (or animal definition), until one has surveyed the reproductive processes that actually take place, and come to understand exactly what it is to be live-bearing and to be egg-bearing and the forms each takes. (*GA* II, for instance, makes much of the distinction between being internally live-bearing and being externally live-bearing while internally egg-bearing, which only a close look at animals and their generation would make one aware of.) *HA* has this purpose, Balme now suggests, before it has the others already mentioned, though it has those as well. It will collect and analyze the differentiae, putting us in a position to give them exact definition. This will facilitate the identification of the significant associations and dissociations of differentiae, which itself will make discovery of causes, and thereby the establishment of causal definition, possible. *HA* itself contains only the first two steps: the collection and analysis of differentiae and their significant groupings; the others are left for other works. Such, as I understand it, is Balme's overall picture of both the immediate and the longer-range purposes of *HA*.[21]

[18] My source for Balme's views in the late 1970s is an unpublished talk, 'The place of biology in Aristotle's philosophy' (different from his 1987a under the same title), given at the Institute for Advanced Study, Princeton, in 1977. In this paper the eventual aim of a systematic classification of the animal world, subsequent upon the discovery of causes, is still present, but it recedes even further into the background than it had in the 1975 revision of the 1961 paper.

[19] Though this paper was never published, its central ideas turned up in subsequent publications. See, for instance, the opening sections of the introduction to the Loeb edition of *HA* VII-X (Balme 1991).

[20] Cf. Balme 1987a and 1987b (Part I). In these papers Balme endorses the thesis argued by Pierre Pellegrin, in ch. 2 of Pellegrin 1986 that *genos, eidos*, and analogy are used in the biology in harmony with their use in the logic, *Metaphysics*, and *PA* I, and that in both sets of places an *eidos* is always a subdivision of some *genos*, but that this relationship may exist at any level, and is not restricted to some fixed taxonomic stage, for there is no taxonomy in Aristotle, and no ultimate aim at one. This involves abandoning the claim of inconsistency between Aristotle's logic and his biology which Balme made in his 1961 and 1962a papers, but sits well both with his own denial of taxonomy and with his new emphasis on definition as an aim of *HA*.

[21] [This account of the development of Professor Balme's view of the aims of *HA* took us through 1983. I noted that he was currently at work on a new edition with commentary of the

I think this picture of the aims of *HA* is substantially right, but recent work by James Lennox and myself suggests an amendment. Professor Lennox and I think that in *HA* itself Aristotle takes the 'significant groupings of differentiae' much further, and in a more systematic way, than Balme has suggested. In a paper that Lennox read to the Williamstown conference on Philosophical Issues in Aristotle's Biology in 1983, he observed that the *HA* statement of purpose from which Balme quoted, at 491a7-14, itself speaks of its connection to the *Posterior Analytics*, and that the latter work contains important clues to the (or, certainly, to *a primary*) purpose of *HA*. The movement from 'grasping the differentiae and attributes' to 'trying to discover their causes', Aristotle says, is the 'natural' (*kata phusin*) way to proceed in this inquiry, for: 'it becomes apparent from [this *historia*] both *about which things* and *from which things* the *apodeixis* must proceed.'

Apodeixis is of course a central subject of *APo.*, and the expressions 'about which things' (*peri hōn*) and 'from which things' (*ex hōn*) are also standard *Posterior Analytics* terminology: the *peri hou* is the feature whose possession by a subject one is attempting to explain; it turns up as a predicate in the conclusion of the explanatory deduction. The *ex hou* is the feature whose possession explains the possession of the former feature; it is the middle term of the deduction (*APo.* I.10 76b21-22).

The theory of *APo.*, Lennox suggested, is lurking in the background.[22] In a recent study he has taken this further.[23] He notes Aristotle's concern throughout *APo.* for producing explanations 'at the right level'. That is to say, if we are to exhibit the cause of a subject's possessing a feature, we must identify that subject at the proper level of generality. The attribute, Aristotle says, must be possessed by the subject 'in virtue of itself' (in one sense of that multi-sensed expression); in fact the attribution must be 'primitively universal' (*APo.* I.4-5 *et passim*). Aristotle's most frequent example is the possession by isosceles triangles of an angle sum equal to two right angles. (Following Lennox, I will call this attribute '2R'.) Isosceles triangles do not possess 2R in virtue of themselves—that is, in virtue of being isosceles—but in virtue of being triangles. After all, *all* triangles possess 2R, and no

whole of *HA* (above, n. 4) and that it would be interesting to see what shape his view takes as a result of that work. Unfortunately, at the time of his death in 1989, Balme had not yet drafted an introduction to his *editio maior* of *HA*. However, see my preface in Balme forthcoming.] The picture I have given of his current views and the story of its development are constructed from the works cited, and not directly from conversation. (In a letter written to me after reading the first published version of this chapter, he described its account of the development of his views as accurate, if smoother than the historical reality.) For Balme's views on the form that the resultant causal definitions will take, their connection with demonstration, and the metaphysical issues involved, see the papers cited in the previous note, and Balme 1987d—which joined to a slightly revised reprint of Balme 1980 two new appendices, 'Note on the *aporia* in *Metaphysics* Z' and 'The snub' (itself reprinted from Balme 1984).

[22] Wolfgang Kullmann also made this suggestion, we later learned, and developed it in ways similar to Lennox, in Kullmann 1974.

[23] Lennox 1987a.

wider class of plane closed figures does. (Thus, going down the scale of generality, 'triangle' is the *first* universal to possess the attribute in all cases.) It is the *triangle* that possesses 2R 'in virtue of itself' and 'primitively universally', and it is that possession whose explanation must be sought. Finding that level, then, is a crucial step in the causal inquiry.

Lennox suggests that one can formulate the identification of the primitively universal level as a sort of explanatory syllogism. In canonical form:

2R belongs to every triangle.
Triangle belongs to every isosceles.

2R belongs to every isosceles.

Lennox calls this an 'A-type explanation', and he distinguishes it from the 'B-type explanation', the type Aristotle is ultimately after. The corresponding 'B-type explanation' would have the form:

2R belongs to every M.
M belongs to every triangle.

2R belongs to every triangle.

Here, M would be the mathematical feature, derived from being a three-straight-sided Euclidean plane figure, which actually explains having 2R, whatever that feature might be.[24] Such an explanation is the mathematician's goal, but he will not achieve it unless he has already brought the feature to be explained to the right subject level, and that is the task, when needed, of the 'A-type explanation'.

Now, in his paper, Lennox shows the pervasiveness of Aristotle's concern for the proper level and his willingness to use 'A-type explanations' to get us there. He observes that in one passage in *APo.* II.14, the chapter Balme noted as recommending the grouping of differentiae according to their association and dissociation, Aristotle identifies division as part of the method for getting us to that level, especially if we are coming down from a higher level of generality. Lennox goes on to show that Aristotle makes use of those 'A-type explanations' in the biology's explanatory treatises, but at the time of composing the Williamstown paper he did not have much to say about *HA*'s connection to this.[25]

[24] The full proof, written in canonical form, would involve a sequence of syllogisms. (Cf. *Metaph.* Θ.9 1051a24-26 and Mendell 1984.)

[25] I do not myself think that what Lennox calls 'A-type explanations' are actually offered by Aristotle as explanations. *APo.* II.14 introduces its subject as the question of how we are to formulate the propositions expressing the connections *to be explained*. The syllogisms that Lennox presents as explanations of a certain sort are rather articulations of the steps involved in finding the right level of generality for these *explananda*. Aristotle's point in *APo.* I.4-5 is not that isosceles triangles' possession of 2R is explained by their being triangles. Rather, it is that isosceles

He and I subsequently, however, formed the hypothesis that one important function of *HA* is precisely to find, with respect to each of the animal differences[26] Aristotle has collected, that right level at which causal explanation can take place—not to find the explanation but to find the level at which it can be found. And in fall 1983 we began reading through *HA*, both separately, and at regular intervals together, to see if we were right. We think we are, and I would like to indicate some reasons why.[27] Let me note first some of the power and interest of this hypothesis of ours, that a primary aim of *HA* is to identify for each of the observable animal differentiae the widest class of animals (identified either by name or by other differentiae they possess[28]) that possess that feature.

For one thing, the establishment of this hypothesis would substantiate, and carry further, David Balme's claim that *HA* is a theoretical treatise, aimed at facilitating the discovery of causes. Balme has maintained that *HA* has as one of its aims the grouping of differentiae, but has not said much about the means and the extent to which it does that. We see the grouping done by means of these widest-class generalizations, and find these generalizations systematically and pervasively developed throughout *HA*.[29]

Second, it would make good sense of Aristotle's claim that *HA* not only lays out the data, but allows what is cause and what effect to 'become apparent'. The widest-class generalization does not identify the cause, but takes us beyond the simple laying out of differentiae and puts us in a crucially better position to do so.[30]

Finally, the survey of my evidence for the hypothesis that a primary purpose of *HA* is the laying out of such widest-class generalizations will provide us with a

triangles' possession of 2R is explained by *the very same thing* that explains triangles' possession of 2R (and that one will understand that explanation only or better when the wider subject is used). I do not have space to argue this here, but in any case this reservation does not affect Lennox's main thesis, that Aristotle's practice in *PA* mirrors the insistence in *APo.* that explanation be at the right level, with which I thoroughly agree. (Lennox has taken these concerns on board: see Lennox 1991, 273-74 n. 16 [2001b: 67-68 n. 16].)

[26] Balme's tendency was to translate *diaphorai* as 'differentiae'; Lennox and I, following Pellegrin, have mostly translated it 'differences'. In this chapter I tend to stay with the former when speaking of Balme's views and the latter when speaking of ours.

[27] Lennox presents his own case for our hypothesis in Lennox 1987a and 1991.

[28] See *APo.* II.14 98a3-19 on how to deal with the absence of a name for the subject group in question.

[29] This, of course, does not rule out *HA's* having as a simultaneous aim the informed and analytical laying out of differentiae which Balme has suggested in his more recent work it does (cf. above, p. 312). A full picture will need to include both aims, but here I focus on the case for the 'widest-class' hypothesis that Lennox and I have developed as being part of that picture.

[30] There are also implications of our hypothesis for the relationship between the biological works and the 'philosophical treatises' (and more generally for, in Balme's words, the place of biology in [*sic*] Aristotle's philosophy), which cannot be dealt with here. On the connections between the biology and the philosophy, see the various essays in Gotthelf and Lennox 1987.

standard by which we can measure whether the *Historia Plantarum* has as any part of its aim the same purpose.

Let us, then, have a look at *HA*, and at the case for this hypothesis.

As Balme has shown, *HA* is a study of animal *differences*. It runs for nine books. Book I opens with a review of the way in which the parts—that is, tissues and organs—of animals can differ or be the same, then announces that 'animals differ in respect of lives (*bioi*), activities (*praxeis*), and characters (*ēthē*), as well as parts.' ('Activities' include cooling, keeping warm, procuring food, copulating, rearing young. 'Lives'—or 'modes of life', 'lifestyles' as some now say—include habitat [being a land, water, marsh-dwelling, or amphibious animal, for instance, and if land, living above ground or underground, in or not in a shelter] and sociability [living alone or in groups of various sorts]: generally, perhaps, 'activities in relation to environment broadly construed'. Typically, Aristotle refuses to discuss these two categories separately, speaking of 'lives and activities', or 'the activities that make up the life'. 'Characters' include degrees of intelligence, cunning, etc., and dispositions—fierceness, timidity, etc. These are not unconnected with activities and lives, as Aristotle notes, but he does distinguish them.)

These categories of difference are broken down into many more subcategories and given many more examples than my own illustrations suggest, and under 'parts' Aristotle also reviews the sorts of parts that all or most animals must have, and their differences—proceeding, roughly, across soul capacities, from nutrition to reproduction to sensation to locomotion. This review of differences takes five of our chapters.[31] Then follows, as chapter 6, the methodological discussion from which I have quoted, after which the main body of the treatise begins, in chapter 7. The main body divides as follows. Books I–IV, on parts; Books V–VII on activities and lives (including generation), and Book VIII on characters. (Book IX returns to generation, to treat the special case of humans.) Though a full case for the 'widest-class' hypothesis will have to include discussion of Books V–IX, I will focus here on the first section, the books on animal parts.

[31] I say 'our chapters' because the chapter divisions have no ancient authority, and are not found in the manuscripts. There are no chapter divisions for the biological works in the first printed edition (the Aldine of 1495-98), and none appear in Greek texts until the second Basel edition of 1539. These appear to derive from Theodorus Gaza's fifteenth-century Latin translation of the biology, *De Animalibus* (*HA, PA, GA*). The first printed edition (Venice, 1476) divides the Books via large initials into sections which correspond in a great majority of cases (but *not* in all) with our current divisions. The first edition which does almost fully correspond is the 1492 Venice edition printed by Iohannes et Gregorius de Gregoriis fratres, on the title page of which we are told that 'Sebastianus Manilius Romanus recognouit & per capita disposuit.' (My thanks to the British Library for making all these wonderful editions so easily available.) It is worth remembering that our chapter divisions are the construct of late medieval or Renaissance editors and not Aristotle's own, and worth reading the text right through them to see if its import changes. Sometimes it does. For more details on the origins of our chapter divisions, see chapter 12 above, n. 15, and esp. Beullens and Gotthelf 2007, 477-83 and its Appendix 1.

That methodological chapter in Book I—chapter 6—begins by identifying some commonly recognized large kinds of animals: birds, fishes, cetacean—these, he notes, are all 'blooded' (that is, red-blooded, pretty much our vertebrates). And among the bloodless (pretty much our invertebrates), there are the hard-shelled ones (clams, oysters, etc.), the soft-shelled ones (lobsters, crabs, etc.), the 'softies' (in Peck's literalizing of *malakia*—octopuses, squids, etc.), and the insects. Then to the blooded category he adds two groupings of his own, the live-bearing four-footed animals (our mammals less man) and the egg-bearing four-footed animals (our amphibians and reptiles excluding snakes), which he argues are authentic *genē*. This is not an attempt at exhaustive taxonomy—he notes that groups like man and the snakes do not fall under any wider natural kinds, and that there are very few intermediate kinds, and later on he notes numerous animal forms that share features of more than one large group. But these large groups are true kinds (genera, not just 'differentiae-classes'), in that members of each group share many features in common, with differences often just a matter of quantity or degree: 'bird differs from bird by the more [and the less] or degree (one is long-feathered, another is short-feathered)', he says elsewhere (*PA* I.4).

The subsequent discussion, in the main body of *HA*, treats blooded animals first, then bloodless. Within the blooded, he explores external parts first, then internal. The external parts tend to be reviewed large kind by large kind. After man, as a standard of comparison, come the live-bearing four-footed animals, then the egg-bearing four-footed animals, birds, fishes, and then snakes. This large-kind-by-large-kind review has often been appealed to as evidence of taxonomic and natural history concerns. But note first of all the division into external and internal parts. That takes precedence over the division into animal kinds. Furthermore, I said Aristotle 'tends' to survey large kind by large kind. As I will illustrate in a moment, where he finds that a part transcends the subject class under consideration—say, live-bearing four-footed animals—he happily indicates the wider class (here, four-footed animals)—often, I will argue, the widest class (say, footed animal)—that possesses the feature. The division of animals into large classes here is a working one for convenience of achievement of his actual goals.

Now, as for our hypothesis regarding those actual goals, notice that if in fact one's concern *were* with identifying the wider class possessing a given feature, on the assumption, say, that locating that would be a key to explaining the feature, then it *would* make great sense—i.e., save much time—to look at groupings already made on the basis of possession of many (significant) features in common. The induction from low-level animal forms has already been done, in the development of the language, or by previous study. From there one is in an ideal position to look across these large groupings to check for inter-generic features—which, Aristotle thinks, will often be there, frequently as analogues, though not nearly so frequently as there will be differences.

A closer look at Aristotle's text confirms, I think, that at least one main purpose of *HA* is to identify the widest class that possesses the feature under discussion. My own examination of the text of *HA*, even of the first four books, is by no means complete—this is a report of work in progress—but it strongly suggests the correctness of this 'widest class' hypothesis. One could group the evidence into five categories. No one piece of evidence or argument by itself will be conclusive, but taken together they are impressive.

(1) The large-kind-by-large-kind surveys throughout *HA* make good sense on this hypothesis, as I have already explained.
(2) More strikingly, as I have hinted, the departures from kind-by-kind survey—the trans-generic generalizations—make good sense too.
(3) There is a body of linguistic evidence, a peculiar form of words common to quite a few generalizations Aristotle reports in these chapters, which once in a while confuses translators. This form of expression seems nicely explained by the hypotheses under consideration (though I hasten to acknowledge that weaker hypotheses could accommodate it also).
(4) There is the structure of the discussion of features common only to some members of a large kind, where widest-class generalizations seem very much what is aimed at.
(5) There is a natural account one can give of the frequent-enough passages that are clearly not identifying widest subject classes, an account which makes these passages natural parts of an endeavor in which the overall purpose is precisely to arrive at such subjects.

I have already said something about the first two categories. In the remaining space I can devote to *HA*, I will just illustrate the claims I have already made and then go right to the other three categories of evidence.

(1) and (2): After the introductory chapters I have already summarized, Aristotle begins the main body of *HA* with a careful and systematic delineation (from top to bottom) of external parts in man, which by virtue of their greater familiarity can serve as a standard of comparison and an aid to not missing anything in the other animals. The internal parts, by contrast, are more familiar in the (other) *animals*, and beginning with the blooded animals, he again proceeds from top to bottom. The delineation of parts precedes the kind-by-kind survey. The latter begins at the start of Book II, where Aristotle discusses external parts, focusing first on those in live-bearing four-footed animals. The opening words of Book II are:

> Some of the parts of animals are common to all, and we said earlier, some only to certain kinds, and the ways in which they can be the same as or different from each other we have often stated.[32]

[32] *HA* II.1 497b6-9. The Greek text reads: *τῶν δ' ἄλλων ζῴων τὰ μόρια τὰ μὲν κοινὰ πάντων ἐστίν, ὥσπερ εἴρηται πρότερον, τὰ δὲ γενῶν τινων. τα αὐτὰ δὲ καὶ ἕτερά ἐστιν ἀλλήλων τὸν ἤδη*

The subsequent discussion, though focused around live-bearing four-footed animals, ranges over features common to wider classes of animals as well, sometimes four-footed ones generally, sometimes footed animals generally, sometimes live-bearing four-footed animals and man—for example in the possession of hair, though Aristotle focuses on the differences between live-bearing four-footed animals and man in this respect.

After hair, Aristotle moves to legs and feet, another set of features by which live-bearing four-footed animals differ from other animals. Here (499a31) he moves right to the four-footed animal level, then immediately further up as follows:

> All four-footed animals have legs which are bony, sinewy, and fleshless. So too have all footed animals, apart from man. Also, they lack buttocks—this absence is even more marked in birds. With man the reverse is true: his buttocks, thighs, and calves are pretty much the fleshiest parts of the whole body.[33]

Apart from gross bodily configuration and number of legs, there is not a whole lot that all and only live-bearing four-footed animals have in common. Most of their common features are internal. The chapters on birds and fishes, on the other hand, offer good examples of descriptions of features all and only birds—or all and only fishes—have. In the chapter on birds, for instance:

> In some of their parts birds resemble the animals we have just been discussing: they all have two legs, like man—in this respect resembling him more than any other animal does—except that birds bend their legs backwards, like four-footed animals, as we remarked earlier. Birds have no forefeet or hands, but wings—a peculiarity distinguishing them from other animals. Their haunch-bone is long, like a thigh, and attached to the body as far as in the middle of the belly: the result is that when taken separately it looks as though it is a thigh, although the thigh is actually a different part, extending from it to the shin. . . . All birds have numerous claws, and all in one way or another have numerous toes. . . .
>
> Birds have a mouth, but a peculiar one: they have neither lips nor teeth, but a beak. They have no ears or noses either, but passages serving these sensory faculties: the passages for the nostrils are in the beak, those for hearing in the head. . . .
>
> Furthermore, birds have neither horny scales nor hair, but feathers, and the feathers always have quills. Birds have no actual tail, but a rump.[34]

πολλάκις εἰρημένον τρόπον. Here and in the notes that follow the Greek for passages from *HA* is taken from Balme 2002.

[33] *HA* II.1 499a31-499b4. The Greek text reads: *πάντα δὲ τὰ τετράποδα ὀστώδη τὰ σκέλη ἔχει καὶ νευρώδη καὶ ἄσαρκα, ὅλως δὲ καὶ τὰ ἄλλα ζῷα ἅπαντα ὅσα ἔχει πόδας, ἐκτὸς ἀνθρώπου (ἔστι δὲ καὶ ἀνίσχια)· καὶ γὰρ οἱ ὄρνιθες ἔτι μᾶλλον τοῦτο πεπόνθασιν. ὁ δ' ἄνθρωπος τοὐναντίον· σαρκώδη γὰρ ἔχει σχεδὸν μάλιστα τοῦ σώματος τὰ ἰσχία καὶ τοὺς μηροὺς καὶ τὰς κνήμας.*

[34] *HA* II.12 503b29-504a3, 5-6, 19-23, 30-32; tr. Peck. The Greek text reads: (503b29) *ὁμοίως δ' ἔνια μόρια καὶ οἱ ὄρνιθες τοῖς εἰρημένοις ἔχουσι ζῴοις· καὶ γὰρ κεφαλὴν καὶ αὐχένα πάντ' ἔχει καὶ νῶτον καὶ τὰ ὕπτια τοῦ σώματος καὶ τὸ ἀνάλογον τῷ στήθει· σκέλη δὲ δύο καθάπερ ἄνθρωπος μάλιστα τῶν ζῴων πλὴν κάμπτει εἰς τοὔπισθεν ὁμοίως τοῖς τετράποσιν,*

(3) The discussion of teeth in live-bearing four-footed animals leads us to our third category of evidence: the linguistic form in which Aristotle often reports his findings. He begins (501a8):

> With regard to teeth, the animals exhibit great differences both from each other and from man. For whichever are four-footed and blooded and live-bearing, all have teeth, but to begin with some are dual-toothed [i.e., toothed in both jaws], others are not. For whichever are horned are not dual-toothed.[35]

In the discussion of the parts in birds, Aristotle notes that: 'Whichever birds rise well into the air, all have four toes' (504a8-9). Several pages later he opens his survey of internal parts of blooded animals by noting that: 'Whichever animals are four-footed and live-bearing, all have an esophagus and a windpipe' (505b32-33). And so on. Whichever are X, all these are Y. In Greek: *hosa* X, *panta echei* Y.

This redundant use of quantifiers is rare outside the biological works (and less frequent in the biological works outside *HA*), and though it could be just a quirk of style, it suggests to me a concern with levels of generality. The *hosa* ('whichever', lit. 'as many as') typically picks out a subclass of some wider class and *suggests*—though it by no means entails—that *only* this subclass has, or is known to have, the feature. I stress 'does not entail' for two reasons. First, the separative force of *hosa* can simply be directing our mental attention to the subject class, as if to say, 'Focusing now on X, all have Y'. Second, even when we distinguish our subject class from a wider class the rest of which does not have the feature (imagine, for instance, that no other four-footed animals had teeth), that would not rule out yet another class (e.g., fishes) from having the feature.

Still, the redundancy of the quantifier suggests to me a focus on finding universal propositions, at a reasonably high level of generalization. The *hosa* propositions *are* often convertible, and if not are ripe to serve as premises for so-called perfect induction to that desired level. By itself, this usage would not count for much, but together with the rest of the evidence, it is strongly suggestive.

ὥσπερ εἴρηται πρότερον. χεῖρας δ' οὐδὲ πόδας προσθίους ἔχει, ἀλλὰ πτέρυγας ἴδιον πρὸς τὰ ἄλλα ζῷα. ἔτι δὲ τὸ ἰσχίον ὅμοιον (504a1) *μηρῷ μακρὸν καὶ προσπεφυκὸς μέχρι ὑπὸ μέσην τὴν γαστέρα, ὥστε δοκεῖν διαιρούμενον μηρὸν εἶναι, τὸν δὲ μηρὸν μεταξὺ τῆς κνήμης, ἕτερόν τι μέρος....* (a5) *πολυώνυχοι δ' εἰσὶ πάντες οἱ ὄρνιθες, ἔτι δὲ πολυσχιδεῖς τρόπον τινὰ πάντες...* (a19) *στόμα δ' οἱ ὄρνιθες ἔχουσι ἴδιον· οὔτε γὰρ χείλη οὔτ' ὀδόντας ἔχουσιν, ἀλλὰ ῥύγχος, οὔτ' ὦτα οὔτε μυκτῆρας, ἀλλὰ τοὺς πόρους τούτων τῶν αἰσθήσεων, τῶν μὲν μυκτήρων ἐν τῷ ῥύγχει, τῆς δ' ἀκοῆς ἐν τῇ κεφαλῇ....* (a30) *ἔτι δ' οὔτε φολίδας οὔτε τρίχας ἔχουσιν, ἀλλὰ πτερά· τὰ δὲ πτερὰ ἔχει καυλὸν ἅπαντα. καὶ οὐρὰν μὲν οὐκ ἔχουσιν, ὀρροπύγιον δέ ...*

[35] *HA* II.1 501a8-12. The Greek text reads: *ἔστι δὲ καὶ περὶ τοὺς ὀδόντας πολλὴ διαφορὰ τοῖς ἄλλοις ζῴοις καὶ πρὸς αὐτὰ καὶ πρὸς ἄνθρωπον. ἔχει μὲν γὰρ πάντα ὀδόντας ὅσα τετράποδα καὶ ἔναιμα καὶ ζωοτόκα, ἀλλὰ πρῶτον τὰ μέν ἐστιν ἀμφώδοντα, τὰ δ' οὔ. ὅσα μὲν γάρ ἐστι κερατοφόρα, οὐκ ἀμφώδοντα ἐστιν.*

Now, Aristotle's interest is of course not only in features common to very wide classes, but also in the way subclasses of these kinds differ from each other—the bulk of *HA* in fact is devoted to differences at the intermediate and lower levels—which brings us to our fourth category of evidence.

(4) Aristotle is particularly interested in exceptions to broad generalizations, for the reason, as David Balme has also suggested, that they imply other differences at that same level, correlated causally with these. For instance, in the chapter on external parts in egg-bearing four-footed animals, II.10, Aristotle writes:

> [They] . . . have a head, a neck, a back, under and upper parts, also front and hind legs and an analogue of the chest, just as the live-bearing four-footed animals do [note the move to the wider class with regard to these external parts], and most of them have a fairly large tail, though a few have a smallish one. All of them are polydactylous, i.e., their feet are many-cloven [split into many digits]. In addition they all have sensory organs including a tongue, except the Egyptian crocodile.[36]

This distinctive feature of the crocodile might imply that there is something unique to the crocodile which explains that feature's distinctness, but note the interesting twist:

> In this respect it resembles certain fishes: for generally speaking, the tongue of fishes is prickly and not properly separate, while some fishes exhibit in that position a quite smooth and unarticulated surface unless you pull their mouth well open.[37]

In *PA*, Aristotle offers a two-part functional explanation for the crocodile's situation, one part of which depends on certain resemblances to fishes, one part on something unique to the crocodile (*PA* II. 660b11-34). Reaching these explanations is much aided by knowing at what subject level or levels to look for the explanation, and our suggestion is that finding those levels is a major part of Aristotle's purpose in *HA*.

Again, in the chapter on external parts in birds, II.12, after some lines on features uniquely common to birds, he introduces his first set of differences among these animals:

> All birds have numerous claws, and all *in one way or another* have numerous toes: in most birds the toes are quite distinct, and the swimmers, although web-footed, have separately

[36] *HA* II.10 502b28-503a1. The Greek text reads: *τὰ δὲ τετράποδα μὲν ᾠοτόκα δὲ καὶ ἔναιμα [οὐδὲν δὲ ᾠοτοκεῖ χερσαῖον καὶ ἔναιμον μὴ τετράπουν ὂν ἢ ἄπουν] κεφαλὴν μὲν ἔχει καὶ αὐχένα καὶ νῶτον καὶ τὰ πρανῆ καὶ τὰ ὕπτια τοῦ σώματος, ἔτι δὲ σκέλη πρόσθια καὶ ὀπίσθια καὶ τὸ ἀνάλογον τῷ στήθει, ὥσπερ τὰ ζωοτόκα τῶν τετραπόδων, καὶ κέρκον τὰ μὲν πλεῖστα μείζω, ὀλίγα δ' ἐλάττω. πάντα δὲ πολυδάκτυλα καὶ πολυσχιδῆ ἐστι τὰ τοιαῦτα. πρὸς δὲ τούτοις τὰ αἰσθητήρια καὶ γλῶτταν πάντα, πλὴν ὁ ἐν Αἰγύπτῳ κροκόδειλος.*

[37] *HA* II.10 503a1-5. The Greek text reads: *οὗτος δὲ παραπλησίως τῶν ἰχθύων τισίν· ὅλως μὲν γὰρ οἱ ἰχθύες ἀκανθώδη καὶ οὐκ ἀπολελυμένην ἔχουσι τὴν γλῶτταν, ἔνιοι δὲ πάμπαν λεῖον καὶ ἀδιάρθρωτον τὸν τόπον μὴ ἐγκλίναντι σφόδρα τὸ χεῖλος.*

> articulated toes. Whichever birds rise well into the air, all these have four toes; most of them have three toes in front and one behind, instead of a heel: a few have two in front and two behind, such as the bird called the wryneck.[38]

The wrynecks are thus fairly peculiar. Often, on noting a peculiarity, Aristotle goes on, as he does here, to describe other peculiarities of the class under discussion. He does this with the elephant and the camel, for instance. Once again, on the 'widest-class' hypothesis this makes perfect sense: the explanation of a peculiarity is likely to be found on the level of that peculiarity.

But Aristotle is interested not only in exceptions, but in variation—difference—quite generally. After a few more lines on features uniquely common to birds—the beak and feathers—he goes on:

> Birds have no actual tail, but a rump; the long-legged and web-footed birds have a short one, the others a long one. The latter fly with their feet close up under their belly, whereas the small-rumped birds fly with their legs stretched out.[39]

Here again are correlations. Among birds, all and only long-legged and web-footed ones have a short rump and fly with their legs stretched out. This does not give us the explanation for any of these features, but it does put us in a better position to do so, and in Aristotle's explanations in *PA* (694b12-25) these features *are* causally connected.

(5) My final category of evidence encompasses a range of apparent exceptions to our hypothesis, namely the many passages in *HA* where Aristotle is not, so far as I can see, offering widest-class correlations. What else might he be doing, compatible with that being his wider project or aim?

Here I will have to be brief, but notice first that in order to find the widest subject class possessive of each animal feature, one must first have the features before one. Thus, sometimes, I suggest, Aristotle is just listing features possessed by some readily observable class, not yet having asked what the widest class possessive of that feature is. Sometimes he has already moved to a higher level—all fishes of a certain type have a certain feature, say—but he has not yet looked to see how many other fishes, or even non-fishes, have it as well.

Sometimes there seems to be an allusion to the wider class, without it being said how wide the class is—as with the pervasive possession of heads, mentioned in several

[38] *HA* II.12 504a5-12. The Greek text reads: *πολυώνυχοι δ' εἰσὶ πάντες οἱ ὄρνιθες, ἔτι δὲ πολυσχιδεῖς τρόπον τινὰ πάντες· τῶν μὲν γὰρ πλείστων διῄρηνται οἱ δάκτυλοι, τὰ δὲ πλωτὰ στεγανόποδά ἐστι, διηρθρωμένους δ' ἔχει καὶ χωριστοὺς τοὺς δακτύλους. εἰσὶ δ' ὅσοι αὐτῶν μετεωρίζονται πάντες τετραδάκτυλοι· τρεῖς μὲν γὰρ εἰς τὸ ἔμπροσθεν ἕνα δ' εἰς τὸ ὄπισθεν κείμενον ἔχουσιν οἱ πλεῖστοι ἀντὶ πτέρνης· ὀλίγοι δέ τινες δύο μὲν ἔμπροσθεν δύο δ' ὄπισθεν, οἷον ἡ καλουμένη ἴυγξ.*

[39] *HA* II.12 504a31-35. The Greek text reads: *καὶ οὐρὰν μὲν οὐκ ἔχουσιν, ὀρροπύγιον δέ, οἱ μὲν μακροσκελεῖς καὶ στεγανόποδες βραχύ, οἱ δ' ἐναντίοι μέγα. καὶ οὗτοι μὲν πρὸς τῇ γαστρὶ τοὺς πόδας ἔχοντες πέτονται, οἱ δὲ μικρουρροπύγιοι ἐκτεταμένους.*

of the passages I have quoted. It would be no surprise if Aristotle has not yet decided how wide the class is, since he rightly finds it a tricky question as he moves down towards and into the bloodless animals. Alternatively, he might be perfectly clear how wide the class of headed animals is, but have planned to discuss it when he got to the bloodless cases; sometimes he discusses features at places one may not quite have expected, as he might well discuss what it is to be a head—and thus how wide the class of headed animals is—when he talks about sense organs. This is a made-up case. I have not stopped to look at how he actually handles heads, but the possibilities are clear, and they go a long way towards explaining the apparently exceptional texts.

Also, as I have mentioned earlier, *HA* is a work in progress, and quite long as it is, thank you. It is no surprise then if things one might expect him to say are omitted, especially as *HA* is probably a set of lecture notes, or study reminders, fillable in as needed.

This, incidentally, points up the core of truth in the notebook hypothesis discussed earlier. *HA* has a basic structure, I have been arguing, a specific organization for a specific theoretical purpose. But within that structure Aristotle is clearly adding information as he acquires it. David Balme's current work on *HA* has established that the collection and organization of the information that constitutes *HA*, though substantially by Aristotle after all (including *HA* VII and VIII), and substantially established during his middle period, was added to throughout the rest of his life.[40]

Some caveats, though. One feature of *HA*, on which Balme has much commented, is the selectivity of its reports: Aristotle's descriptions, even in the preliminary, as yet undistributed reports (above, n. 40), are highly selective. Much that he surely has observed is not stated. I have suggested ways in which one might explain what is present and what is not; and much that is left out might well be features that are covered by generalizations already made about some wider class under which the animal in question falls. But until one can tell a full story that makes thoroughly natural and unproblematic both what is present and what is absent in *HA*, one cannot be fully confident of one's hypothesis. More generally, one needs to walk this interpretation, in all its ramifications, line by line through *all* of *HA*, and then

[40] A dramatic set of such cases, to which Balme has already drawn our attention, are the occasional sustained discussions of single low-level animal types, such as the chapter on the chameleon (II.ll). In this discussion, for instance, Aristotle describes both the external and the internal parts of the chameleon together, and even includes information about its activities and mode of life, and this survey seems to make comparisons to other animals only for purposes of identification. It is clear from the passage (503b19-28) that Aristotle (or his source) has dissected the chameleon. Balme very plausibly suggests that what we have here is a preliminary stage of the Aristotelian investigation, material which has not yet been distributed to its proper places, not yet integrated into the separate discussions of external and internal parts of egg-bearing four-footed animals, but merely tacked on to the end of one of the discussions into which its material will be integrated. But this sits well with our 'widest class' hypothesis. After all, one cannot reach widest-class correlations before one knows *how* animals differ from each other, but one cannot know *that* before learning what features they in fact *have*, and these passages reflect just such an earlier stage of investigation.

run it through, so to speak, to be sure it fits the *spirit* of the whole as well. I invite readers to follow me in that effort: the evidence offered here, and in James Lennox's papers (above n. 27), already makes it unquestionably clear that our 'widest class' hypothesis is at least one part of the full picture, and I offer it in that spirit; a fuller read of *HA* will only confirm that picture.

Historia Plantarum

Diogenes Laertius reports for Theophrastus a *Peri Phutikōn Historiai* in ten books. Our *Peri Phutōn Historia* is in nine—perhaps, scholars have argued, because the ancient division was different, perhaps because a book was lost.[41] *HA* is in ten, or nine books. Do the affinities stop there, or was *HP* perhaps modeled after *HA*?

HA begins abruptly with a distinction between homoiomerous and anhomoiomerous animal parts and a discussion of the various ways in which animal parts may be the same as or different from each other. This has appeared so abrupt to some commentators that they have wondered if the passage did not belong later in the treatise,[42] especially since if one moved it, Aristotle's first words would then be expressive of the treatise's focus and structure, as follows:

αἱ δὲ διαφοραὶ τῶν ζῴων εἰσὶ κατά τε τοὺς βίους καὶ τὰς πράξεις καὶ τὰ ἤθη καὶ τὰ μόρια, περὶ ὧν τύπῳ μὲν εἴπωμεν πρῶτον, ὕστερον δὲ περὶ ἕκαστον γένος ἐπιστήσαντες ἐροῦμεν.

Differences in animals are with respect to their modes of life and their activities and their characters as well as their parts, concerning which let us first speak in outline, and later discuss with attention to each kind [sc. of difference under these four heads]. (Tr. mine)

Here are the opening words of *HP*:

τῶν φυτῶν τὰς διαφορὰς καὶ τὴν ἄλλην φύσιν ληπτέον κατά τε τὰ μέρη καὶ τὰ πάθη καὶ τὰς γενέσεις καὶ τοὺς βίους· ἤθη γὰρ καὶ πράξεις οὐκ ἔχουσιν ὥσπερ τὰ ζῷα. εἰσὶ δ' αἱ μὲν κατὰ τὴν γένεσιν καὶ τὰ πάθη καὶ τοὺς βίους εὐθεωρητότεραι καὶ ῥᾴους, αἱ δὲ κατὰ τὰ μέρη πλείους ἔχουσι ποικιλίας.

[41] Zeller 1897, II.354; Keaney 1968; and Sollenberger 1988, 15-16. (For more on the work's title, see Amigues 1988-96, vol. I (1988), xvi-xviii, and on the status of the ninth book, see her vol. V (1996). Notice esp. i-xxvii.)

[42] One such scholar, at the turn of the twentieth century, is C. R. Burckhardt (see Peck's note on him in the introduction to his Loeb edition of *HA*, p. li), who would place 486a5-487a10 at the end of the first chapter (488b28) (Burckhardt 1905). David Balme argues for its retention at the beginning, on the grounds that the treatise properly begins with a survey of the different ways animal features may be the same or differ, as a necessary prelude to a discussion of these samenesses and differences themselves, and notes that these modes of difference are explicitly applied in *HA* VII and VIII to features other than parts, but acknowledges the awkwardness of the fact that this application is not even hinted at in the present opening passage. I am myself inclined to Balme's view, and note only that, as we will see, this diminishes very little the parallel between *HP* and *HA*.

In plants the differences and their nature generally must be understood with respect to their parts and qualities and modes of generation and modes of life; for characters and actions they do not have as animals do. Now, while their modes of generation and qualities and modes of life are more discernible and easy to understand, differences with respect to parts can be very complex. (Tr. mine)

The parallel is striking, and it does not end there. Like *HA, HP* gets its broad structure (at least initially) from this initial division of differentiae. We will review that broad structure and its parallels (or lack thereof) to the overall structure of *HA*, but it is worth noting first some other close parallels in the first books of the two treatises. Then, after a look at the overall structure, we will turn to the question of the purpose of *HP*.

Theophrastus continues (from where the above quotation left off) by exploring an *aporia* concerning what counts as a 'part'. What that difficulty is, is not important for our purposes, but his way out of it is:

But perhaps we should not expect to find in plants a complete correspondence with animals in regard to those things which concern reproduction any more than in other respects.[43] (I.1.3; repeated at the start of I.1.4)[44]

As in the opening passage just quoted, this has the ring of announcing a departure from a standard practice, and a long passage in I.2.3-4 makes it clear that (and why) a comparison with animals will be standard:

'Fibre' and 'veins' have no special names in relation to plants, but, because of the resemblance, borrow the names of the corresponding parts of animals. It may be however that, not only these things, but the world of plants generally, exhibits also other differences as compared with animals: for, as we have said, the world of plants is manifold. However, since it is by the help of the better known that we must pursue the unknown, and better known are the things which are larger and plainer to our senses, it is clear that it is right to speak of these things in the way indicated: for then in dealing with less known things we shall be making these better known things our standard, and shall ask how far and in what manner comparison is possible in each case.[45]

[43] Theophrastus, *HP* I.1.3 (1-2). The Greek text reads: *τάχα δὲ οὐχ ὁμοίως ἅπαντα ζητητέον οὔτε ἐν τοῖς ἄλλοις οὔθ' ὅσα πρὸς τὴν γένεσιν . . .*

[44] All translations of passages from *HP*, and all citations of the original Greek, are from Hort's Loeb edition (Hort 1916), unless otherwise indicated.

[45] Theophrastus, *HP* I.2.3-4 (3.4-4.3). The Greek text reads: *ἶνες δὲ καὶ φλέβες καθ'αὑτὰ μὲν ἀνώνυμα τῇ δὲ ὁμοιότητι μεταλαμβάνουσι τῶν ἐν τοῖς ζώοις μορίων. ἔχει δὲ ἴσως καὶ ἄλλας διαφορὰς καὶ ταῦτα καὶ ὅλως τὸ τῶν φυτῶν γένος· πολύχουν γὰρ ὥσπερ εἰρήκαμεν. ἀλλ' ἐπεὶ διὰ τῶν γνωριμωτέρων μεταδιώκειν δεῖ τὰ ἀγνώριστα, γνωριμώτερα δὲ τὰ μείζω καὶ ἐμφανῆ τῇ αἰσθήσει, δῆλον ὅτι καθάπερ ὑφήγηται περὶ τούτων λεκτέον· ἐπαναφορὰν γὰρ ἕξομεν τῶν ἄλλων πρὸς ταῦτα μέχρι πόσου καὶ πῶς ἕκαστα μετέχει τῆς ὁμοιότητος.*

A similar point is made in passing in I.1.5. Specific comparisons and contrasts with animals are made at least in the following additional places in Book I: 1.3 (four or five times more), 1.9, 1.12 (twice), 2.5 (three times), 4.2, etc.

Most of the facts about animals with which those about plants are compared or contrasted are commonplaces, and do not necessarily reflect a careful study of *HA*, much less a borrowing of its structure, but the terminology is consistently Aristotelian, as is such doctrine that is not commonplace, and both of these reflect an affinity of general theoretical approach. Other passages more clearly suggest a modeling after *HA*.

The passage that immediately follows the resolution of the *aporia* about parts identifies several general ways in which parts may differ, and Theophrastus says in outline (*hōs tupōi labein*, I.1.6) much the same things, frequently in much the same words, that Aristotle says in the opening passage of *HA*.[46]

This opening passage of *HA* is followed by the sentence I quoted above, after which Aristotle outlines the *bioi kai praxeis* and the *ēthē* that he expects to study in more detail later.[47] This is followed, in chapters 2-5, as I mentioned earlier, by a survey of the more common parts of animals, proceeding soul-faculty by soul-faculty. In chapter 6 Aristotle identifies the *megista genē* into which animals are divided, notes difficulties for any comprehensive classification scheme, gives a brief statement of purpose, and explains why he will use man as the standard in surveying the external parts. Chapter 7 begins the survey as follows:

μέγιστα μὲν οὖν ἐστι τάδε τῶν μερῶν εἰς ἃ διαιρεῖται τὸ σῶμα τὸ σύνολον·

Now, these are the main parts into which the body as a whole is divided;

Theophrastus does not survey the *geneseis, pathē*, or *bioi* he expects to discuss, but turns directly to parts, with the words:

αὐτὰ δὲ τὰ μέρη διαριθμησαμένους πειρατέον περὶ ἑκάστου λέγειν. ἔστι δὲ πρῶτα μὲν καὶ μέγιστα καὶ κοινὰ τῶν πλείστων τάδε,

Let us try to give a delineation of the parts themselves, speaking about each. First and main and common to most are these: (I.1.9; tr. mine)

He begins with the visible parts, 'regarding them as members corresponding to the members of animals' ('members': *melē*; cf. *HA* 486a8-13)—root, stem, branch, and twig. From there he moves to 'the things of which such parts are composed,

[46] Thus, plants may differ in respect of parts, Theophrastus says, in that (1) one plant has it, another not (cf. *HA* 486b11-14—wording very close); (2) they are unlike in shape, color, compactness, texture, and the other πάθη (cf. *HA* 486a25-b6: many differences 'lie in the contrasting opposition of their παθήματα, e.g. color and shape'), or in 'excess or defect' (ὑπεροχή καὶ ἔλλειψις) in number or size, 'or, to speak generally, all the above mentioned differences are included under excess and defect: for the more and the less are the same thing as excess and defect' (τὸ γὰρ μᾶλλον καὶ ἧττον ὑπεροχή καὶ ἔλλειψις; cf. *HA* 486b16-17: τὸ γὰρ μᾶλλον καὶ ἧττον ὑπεροχὴν ἄν τις καὶ ἔλλειψιν θείη); (3) they are differently arranged, e.g., differently positioned, or symmetrical or not (cf. *HA* 486b23-487al).

[47] περὶ ὧν τύπῳ μὲν εἴπωμεν πρῶτον (487al2-13). Cf. Theophrastus' use of τύπῳ cited in the sentence to which the previous note is attached.

namely bark, wood, and core . . . , and these are all *homoiomerē*' (I.2.1; *HA* begins with a distinction between *asuntheta*, 'which divide into *homoiomerē*' and *suntheta* 'which divide into *anhomoiomerē*'). Theophrastus continues the analysis down to the elementary powers, bringing in a variety of parts-related features on the way, showing for instance that various *pathē* and *dunameis* (e.g., being easily splittable) are due to features of the plant's constituent materials.

Ready now (I.3.1) to state the actual differences between parts of plants in these various ways, Theophrastus observes that this study (*mathēsis*) will proceed more easily if we divide according to forms (*diairoumenon kata eidē*) wherever possible. Then he says:

πρῶτα δέ ἐστι καὶ μέγιστα καὶ σχεδὸν ὑφ' ὧν πάντ' ἢ τὰ πλεῖστα περιέχεται τάδε, δένδρον θάμνος φρύγανον πόα.

First and 'very large'—those by which pretty much all or most <of the forms> are embraced—are these: tree, shrub, undershrub, herb.[48]

And Theophrastus too spends some time (I.3.2-6) discussing the limitations of this, or any, attempt to produce a comprehensive classification of the plants. (Theophrastus, more it seems than Aristotle, is bothered by *epallaxis* ['overlapping'] even with respect to differentiae themselves,[49] and he continues on about this through chapter 4.)

The parallels, in short, are overwhelming, and there cannot be any question but that Theophrastus had *HA* before him as a model for his work so far as his distinct subject matter permitted it. How closely, then, did he follow *HA*? We see that though the tools are much the same, the strategies of deploying them are somewhat, though not terribly, different. What about the overall structure of the two treatises? A brief look at that will put us in a position to explore *HP*'s purpose.

HP I, as a whole, seems to correspond to *HA* I–IV. It is, as they are, concerned with parts (containing on the way various methodological remarks that bear on the entire enterprise). *HP* II–III, like *HA* V–VI (and IX), seem to focus on *geneseis*—modes of generation, including types of spontaneous generation, the effects of cultivation as against germination in the wild, and so forth. In Book IV Theophrastus turns to trees typical of, or peculiar to, various regions, themselves usually distinguished by their climates. This seems to mirror in certain ways some of Aristotle's discussions in *HA* VII and VIII, and to fall under *bioi*. Theophrastus, like Aristotle, includes a review of the sorts of diseases various organisms are subject to, probably because these reveal something about the organism's nature.[50]

[48] *HA* I.6 begins: γένη δὲ μέγιστα τῶν ζῴων, εἰς ἃ διῄρηται τἆλλα ζῷα, τάδ' ἐστίν ('Very large kinds of animals, into which animals have been divided, are these:'). A few lines later (490b16-17) they are in effect characterized as cases where one εἶδος περιέχει many εἴδη.

[49] 1.3.2ff; cf. above, n.12, for evidence of Aristotle's concern.

[50] Cf. Arist., *Resp.* 480b22-24: 'As for health and disease it is the business not only of the physician but of the natural philosopher to discuss their causes up to a point' (tr. Hett). The

HP V, which 'endeavour[s] to speak of timber (*hulē*), saying of what nature (*poia*) is that of each tree, what is the right season for cutting it, which kinds are hard or easy to work, and anything else which belongs to such an inquiry (*historia*)' (V.1.1. *init.*), might strike a reader of *HA* as peculiar: should *HA* have discussed the types of meat that different animals provide, 'and which kinds are hard or easy to work'? But, as David Balme pointed out to me, the answer would be yes *if* such an investigation revealed facts about the nature of those animals that were not best revealed in any other way, as Theophrastus apparently thinks is the case for wood (and perhaps makes clear at I.5.4-5, where he states that the various *dunameis* of plants belong to their nature, in that they have these *dunameis* because [*dia*] of certain features of their material makeup).

As I mentioned twice before, Aristotle's survey of the differences among the external parts of animals begins with man: 'Every nation reckons currency with reference to the standard most familiar (*pros to . . . gnōrimōtaton*) to itself; and we must do the same in other fields: man is, of necessity, the animal most familiar to us' (*HA* I.7 491a20-23, tr. Peck). For Theophrastus the standard is provided by trees—the *genos* of plants most familiar to us; and they (and to a lesser extent, shrubs) get primary attention in the first five books: 'However, as has been said, these <parts> belong especially to trees, and this way of dividing them up (*merismos*) applies more properly to these; and it is right to make <trees> the standard (*pros ha . . . tēn anaphoran*) in treating of the others' (*HP* I.1.11, tr. Hort, much modified).[51]

In Book VI Theophrastus begins his look at the remaining *megista genē*, undershrubs and herbaceous plants (including grains), and this continues through Book VIII. One difference from *HA* is that while Aristotle firmly structures the work by differentiae, reviewing the *megista genē* once for parts, again for generation, and again for modes of nutrition and characters,[52] Theophrastus tends to discuss, for the undershrubs and herbs, parts and generation together.

Book IX is about the 'juices' (or better, the fluids—moisture, sap, etc.) in plants; the tastes plants have; and the medicinal uses of herbs. Taste and degree and kind of moisture were mentioned as differentiae in Book I (chapter 12), where the latter explicitly (and the former, perhaps implicitly) was said to derive from (and therefore be indicative of) the peculiar blend (*krasis*) or elemental mixture (*mixis*) which constitutes the material nature of the plant (as is surely true as well for its medicinal

Greek text (from Ross 1955) reads: *περὶ δὲ ὑγιείας καὶ νόσου οὐ μόνον ἐστὶν ἰατροῦ ἀλλὰ καὶ τοῦ φυσικοῦ μέχρι του τὰς αἰτίας εἰπεῖν.*

[51] Theophrastus, *HP* I.1.11 (6-9): *ἀλλὰ μάλιστα ταῦτα ὑπάρχει, καθάπερ εἴρηται, τοῖς δένδροις κἀκείνων οἰκειότερος ὁ μερισμός. πρὸς ἃ καὶ τὴν ἀναφορὰν τῶν ἄλλων ποιεῖσθαι δίκαιον.*

[52] And that is an oversimplification, as the books on parts review them first for external, then again for internal parts; those on generation first for copulation, then for mode of generation; and there are at least three distinct reviews in *HA* VII–VIII.

power, where it has any). The book thus takes its place in *HP* and, at least in the text as we have it, brings it to a close.

What can we say in general? Aristotle has structured *HA* almost entirely around features (parts, activities, lives, characters), with groupings by *megista genē* only under these features, with rare exceptions that are explainable as early stages in the production of such a structure. Theophrastus' structure seems, to this reader at least, less sure. As just mentioned, in the case of the less familiar *megista genē*, he tends to group all differentiae of a *megiston genos* together under that *genos*, though he has said *explicitly* (where we have had to infer it, against the tradition, in the case of Aristotle) that the groupings in terms of *megista genē* are only for the purpose of facilitating the study of differentiae. Now, what appears to me as unsureness may be just that, a sign of an inability—either an inability to master the material, shaping it to the intended structure, or an inability fully to master the theoretical structure itself—not completely surprising even in the brightest of inheritors of a once hard-won approach, now taken for granted. *Or*, it might be a sign simply of *HP*'s being an earlier stage in Theophrastus' work than *HA* was in Aristotle's.

Alternatively, though this is less likely, there may be some theoretical purpose in doing it that way, just as Aristotle appears to have one in discussing the external then internal parts of each bloodless kind together under the kind, rather than surveying the external parts of each kind first, and then the internal.[53] Finally, the structure might be cleaner and clearer than I am making out, its invisibility to me due simply to my relative unfamiliarity with the full text of *HP*.[54] But whichever it is, it's clear that Theophrastus is working at least within the same broad theoretical structure as was Aristotle, and that we can expect the broad theoretical purposes of the treatises to be much the same, or at least to have more or less the same methodology, though we might well find differences of emphasis. What, then, can we say about the purposes of *HP*?

With *HA*, as we recall, we distinguished an immediate and a more ultimate purpose. The immediate purpose, as Balme established in 1961, and which there has never been reason to doubt, is, as he put it, 'a collection of all observable differentiae, collected *qua* differentiae. The animals are called in [only] as witnesses, taken primarily in the order in which the differentiae are taken' (above, p. 311). *HA* is neither taxonomy, nor natural history, nor 'a rag-bag of observations', as Balme in 1961 put one view he was criticizing. The more ultimate purpose of that

[53] For a view of what Aristotle's purpose might be, cf. Lennox 1987a, 104 n. 30 (2001b, 36 n. 29). Alternatively, the relatively limited amount of information available about the bloodless kinds, and/or the way the widest-class generalizations fall out, might not call for separate surveys.

[54] In fact, I believe it is a matter of 'unsureness' or inability and not a deeper theoretical purpose. H. B. Gottschalk (below, n. 90) has commented to me that 'the unsureness you find in Theophrastus [in *HP*] is characteristic of his thinking', and refers to D. E. Eichholz's introduction to his *Theophrastus: De Lapidibus* (Eichholz 1965, 5ff.; cf. esp. 7 n.1), and W. D. Ross's introduction to the *Metaphysics* (Ross and Fobes 1929, xxv f.).

collection has been for Balme, 'the significant causal grouping of differentiae' and thereby the discovery of causes, and ultimately, or in the process, the explanatory definition of animal kinds (however exactly the form, and object(s), of definition are to be understood). James Lennox and I have argued that this grouping of differentiae occurs in the form of 'widest-class generalizations' and in fact is part of the immediate purpose, being typically the very *form* in which the 'collection of observable differentiae' is presented. This entails that the discovery of causes is certainly an essential part of any more ultimate purpose of *HA*, though it of course does not exclude any wider purpose, such as definition.

So there are two levels at which we can ask the aim of *HP*: its immediate purpose and its more ultimate one. Let us start with the immediate purpose. First, as we have seen, there is clearly a structure to *HP*, even if it is not as luminous, or as consistently articulated, as *HA*'s structure was in *HA*. No 'rag-bagger' Theophrastus.

Theophrastus, more firmly even than Aristotle, seems to reject the possibility of any rigorous classification scheme, and (though this deserves a more careful examination than I have given it) his biological corpus, like Aristotle's, is largely free of intermediate genera. There is, in short, no sign even of an ultimate, let alone an immediate taxonomic aim. A mere natural history? A quick review of the structure of *HP* will make clear that it is not, that Theophrastus has not grouped his material by plant type.

Books I and II–III, as we have noted, treat parts and modes of generation, respectively, and even within these units, the division is by differentiae. Consider, first, the structure of Book I from the point where we left off with it earlier. As we saw, the book begins with a division of plant differentiae into parts, *pathē, geneseis, bioi*; resolves a difficulty in what we are to count as 'parts'; and offers an *arithmēsis*, with preliminary definitions, of the major parts, noting that parts of *trees* are to be used as the standard for those in other plants, followed by an *arithmēsis* of the constituents of these parts, down to the elementary powers. Theophrastus then suggests that a division of plants into *megista genē* will be useful towards this end, and provides one, urging however that this division, and in fact all the divisions of plants by one or another pair of major differentiae, is only approximate, though still natural and scientifically important. That is where we left him.

At this point (I.4.1), Theophrastus incorporates into the discussion of parts a discussion of a plant's physical form (*morphē*), here meaning overall presentation—that is, its shape and those features it possesses as a whole (or throughout the whole of itself)[55]—thereby completing the list of differentiae (here: parts and *morphai*) to be discussed. We are then (I.5.1) told that:

[55] Theophrastus, *HP* I.4.1 (1-6): *οἷον λέγω μέγεθος καὶ μικρότης, σκληρότης μαλακότης, λειότης τραχύτης, φλοιοῦ φύλλων τῶν ἄλλων, ἁπλῶς εὐμορφία καὶ δυσμορφία τις, ἔτι δὲ καὶ καλλικαρπία καὶ κακοκαρπία* ('I mean differences such as those in size, hardness, smoothness, or their opposites, as seen in bark, leaves, and the other parts; also, in general, differences as to

Next we must try to give the differences as to particular parts, in the first instance speaking broadly of those that are universal, and then of special differences between individual kinds; and after that take a thorough look more widely.[56]

And so the remainder of the book proceeds. We get a full survey of the differences (sometimes labeled divisions) and subdifferences, with at least one example for each difference of a tree or other plant which possesses it, which Theophrastus has found or noted in cores, roots, knots, direction and pattern of growth, leaves and their composition, seeds, taste, shape, and the whole form (*hē hole morphē*), degree and kind of plant moisture, miscellaneous others (downy-ness, leafiness, etc.), flowers, and fruit production, concluding with a general account of differences that apply to a plant as a whole.

Books II–III appear to be organized in much the same way. The focus is on methods of propagation, and related facts about generation, and plant-types are brought in only or largely as examples, except where those exemplary types are sufficiently unfamiliar to require some identificatory description, or where a familiar type has subtypes which, though not usually or easily distinguished, require distinguishing here because some aspect of the mode of generation is different in each.[57] And even in the later books where, as I have noted, almost all the differentiae of the undershrubs and of the two types of herbaceous plants are treated together, the review within that discussion, after the initial division into subkinds, seems from a quick look to be substantially by differentiae or at least struggling towards such.

HP is not, then, organized by plant type, but by difference, as is *HA*. In fact, while the word *diaphora* appears only a few times, I think, in *HA* I, and there is only that one statement of purpose that involves it,[58] *HP* I is riddled with the word, with some form of it appearing almost once a Loeb page of Greek, and surely more than once a page if

comeliness [*eumorphia*] or its opposite and as to the production of good or of inferior fruit'). On *hē holē morphē* in I.1.8, see n. 81 below.

[56] Hort's tr. modified. This last clause seems a reference to features falling under *morphē*, and any others that belong best with parts but are not themselves parts. The Greek text reads: πειρατέον δ' εἰπεῖν τὰς κατὰ μέρος διαφορὰς ὡς ἂν καθόλου λέγοντας πρῶτον καὶ κοινῶς, εἶτα καθ' ἕκαστον, ὕστερον ἐπὶ πλεῖον ὥσπερ ἀναθεωροῦντας.

[57] In later books Theophrastus sometimes falls into such identificatory surveys by type and subtype, especially, I think, for the wilder kinds, which would be more unfamiliar to his listeners. Aristotle, too, falls into this at least once in *HA* VIII with birds (cf. chs. 19-28). But the connection of the type and subtype survey to the review of animal *ēthē* which is going on in the surrounding context seems clear (compare VIII.19-28 with the preceding and succeeding chapters). The bird names (*cum* minimal descriptions) are brought in as headings under which facts about their *ēthē* are to be classified. These bird-types eventually are to be grouped as they agree, or disagree, in *ēthē* (as has already been done, or is about to be done), for the other birds, and more generally for the other animals. But however this may be, the immediate point is that such kind-by-kind surveys, not clearly exemplifying particular differentiae, are rare, certainly in Aristotle, and, I think, also in Theophrastus.

[58] 491a9, see above, n. 14.

one includes forms of *diairein* and *diairesis* (which nearly always imply 'of differentiae'). And it abounds in statements of immediate purpose or subpurpose. Consider:

I.1.1 (opening words): *τῶν φυτῶν τὰς διαφορὰς καὶ τὴν ἄλλην φύσιν ληπτέον . . .*

In considering the differences of plants and their nature generally) . . .[59]

I.1.6: *αἱ δὲ τῶν μερῶν διαφοραὶ σχεδὸν ὡς τύπῳ λαβεῖν εἰσιν ἐν τρισίν . . .*

Now the differences in regard to parts, to take a general view (*ὡς τύπῳ λαβεῖν*), are of three kinds.

I.2.2: *καὶ τοῖς μὲν δένδρεσιν ἔστιν οὕτως διαλαβεῖν.*

Let these things be divided then in this way.[60]

I.5.1: *πειρατέον δ' εἰπεῖν τὰς κατὰ μέρος διαφορὰς ὡς ἂν καθόλου λέγοντας πρῶτον καὶ κοινῶς, εἶτα καθ' ἕκαστον, ὕστερον ἐπὶ πλεῖον ὥσπερ ἀναθεωροῦντας.*

Next we must try to give the differences as to particular parts, in the first instance speaking broadly of those that are universal and common, and then of special differences between individual kinds; and after that take a thorough look more widely.[61]

I.14.5 (final words of Book I): *ἀλλὰ δὴ τὰς μὲν τῶν μορίων διαφορὰς καὶ τῶν ἄλλων οὐσιῶν ἐκ τούτων πειρατέον θεωρεῖν. περὶ δὲ τῶν γενέσεων μετὰ ταῦτα λεκτέον· τοῦτο γὰρ ὥσπερ ἐφεξῆς τοῖς εἰρημένοις ἐστίν.*

However we have said enough for study of the differences between parts and between general characters. We must now speak of the methods of growth, for this subject comes naturally after what has been said.[62]

[59] Tr. Hort, with 'differences' for *diaphorai.*

[60] My own translation. The 'things' are the broad category of parts.

[61] Cf. n. 70 below.

[62] See p. 334 below for a more literal translation of the first sentence. For similar indications of purpose, cf., e.g., I.5.4, I.6.1, I.6.3, I.6.7, I.6.12, I.7.3, I.10.8, I.14.2. This study of differences, Theophrastus makes clear in passing in a remark in Book VII.I5.3, is *not* a study of oddities: *πολλὰ δ' ἄν τις ἴσως λάβοι καὶ ἕτερα περιττά. χρὴ δέ, ὥσπερ πολλάκις εἴρηται, τὰς ἰδιότητας θεωρεῖν καὶ τὰς διαφορὰς πρὸς τὰ ἄλλα* ('Possibly one might mention many other eccentricities [*peritta*]. But as has been repeatedly said, we must only study the peculiarities [*tas idiōtētas*] and differences which one plant has as compared with others'). (I have replaced Hort's translation of *theōrein* as 'observe' with 'study', which it seems most often to mean in *HP*.) The concern to study differences in order to make possible explanation, which I attribute to *HP*, can be shown to be manifested in Theophrastus' other scientific works as well. I thus take issue with the thesis maintained by John Vallance (in Vallance 1988) that Theophrastus has a special interest in *perrita.* Cf., e.g., *De Lapidibus* 48 *init.* (brought to my attention by H. B. Gottschalk) with the introduction to Caley and Richards's translation of *De Lapidibus* (Caley and Richards 1956, 9) and *De Igne* §§ 44 *fin.*, 57 *init.* Dr. Gottschalk suggests that the method of arranging data by differentiae may be found even in the *Phusikōn Doxai.* The fragment quoted by Athenaeus (IX.387B) from Theophrastus' '*Peri Zōiōn* III' is also suggestive: 'The birds have this difference: while some are heavy and do not fly, e.g., francolin, partridge, cock, pheasant; [those] immediately can walk and have their feathers.'

So, *HP* is in its immediate purpose a collection, an analysis, of differentiae. But what is its more ultimate purpose? What is the survey of differentiae *for*? Causes? Definitions? Differentiae for their own sake, with no further purpose ('The man just liked plants . . . ')?

Arthur Hort, the Loeb editor and translator, seemed to have thought the purpose was 'the principle of classification'—that is, the idea of defining the true nature of the real objects in the world. This had begun with Plato, and was absorbed by Theophrastus:

> Throughout his botanical works the constant implied question is 'What is its *difference*?' 'What is its essential nature?', viz. 'What are the characteristic features in virtue of which a plant may be distinguished from other plants, and which make up its own "nature" or "essential character"?'.[63]

Hort says nothing of causes, not even of how their investigation might serve definition. Benedict Einarson, in his introduction to the Loeb *De Causis Plantarum*, says that the aim is the discovery of *causes*, but he says nothing about definition of essential nature:

> Theophrastus first classifies and identifies plants and gathers information about them in the *Historia Plantarum*, then proceeds to account for certain common or distinctive characters in the *De Causis Plantarum*.[64]

What does Theophrastus himself say? While *HA's* one statement of immediate purpose is followed directly by a statement of more ultimate purpose ('to try to discover the causes'[65]), *HP*'s many avowals of concern for differentiae say little of the use to which the survey might be put, though it is certainly clear that some further use is in mind. Consider again, for instance, the closing words of Book I:

> However, we have said enough for study (*theōrein*) of the differences between parts and between general characters. We must now speak of the methods of growth, for this subject comes naturally after what has been said.

[63] pp. xxi-xxii (xvii-xviii in first printing).

[64] p. ix. In a 67-page introduction, Einarson says nothing more about the aims of these treatises than this.

[65] ἵνα πρῶτον τὰς ὑπαρχούσας διαφορὰς καὶ τὰ συμβεβηκότα πᾶσι λαμβῶμεν. μετὰ δὲ τοῦτο τὰς αἰτίας τούτων πειρατέον εὑρεῖν ('in order that we may grasp the differences and <in general> the features that belong to all. After this we must try to discover the causes') (491a9-11; for a fuller discussion of this passage, see chapter 12 above, pp. 271-74). David Balme has suggested that the entire methodological passage in I.6 of which this is a part may well have been added to *HA* when it was prepared as a lecture course in the Lyceum, to be delivered before the *PA* and *GA* courses, although actually written after them.

More literally, the first of these sentences reads:

However, we must [go on now to] try *to study the differences* of the parts and of the wholes themselves *from these things*.

But 'these things' *are* the differences that have been studied across Book I.[66] That suggests that the present study of differences is to be followed by another study of these differences that is to be built on the present study. And this is most likely to be the study of causes. Consider the following passage:

I.3.6: *ἀλλ' ὅμως τοιαῦτα διαιρετέον· ἔχει γάρ τι τῆς φύσεως κοινὸν ὁμοίως ἐν δένδροις καὶ θάμνοις καὶ τοῖς φρυγανικοῖς καὶ ποιώδεσιν· ὑπὲρ ὧν καὶ τὰς αἰτίας ὅταν τις λέγῃ περὶ πάντων κοινῇ δῆλον ὅτι λεκτέον οὐχ ὁρίζοντα καθ' ἕκαστον· εὔλογον δὲ καὶ ταύτας κοινὰς εἶναι πάντων.*

Nevertheless, we are bound to use such distinctions.[67] For there is a certain common character alike in trees, shrubs, undershrubs, and herbs.[68] Wherefore, when one mentions the causes also, one must take account of all alike, not giving separate definitions for each class (*οὐχ ὁρίζοντα καθ' ἕκαστον*), it being reasonable to suppose that the causes too are common to all.[69]

This passage is speaking to the value of a division into *megista genē*, and of the broad divisions of differentiae, notwithstanding the fuzzy edges, and suggests that this division will play an important role in the exposition of causes, quite in the spirit of Aristotle's insistence that explanation be at the right level, discussed above.[70] (I will

[66] *ἀλλὰ δὴ τὰς μὲν τῶν μορίων διαφορὰς καὶ τῶν ἄλλων οὐσιῶν ἐκ τούτων πειρατέον θεωρεῖν.* (We will return to this occurrence of *ousia* shortly.) Note the similar passage at I.14.2: *τὰς μὲν οὖν κατὰ τὰ μέρη διαφορὰς πειρατέον ἐκ τούτων θεωρεῖν*, where *ταῦτα* are themselves the differences of the parts.

[67] Sc. both those between tree, shrub, undershrub, herb (cf. I.3.2-3) and those between wild/cultivated, flowering/flowerless, evergreen/deciduous, etc., which he has just been arguing have fuzzy edges.

[68] Sc. (from the sentence that follows): and a natural difference between wild and cultivated.

[69] I have not replaced 'distinctions' with 'divisions', in the first sentence, as I have usually been doing, since as Hort notes *τοιαῦτα διαιρετέον* is only a twofold conjecture, part by Schneider part by Wimmer, for the Aldine *τοῖς αὐτοῖς αἱρετέον*. (Neither says what the MSS readings are.) Hort's text is full of conjectures by Wimmer and Schneider, many apparently without direct basis in the MSS. A new edition of *HP* surely is a desideratum. [*2010:* Ask and ye shall receive: the first volume of Suzanne Amigues' eventual five-volume edition, in the Budé series, appeared in 1988, the fifth and final volume in 2006 (Amigues 1988-2006). As for the opening sentence of I.3.6, Amigues reports that the Aldine *τοῖς αὐτοῖς αἱρετέον* is found likewise in all the manuscripts; like Hort, though, she prints Wimmer's *τοιαῦτα διαιρετέον*.]

[70] pp. 313-14. Compare in particular Aristotle's injunctions at *PA* I. 639a15, repeated at 644b1, 645b10, and note 645b1: 'It is necessary first to divide off, in relation to each kind, the attributes that belong essentially to all the animals <of that kind>, and then to try to divide off their causes.' (tr. Balme, additional clarifying phrase my own). In his note to 645a36-b14, which includes this passage, Balme writes: 'This recapitulates 644a23-b7 and 639a15-b10, but adds that we should next state the causes of the common attributes, i.e. before going on to those peculiar to genera or to species.' Whether Theophrastus' practice in *CP* is in agreement with this

come back to the mention of definition very shortly.) It seems to suggest that a causal inquiry will follow the study of differentiae, and it suggests that at that point we will need something like our 'widest-class generalizations'. And it does seem to announce that we are to do certain things now that will put us in a position more easily to secure them. However, the passage does not clearly say that the entire thrust of the current enterprise is to enable us better to seek causes, as Aristotle in fact seems to say of his *historia*.[71] The same seems true for a second passage that mentions causes, in fact final causes:

> Now the nature of the major parts has been discussed already—I mean such parts as the root, the stem, and the rest. Their functions and those things for the sake of which they each are will be discussed later.[72]

Theophrastus has just digressed from his enumeration of the major parts of plants and their constituents to make a methodological point, which leads him to observe that his next step after this enumeration will be to discuss the differences of these parts, a discussion which he says (in a passage we will look at in a moment) will make clear their basic natures. As he is about to resume the enumeration, it seems to occur to him that someone will wonder where the inquiry into causes fits in. It seems clear from the brevity of the mention that his listeners would take for granted that a causal inquiry belongs somewhere in all this, but perhaps would not be certain where. That seems to suggest, again, that the present inquiry will be followed by a causal one, but again does not make clear that the former one is for the sake of, or directed immediately at, the latter.

The two passages just discussed are, I think, the only ones in *HP* I that mention an inquiry into causes. Now an inquiry into the causes of plant features has come down to us, and one might wonder if that is not the inquiry referred to, and if it might not contain any clues as to the purpose of *HP*. Strangely, the answer to the first question seems to be no. Both these passages speak of an inquiry into the causes of *parts*, and so far as I can see, there is no such inquiry in *CP*. *CP* is rather an inquiry into plant *generation*, roughly on the model of Aristotle's *GA*, and a much more accurate title for *CP* would be *De Generatione Plantarum*. *CP* discusses parts only so far as they bear on generation, as *GA* itself does (and announces itself, in its opening

injunction at I.3.6 is a matter requiring separate investigation, but such passages as this, and, for instance, I.5.1 (quoted above, p. 332), make clear that he is sensitive to the issue. (For a fuller discussion of the significance of the Aristotle passage [645a36ff.], see chapter 9 above.)

[71] At 491a11-14, on Lennox's analysis (above, pp. 313-14.). The text there, following immediately upon that quoted in n. 65 above, reads: *οὕτω γὰρ κατὰ φύσιν ἐστὶ ποιεῖσθαι τὴν μέθοδον, ὑπαρχούσης τῆς ἱστορίας τῆς περὶ ἕκαστον· περὶ ὧν τε γὰρ καὶ ἐξ ὧν εἶναι δεῖ τὴν ἀπόδειξιν, ἐκ τούτων γίνεται φανερόν.*

[72] I.2.4. The second sentence reads in the original: *αἱ γὰρ δυνάμεις καὶ ὧν χάριν ἕκαστον ὕστερον ῥηθήσονται.* The *γὰρ* seems important to our discussion, but I am not sure what to make of it.

words, as doing).[73] Nowhere in *CP* does Theophrastus take the differences in *parts* laid out in *HP* I and ask why *these* plants, for instance, have this type of root, and *those* that type, telling us thereby 'what its function is and that for the sake of which it exists', as Aristotle's *PA* does relentlessly for animal parts. *CP*—more properly '*GP*'—may well have been intended as the first in a series of inquiries which was eventually to contain the non-existent '*PP*' ('*De Partibus Plantarum*'). My overall impression even of *HP* alone is that Theophrastus had a special preference for generation. It must be left for a separate discussion to what extend the causal inquiry into generation makes use of the laying out of differences in *HP* II ff., and to what extent the manner in which they have been laid out in *HP* is geared particularly to the explanation in *CP*. As for *HP* I, there are very few 'widest-class generalizations' to be found there.[74] For the most part, the review is differentia by differentia, and if a widest-class inquiry is a stage in the overall study, it is a stage well down the road, as the first of our two passages just above suggests.

Is *HP* then more immediately a study of definition? It was the connection between the opening of *HA* and the discussion of definition by division in *PA* I.2-4 that led David Balme to see definition as a primary purpose of *HA*. The *PA* passage argued that division should begin with a natural *genos, eidē* of which differ primarily in 'the more and the less' (and gave as examples some of those listed as *megista genē* in *HA* I.6). It argued that we should take the multiple differentiae of this *genos* all at once and divide down simultaneously and successively, that is by further specification of the 'mores or lesses' of those initial differentiae. This would tie the distinction between *eidos* and *genos* to the distinction between sameness differing only in more and less and sameness only by analogy. The opening passage of *HA* introduces this apparatus of levels of sameness and difference, elaborates on the notion of 'the more and the less', and proceeds, at great length, to collect all the differentiae Aristotle can find—for the purpose, Balme inferred, of producing such definitions by division.

[73] The opening words of *CP* speak only of a study of generation: 'That the modes of generation are several, and how many and what they are, is stated in the *History*. Since not all occur in all [plants], it is proper to divide off those that belong to each and on account of what sort of causes, using principles drawn from their peculiar *ousiai*. For, our accounts must first of all be in harmony with what is said there' (tr. mine). Compared to *GA*, it is worth noting, the limited explicit attention to the final cause in *CP* is no great surprise, though to many readers Theophrastus has seemed to be one more person who, with respect to any cause beyond material necessity, merely 'touched on it [in *CP* I.1.1] and said good-bye' (cf. Arist. *Phys.* II.8 198b15). A satisfactory analysis of the role of teleology in *CP* will need to take account of its character as a '*GP*' and not a '*PP*'.

[74] A clear example is the first paragraph of I.10.8. There are a few others, but they do seem few and far between. There are no occurrences of the 'redundant quantifiers' discussed earlier (p. 320), and none of the other evidences for Aristotelian preoccupation with widest-class generalizations is in evidence in *HP* I.

Has *HP* that tie to the *PA* theory? As we have seen, Theophrastus does introduce early on differing ways of being the same and different, and makes significant use of the notion of analogy. But the tie to the notions of *genos* and *eidos* seems less in evidence, and no mention is made of definition. What Balme has observed about *HA* was at least equally true of *HP*: division is nowhere used for definitional purposes, and no definitions by division of plant kinds are produced. As we have noticed, there are frequent divisions of *differentiae* in *HP* I (as there are throughout *HA*)—e.g., of roots, some are woody, some fleshy, some are both, etc.—but all in the service simply of laying out the differentiae, not of defining anything. Nor does it seem to me that the definition of plant kinds or forms ever becomes a distinct purpose of *HP*. There are a few passages in *HP* I that might seem to suggest this, however, so we had better have a brief look at them.

In the lines that immediately precede the reference to the final cause quoted just above,[75] in the midst of his overview of plant parts and their constituents, Theophrastus reflects on his project:

> And when we have taken the parts, we must next take the differences which they exhibit, for thus will their essential nature become plain, and at the same time the general differences between one kind and another (*οὕτως γὰρ ἅμα καὶ ἡ οὐσία φανερὰ καὶ ἡ ὅλη τῶν γενῶν πρὸς ἄλληλα διάστασις*).[76] (I.2.4)

On first reading this might suggest that the making plain of the *ousia* was the aim of the inquiry, especially if with Hort one takes the *genē* to which Theophrastus refers to be those of plants. The *hē* that begins the very next line, however, clearly picks up our *ousia* and is explicitly referred to the parts.[77] A *diastasis* need be only a firm distinction, or set of distinctions,[78] and implies no more than Hort's 'general differences between' does. Still, I wonder if the *genē* are not of *parts* rather than of plants, even though plants in general, as compared and contrasted with animals, have just been mentioned (I.1.3)—since the wider context, for some length in either direction, is the *diarithmēsis* (I.1.9) of parts and their constituents.

Hort's reading might be thought to be supported by the closing lines of that *diarithmēsis*, at I.2.6.:

> Having now, we may say, taken all the parts, we must endeavour to give the differences between them and the essential characters (*ousiai*) of trees and plants taken as wholes.[79]

[75] p. 335 (and n. 73).

[76] *εἰλημμένων δὲ τῶν μερῶν μετὰ ταῦτα ληπτέον τὰς τούτων διαφοράς· οὕτως γὰρ ἅμα καὶ ἡ οὐσία φανερὰ καὶ ἡ ὅλη τῶν γενῶν πρὸς ἄλληλα διάστασις.*

[77] *ἡ μὲν οὖν τῶν μεγίστων σχεδὸν εἴρηται· λέγω δ' οἷον ῥίζης καυλοῦ τῶν ἄλλων.* The next line of the Greek is quoted above in n. 72, which is attached to Hort's translation of the whole passage.

[78] LSJ and Bonitz, *Ind. Arist.*, s.v.

[79] *εἰλημμένων δὲ πάντων τῶν μορίων ὡς εἰπεῖν τὰς τούτων διαφορὰς πειρατέον ἀποδιδόναι καὶ τὰς ὅλων τῶν δένδρων καὶ φυτῶν οὐσίας.*

However, Theophrastus' use of *ousia* is rather loose, and in I.2.6 seems to mean no more than 'the features that trees, and plants, have as a whole'. We have seen earlier (above, p. 306) how Theophrastus often uses *hē morphē* or *hē holē morphē* with this meaning, and that he in fact goes on to review these differences just after reviewing those of the parts. What is striking is the way in which, apparently, *ousiai* is used later on (I.14.3) for plants as the subject of (at least a subset of) just such features:

> Thus we must endeavour to study in the light of the instances mentioned the differences seen in the various parts of the plants [i.e. enough, then, for the differences in parts]. But there appear to be the following differences which affect the plant's whole being.[80]

'The plant's whole being' here surely means no more than 'the plant as a whole', as is certainly clear from the next-to-last sentence of *HP* I (the translation of which appears above [p. 309], and the Greek in n. 66). The differences of *hai allai ousiai* in that passage must mean (which is not to say it translates into) 'those differences discussed, over and above those of the parts, which belong to the *ousiai*—i.e., to the plants as wholes'.[81] If *ousia* in this use has any tie with its technical meaning, it is as designating one (or more) of the *ousiai* in the world (in the sense of Aristotle's *Metaphysics Z* 2), not the *ousia* of any of those *ousiai*.

Some occurrences of *ousia* in *HP* I do, of course, bear the latter, more definition-centered meaning, but I do not think any of them occur in passages that speak to the purpose of the enterprise. So we must conclude that the production of definitions of essential nature of plant kinds is not a distinct purpose of *HP*, not even ultimately. But I say 'distinct' because definitions also have a role as first principles of demonstration, or explanation, and in this role there is some evidence that Theophrastus had an interest in them. The evidence is unfortunately outside of *HP*, but is not without some use. Consider the opening lines of *CP*, quoted earlier (n. 73) for another purpose, and a passage later on in that work:

[80] I say 'at least a subset' because, in his actual review of differentiae, Theophrastus makes a clear distinction between features that fall under *hē morphē* (or *hē holē morphē*), which are reviewed in 1.12, before the present passage, and those that 'affect the plant's whole being'. However, in the preliminary overview in 1.4, from which I quoted earlier and to which I just referred, the distinction is less clear (and cf. the summary remark at the end of I.4.3), but is made nonetheless. However, it is made without the use of *ousia*, and seems clearly to be between differentiae on the same level, supporting my suggestion that *ousia* in the present passage has no metaphysical (or definitional) significance.

[81] I do not think *hē holē morphē*, in the interesting line at the close of I.1.8, means anything stronger than we have found it to mean elsewhere (above, p. 306). It has the same position in the preliminary survey of differences ending there that the discussions of overall features have in the full review, and I wonder if a remark in I.14.3, soon after our first *ousia* passage, might not bear that out: *καί πως τά γε τοιαῦτα ἐν τοῖς μέρεσιν ἢ οὐκ ἄνευ τῶν μερῶν ἐστιν* (Hort: 'Now it may be said that such differences are seen in the parts, orat least that particular parts are concerned in them.'). But I am not sure, and I am not able to explore it here.

That the modes of generation are several, and how many and what they are, is stated in the *Historia*. Since not all occur in all [plants], it is proper to divide off those that belong to each and on account of what sort of causes, using *archai* drawn from (*kata*) their *idiai ousiai*. For, our accounts (*logous*) must first of all be in harmony with what is said there.[82]

In dealing with all other occurrences in trees or plants, we must endeavour to start with the study of trees, taking as our bases (*lambanontas*) the *idia ousia* of each and the nature of its countryside, since from these the common attributes (*koina pathē*) and the variations between kinds (*hai kata genē parallagai*) and what [sort of countryside] comports with and is appropriate to each of them will become evident. We must also be able to discern (*theōrein*) what is merely similar and what is the same, since many things which differ are not thought to differ, as in other matters.[83]

These passages make clear that definitions which state the *ousia* of plant kinds will play a crucial role in the causal explanations that explanatory treatises such as *CP* are aimed at offering, and a study is needed of how they will do that.[84] There is no indication in either of these passages, however, that the study of differences, which *HP* clearly is, will help us towards that end. But I think there is such an indication in the fascinating passage from Theophrastus' little tract we call his *Metaphysics* that I have quoted at the head of this chapter.

As Ross explains in his commentary (p. xxii), this passage 'develops the notion that difference is an essential feature of the universe', and builds on this 'the thesis that it is essential that knowledge should take account of the differences between things'. The passage invites several remarks germane to our inquiry.

(1) It defines the task that perception passes on to thought as 'studying the differences and searching out the causes'. This characterization (which incidentally mirrors *HA*'s description of the *methodos* of which it aims at completing the first half)[85] makes no mention of definition as a basic task of thought. Once armed with the differences, it 'searches out the causes'.

(2) Since 'therefore' translates *ara*, the task of 'studying the differences and searching out the causes' is central to the achievement of *understanding*

[82] I.1.1, tr. mine. The Greek text reads: *τῶν φυτῶν αἱ γενέσεις ὅτι μέν εἰσι πλείους, καὶ πόσαι καὶ τίνες, ἐν ταῖς ἱστορίαις εἴρηται πρότερον·ἐπεὶ δὲ οὐ πᾶσαι πᾶσιν, οἰκείως ἔχει διελεῖν τίνες ἑκάστοις καὶ διὰ ποίας αἰτίας, ἀρχαῖς χρωμένους ταῖς κατὰ τὰς ἰδίας οὐσίας· εὐθὺ γὰρ χρὴ συμφωνεῖσθαι τοὺς λόγους τοῖς εὑρημένοις.* Here and elsewhere in this chapter, the Greek of *CP* is taken from Einarson and Link 1976–90.

[83] II.19.6; Einarson and Link's tr., modified. The Greek text reads: *ὑπὲρ δὲ τῶν ἄλλων ὅσα συμβαίνει τοῖς δένδροις ἢ φυτοῖς πειρατέον ἐκ τῶν δένδρων μετιέναι καὶ θεωρεῖν, τὴν ἰδίαν οὐσίαν ἑκάστου λαμβάνοντας καὶ τὴν τῆς χώρας φύσιν· ἐκ τούτων γὰρ τὰ κοινὰ πάθη καὶ αἱ κατὰ γένη παραλλαγαὶ καὶ τὸ ἁρμόττον καὶ τὸ οἰκεῖον ἑκάστοις γίνεται φανερόν. δεῖ δὲ καὶ τὸ ὅμοιον καὶ τὸ ταὐτὸ δύνασθαι θεωρεῖν, πολλὰ γὰρ δοκεῖ διαφέροντα οὐ διαφέρειν, ὥσπερ καὶ ἐπὶ τῶν ἄλλων.*

[84] I have attempted this for Aristotle in chapter 7 above. See also chapter 10.

[85] Quoted in nn. 65 and 71 above.

(*epistēmē*) which, from its role in the passage, seems the overriding aim of thought. That is not, however, the main thrust of the second paragraph.

(3) The point of the second paragraph is rather that the grasp of differences is necessarily central to this overriding aim of understanding. This is argued first from the plurality of the things that are and from the nature of universals, but then from a thesis about understanding, namely that it is almost always of *idia*.

(4) This thesis is itself argued for—to interpret—from a premise about the starting points of understanding and a premise about its objects, the attributes (*kath' hauta kata sumbebēkos*) whose possession by their subjects we are to understand.

(5) The premise about the starting points is actually a premise about 'the *ousia* and the *ti ēn einai* of each thing', namely that it is *idion* to its possessor; and as a premise (*gar*) in the argument for the centrality to understanding of grasping difference, it seems to imply that the grasp of 'the *ousia* and the *ti ēn einai*' depends on, is built upon, a grasp of difference.

(6) Perhaps as a corrective to an overemphasis on difference, the passage reminds us that understanding also involves, or perhaps in some sense more centrally involves, a grasp of sameness (which can come in various forms). This may suggest that of the various tasks that are involved in the achievement of understanding—the detection of differences, the discovery of causes, the development of definitions—all or some of them involve a grasp of sameness.

(7) As is clear from the survey of the types of sameness, any grasp of sameness involves as well a grasp of differences, and the achievement of understanding starts from a study of differences (which itself aims at the achievement of that understanding). It would seem, in sum, that the study of differences has the achievement of definitions as a part of its aim.

My deliberateness here stems from the complexity of this passage, which I find fascinating for the amount of theory, or its intimation, that has been packed into such a small space (and of course for the theory so intimated). And perhaps it is indicative too of a well of uncertainty (in my own mind, I mean) that my analysis has now reached.

It seems clear that the study of differences, which *HP* certainly is, is aimed more ultimately at the discovery of causes and at the development of the definitions that play a fundamental role in the exposition of that discovery. My reading of this *Metaphysics* passage brought into focus for me an obvious fact, but one worth noting because the standard texts that we have used to explore the problem of the aims of a *Historia* do not I think make mention of it, namely that the 'grasping of differences' and the 'search for causes' are themselves components of the striving for *understanding*.[86] My uncertainty concerns the exact formulation of how the development of definitions and the search for causes are connected, and why it should be natural in the Aristotelian tradition to speak of science as aimed at differences and causes, not at definitions, yet of the grasp of definitions as central to understanding. We know all the pieces, but I do not know how to sort them out. I take comfort, however, in the fact, or what I take to be a fact, that one can go a long way towards understanding the aims of an Aristotelian *historia*, whether *plantarum* or *animalium*, without resolving this (what I at least take to be) deep problem in the interpretation of Aristotelian philosophy.[87]

For, I think what our discussion shows is that the aims of the *Historia Plantarum* are the same as those of the *Historia Animalium*, in both the short term and the long; that the immediate aim is the laying out of all the differences that are to be found among the kinds of things under study; that this is to be done systematically, by major categories; that the more ultimate aim is the discovery of causes; and perhaps *the* ultimate one is best expressed as the achievement of understanding; that these more ultimate aims involve, in some way that is not really clear, the development of definitions of essential nature; and that the arrival at what Lennox and I have called 'widest-class generalizations' plays an essential role in the transition from the laying out of differences to the discovery, and laying out, of causes.

I think it is also the case that Aristotle was farther along in *his Historia*—at least so far as the *parts* of their respective organisms were concerned—than Theophrastus was in his, in that *HA* has much more in the way of these widest-class generalizations that both men thought necessary to the establishment of causes than does *HP*. Finally, I do not think I have found anything in Theophrastus to suggest that the picture of *HA* that David Balme, James Lennox, and I (in our various ways) have been building up needs important modification; in fact, I think rather that it confirms that picture.

Our investigation of *HP* confirms, as well, two remarks made (if our ancient source is correct) some twenty-two hundred years apart. In his magisterial *Die Philosophie der Griechen*, Eduard Zeller wrote that 'In creative power of intellect [Theophrastus] is not

[86] Though the famous passage in *PA* I.5 aimed at motivating students to the study of animals, even the 'less-valued ones', even their insides, argues in part from the *epistēmē* such a study will engender.

[87] Some important pieces of the puzzle can be found, I think, in Bolton 1987. One wishes, of course, that we had some of Theophrastus' own *Posterior Analytics*. (Zeller mentions an anonymous scholium [*Schol. in Ar.* 240a47] according to which for Theophrastus 'definition is embraced under demonstration', and refers to two or three other 'equally unimportant' testimonia from the commentators concerning definition and/or demonstration. His characterization of their usefulness is unfortunately apt. Cf. Zeller 1897, II. 363 n. 1. (See also Fortenbaugh et al. 1992, Part I, 254-60, and the brief commentary on these texts in Huby 2007, 154-60.)

indeed to be compared with Aristotle.... The fundamental principles of the system suffer no changes and are not infrequently stated in the very words of Aristotle.'[88] And what did Aristotle himself think of this? Diogenes Laertius tells us that 'It is said that Aristotle applied to him and Callisthenes what Plato had said of Xenocrates and himself (as already related), namely that the one needed a bridle and the other a goad; for Theophrastus interpreted all his thought with preeminent cleverness.'[89] If Aristotle had hoped for more, as we might have, his characteristic generosity did not permit him to say so, at least not for posterity.[90]

[88] Zeller 1897,II. 351, 355 (and cf. 355 n. 4).

[89] Diogenes Laertius, *Vit.* V.39, tr. after Hicks). The Greek text (from Long 1964) is: *λέγεται δ' ἐπ' αὐτοῦ τε καὶ Καλλισθένους τὸ ὅμοιον εἰπεῖν Ἀριστοτέλην, ὅπερ Πλάτωνα, καθὰ προείρηται, φασὶν εἰπεῖν ἐπί τε Ξενοκράτους καὶ αὐτοῦ τούτου· φάναι γάρ, τοῦ μὲν Θεοφράστου καθ' ὑπερβολὴν ὀξύτητος πᾶν τὸ νοηθὲν ἐξερμηνεύοντος, τοῦ δὲ νωθροῦ τὴν φύσιν ὑπάρχοντος, ὡς τῷ μὲν χαλινοῦ δέοι, τῷ δὲ κέντρου.*

[90] An earlier draft of this chapter was presented to the 1985 Project Theophrastus conference, 'Theophrastus on Science', at the University of London, and was commented on by William Fortenbaugh. My thanks to Professor Fortenbaugh, James Lennox, and Mario Mignucci, for very interesting discussion, and to Dr. H. B. Gottschalk for kindly providing me afterwards with valuable written comments.

PART V

Aristotle as Theoretical Biologist

15

Darwin on Aristotle

> Linnaeus and Cuvier have been my two gods, though in very different ways, but they were mere school-boys to old Aristotle.

This famous line, from a letter to William Ogle, the English translator of Aristotle's *Parts of Animals*, has often been quoted as evidence of Charles Darwin's respect for Aristotle's biological writings, and thereby as witness to the quality of Aristotle's work as biologist. More recently, however, a few scholars have sought to dampen our enthusiasm for this letter, suggesting that Darwin was familiar with very little of Aristotle and understood less, and that the vaunted letter was merely an extravagantly worded gesture of politeness from Darwin to his friend Ogle upon receipt of a gift copy of Ogle's just-published translation.

In this chapter I redeem the famous Darwin letter, and explore the Victorian master's actual acquaintance with, and estimate of, the ancient philosopher's biological work.

Making use of two additional letters from Darwin in which Aristotle is mentioned, the full text of the Ogle–Darwin exchange, and the occasional references to Aristotle in Darwin's earlier work, I argue (1) that the famous letter was *not* the 'gesture of politeness', but was a *second* one, sent unsolicited more than a month after a polite but unenthusiastic first letter; (2) that it was carefully formulated and meant quite literally; and (3) that it reflected an authentic, and substantial, *increase* in Darwin's respect for Aristotle as biologist, based on a careful reading both of Ogle's

This material, in very preliminary form, was first presented to the 1988 NEH Summer Institute on Aristotle's Metaphysics, Biology, and Ethics, at the University of New Hampshire, directed by John M. Cooper, Michael Frede, and myself. A substantially developed version was read to a conference on 'Aristotle's Animals in the Middle Ages and the Renaissance' in Leuven in May 1997; a sketch of the argument appeared as part of my contribution to the published proceeding of the conference, 'Closing Words', in Steel et al. 1999. The penultimate version was presented at a History and Philosophy of Science Faculty Colloquium at Cambridge University in May 1998, where I received helpful comments. It was published in *The Journal of the History of Biology* in 1999. (For further acknowledgments, see the final note below, p. 369; see also the Preface above, p. ix). The online version of this chapter is provided without the images of the Darwin letters that are available in the print version, in the full book.

Introduction and of the portion of Ogle's translation which Darwin says he has read.

I go on to hypothesize (4) what it was in Ogle's Introduction, and in Aristotle's text, that led to Darwin's increased respect; and I close by suggesting (5) that, while Darwin had indeed read very little of Aristotle's biological work (sadly, he died just two months after this first real encounter with Aristotle's text), his famous remark may have reflected some real insight into the teleological aspect of Aristotle's thought, and indeed more insight into that aspect than Ogle had achieved, as evidenced by a portion of their exchange that is rarely cited.

Let us begin with the famous letter. It reads in full as follows:

Feb. 22, 1882
My dear Dr. Ogle,

You must let me thank you for the pleasure which the Introduction to the Aristotle book has given me. I have rarely read anything which has interested me more; though I have not read as yet more than a quarter of the book proper. From quotations which I had seen I had a high notion of Aristotle's merits, but I had not the most remote notion what a wonderful man he was. Linnaeus and Cuvier have been my two gods, though in very different ways, but they were mere school-boys to old Aristotle.—How very curious, also, his ignorance on some points as on muscles as the means of movement—I am glad that you have explained in so probable a manner some of the grossest mistakes attributed to him.—I never realized before reading your book to what an enormous summation of labour we owe even our common knowledge. I wish old Aristotle could know what a grand Defender of the Faith he had found in you.

Believe me my dear Dr. Ogle,
Yours very sincerely,
Ch. Darwin[1]

The detractors' argument has been set forth most fully by Simon Byl,[2] who makes the following three points:

[1] Cal. 13697; DAR 261.5:19. See Figure 1. Darwin letters are cited herein by number from F. Burkhardt et al., eds., *A Calendar of the Correspondence of Charles Darwin, 1821-1882, with supplement* (Burkhardt et al., 1994) ('Cal.'), and then by classmark in the Cambridge University Library collection ('DAR'). In two cases the originals are currently held in the Huxley collection in the Archives of the Imperial College of Science, Technology and Medicine, London, and are cited by that institution's classmark ('Imperial College, Huxley'). My thanks to Pat Corvini for spotting several errors of transcription in the version of this letter I originally printed in *Journal of the History of Biology* 32 (1999), 4. I have also found, and corrected, some minor errors of punctuation, and transcription of the endings, in my earlier transcriptions of subsequent letters.

[2] Byl 1973; modified version in Byl 1980, xxx-xxxii. Byl's discussion is endorsed in Pellegrin 1986, 170 n. 10. Byl will be quoted from the 1980 book version.

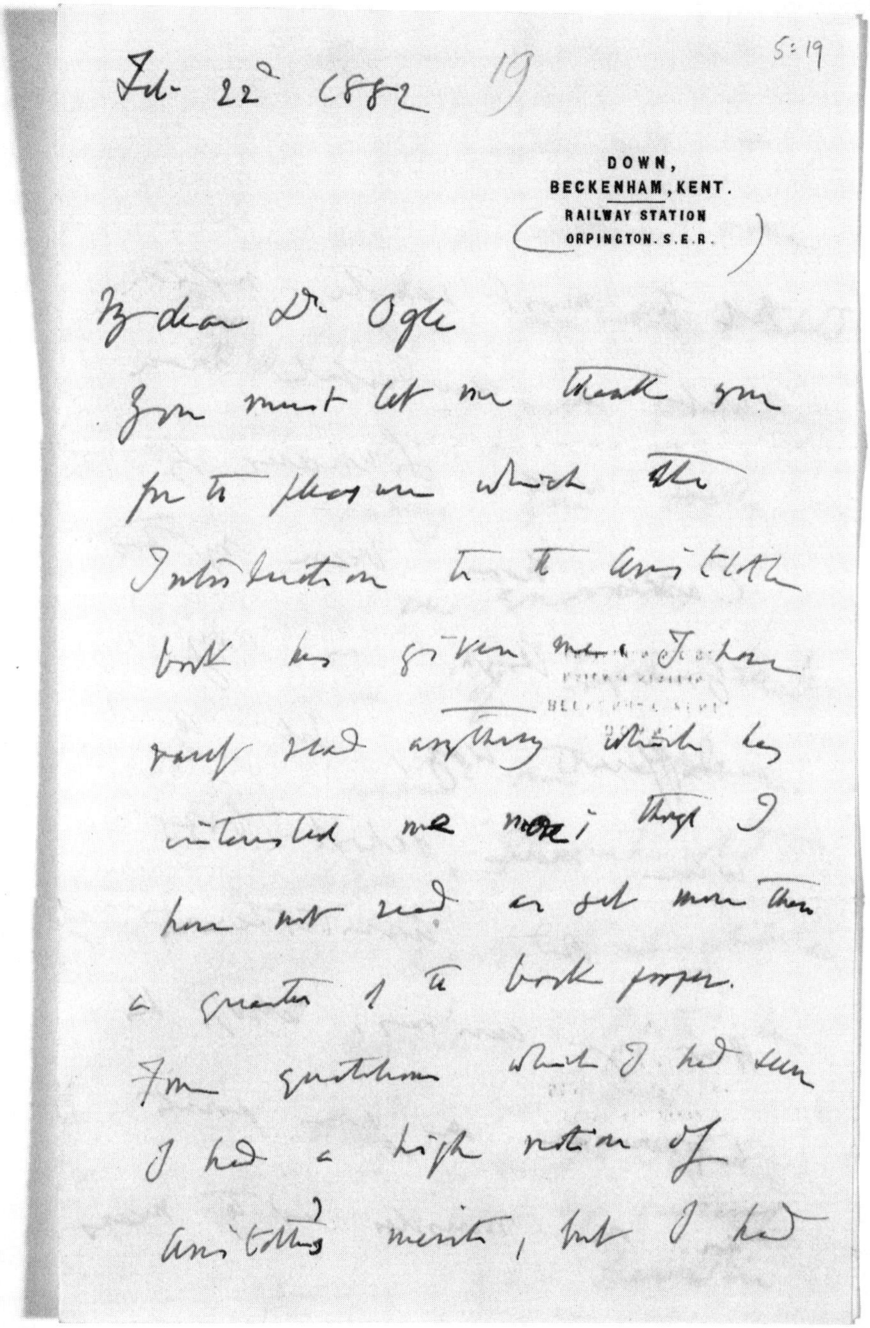

Feb. 22d 1882

DOWN,
BECKENHAM, KENT.
RAILWAY STATION
ORPINGTON. S.E.R.

My dear Dr Ogle

You must let me thank you for the pleasure which the Introduction to the Aristotle book has given me. I have rarely read anything which has interested me more; though I have not read as yet more than a quarter of the book proper.

From quotations which I had seen I had a high notion of Aristotle's merits, but I had

Figure 1

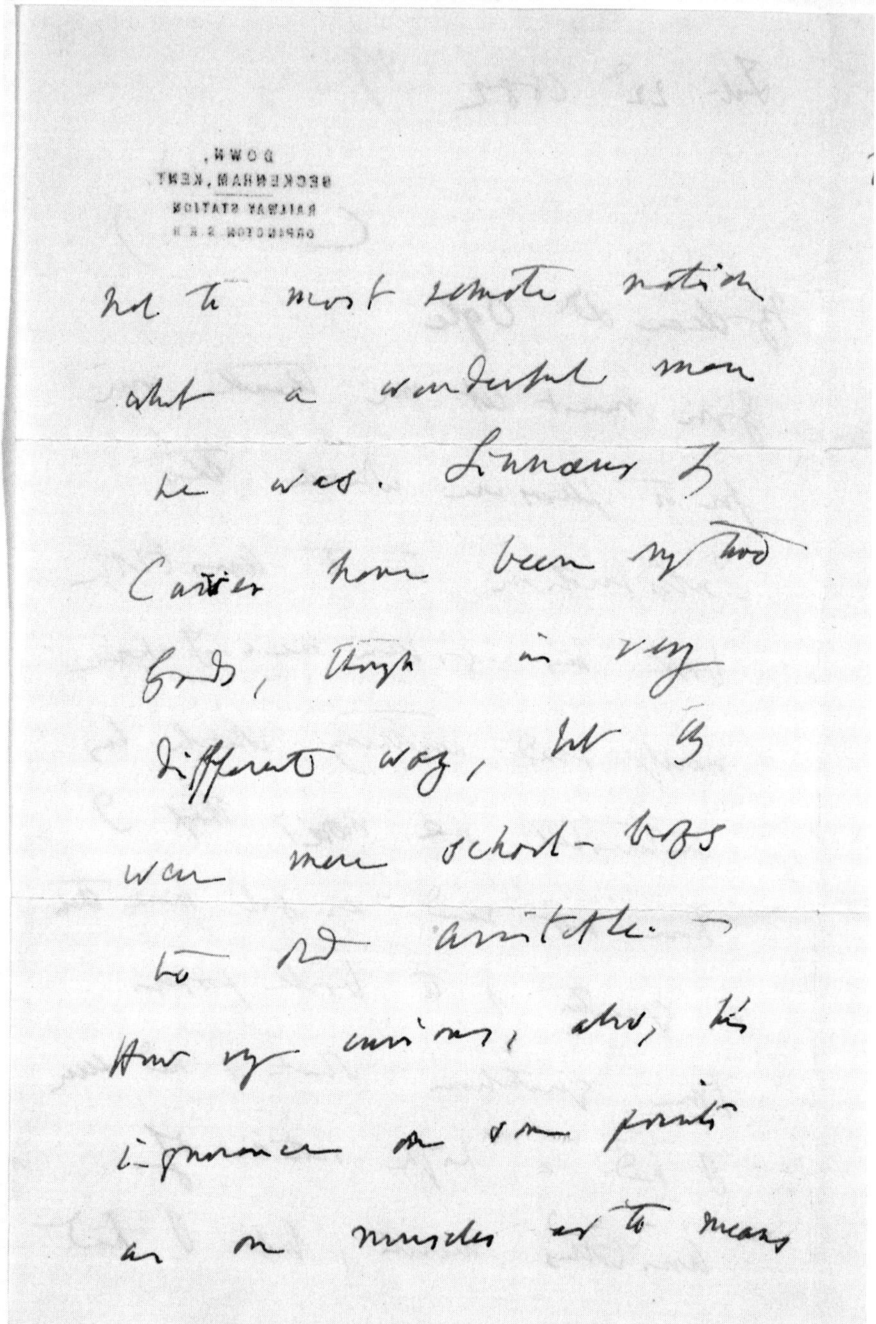

not the most remote notion
what a wonderful man
he was. Linnaeus &
Cuvier have been my two
gods, though in very
different ways, but they
were mere school-boys
to old Aristotle.
How very curious, also, his
ignorance on some points
as on muscles as the means

Figure 1. Continued

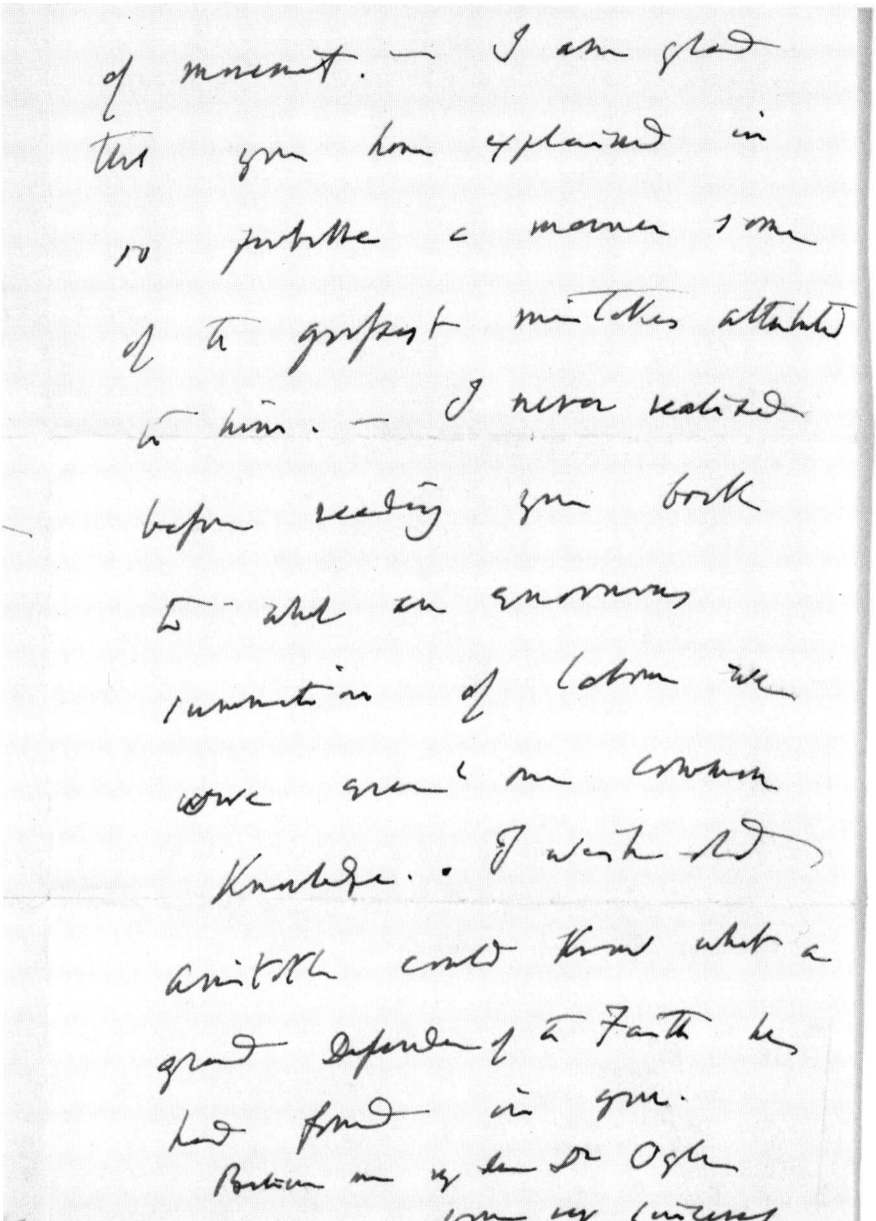

of moment. — I am glad that you have explained in so probable a manner some of the grossest mistakes attributed to him. — I never realized before reading your book to what an enormous summation of labour we owe even our common knowledge. I wish old Aristotle could know what a grand Defender of his Faith he had found in you.

Believe me my dear Dr Ogle

yours very sincerely

Ch. Darwin

Figure 1. Continued

1. This is a 'lettre de remercîments et de compliments', with a 'caractère dithyrambique' appropriate to such an occasion (he notes the reference to Ogle as 'Defender of the Faith'). As such it cannot be counted on to be a reliable account of Darwin's actual views.
2. Darwin's judgment is based mainly on Ogle's introduction. Darwin reports in our letter that he has read only a small portion of the 'book proper', and in a 'Historical Sketch' added to the Preface in the 3rd edition of *On the Origin of Species* (1861), Darwin had said straightaway that he is not familiar with the writings of biologists before Buffon, including in that group 'authors from the classical period'. Indeed, in the notorious footnote to this remark, added in the 4th edn. (1866), Darwin, referring to Aristotle's presentation in *Physics* II.8 of the Empedoclean position on the formation of teeth, attributes the Empedoclean position to Aristotle himself.[3] Thus, Darwin's judgment, even where it is sincere, has little value.
3. And, insofar as Darwin's judgment has any value, we must not forget that in this letter he speaks of 'Aristotle's ignorance on some points', and of his 'grossest mistakes' in others.

So, Byl concludes, we may cite 'le mot de Darwin' only with great 'prudence' and 'circonspection'.

Oddly, this assessment itself suffers from insufficient attention to the full content, and actual context, of the correspondence in question. Let us try to remedy that, examining first the circumstances of the famous letter to Ogle.

William Ogle (1827–1912) was a physician and naturalist, with classical training, whose small but interesting set of publications earned him inclusion in the *Dictionary of British and Irish Botanists and Horticulturists*.[4] Darwin cites Ogle at several

[3] The specific mention of Darwin's unfamiliarity with ancient authors occurs only in this first appearance of the 'Historical Sketch', but its removal from subsequent editions was certainly not due to any additional reading. Darwin's citation of the Empedocles passage in the 4th edition footnote was supplied by a correspondent, C. J. Grece, as Darwin there indicates; there is no reason to think that Darwin himself opened a text of the *Physics*. The letter from Grece does not seem to have survived, so that we cannot tell if Grece himself attributed the Empedoclean theory to Aristotle; in a November 1866 letter (Cal. 5276; DAR 165:220), Grece reintroduces himself as the person who 'a year or two ago' brought the Aristotle passage to Darwin's attention.

[4] Ed. R. Desmond, 3rd edn. (London: Taylor & Francis, 1977). The *Dictionary* (followed by the Darwin Correspondence Project's *Calendar* (above, n. 1)) incorrectly gives Ogle's year of death as 1905, basing its information on *Who Was Who: Supplementary Volume 1897-1916*, which cites 16 May 1905 as the date of Ogle's death. This citation was itself apparently based on a death notice which appeared on 17 May in the *Times* of London for one William Ogle, M.D., F.R.C. P. But this notice describes the deceased Dr. Ogle as 'sometime fellow of St. Catherine's College, Cambridge', while the translator of Aristotle's *Parts of Animals* (*PA*) is identified in all its editions as 'sometime fellow of Corpus Christi College, Oxford'; indeed, the character of the Preface to the 1912 edition of the *PA* translation, in volume 5 of the Oxford *Complete Works*, suggests that the translator was alive at the time the volume (or at least the 1911 fascicle) went to press. The

places in his writings of the late 1860s, and their correspondence appears to have begun in that period. The younger man by eighteen years, Ogle was clearly an enthusiast of Darwin's work, and in these letters they briefly discuss evidence of Darwin's theories, they recommend books to each other, and Ogle occasionally passes on to Darwin information about classical authors—both their discoveries and their mistakes on various matters of biological detail. Ogle's letters sometimes have a deferential quality—he knew he was dealing with a greater mind than his own; Darwin in his letters always projects a quiet respect for Ogle as a skilled observer and a source of valuable biological and medical information, classical and contemporary.

In January 1882, several months before Darwin's death, Ogle published a work entitled *Aristotle on the Parts of Animals*,[5] a translation of the Aristotelian treatise (hereafter abbreviated '*PA*') together with a 35-page Introduction and over 100 pages of endnotes.[6] On January 17, presumably just days after publication, Ogle sent a copy of the book to Darwin with the following cover letter:

suspicion that there were two Dr. William Ogles living in England in 1905 is confirmed by the *Alumni Cantabridgienses*, which reports that the Cambridge Ogle was born in 1824, while our Ogle, according to all sources, including the *Alumni Oxonienses* and the *Times* of London (see below), was born in 1827. To make matters worse, the *Dictionary of National Biography* reports a John William Ogle, M.D., F.R.C.P., an Oxford man who, as our William, resided in London, and whose years of birth and death were the very same as those of the other William Ogle (1824-1905), although John William died on August 8 as against William's May 16. Unfortunately, some biographical notes regarding our William Ogle attribute some of John William's accomplishments to our William in place of some of his own (in part because both were associated with St. George's Hospital). It may be thought astonishing that there were three English physicians with surname Ogle and one forename William, two residing in London in early 1905 and all within three years of age of each other, but so it was. That our Dr. William Ogle is distinct from the other two, and outlived them both by almost seven years, is confirmed by two sentences of an obituary which appeared on the first page of the *Times* of London of Monday, April 15 1912: 'We regret that Dr. William Ogle died at his residence, 10, Gordon-street, on Friday. . . . In 1882 Dr. Ogle published a translation, with introduction and notes, of Aristotle's "De Partibus Animalium".'

(The second of the two sources the Darwin Project *Calendar* cites for information about Ogle, R. B. Freeman's *Charles Darwin: A Companion* (Folkstone: Wm. Dawson and Sons, 1978), gives 1912 as the year of his death, relying very likely on P. Thomas Carroll, *An Annotated Calendar of the Letters of Charles Darwin in the Library of the American Philosophical Society* [Carroll 1976], which also gives 1912.)

[5] Ogle 1882. The only previous translation into English was Thomas Taylor's, published in 1810.

[6] Many of these notes are still valuable; this translation, somewhat revised, with many of the notes shortened or omitted, was one of the very first to appear in the Oxford translation series, edited by J. A. Smith and W. D. Ross, as a separate fascicle in 1911, and as part of vol. v in 1912. It was revised yet again by Jonathan Barnes for inclusion, although without the notes, in *The Complete Works of Aristotle: The Revised Oxford Translation* (Barnes 1984).

Dear Mr. Darwin,

I have given myself the pleasure of sending you a copy of a translation of the 'De Partibus' of Aristotle; and I feel some self-importance in thus being a kind of formal introducer of the father of naturalists to his great modern successor. Could the meeting occur in the actual flesh, what a curious one it would be! I can fancy the old teleologist looking sideways and with no little suspicion at his successor, and much astounded to find that, while there was actually no copy of his own works in the house and while his views were looked on as mere matters of antiquarian curiosity, Democritus whom he thought to have effectually and everlastingly squashed, has come to life again in the man he saw before him! I have, however, such faith in Aristotle as a real honest hunter after truth, that I verily believe, that, when he had heard all you have to say on your side, he would have given in like a true man, and have burnt all his writings; and this pray do, if it so please you, with the one volume of them which I send you.

Believe me,
Yours truly,
W. Ogle[7]

I will comment on Ogle's picture of Darwin as a Democritean anti-teleologist later, but I want to turn immediately to Darwin's letter of response, which is dated the very same day as Ogle's letter, and which preceded the famous 'school-boys' letter by more than a month:

Jan. 17th, 1882
My dear Ogle

I am very much obliged to you for your gift of your Aristotle. By turning over the pages I suspect that your Introduction will interest me more than the text, notwithstanding that he was such a wonderful old fellow.—

Pray believe me,
Yours sincerely & obliged,
Charles Darwin[8]

This is the 'lettre de remercîments', the 'gesture of politeness' as I called it above. The famous 'school-boys' letter is dated February 22, some five weeks later, and, so far as we can tell, is unprompted: the Darwin Archives contain no letter from Ogle to Darwin in the intervening period.

[7] Cal. 13621; DAR 173:10. See Figure 2.

[8] Cal. 13622; DAR 261.5:18. See Figure 3. There is some uncertainty whether Darwin would have received, and thus been able to respond to the receipt of, Ogle's book and letter on the same day they were sent. So it is possible that one or the other of the January 17 dates is incorrect. On the other hand, Adam Perkins, Royal Greenwich Observatory Archivist at Cambridge University Library [and now also the Keeper of Manuscripts and University Archives], commented that 'In the contemporary Royal Observatory correspondence of the Astronomer Royal George Airy, I have noted on some occasions just this happening, and I have inferred that postal deliveries could be rapid and frequent at this period. Airy, a "workaholic", on one or two occasions could write, receive an answer and reply to that on the same day, though I think that this only happened when he was writing to people at points on the railway network' (personal communication, October 7, 1997). The matter of the two January 17 dates will be addressed by the Darwin Correspondence Project when the series reaches 1882.

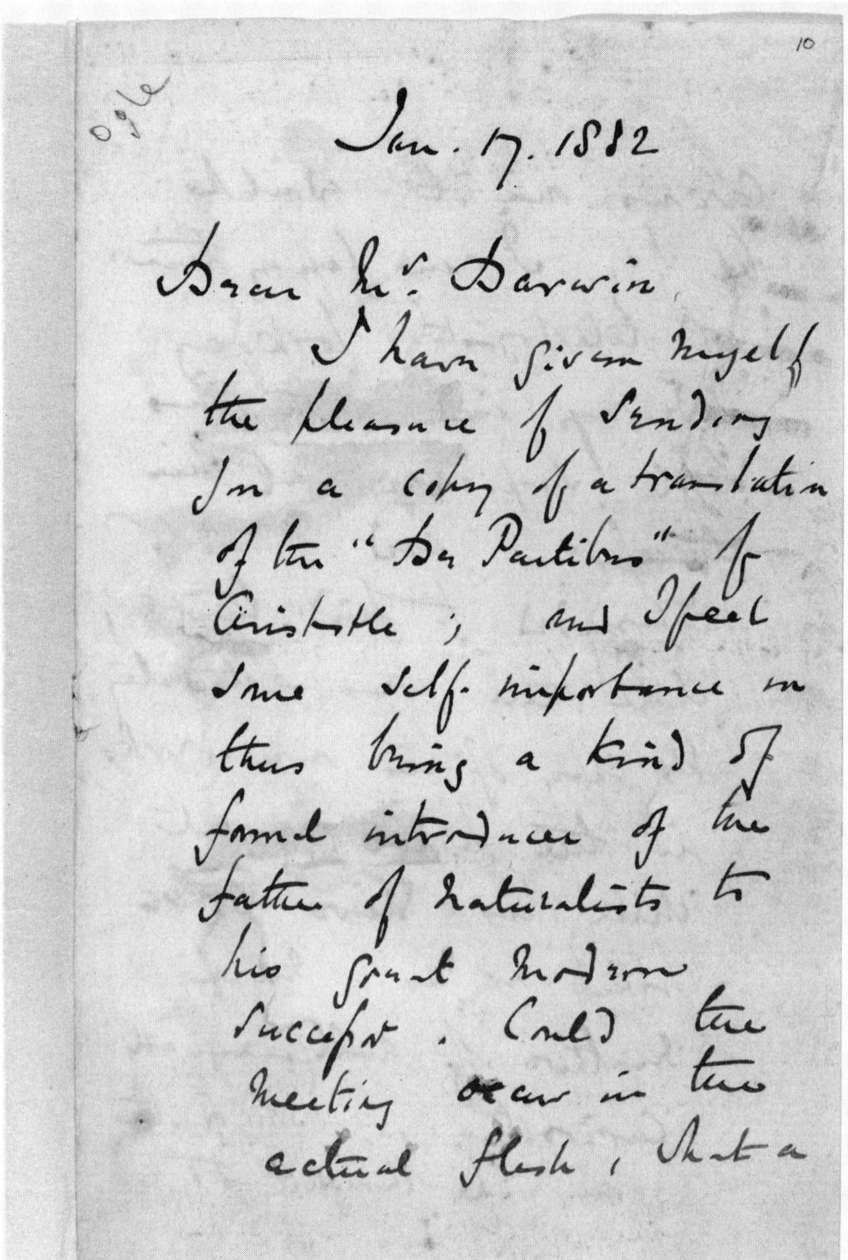

Ogle

10

Jan. 17. 1882

Dear Mr. Darwin,

I have given myself the pleasure of sending you a copy of a translation of the "De Partibus" of Aristotle; and I feel some self-importance in thus being a kind of formal introducer of the father of naturalists to his great modern successor. Could the meeting occur in the actual flesh, that a

Figure 2

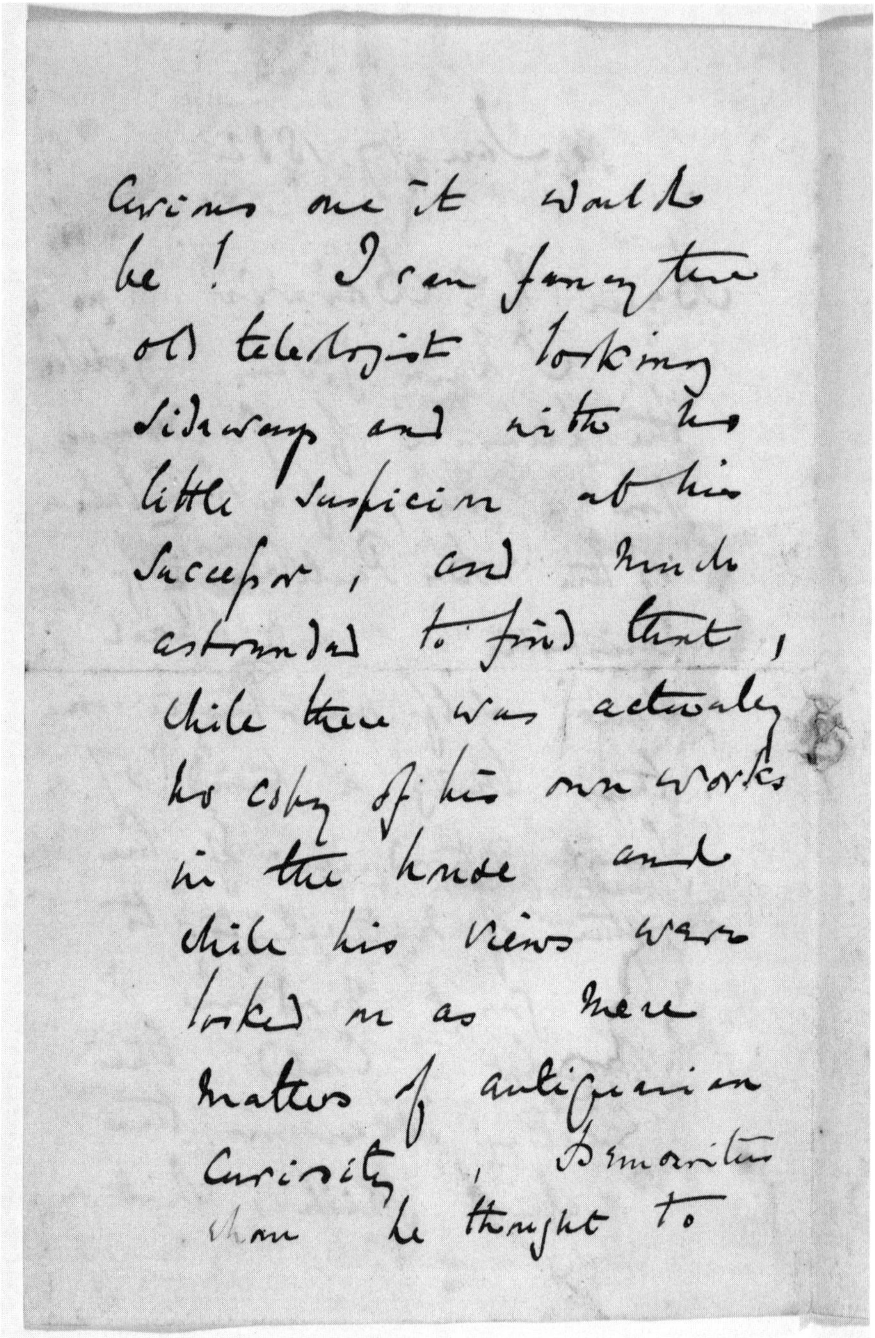
curious one it would be! I can fancy the old teleologist looking sideways and with his little suspicion at his successor, and much astounded to find that, while there was actually no copy of his own works in the house and while his views were looked on as mere matters of antiquarian curiosity, Democritus whom he thought to

Figure 2. Continued

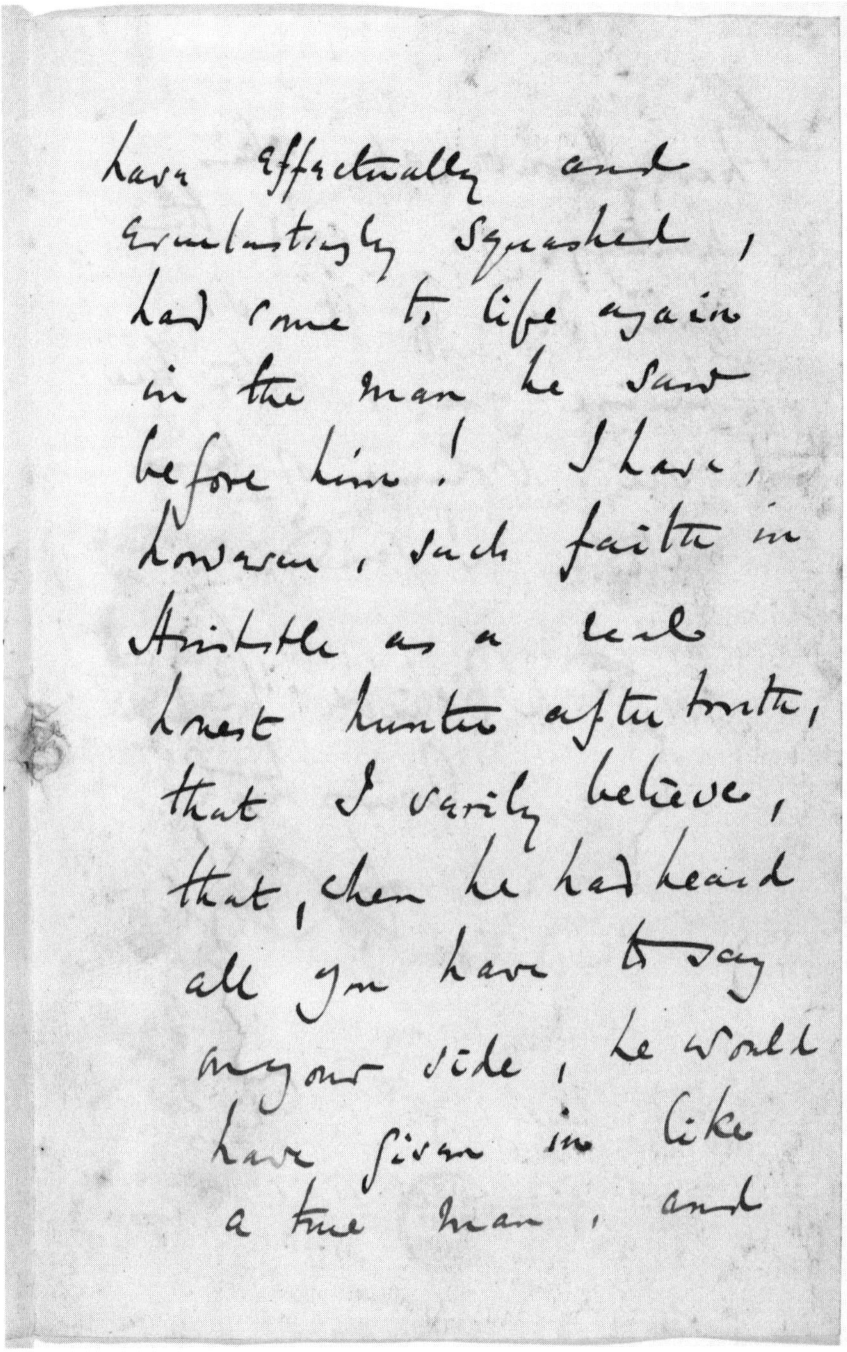

have effectually and everlastingly squashed, had come to life again in the man he saw before him! I have, however, such faith in Aristotle as a real honest hunter after truth, that I verily believe, that, when he had heard all you have to say on your side, he would have given in like a true man, and

Figure 2. Continued

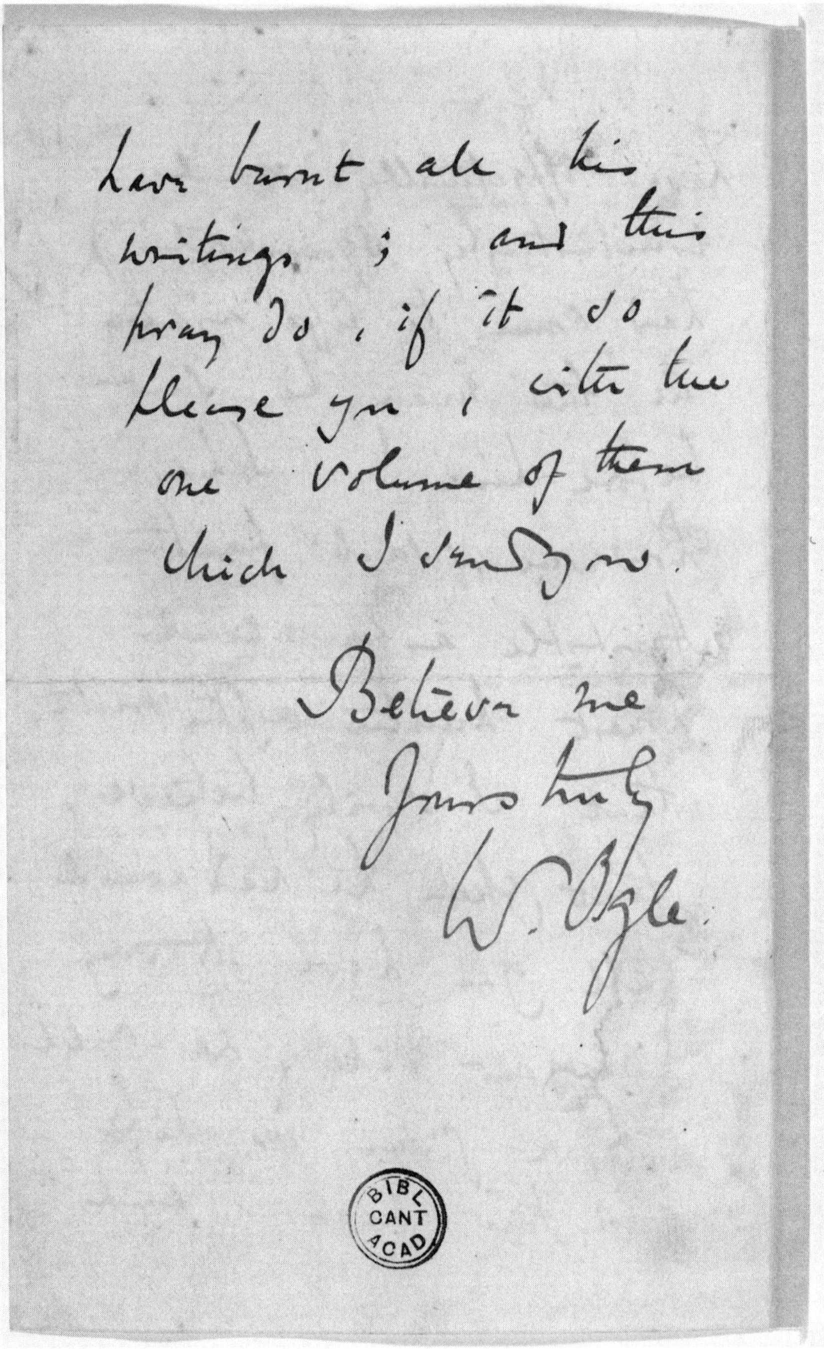
have burnt all his writings; and this pray do, if it so please you, with the one volume of them which I send you.

Believe me
Yours truly
W. Ogle.

Figure 2. Continued

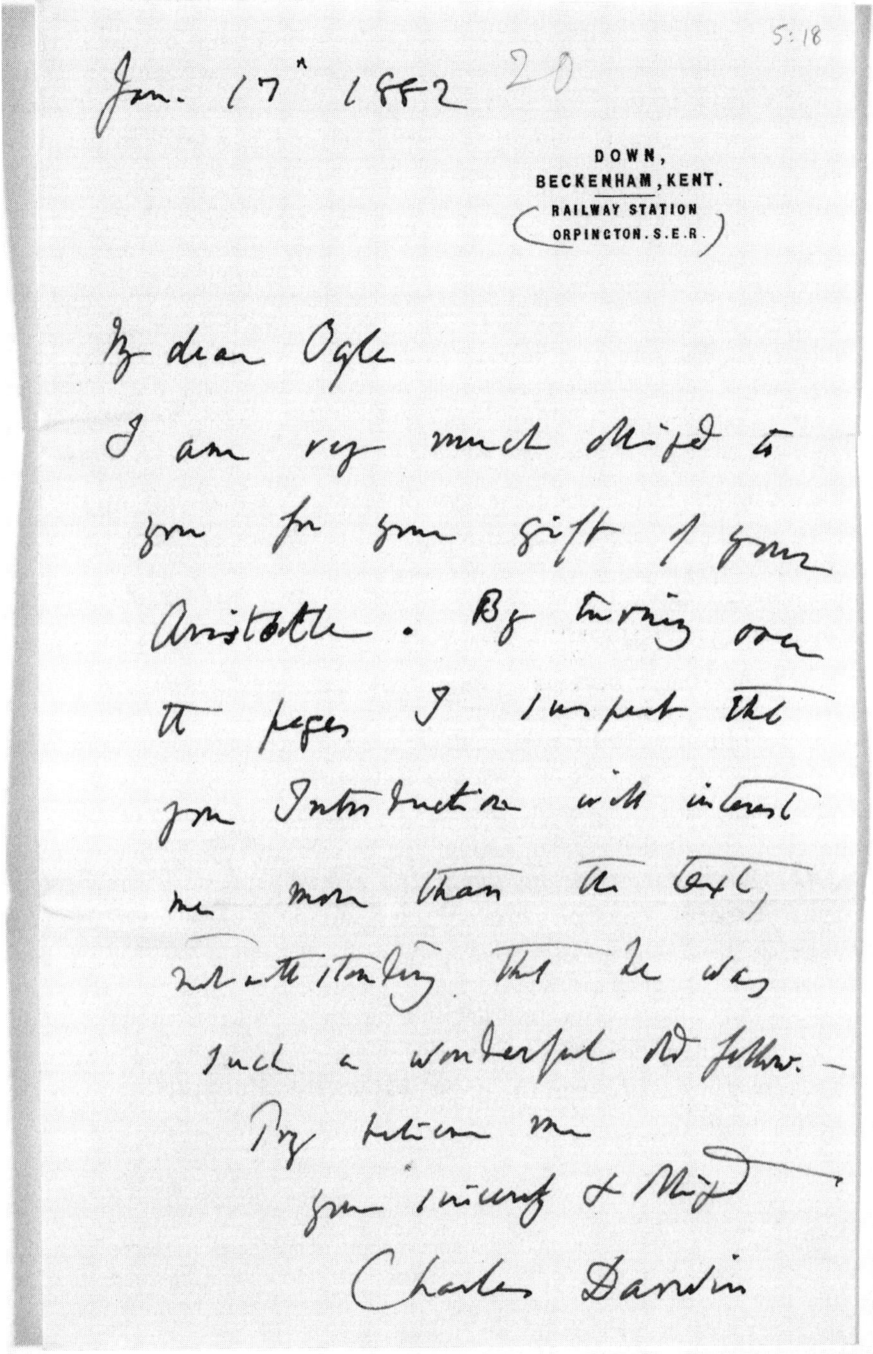

5:18

Jan. 17th 1882 20

DOWN,
BECKENHAM, KENT.
RAILWAY STATION
ORPINGTON. S.E.R.

My dear Ogle

I am very much obliged to you for your gift of your Aristotle. By turning over the pages I suspect that your Introduction will interest me more than the text, notwithstanding that he was such a wonderful old fellow.—

Pray believe me
yours sincerely & obliged
Charles Darwin

Figure 3

So Darwin, having written Ogle a polite response, begins reading the Introduction. He finds it so fascinating that, contrary to his original expectation, he begins reading the translation proper, and about one-quarter of the way through he has been so enjoying himself that it occurs to him to write Ogle a second letter, thanking him for that pleasure. The letter is authentically felt, and not only expresses Darwin's pleasure and gratitude, but also describes a change that has taken place in Darwin's view of Aristotle. I repeat the opening portion:

> My dear Dr. Ogle,
> You must let me thank you for the pleasure which the Introduction to the Aristotle book has given me. I have rarely read anything which has interested me more; though I have not read as yet more than a quarter of the book proper. From quotations which I had seen I had a high notion of Aristotle's merits, but I had not the most remote notion what a wonderful man he was. Linnaeus and Cuvier have been my two gods, though in very different ways, but they were mere school-boys to old Aristotle.

Darwin does not identify why his respect for Aristotle has increased, but it clearly has, and in light of that increased respect, Darwin expresses surprise at the 'wonderful man's' ignorance of certain matters, mentioning the muscles and their role in making movement possible.

Let us ask, then: how accurate is Darwin's letter in reporting his reading and his change of view?

Assuming Darwin read straight through from the beginning, we can certainly say that he had read at least 14 percent of the translation itself, and read (at least some of) it closely. His mention of Aristotle's failure to understand the role of the muscles in facilitating movement provides the evidence for this. For, the term 'muscles' appears neither in the translation nor in Ogle's Introduction, but only in three of Ogle's endnotes, the first two of which are consecutive and are together likely to have been Darwin's source. The notes are to the opening paragraphs of Book II ch. 1, around 646b19, where Aristotle is explaining the final cause of homogeneous parts of animals such as flesh (under which, as Ogle points out in his notes, Aristotle subsumes the muscles), and an easy calculation shows that this passage appears almost 15 percent of the way into the translation itself. Darwin may well have read beyond that point, or he may have overestimated the length of what he had read so far.[9]

[9] The translation appears on pp. 1–140 of the 1882 edition. The notes in question, nn. 9 and 10 on p. 153, are attached to pp. 20 and 21 of the translation. The less likely third note, n. 20 on pp. 196–97, is attached to p. 68 of the translation, nearly 50 percent of the way into 'the book proper'. The copy Ogle sent Darwin is still in the Down House Library; short of forensic-type study, it provides no guidance on this question (or others), since it has no annotations. On the highly methodic character of Darwin's reading habits, see parts (i) and (ii) of the Introduction to *Charles Darwin's Marginalia*, vol. 1 (Di Gregorio 1990), xii–xvii.

What about the change in Darwin's view of Aristotle? In Darwin's words again: 'From quotations which I had seen I had a high notion of Aristotle's merits, but I had not the most remote notion what a wonderful man he was.' Is this an accurate statement of Darwin's prior view, or just rhetorical excess? What *was* Darwin's view of Aristotle prior to reading Ogle's Introduction and the first quarter or so of *PA*?

We have already seen that in the historical sketch added to the third edition of *On the Origin of Species* (1861) Darwin says he is not familiar with the 'writings' of the classical authors, including of course Aristotle. Indeed, in one of his early notebooks, in an entry dated to 1838, listing things 'To be read', Darwin writes, 'Read Aristotle to see whether any my views very ancient?'.[10] Unfortunately, by the time of this edition of the *Origin*, he still had not done so. But of course Darwin never says otherwise. In the sentence to Ogle just quoted, he says that '*From quotations which I had seen* I had a high notion of Aristotle's merits'.[11] But, was this true?

Yes, very much so, as confirmed by an unpublished letter written three years before he received Ogle's book. Dated February 12, 1879, it is in response to a letter (which has not been found) from one J. A. Crawley, a Cambridge BA of about thirty-five years of age,[12] who had apparently asked something about Aristotle's biology as found in the original Greek. Darwin writes back as follows:

Dear Sir,
I am sorry to say that I can give you no information.

I have forgotten the very little Greek which I once knew. Nor have I ever read, to my shame be it spoken, the works of Aristotle. From extracts, which I have seen, I have an unbounded respect for him, as one of the greatest, if not the greatest observers, that ever lived.—

Dear Sir
Yours faithfully,
Ch. Darwin[13]

Here we have a sincere private letter to someone Darwin apparently does not know well, if at all, written on no special occasion, in which the expression of

[10] Barrett et al. 1988, 325 c267.

[11] My emphasis.

[12] Crawley was inadvertantly omitted from the *Calendar*'s Biographical Register; this information was kindly supplied by Jonathan Topham of the Darwin Correspondence Project. A fuller discussion of Crawley and his letter will accompany the publication of the letter when the series reaches 1879.

[13] Cal. 11875; DAR 143:302. See Figure 4.

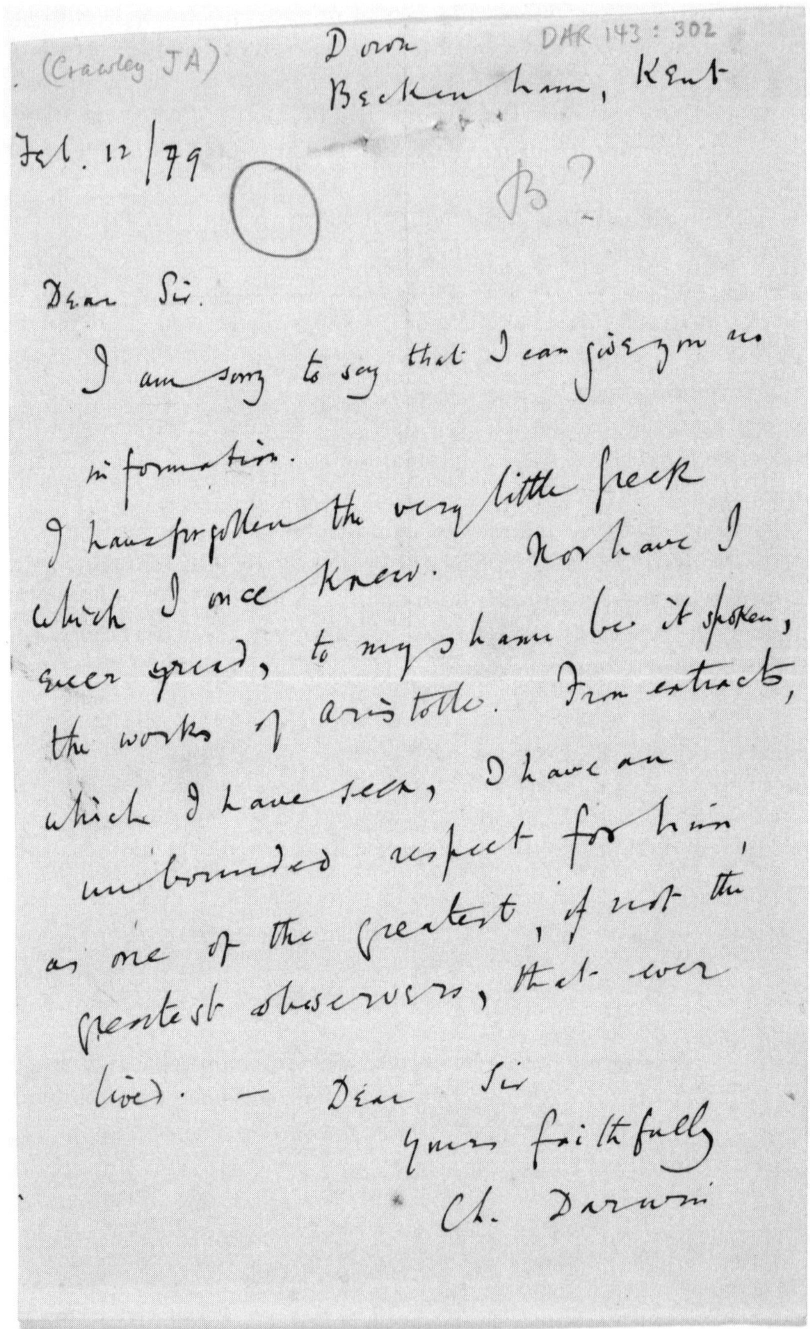

(Crawley JA)

Down
Beckenham, Kent

DAR 143 : 302

Feb. 12/79

B?

Dear Sir.

I am sorry to say that I can give you no information.

I have forgotten the very little Greek which I once knew. Nor have I ever read, to my shame be it spoken, the works of Aristotle. From extracts, which I have seen, I have an unbounded respect for him, as one of the greatest, if not the greatest observers, that ever lived. — Dear Sir

Yours faithfully
Ch. Darwin

Figure 4

praise for Aristotle is structurally and verbally very similar to the corresponding one in the 1882 'school-boys' letter. The thought is very likely to be the same, and so the greater specificity of the 1879 letter should help us to understand the force of the famous 1882 letter:

1882: 'From quotations which I had seen I had a high notion of Aristotle's merits . . . '

1879: 'From extracts, which I have seen, I have an unbounded respect for him, as one of the greatest, if not the greatest observers, that ever lived.'

So, prior to reading Ogle, Darwin had not read any of Aristotle's biology directly, but he had seen enough 'quotations' (or 'extracts') from Aristotle in the works of authors he *had* read, to form the impression that Aristotle was an outstanding observer. I have tried to confirm that Darwin had formed such an impression, by checking the indexes to assorted Darwiniana, including the *Calendar* of letters, and the recent editions of Darwin's published works, notebooks, and marginalia. There are not many references to Aristotle, but those that are there do note other authors' reports of Aristotelian observations, more often to show that some fact that Darwin is reporting was known even to the ancients, occasionally to note some detail that Aristotle did not know.[14] These references confirm what is already clear from the 1879 letter to Crawley, that the 1882 letter to Ogle accurately states Darwin's view of Aristotle prior to reading Ogle.

But 'Linnaeus and Cuvier have been my two gods'. Interestingly, being 'one of the greatest, if not the greatest observers, that ever lived' (1879) was not sufficient in Darwin's eyes to place Aristotle in the pantheon. What new information was provided, then, by Ogle's Introduction, and the first quarter or so of the *PA* together with Ogle's notes, that resulted in Aristotle's apotheosis?

[14] e.g.: 'From the time of Aristotle to the present time solid hoofed swine have occasionally been observed in various parts of the world.'—*Variation of Animals and Plants under Domestication* (Barrett and Freeman 1988, vol. 19, p. 67). (Cf. Aristotle, *HA* II. 499b12-13; *GA* IV. 774b19. Darwin does not here indicate his source.)

'Tame ducks were not known in Aristotle's time, as remarked by Volz, in his *Beitrage zur Kulturgeschichte* (1852), p. 78' (Barrett and Freeman 1988, vol. 19, 256 n. 7).

'It appears that Aristotle was well aware of the change in mental dispositions in old hens' (Barrett and Freeman 1988, vol. 20, 22 n. 56). (Cf. with Darwin's entire discussion here *HA* VIII. 631b5-20; any of the several authors Darwin cites in this note, including C. Waterton and I. Geoffroy Saint Hilaire, might have been his source.)

Darwin read widely in many authors who frequently cite Aristotle, including Cuvier, Owen, and I. Geoffroy Saint Hilaire among others, and Aristotelian references, as the few citations above suggest, are sprinkled throughout the nineteenth-century biological literature. The list of Names and References in Di Gregorio 1990 includes some ten places where Darwin comments on or marks a report of an Aristotelian observation.

Aristotelian information was occasionally also brought to Darwin's attention in correspondence, primarily (but not only) by Ogle; cf., e.g., Cal. 1818 (E. Blyth, 1856; DAR 98: 112-113 & 117-120), 8705 (Ogle, 1873; DAR 173: 7), 10167 (Ogle, 1875; DAR 46.2 (series 3): 63-64).

Before I try to answer this question, let me insist that we are to take *these* words from the 1882 letter very seriously as well. Note first the care with which the view is expressed: 'Linnaeus and Cuvier have been my two gods, *though in very different ways*'. The adverbial phrase is no dithyrambic excess. On the contrary, it is quite careful and subtle.

In what different ways, then? What were Darwin's estimates of Linnaeus and Cuvier? To start, I think we must take 'my two gods' to mean, 'my two heroes', 'the two earlier giants of the field, whose achievements I most admire and have wanted to emulate'. The question, then, is why Darwin might have granted Linnaeus and Cuvier (and no one else) this status. Speculation will be necessary, but let us try to ground it in whatever limited concrete evidence is available.

In the case of Linnaeus, the answer seems reasonably clear. His classification system brought significant order to a vast sea of biological data, providing the basis for virtually all subsequent work, and it is hard to imagine anything else in Linnaeus that could have earned him the exalted status in question.[15] Cuvier is the more difficult case, since he did significant work in several areas, including systematics, descriptive comparative anatomy, and comparative functional explanation. Although Cuvier's work in systematics was of importance to Darwin,[16] I venture to suggest that his work in the latter two areas, especially the third, was the focus of Darwin's admiration.

First of all, to idolize Cuvier and Linnaeus for their respective achievements in systematics is not to idolize them 'in very different ways'. In addition, I have not come across any explicit praise by Darwin of Cuvier for his systematics. On the contrary, there is the following exchange with Huxley in 1857. In response to a September 26 letter from Darwin proposing that classification should be essentially genealogical, Huxley wrote:

> Cuviers definition of the object of Classification seems to me to embody all that is really wanted in Science—it is *to throw the facts of structure into the fewest possible general propositions*—This of course leaves out of view & passes by all questions of pedigree & possible modifications—dealing with existing animals as faits accomplis[17]

[15] Indeed, see n. 19 below for some specific evidence that Darwin did prize aspects of Linnaeus's systematics.

[16] So that Ernst Mayr could write: 'Personally, I think, Darwin admired Cuvier as the destroyer of the linear scala naturae. He made it ever so much easier for Darwin to construct a branching phylogeny. Accepting evolution it was easy for Darwin to break down the branches more finely. [And] Cuvier's five embranchements were indeed branches in the Darwinian phylogeny' (personal communication, August 16, 1997).

[17] Burkhardt et al. 1990, 461; punctuation and emphasis as printed there. The letter is not dated. (Cal. 2144; DAR 205.5 (Letters). Darwin's Sept. 26th letter is Cal. 2143; Imperial College, Huxley 5:54.)

On October 3 Darwin replied:

I knew, of course, of the Cuvierian view of Classification, but I think that most naturalists look for something further, & search for 'the natural system',—'for the plan on which the Creator has worked' &c &c.—It is this further element which I believe to be simply genealogical.[18]

There is no sign of admiration here, and perhaps none to be expected for a system that, as Darwin here views it, fails to reflect clearly the *causes* of the groupings found in nature.[19]

By contrast, on the one occasion I have found where Darwin lauds Cuvier on a matter of theory, it is for his position in his dramatically important debate with E. Geoffroy Saint Hilaire over the explanatory primacy of morphology versus function. In the last section of ch. 6 of *Origin*, entitled 'Summary: The Law of Unity of Type and of the Conditions of Existence embraced by the Theory of Natural Selection', Darwin spoke of 'The expression of conditions of existence, so often insisted on by the illustrious Cuvier' and inferred from the considerations which establish that both Unity of Type and Conditions of Existence are explained by natural selection that:

the law of the Conditions of Existence is the higher law; as it includes, through the inheritance of former variations and adaptations, that of Unity of Type.[20]

This endorsement of Cuvier over Geoffroy mirrors a very interesting set of marginalia in Darwin's copy of Whewell's *History of the Inductive Sciences*, made nearly twenty years earlier. As Edward Manier writes:

[18] Burkhardt et al. 1990, 463 (Cal. 2150; Imperial College, Huxley 5:139). Compare the opening section, 'Classification', of *Origin* ch. 14, esp. the third paragraph, where however Cuvier is not named.

[19] By contrast, in the corresponding passage in *Origin* (see previous note), *Linnaeus* comes in for some implicit praise. In support of the view that classification ought to reflect what is explanatorily prior, Darwin writes: 'Expressions such as that famous one by Linnaeus . . . that the characters do not make the genus, but that the genus gives the characters, seem to imply that some deeper bond is included in our classifications than mere resemblance.' This might well suggest that Darwin's admiration for Linnaeus is based not only on the breadth of his classification system, but on his grasp of a fundamental fact about how classification ought to be done. (I am indebted to James Lennox for bringing the significance of this passage to my attention.) Whether the criticism of Cuvier in the letter to Huxley is entirely fair is of course a separate matter.

[20] He is again 'the illustrious Cuvier' in ch. 14 (in the section, 'Development and Embryology'), even when he misses that barnacles are crustaceans. The only other eminent to earn that appellation in *Origin* is von Humboldt, although Darwin uses the terms of several others in other works. The ironic note occasionally implicit in today's use of 'illustrious' seems completely missing from the nineteenth-century citations in the OED, s.v. illustrious, and each of Darwin's uses seem a straightforward acknowledgment of deserved high standing.

Darwin's annotations indicate his flat rejection of Geoffroy's general position as it was described by Whewell. He marked as 'clearly wrong' the passage where Geoffroy was characterized as holding that

> the structure and functions of animals are to be studied by the guide of their analogy only; our attention is to be turned, not to the fitness of the organization for any end of life or action, but to its resemblance to other organizations by which it is gradually derived from the original type.

Geoffroy was portrayed as making no use of the concept of adaptation and as not taking into account the interaction of organisms with their environment. Darwin circled the reference to Cuvier's account of the 'part which the animal *has to play* in nature,' and drew a line to the marginal comment: 'this qualified is correct. Owing to external contingencies and numbers of other allied species and not owing to mandate of God.' Whewell subsequently claimed 'That the parts of the bodies are made⋆ in order to discharge their respective offices⋆⋆ is a conviction which we cannot believe to be otherwise than an irremovable principle of the philosophy of organization.' Darwin inserted the comments (at ⋆) 'born, altered', and (at ⋆⋆) 'under changing circumstances' . . .[21]

Darwin always prized getting the explanatory priority right, and it was Cuvier who stood in Darwin's mind for the primacy of fitness over structure. That Darwin also prized descriptive precision is evident everywhere in his work,[22] and his citations from Cuvier on matters of detail are frequent enough, so it would not be unreasonable to imagine that Cuvier's descriptive achievements, and those aspects of his work in systematics which Darwin did appreciate,[23] contributed to Darwin's overall admiration for Cuvier; but I would suggest that it was Cuvier's grasp of the fundamental causal priority in biological thought that was foremost in Darwin's mind.[24]

We may hypothesize, then, that Linnaeus had earned his exalted standing in Darwin's eyes for his wide-ranging and in some ways philosophically sensitive system of classification, and Cuvier his primarily for the extensive, descriptively precise and explanatory functional analysis of animal characters. But now Darwin reads Ogle's Introduction, and *PA* Book I and the beginning of II (with Ogle's notes) and what does he find? First of all, I propose, he finds, centuries before

[21] Manier 1978, 53. The Whewell annotations are to Whewell 1837, III. 462, 467: cf. *Charles Darwin's Marginalia*, vol. I (Di Gregorio 1990), p. 868.

[22] Cf. esp. the Introduction to *Charles Darwin's Marginalia*, vol. I (Di Gregorio 1990, xii–xiv).

[23] Above, n. 16.

[24] In these features of his work Cuvier is himself apparently influenced by an extensive and careful study of Aristotle, on whose biological writings he heaped praise. Speaking, for example, of Aristotle's *Historia Animalium*, Cuvier wrote: 'I cannot read [this work] without being ravished with astonishment. Indeed it is impossible to conceive how a single man was able to collect and compare the multitude of particular facts which underlie the numerous general rules and aphorisms contained in this work and of which his predecessors never had any idea' (Cuvier 1841, I. 146–147 [translation by G. H. Lewes 1864, 269]). See also Cuvier 1841, 147ff.

Linnaeus or Cuvier, both that Aristotle had a systematic and plausible scheme of animal classification, *and* that he sought for each part, wherever possible, to identify its function in the life of, and so to explain its presence in the body of, the animals that possess it. For Ogle's Introduction is in large part a discussion of just those two features of Aristotle's work, and the first book of *PA* is largely concerned both with teleological explanation of animal parts, in chs. 1 and 5 (as is II.1), and with classification, in chs. 2-4.[25]

And one of the striking features of Ogle's Introduction, especially to a nineteenth-century reader, is the fresh sense it gives of the breadth—the immensity—of Aristotle's undertaking, and of the way he has brought an incredible mass of data under theoretical control. No surprise then that a careful reader and a great biological mind would write: 'I never realized before reading your book to what an enormous summation of labour we owe even our common knowledge.'[26]

This 'enormous summation of labour' will inevitably have missed some important facts, such as the role of the muscles in facilitating movement, but Darwin is focused not on the mistake as such, but rather on his surprise that someone of such genius would make that mistake; that very surprise is testimony to respect: 'How very curious . . . his ignorance' on such points.[27] So, Darwin takes interest in Ogle's attempt, across several pages of the Introduction, to explain, or explain away, the 'grossest mistakes attributed to him'—the history of Aristotle's text, Ogle suggests, has permitted much interpolation, and Aristotle's limited access to certain animals and his 'habit of hasty generalization' have allowed in some mistakes—and Darwin takes pleasure in Ogle's success in explaining all this 'in so probable a manner'.[28]

All of this, to sum up, suggests an authentic respect on Darwin's part for Aristotle—even perhaps a growing *affection* for Aristotle—and one that is well

[25] Strictly, these chapters are concerned with the method of establishing definitions rather than with any method of establishing a hierarchical classification scheme; but it was standard in Ogle's time to read them the latter way, and Ogle indeed does so in his notes (as he did in his Introduction); Aristotle's text is so abbreviated, and otherwise difficult, that Darwin would certainly have relied heavily on Ogle's Introduction and notes.

[26] In a letter to Ernst Mayr, commenting on the issue raised in this chapter, Michael Ghiselin writes, 'I think that Darwin admired Cuvier, Linnaeus, and Aristotle for their capacity to reduce large bodies of material to intelligible order. Lyell, whom Darwin also admired, had the same ability' (August 6, 1997, cited by Mayr, above n. 16; quoted here with Ghiselin's permission). This is very likely part of the overall picture, as I suggest above in the case of Linnaeus.

[27] Indeed, perhaps it was Darwin's very surprise at the thought that someone of such impressive genius could have missed the function of the muscles that made him realize 'what a great summation of labour we owe even our common knowledge'. Perhaps, too, this very sequence of thought, stimulated by his reading of Ogle's notes on muscles, almost one-quarter of the way into 'the book proper', began the reflections which sparked Darwin to write the famous letter.

[28] Ogle 1882, xi-xix. Byl is confused in thinking that Darwin simply accepts as Aristotle's 'the grossest mistakes attributed to him'; on the contrary, the attribution of most of these is disputed by Ogle (xi-xiii), and this is part of what Darwin finds 'so probable'. So, based on my argument so far, Byl's first and third points (above, p. 350) fail. I comment on his second point below.

thought out. The letter is fresh, eager, excited. Darwin had indeed read little of Aristotle even to this point, and sadly his death only two months later prevented him from reading much further 'to see whether any my views very ancient'; but his limited exposure to Aristotle was not perfunctory, and he made very good use of it indeed. While Darwin's exposure was too limited to provide by itself strong evidence of Aristotle's greatness, those of us who *have* ample evidence of that greatness from our own study of the text can see how acute Darwin's perception was. How unfortunate, not only for Darwin, but also for us, that he never got to read very much of Aristotle—it would have been a wonderful encounter.

For, I think myself, to borrow a term from the mathematicians, that there is an *isomorphism* between Darwin's biological vision and Aristotle's—that there is much in Darwin's biological theorizing that reflects what one gets if one imagines Aristotle having to accept into his system (i) full evidence of the evolution of species and (ii) a more powerful biochemistry than was projectible in his time. This is particularly true, I believe, on the matter of biological teleology—and I want to conclude this chapter by conjecturing—and it is only a conjecture—that Darwin may indeed have seen this while reading Ogle's Introduction and the first quarter or so of *PA*, especially I.1 and the opening of II.1, which he had recently reached when he was inspired to write the famous letter. For, if my suggestion is right that Darwin exalts Aristotle in part for anticipating Cuvier in studying the way function (within environment) determines structure, then Darwin has understood something about Aristotelian teleology that Ogle did not. We recall that in his first letter Ogle spoke of Darwin as a Democritean anti-teleologist to whose work in this area Aristotle, as 'a real honest hunter after truth', would respond by 'burn[ing] all his [own] writings'.

If one reads Ogle's Introduction, one understands how Ogle could think that. For he sets out a fundamental contrast between teleology and mechanism, putting Aristotle largely on the side of teleology, and Democritus (and in effect the Darwinian moderns) on the side of mechanism. The contrast is stark; as Ogle writes, starting with the mechanists: 'One group of philosophers there was, who fancied that they found an adequate cause in the necessary operations of the inherent properties of matter; while another sought a solution in the *intelligent action of a benevolent and foreseeing agent*, whom they called God, *or Nature*, as the case might be.'[29] The latter would ask, having come upon the marvelous character of, for example, the system of blood vessels, 'What but foresight and intelligent purpose can have made these channels throughout the body . . . ?'[30] Ogle's Aristotle combines elements of the two approaches, since he recognizes the limitations material necessity places on teleology, but he is still fundamentally a teleologist, who views 'Nature' as an 'intelligent . . . foreseeing agent'.

[29] Ogle 1882, i; my emphasis.
[30] Ogle 1882, i.

But, contrary to Ogle's sentiment, Aristotle's natures are *not* literally intelligent, not literally planners, and arguably the philosopher's genius lies in plotting a *third* course between these two pictures, and defending it as scientifically legitimate.[31] This third course involves the postulation not of an 'intelligent Nature' but of inherent *natures*—that is, of natural capacities (in Greek, *dunameis*) *directed at form*, capacities for the production of a living organism of a particular type that are irreducible to the capacities of the elements that constitute such an organism.[32] From that perspective the isomorphism with Darwin becomes clear, and there is some evidence that Darwin himself saw it.

For it is Aristotle's view that living processes are irreducibly actualizations of capacities for self-development and self-maintenance, and actualizations of ends on that scale of course bring along whatever intermediate states are necessary (or best[33]) for such ends; the intermediate states thus occur *because* they stand in that relation to their end, and not as the result of material interactions alone. Animal parts, for example, come to be, and are present, *because they are needed for the life of the organism in question, or because it is better that organisms of that type have this part than not.*[34] But as various scholars have pointed out, natural selection makes the very same thing true of animal parts in Darwinian theory. For, when they are due to natural selection, parts are present in an organism, in just the form they are, precisely *because* their presence has made survival possible or has provided for animals that have these parts a survival advantage.[35] The difference is that, while Aristotle takes it as a basic fact of nature that animals come well-adapted, with a capacity to reproduce themselves, Darwin offers a mechanism—viz., natural selection—by which well-adaptedness is established and maintained.[36]

[31] Indeed, as James Lennox has reminded me, Aristotle was aware of, and clearly rejected, a teleological picture of just the sort Ogle describes, viz., that of Plato's *Timaeus*. See Lennox 1985d.

[32] Cf. chapter 1 above. In the next paragraph I drastically summarize key aspects of my account, presented much more fully there, especially in the chapter's 'Postscript' 1986, secs. I and III.

[33] Given that Aristotelian ends are optimally structured: see chapter 8 above, pp. 191-92.

[34] *PA* I.1 640a33-b1, II.2 648a13-16, III.8 670b23-24, *GA* I.4 716a12-21 and ff., etc.

[35] On the tense shift, from 'has made' to 'makes', see Wright 1976, 87-90 and Binswanger 1990, 120-26.

[36] Darwin differs also from Aristotle in (i) insisting on a greater variability in environments for a given population than Aristotle was aware of, and (ii) recognizing, as Aristotle did not, that a slight variation in the structure and functioning of animal parts within populations in a given environment frequently make a difference to survival and reproduction.

An even closer connection between Aristotelian and Darwinian teleology has been argued by David Depew in Depew 1997, n. 3, although I think his account gives insufficient attention to what is for Aristotle the *basic fact* that animal forms *come* well-organized. In Aristotle's account of generation (which Depew rightly praises), the reproductive process operates to replicate these basic forms, so that it is these well-organized forms' *actual* presence in the parents (e.g. the chicken) which explains both their presence as potentialities in the developmental stages of the offspring (e.g.

What is so interesting for us about this isomorphism (or deep underlying similarity) between traditional Aristotelian teleology properly understood and the Darwinian version, is that Darwin himself may have noticed it. James Lennox, in a very interesting paper entitled 'Darwin *was* a Teleologist',[37] cites a letter from Darwin to Asa Gray, in response to 'a brief appreciation of Darwin published in *Nature*, in June of 1874'.[38] Gray had remarked that Darwin did not destroy teleology, as many friends of Darwin, as well as many enemies, had thought, but rather put it on a scientific footing.[39] Darwin writes to Gray in response:

> What you say about Teleology pleases me especially and I do not think anyone else has ever noted that.[40]

While the emergence of a new species is the *result* of very many individual cases of natural selection, and not a goal *aimed at* by the process, selection nonetheless does make it the case that the character selected is present *because* it contributes to the life of the organisms possessing it, which is the core claim of any teleological theory. Aristotle's teleology is a *stronger* teleology than Darwin's, since Aristotle postulates a primitive or basic *directiveness on an end*, an irreducible (yet non-conscious) directiveness built into each ontogenic process, which directiveness cannot be explained by reference to anything simpler (such as a series of selections leading to parent organisms which then reproduce themselves);[41] but against the background of a pure Democritean mechanistic world-view, it is the *similarity* between Aristotle and Darwin, not the *difference*, that one notices.[42]

the egg), and the elemental interactions involved in their subsequent re-actualization, and not the other way round (*PA* I. 640a22-26). But Depew is right to embrace the view that it is the developmental base of Aristotelian teleology that links Aristotle with Darwin here.

[37] Lennox 1993.

[38] Ibid., 409.

[39] 'let us recognize Darwin's great service to Natural Science in bringing back to it Teleology; so that instead of Morphology versus Teleology, we shall have Morphology wedded to Teleology' (from Gray's 'Charles Darwin: A Sketch', in Dupree 1963, 237).

[40] Cal. 9483, June 5, 1874. Cited by Lennox (1993, 409) from Francis Darwin's *The Life and Letters of Charles Darwin*, 3 vols. (F. Darwin 1887, III. 189; Lennox's '308' is a reference to *The Autobiography of Charles Darwin and Selected Letters*, the 1958 Dover reprint of F. Darwin's 1892 abridged edition of *Life and Letters*.) It is not clear, however, that either Gray or Huxley (whose remark similar to Gray's is cited by Francis Darwin in F. Darwin 1887, ii. 201) understood the connection between teleology and selection as Darwin did; indeed it is quite clear from the latter citation that Huxley did not. Gray's statement remains, however, something that Darwin could endorse.

[41] Cf. chapter 3 above, pp. 71, 73, 83, 88.

[42] The dispute between James Lennox and Michael Ghiselin whether Darwin is to be counted a teleologist seems to turn on this issue. (Lennox 1993, Ghiselin 1994, Lennox 1994). Ghiselin appears to conceive teleology as involving essentially this sort of primitive directiveness—either by a conscious agent or by an irreducible nature; he wants to stress that, by contrast, the better-adapted organisms that result from natural selection are *not*, in that sense, *aimed at*. Lennox's view seems to be

And my suspicion is that Darwin noticed it too, as he read Ogle's Introduction and *PA* I.1 and II.1—and I suspect as well that although he disagreed with Ogle's view of his own relationship to Aristotle, he declined to express that disagreement in a letter to Ogle aimed at thanking him so much for what had turned out to be so lovely a gift.

Ogle was very pleased with Darwin's letter, and with the 'mere school-boys' reference. In his next letter to Darwin, nearly two months later (and his last, for Darwin died a few days after that), Ogle wrote: 'Thank you for your kind and eulogistic letter re 'the parts of animals'. It gave me much pleasure. I am glad also to have added a third person to your gods and completed the Trinity . . . '[43]

William Ogle believed that Darwin meant what he said, and that Darwin knew of what he spoke; my argument here has been that *we* should believe these things of Darwin too. He did mean exactly what he said, and even on limited reading knew quite much of what he spoke. The famous 'school-boys' letter is a great tribute, and a theoretically sensitive one, across 2,200 years, from one great biological master to another.[44]

that so long as features are present *because* they render the organism more fit, teleology is present. I think myself that we need to distinguish a stronger and a weaker (but still philosophically and scientifically substantial) sense (or type) of teleology. In the stronger sense, Aristotle is, but Darwin isn't, a teleologist; in the weaker (but still substantial) sense, both Aristotle and Darwin are, while Empedocles and Democritus aren't, teleologists. (Ghiselin seems to hold that the weaker sense has no legitimacy as a conception of *teleology*, but in my view the literature Lennox cites in his paper makes clear that it is a legitimate, and important, conception.)

This issue is complicated by the fact that various opponents of Darwin, such as von Baer, who thought natural selection unable to account for the extensive 'correlation of variations' to be found in nature, and opted for a teleological thesis according to which the results of evolution are indeed *aimed at*, characterized Darwin as having abandoned teleology *altogether*. But Darwin clearly did not think he had, and this has generated much confusion (on which see also Beatty 1990). On von Baer's critique of Darwin, cf. Russell 1982, 238-45 (and ch. 13 generally), and Lenoir 1989, ch. 6. Distinguishing stronger and weaker senses (or types) of teleology would help to sort out both the historical and the philosophical issues here.

[43] Cal. 13767; DAR 173:11.

[44] [2011:] The Darwin letters are quoted from the Darwin Archives with the kind permission of the Syndics of Cambridge University Library. I thank Adam Perkins, now Curator of Scientific Manuscripts, CUL Department of Manuscripts and University Archives, for his advice and assistance in this matter.

[1999]: For research assistance and comments, I would particularly like to thank Frederick Burkhardt and Jonathan Topham of the Darwin Correspondence Project, and Adam Perkins of the Department of Manuscripts at Cambridge University Library, who were all most generous with their time and expertises. Alan C. Bowen and Kelly Smith provided useful comments and/or research assistance as well. James Lennox, David Depew, Ernst Mayr, and Michael Ghiselin, each of whom knows vastly more about Darwin than I do, were very generous in supplying comments to an *amateur*—but perhaps they recognized that I have indeed found Darwin, as I have always found Aristotle (if in a much deeper way), extraordinarily lovable.

Coda

Aristotle as Scientist: A Proper Verdict

(with emphasis on his biological works)

THESIS

Aristotle was not an 'armchair theorist', spinning theories out of his head, who 'held back the course of science for two thousand years'. And though he was a brilliant and careful observer, some of whose findings were not rediscovered until the nineteenth century, his greatness as a scientist does not lie in that. His greatness lies rather in the *systematic* and *explanatory* character of his work—in, broadly speaking, the *methodology* he practiced. It lies, specifically, in:

- the *range* of data he collected, and the *care* with which he collected it;
- the systematic way he *organized* that range of data;
- the way he *explained* the data he collected and organized; and
- the way he *organized* his explanations into a comprehensive body of scientific *understanding*, which was empirically based and revisable as new knowledge was discovered.

Aristotle's biological works comprise over 25 percent of his surviving corpus. I have spent a good part of the last thirty-five years with those treatises. What first brought me to Aristotle's biological works was the question of what can be learned from them about Aristotle's *philosophy*—especially his metaphysics, natural philosophy, and philosophy of science. But a dramatic side benefit of this study has been an awareness of the power of these works as *science*.

That power comes I think from a powerful vision on Aristotle's part of what science should be, plus a brilliant, persistent, probing, analytical, and synthesizing

This chapter was first prepared in 1987 as a talk to the regular and visiting fellows at Clare Hall, Cambridge, where I had become a life member in 1985 (see the preface above, p. ix). I have since given it often, at colleges and universities in the United States, Europe, and Japan and at the Institute for Advanced Study, Princeton, when I was a visiting member in the first half of 2001. The most recent occasion was in late 2010, at Concordia University. Andrea Falcon on that occasion, and David Charles previously, provided me with many valuable comments, which have improved this chapter. Items from the handout which I standardly provided on these occasions are here placed in boxes or as figures, or are quoted within the present chapter itself.

mind which has allowed him to put a major part of that vision into practice. It's the picture of Aristotle's practice as a scientist, as manifested especially in his biological works, that I'd like to talk with you about now. Against a common view that Aristotle was, as it's said, an 'armchair theorist', who—it's also said—'held back the course of science for 2,000 years' (not to mention other faults of late ascribed to him), I want to offer a very different verdict.

THE SURVIVING BIOLOGICAL CORPUS

The 'big three':

1. *History of Animals* (*HA*)
2. *Parts of Animals* (*PA*)
3. *Generation of Animals* (*GA*)

Minor works:

4. *Progression of Animals* (*IA*)
5. *Motion of Animals* (*MA*)

Studies of the basic life-capacities:

6. *On the Soul* (*An.*)
7. *Parva Naturalia* ('short nature studies') (*PN*)
   (a) *Sense and Sensible Objects*
   (b) *Memory and Recollection*
   (c) *Sleep and Waking*
   (d) *Dreams*
   (e) *'Divination' in Sleep*
   (f) *On Length and Shortness of Life*
   (g) *On Youth and Old Age, Life and Death, and Respiration*

[Some manuscripts place 4 and 5 between 7e and 7f.]

The overview I provide will, I hope, be of use to those interested in the question of Aristotle's rightful place in the history of science—but also to those interested in his philosophic thought, for suggestions regarding the relationship of his scientific practice to the theory of science he presents in the *Posterior Analytics* and elsewhere. My own view is that the connection between the biology and the *Posterior Analytics* is very close. I believe that there are close connections as well between the biology and the *Metaphysics*—and between the biology and the theory of soul in the *De Anima* (which literally means 'On the Soul'); indeed the *De Anima* is, among other things, a study of the basic life-capacities. I can only touch on these more philosophical connections here. For those who wish a fuller picture, perhaps I may be permitted to recommend my favorite book on the subject: *Philosophical Issues in Aristotle's Biology* (Gotthelf and Lennox 1987). Our focus today is on Aristotle as scientist. So, let us begin.

There are actually *several* traditional pictures of Aristotle as scientist that I want to reject, so let me begin with those. The first and perhaps most common view is the one I've already mentioned, that of Aristotle as *non-scientist*—as an 'armchair theorist', who spun theories out of his head—or out of his philosophy—or out of the pre-philosophical beliefs and biases of his culture. In all its different forms, this is a picture of a thinker not properly constrained by evidence. Sometimes the complaint is directed towards later thinkers under the sway of Aristotle's authority. The symbolic image here is that of the Renaissance schoolmen, supposedly Aristotelian, who refused to look through Galileo's telescope at the moons of Jupiter, which were supposedly not permitted by Aristotle's astronomical science.[1]

[1] Most often cited as refusing to look through Galileo's telescope is Galileo's Paduan colleague, the Aristotelian Cesare Cremonini, though others are mentioned as well. One source of the Cremonini attribution is a letter to Galileo from Paolo Gualdo. The relevant portion is quoted in (e.g.) Muir 2007, 17-18. Mario Biagioli (Renn 2001, 301 n. 62) points to passages in Galileo's own writings: 'On Cremonini's refusal to confront Galileo's discoveries see Galilei 1890-1909, XI 526, 100 and especially XI 564, 165.' (For a more positive view of Cremonini, as a prominent and talented Aristotelian, see Schmitt 1980 and Schmitt 1983, 11-12. My thanks to Andrea Falcon for these references.)

The 'armchair theorist' charge, in one form or another, goes back at least to Francis Bacon. Bacon is aware of, and to some extent impressed by, Aristotle's empirically based efforts in *HA* (*Novum Organum* [*NO*] I. xcviii), but he thinks Aristotle's results vague and inadequate (and a 'natural history' in the sense I go on to define), and not suitable to facilitating the discovery of causes (*NO* I. xcviii, II.xxvii, *et passim* for Aristotelian science generally). (What is amusing—or sad—is that what Bacon says the preliminary study of animals *should be* is, as we'll see, far closer to what *HA* actually *is*.) Thinking that Aristotle has no methodology for reaching a grasp of actual causes, and only a formal, demonstrative logic, Bacon views Aristotle's working methodology as essentially deductive, from first principles grasped 'somehow'. (*NO* I. lxiii *et passim*.) Thus the 'armchair theorist' picture.

The spread through twentieth-century English-speaking culture of the familiar slogan regarding the retardation of science is probably due to Alfred North Whitehead, Bertrand Russell, and Karl Popper. Whitehead is reported to have said just those words in a Harvard lecture (claiming only '1,000 years'; cf. Kniseley 1964, 263), but that report is just a recollection. At Whitehead 1925, 41, he ties a complaint that sounds much like Bacon's to the retardation of physics in the Middle Ages (my thanks to John Wilkins for this passage). In his widely read *History of Western Philosophy* (1946), speaking of the authority that Aristotle's logic held for so long, Russell writes that '[b]y the time that logical originality revived, a reign of two thousand years had made Aristotle very difficult to dethrone. Throughout modern times, practically every advance in science, in logic, or in philosophy has had to be made in the teeth of opposition from Aristotle's disciples' (Russell 1946, 202; cf. 159-60). See also Russell's nasty crack about 'Mrs. Aristotle' (Russell 1950, 135-36) and for a reply Mayhew 2004, 81-89.

A somewhat differently focused retardation complaint appears at Popper 1998, 5, building on criticisms of Aristotle's essentialism and its consequences for scientific method in Popper 1945 and 1963. Referring to Popper, David Hull, for instance, titles a critique of Aristotle's alleged aprioristic essentialism, 'The Effect of Essentialism on Taxonomy: Two Thousand Years of Stasis' (Hull 1965; for a reply see Lennox 1987b). A more (post-)modern ascription of apriorism to Aristotle may be found, in part, in the work of G. E. R. Lloyd, when he insists that important theses in Aristotelian science have been unreflectively absorbed by Aristotle from the surrounding culture. See esp. Lloyd 1983; two penetrating reviews that take Lloyd to task for that are Lennox 1985c and Parker 1984. (See also my criticism of other aspects of Lloyd's approach to

A second picture, arising more in reaction to Aristotle's biological work than to his physics and astronomy, is of Aristotle as *pre-scientist*. His forte, we are told, was the collection of data about animals, and the classification of animals based on that data—that is, the groupings of animals into a hierarchy of classes: as we now call them (in English), species, genus, family, order, etc. This is a view common to older philosophy of science textbooks, in which Aristotle's practice is offered as an example of the preliminary stages of science. Preliminary—because real science begins with the producing of theories about the data, theories that attempt to explain the data. Aristotle offers no such theories, we are told, and thus leaves us in the anteroom of science (which is at least better than the former picture, which has Aristotle diverting us at the door and leading us into the muck of a priori non-science).[2]

A more positive-spirited verdict, especially among biologists interested in the history of their discipline, represents a third picture of Aristotle as scientist: this is the picture of Aristotle as *brilliantly sensitive observer*. According to this picture, Aristotle is praised for particular observations which were not rediscovered until quite late in modern biology, in some cases not until the nineteenth century. Occasionally the praise extends to Aristotle's descriptive method quite generally, or to his anticipation of particular principles (not just particular observations) of contemporary biology. This picture goes back at least to the great early nineteenth-century French biologist, Georges Cuvier, who once wrote:

(T1) I cannot read this work [Aristotle's *History of Animals*—the largest of his biological treatises] without being ravished with astonishment. Indeed it is impossible to conceive how a single man was able to collect and compare the multitude of particular facts implied in the numerous general rules and aphorisms contained in this work and of which his predecessors never had any idea.[3]

Aristotle's scientific theorizing in chapter 8, and in the penultimate paragraph of the Additional note to chapter 7, above).

My thanks to fellow members of the History of Philosophy of Science list, HOPOS-L (Yahoo groups), for online list discussion both of Bacon's attitudes towards Aristotle and of the sources of the retardation of science slogan.

[2] Few writers are completely unaware that Aristotle wrote explanatory treatises, so this view of Aristotle's merits as pre-theoretical scientist is usually combined with a picture of his theoretical work as 'armchair' rationalism, as per the first picture just mentioned. That's effectively true of Francis Bacon's view (see the previous note), though Bacon is less interested in this pre-theoretical work apart from its role in facilitating the more theoretical enterprise. In his very influential *The Story of Philosophy* (1926), Will Durant writes that '[Aristotle's] specialty is the collection and classification of data; in every field he wields his categories and produces catalogues. But side by side with this bent and talent for observation goes a Platonic addiction to metaphysics; this trips him up in every science, and inveigles him into the wildest presuppositions' (102).

[3] Cuvier 1841, I. 146–47 (tr. Lewes 1864, 269).

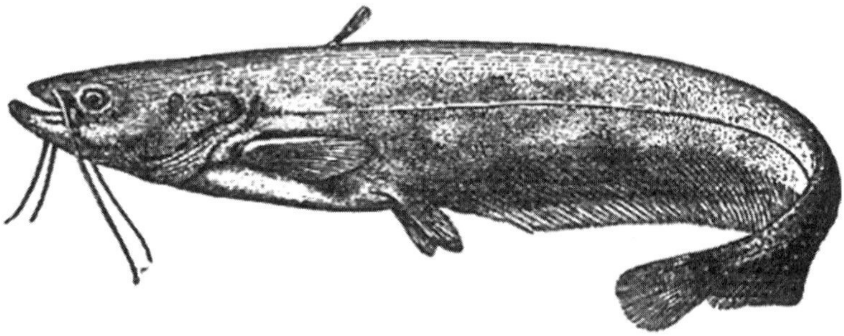

Figure 1. *Parasilurus aristotelis.*

Aristotle *does* deserve praise for such observation, and for his careful selection from reports by others of their observations, so let me take time out to give you some examples. The three most famous have to do with: (i) the catfish, (ii) the octopus, and (iii) the placental dogfish.

(i) The catfish

(T2) Among the river fishes, the male glanis [a species of catfish] takes great care of the young. For the female separates from them after giving birth, but the male stays by and guards the eggs wherever the most part of the spawn has set; he confines his defensive efforts to warding off the other small fishes from stealing the embryos; and this he does for as many as forty or fifty days, until the embryos have grown and can make their escape from the other fishes. He is recognised by the fishermen wherever he happens to be on guard, for in warding off the small fishes he darts at them and makes a sound and a murmur. He stays by the eggs with such devotion that the fishermen, on every occasion when the eggs adhere among deep roots, draw them up to as shallow a place as they can; but all the same he does not desert the embryos but, if it so happens, gets caught quickly by means of the hook because he is snapping at the small fishes that approach; but if he is experienced and used to biting hooks he still does not leave the embryos but crushes the hooks with the hardest of his teeth and destroys them.[4]

In this passage, which is from the treatise Cuvier referred to, the *History of Animals*, Aristotle reports that in the catfish (called *glanis* in Greek), once the eggs are deposited the female swims off and the *male* stands guard over the spawn. This is in fact not true of the large European catfish with which early modern zoologists were familiar, so they dismissed Aristotle's claim. But the famous nineteenth-century Swiss biologist, Louis Agassiz, who had emigrated to America, discovered there a species of catfish in which the male did just that. He remembered his Aristotle and contacted a colleague in Greece, whom he asked to look further. Sometime before 1857 this colleague found in a northern Greek river, often

[4] *HA* VIII. 621a20-b2, tr. Balme.

referred to by Aristotle (the Achiloos River), a smaller species of catfish in which the males protected the young in just the way reported, and sent over to Agassiz a specimen of that species. Agassiz was entitled to give the fish its official Latin name, and he chose *Parasilurus aristotelis* (see Figure 1). For that reason, it is often called 'Aristotle's catfish'—which is perhaps why the catfish is smiling![5]

(ii) The octopus

Next, in two passages in the *History of Animals*, Aristotle speaks of the reproductive habits of the octopus family. In one place, in his own voice, he reports that:

(T3) The octopus uses its tentacles both as feet and as hands . . . The last of them, which is very sharp and is the only one which is whitish in color and bifurcated at the tip . . . This one it uses in the act of copulation.[6]

Later, he mentions, more cautiously, a view of others, probably fishermen, about this tentacle:

(T4) Some say that the male octopus has a sort of penis (*aidoiōdes ti*) on one of his tentacles . . . and that the whole of it is attached up as far as the middle of the tentacle, which it thus admits into the 'nostril' [elsewhere called the 'funnel'] of the female.[7]

The subsequent history is what one historian calls 'a comedy of errors'.[8] First, a few facts. (Consult Figure 2 on the following page.) One tentacle *is* in fact specialized

[5] See Thompson 1947, 43-50, from which I have taken our illustration (courtesy of University of St. Andrew's Library, which holds the D'Arcy Thompson papers). The classic discussion, from which Thompson took the illustration, is Gill 1906. The drawing was originally made for Gill 1905 on which Gill 1906 is largely based. Agassiz's own report of his find appeared in Agassiz 1856. Valuable resources for the identification of animals Aristotle mentions, and sometimes for the accuracy of what he reports, include, in addition to Thompson's *Glossary of Greek Fishes* (Thompson 1947): Thompson's *Glossary of Greek Birds* (Thompson 1936) with W. G. Arnott's *Birds in the Ancient World From A to Z* (Arnott 2007), and I. C. Beavis's *Insects and Other Invertebrates in Classical Antiquity* (Beavis 1988). (Beavis 1988 is a superior work to Davies and Kathirithamby 1986, though the latter is worth consulting on occasion.)

[6] *HA* IV. 524a2-9, tr. Peck.

[7] *HA* V. 541b8-12, tr. Peck. At *GA* I. 720b32-36, Aristotle says in his own voice that the male's insertion of a tentacle in the female is for the sake of *linking* or *uniting* (*sumplokēs charin estin*) the two, but denies the claim of 'the fishermen' who say that they copulate (i.e., that the male transmits *sperma*) that way. If, as Balme thinks, *HA* is later than *GA* (below, p. 397), then this passage, from *HA* V, may be a restatement of the *GA* information, and the prior passage may represent a change of mind on Aristotle's part. However, in his forthcoming commentary on the two *HA* passages (Balme forthcoming), Balme offers a reading of the three passages that renders them all consistent with each other, translating *aidoiōdes ti* not as 'a sort of penis' but as 'something penis-like'. I leave it to readers of Balme's forthcoming commentary to determine if Balme is right about this, and, if they think he is, to adjust my historical narrative accordingly.

[8] Cole 1975, 32: The full story, which I summarize here, is told at 28-32. See also Thompson 1947, 204-209 (s.v. *ΠΟΛΥΠΟΥΣ*) (from which Figure 2 below is taken, courtesy of the University of St. Andrews Library), and 154-58 (s.v. *ΜΑΛΑΚΙΑ*).

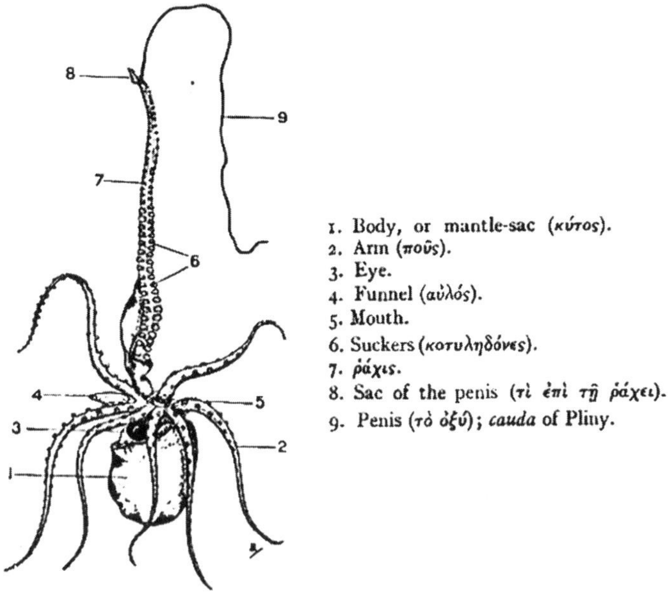

An Octopus, with hectocotylized arm.

Figure 2

for sperm cell transfer in all octopods, and is indeed inserted into the mantle of the female. In several genera, in which the male is much smaller than the female, the arm, which as pictured becomes quite thin in the outer half, actually breaks off and is left there. In 1827 a female was discovered with a small thin body attached to it. The zoologist thought it was a parasite. Ten years later, after looking at five such specimens, Cuvier, whom we have already met, agreed, held it was a new genus of parasite, and even gave it a Latin name. (The name he chose, by the way, was based on its looking like a small version of an octopus' arm!) Fifteen years late a Swiss zoologist decided it wasn't a parasite, and was connected with reproduction, but that it was the whole male of the species, busily engaged in intercourse! A year or so later, a German zoologist, Henri Müller, got it right,[9] and within a year or so after his studies, a colleague showed him Aristotle's report—the earlier notice of which might have saved Müller some time and his predecessors a good deal of embarrassment.

[9] H. Müller 1852.

Figure 3. Dogfish with placenta-like structure.

(iii) The placental dogfish

The third case is the placental dogfish, the so-called 'smooth dogfish' (*Musteles laevis*), a species of shark whose thin eggshell is soon absorbed inside the womb leaving the embryo to develop live inside, attached by a sort of umbilical cord to a placenta-like structure.

(T5) The so-called smooth dogfishes have their eggs between [the parts of] the uterus ... [T]he young are produced with the umbilical cord attached to the uterus, so that ... the embryo's condition appears to be similar to what is found in four-footed animals ... Each embryo has a chorion and membranes of its own round it, just as in the four-footed.[10]

A seventeenth-century zoologist made the same identification but his work was not generally known, and the information was not circulated until another zoologist named Müller, this one Johannes Müller, described the whole structure in

[10] *HA* VI. 565b1-12.

detail in an 1842 monograph, in which he compared favorably Aristotle's report with his own find.[11]

There are many other examples. In the *History of Animals* Aristotle describes the stages in the development of the chick embryo. Some forty years ago, Jane Oppenheimer, then a well-known American embryologist and historian of embryology at Bryn Mawr College, prepared, for an Aristotle conference, an exhibit of eggs, each opened at the day of development Aristotle described, and showed us how astonishingly accurate Aristotle's reports were.[12] And so on. (There are some serious misreports as well—Aristotle sometimes treats as reliable reports that weren't, or is occasionally misled by some point of theory[13]—but biologists who have examined his work as a whole remain astonished by his successes.)

Now, the range, accuracy, and subtlety of data Aristotle collected *is* impressive, and the quality of his method of collection *will* form part of my verdict: Aristotle's work was so important here that much modern zoology from its beginnings in the sixteenth century, and even well into the eighteenth, took off explicitly from Aristotle. But it's wrong, in assessing stature as scientist, to focus on isolated observations, or even just on the quality of descriptive studies—for a collection of observations by itself *is* only science at its beginnings. Aristotle's greatness lies in his *organization* of his data—not just in its range and accuracy—and in his use of his data so organized to *understand* the world of animals, to *explain* why animals are as they are.

For this is always Aristotle's aim: to understand fully *what* animals are (i.e., what features they have), and *why* they are as they are. *How* he thinks we are to do that, we'll discuss, but let me indicate in advance why I reject the traditional pictures.

First, the 'armchair theorist' view. Our friend *Parasilurus aristotelis*, and his octopus and dogfish companions, suggest the implausibility of this picture. In fact, as I'll illustrate shortly, Aristotle's theories in biology are consistently and essentially based on observations—his own or carefully sifted reports of observations by others. (This is true not only in biology but also in the sciences for which Aristotle is less praised, such as astronomy and meteorology and physics generally, although there won't be time for us to get into that.[14]) The 'non-science' charge, as

[11] J. Müller 1842; cf. Thompson 1947, 41-42 (s.v. ὁ γαλεὸς ὁ λεῖος, the Smooth Dogfish), from which Figure 3 is taken (courtesy of the University of St. Andrews Library); Cole 1975, 33-36.

[12] At my invitation, Oppenheimer gave a talk on this same subject to the 1983 Williamstown conference on Philosophical Issues in Aristotle's Biology; she compared Aristotle's reports with photographs she showed us of the developing embryo. I have had the impression that she published an article on the subject containing such photographs, but I cannot find any trace of such a publication.

[13] See Lloyd 1987a for some examples (although Lloyd sometimes overstates his case: see, e.g., above pp. 230 n. 23 and 373 n. 1).

[14] Cf., e.g., Lones 1912, Freeland 1990, and Wilson 2009.

you'll see, is simply absurd, and comes from those who simply have not read Aristotle's biological works.

But so, too, is the 'pre-science' charge—namely, that Aristotle was only collecting data about animals, perhaps as a vehicle for a system of classification—and wrong on two counts. First of all, much of Aristotle's biological writing is clearly *explanatory* in aim: the treatise called the *Parts of Animals* is full of functional explanation of animal organs and tissues; his great work called the *Generation of Animals* is full of explanation, of why animals reproduce to type (deer coming from deer, Aristotle's catfish from Aristotle's catfish), and of why many offspring resemble their particular parents, and why others don't but resemble only a grandparent—or resemble no one in the family at all, and so on. It's only in the *History of Animals* (*HA*), which we have already met, that Aristotle presents *how* things are *without* explanation as to *why* they are that way—and even here the 'pre-science' picture is wrong, in ways worth noting, for these ways point us towards important aspects of Aristotle's methodology and aims in his scientific enterprise.

To start, what we find in *HA* is not a 'natural history' as we have known those in the nineteenth and twentieth centuries: they are descriptive surveys, reporting selected facts of importance or interest about animals, species by species, the species themselves grouped by the larger animal kinds to which they belong. But the *HA* is not organized by animal kind; it's organized rather by animal *feature*. The parts are discussed first, then the activities and broad ways of life, then the dispositions (more about this in a moment). And Aristotle's reports of data are highly selective: he clearly knows more about at least some of the animals than he is saying, so he must have a purpose in mentioning only those features he does mention. What that purpose is we'll have to see; but it's not mere description for description's sake—it's not to produce an animal encyclopedia.

Second, his aim is not 'classificatory' in the sense usually meant, i.e., not *taxonomical*, with levels of groupings, each animal placed under a sequence of ever-wider animals: recent work has shown that Aristotle's works do not contain, or presuppose, such a systematic hierarchical classification of animals, nor do they immediately aim at one. The taxonomical schemes imputed to Aristotle by writer after writer are simply not there. He does speak several times of the 'very large kinds' (in Greek, *megista genē*) of animals—birds; fishes; cetaceans (the water-mammals); 'the four-footed live-bearing animals' (our mammals less man); 'the four-footed egg-laying animals' (roughly our reptiles and amphibians less the snakes); then, the 'soft-bodied' bloodless animals (octopus, squid, etc.); the 'soft-shelled' bloodless animals (lobster, crab, shrimp, etc.); the 'hard-shelled' bloodless animals (oysters, clams, snails, etc.); and the 'insects' (including spiders). And he uses these groupings in his study of animals throughout *HA*. But he sometimes groups animals differently—and in any case does not intend these groupings to be *exhaustive*: man and snakes, for instance, don't fall under any of the 'very large kinds'. Furthermore, Aristotle has no vocabulary for fixed classificatory levels, as we have 'species, genus, family, order', and so forth. His only terms are *genos* and *eidos*, from which (via Latin) the English 'genus' and 'species' come, but he uses these terms at all levels of generality: *genos* is his basic term for an animal group, and can be used at any level, from the *genos* of animals to a certain *genos* of sheep (we might say 'variety'); the best English translation is not 'genus' but *kind*. The term suggests a kinship, a commonality of some sort or other, but it might be a quite loose commonality, and he will even sometimes use it for a group of animals that share a single feature, such as the *genos* of animals with bladders; there it is not what philosophers call a 'natural kind' (a group of animals that possess a common nature) but just a 'class'. The term *eidos*, when it designates a group, virtually always refers to it as a subdivision of some *genos*, and since *genos* can be used at any level of generality, so can *eidos* (below the highest level). But what is most striking is that these intermediate levels are for the most part *not* filled in: Aristotle is not particularly concerned to identify intermediate kinds, and in fact several times comments on the difficulty or even impossibility of doing so.[15]

In sum, then, whatever is going on in the *History of Animals*, it is not descriptive natural history (though it is in part descriptive), and it's not hierarchical classification (though Aristotle does believe that every animal is 'of a specific kind' and does recognize certain larger kinds).[16] The 'pre-science' view is thus wrong, even for *HA*.

What, then, *is* his aim in *HA*, and how does that aim fit into his overall conception of the proper study of animals? Well, what does he *say* about this, in *HA* itself? Aristotle begins the treatise with a survey of the ways in which animal parts can differ or be the same, biologically speaking: for example, 'exactly the same', as human nose to human nose; 'the same in kind but varying in degree or 'the more and the less', as human nose to horse's nose; 'the same only by analogy', when the structure and materials are different but the function is the same, as bird's feathers to fish's scales. So he's interested in how animals differ or are similar. Then he notes that

(T6) Animals differ with respect to their modes of life and their actions and their characters as well as their parts . . .[17]

—that is, the differences fall into one or another of these four categories—and then he goes on to illustrate the sorts of difference, i.e., the sorts of feature, one can find in these various categories. Then comes this passage:

[15] *HA* I.6 490b16, 31-491a6. For other examples, see Balme 1987b, 84-85. See also the next note.

[16] For fuller discussion of the question of whether, and to what extent, *HA* is concerned to establish a system of animal kinds, including 'very large kinds' (the *megista genē* Aristotle speaks of in *HA* I.6 490b7ff.), including a more refined statement of my current view in relation to some recent alternative views, see above, chapters 12 and 13.

[17] *HA* I.1 487a11-12.

(T7) We have stated these things in this way now—in outline form—to provide a taste of the range and sorts of things we must study; later we will have to speak in greater detail, in order that we might grasp the differences and in general the features that belong in every case. After this we must try to discover the causes. For that is the natural way to pursue such an inquiry, once one has completed an investigation (*historia*) concerning each of these: for it becomes apparent from this investigation both about which things the demonstration must be, and from which things it must proceed.[18]

That is, from this *historia* we are about to embark on we will be in a better position to tell which features are causally primary—are those 'from which' the others (the 'about which') are to be explained. Why we will be in a better position to identify causes after we've done the *historia* we will see shortly. '*Historia*', by the way—accented on the next-to-last syllable—is the Greek word from which come the Latin 'historia' and the English 'history', so Aristotle is referring here to the material that is to come in the treatise, the title of which, in English is literally 'The Histories Pertaining to the Animals' (*Ta Peri Ta Zōia Historiai*).

The aim of the treatise is given by the words 'in order that we might grasp the *differences* and in general the features'—the features which animals have and how these features differ from each other. 'After this we must try to grasp the causes'—*why* these features are present in those animals that have them. Why this feature in this animal, that feature in that one? For example, why do camels, cows, deer, etc. have four stomachs (or: a stomach with four compartments), but dogs, pigs, elephants, and we humans, only one? And why do some animals give birth to live young, while others lay eggs? And so on. The *HA* is a study of differences which is preliminary to a study of causes.

Now, for reasons I will indicate later, Aristotle believed that one could not understand what animals are and why, merely by studying the stuff they are made of—the mixture of elements that constitute them. We have to understand them each as a whole—as a living, functioning whole with a definite *mode* of life, and have to see each part and element in relation to the whole: we have to see what the part *does* and what contribution that activity makes to the life of the whole. So the causes, and explanations, which the study of animals must reach are largely functional or *teleological* (meaning: made with reference to *ends* or *goals*). This, incidentally, is what Charles Darwin must have seen when he read the first part of Aristotle's *Parts of Animals* in 1882, three months before he died, and what must have been the basis for his famous praise of Aristotle:

(T8) From quotations which I had seen, I had a high notion of Aristotle's merits, but I had not the most remote notion what a wonderful man he was. Linnaeus and Cuvier have

[18] *HA* I.6 491a7–14.

been my two gods, though in very different ways, but they were mere school-boys to old Aristotle.[19]

Of the three major biological treatises which have survived—*Parts of Animals, Generation of Animals, History of Animals*—the first two are, as I've already noted, explanatory, and contain much functional, or teleological, explanation. The *History of Animals* does not. It is a study of differences—of the features of animals—which will somehow make the discovery of causes possible. The key to how a study of differences might lead to the discovery of causes lies in the *organization* of the data in *HA*. For, the full Aristotelian scientific inquiry must be thought of as having three stages: the *collection* of data, the *organization* of data, and the *explanation* of data. (The explanations themselves will have to be organized into a unified body of understanding proceeding from first principles, but for technical reasons I include that under 'explanation'.) Collection, Organization, Explanation. To get an overall picture of Aristotle's scientific practice we need to take at least a brief look at each of these stages. First, 'Collection'.

(1) The collection of the data

There is no surviving treatise strictly at this stage. This is the notebook stage, where Aristotle records observations and reports, evaluates them, deciding which to accept and which to reject; looks for shared features across different kinds; etc. *HA* has been thought to be such, as we've seen, but it is really at the organization stage, as I will illustrate in a moment. *HA* does, however, contain hints as to what might have been contained in such notebooks. For one thing, on some matters it refers for further details—and for *diagrams*—to a long work called *Dissections* (*Anatomai*), which has not survived. We get a clue as to what this work might have looked like from a reference in *HA* itself to a diagram the earliest manuscripts

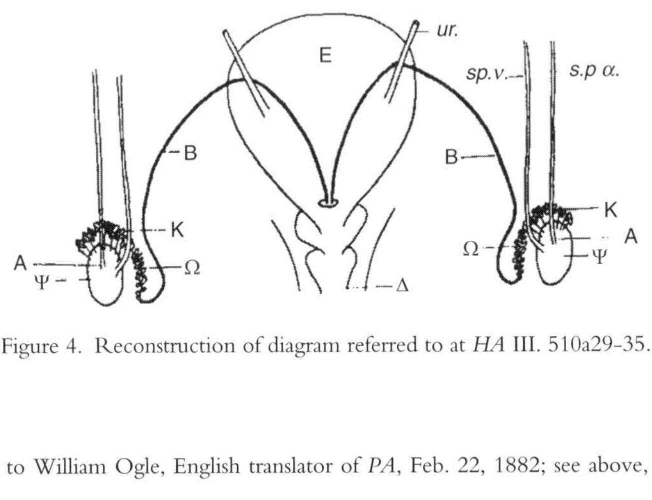

Figure 4. Reconstruction of diagram referred to at *HA* III. 510a29-35.

[19] Letter to William Ogle, English translator of *PA*, Feb. 22, 1882; see above, chapter 15, Figure 1.

must have contained, although none of the surviving ones do. Several editors have reconstructed it; Fig. 4 is one such reconstruction and here is the associated text:[20]

(T9) All this may be studied by the light of the accompanying diagram. The starting-point of the passages that extend from the aorta is marked **A**; the heads of the testicles and the passages descending to them are marked **KK**; the passages extending from these along the testicles are marked **ΩΩ**; the passages turning back, in which the white fluid is, are marked **BB**; the penis **Δ**; the bladder **E**; and the testicles that which is marked **ΨΨ**.

The evidence of *HA* alone suggests that Aristotle dissected over fifty different kinds of animals, and there are several passages, which we don't have time to look at, where he criticizes methods of dissection used by his predecessors, and very sensibly suggests more reliable methods.[21]

Other hints from *HA* as to the Collection stage include some remarks showing how Aristotle dealt with reports he received from others:

(T10) No animal of these kinds has double rows of teeth [top and bottom jaw in front]. There is, however, one such, *if we are to believe Ktesias.*[22]

But *are* we to believe Ktesias? Elsewhere, Aristotle questions Ktesias' accuracy and is unwilling to accept some of his reports; here he seems to accept the report provisionally, but he clearly expresses great caution.[23]

Or consider this passage:

(T11) [In man . . .] common to the upper and lower trunk are the ribs, eight on each side—for concerning the so-called 'seven-ribbed Ligurians' we have heard pretty much nothing worthy of belief.[24]

[20] *HA* III. 510a29-35. Translation (modified) and diagram from Thompson 1910 (with the kind permission of Oxford University Press). For references to the lost *Dissections*, see Gigon 1987, 492-502.

[21] On the number of species *HA* gives evidence Aristotle dissected, see Lones 1912, 102-106. On methods of dissection, good and bad, see *HA* I. 496b4-6, III. 511b10-24, and 513a12-15; and cf. I. 496a9-11. More generally on Aristotle's dissections, see Lloyd 1975 and Cosans 1998. It has been suggested that Aristotle's own manuscripts may have contained illustrations of some of the animals discussed in the text. Manuscripts written in the Byzantine period clearly did, but the argument for an early origin of the illustrations is weak. See Zoltan Kádár's fascinating *Survivals of Greek Zoological Illuminations in Byzantine Manuscripts* (Kádár 1978).

[22] The full passage (*HA* II. 501a24-b1) reads: 'No animal of these kinds has double rows of teeth. There is, however, one such, if we are to believe Ktesias. He asserts that the Indian beast which bears the name of martichoras has a triple row of teeth in both jaws; that it is as big as a lion and just as hairy, with similar feet; that its face and ears resemble a man's, that its eyes are blue and its general color vermilion; that its tail is like that of the land-scorpion; that it has a sting in its tail, and can shoot off like arrows the spines attached to it; that its voice is like a mixture of a shepherd's pipe and a trumpet; that it can run as fast as a deer; and that it is fierce and man-eating.'

[23] *HA* III. 523a26; VII. 606a8.

[24] *HA* I. 493b14-16.

This was said about them somewhere, but by an unreliable source which couldn't be confirmed, showing again Aristotle's caution in accepting reports.

The question of Aristotle's criteria for accepting or rejecting reports needs much more study than it has so far received. There's one treatise in the corpus called *On Marvelous Things Heard*, which is a long list of reports of odd phenomena in various scientific areas. It is usually thought not to be an authentic treatise of Aristotle's, but the pioneer in the recent study of Aristotle's biology, David Balme, who died some twenty years ago (1989), once expressed wonder as to whether it might well be authentic, or derive from a notebook that was authentic, and might be a record of reports which had come in that were awaiting verification or evaluation. Whatever the case about this particular treatise, the enterprise of gathering, and then carefully sifting, reports was a significant part of the Collection stage.

From all this, incidentally, you can see more fully the simplemindedness (to put it generously) of the 'armchair theorist' picture. Even if Aristotle did bring preconceptions to the examination of animals, he was nonetheless immersed in the data, and it is absurd to say otherwise. For a full vindication of Aristotle here, we would need to discuss the evidence offered on the other side case by case—e.g., his claim that women have fewer teeth than men; the lesser role of the female in determining the nature of the offspring in reproduction; etc. That is for another time.[25] But Aristotle's commitment to deriving his theories from observation is clear enough to put the burden of proof fully on the prosecution. As an example of this commitment, consider the following passage, from one of his explanatory works, where we catch him reflecting on the limited data available, and look at what he says. It is in the *Generation of Animals*, and he has just completed his discussion of the reproductive process in bees:

(T12) Such appears to be the truth about the reproduction of bees, judging from argument (*logos*) and what are believed to be the facts (*ta sumbainonta*) about them. The facts, however, have not been sufficiently grasped. If they are grasped, then credence must be given to perceptions more than to arguments, and to arguments if what they affirm agrees with the appearances.[26]

In another explanatory study, a little essay called *On Respiration*, speaking of the inadequate theories of his predecessors as to the role of breathing—and thus the function of lungs and gills—in animals, Aristotle remarks:

(T13) . . . the main reason why they did not give a good account of these things [= various facts about the breathing process] is their lack of experience of the internal parts, and their not grasping that a nature makes everything for the sake of something; for if they had inquired *for the sake of what* respiration is present in animals and sought this in regard to gills

[25] On many such issues, see now Mayhew 2004.

[26] *GA* III.10 760b27-33, tr. Irwin (1988).

and lungs, they would have found the cause [of the presence and functioning, described earlier, of gills and lungs] more easily.[27]

This is a very interesting passage, methodologically. Part of his predecessors' problem, Aristotle says, is that they didn't *look* closely enough—they didn't study the internal parts and their connection to each other with enough care. Probably they didn't *dissect*. In another place, Aristotle speaks of a peculiar theory regarding the function of the windpipe that holds that, just as the esophagus is the passage for solid food, so the windpipe is the passage for liquids. Everyone knew that the windpipe was connected to the lungs, and against that background Aristotle ridicules this liquid-food theory by noting that anyone who dissected would notice that the lungs are not connected to the stomach.[28] That's one failure, then: they didn't look closely enough. But there's a second factor Aristotle has mentioned. In addition to their inexperience, they didn't know what to look *for*. They didn't grasp the fact that every organ is for the sake of something, that is, has a *function*, is there because it contributes in some way to the *life* of the organism as a whole. In particular, they didn't ask what breathing is *for*—what the lungs and gills are *for*—what contribution each makes to the life of the organisms which possess it.

This twofold requirement is generally true, Aristotle held—one needs to know what an organ looks like (what its actual features are) and what it is connected to—but *also* one needs to know what it *does*, what contribution it makes to the life of the organism as a whole. And our view of what sorts of functions there are for a part to fulfill might even lead us to look for particular connections and non-connections, as in the windpipe example I mentioned.

That suggests how Aristotle's collection of data might be guided—as I think it to some extent is—by a broad empirically based conception of the animal and its nature. In that way, for instance, Aristotle's more general philosophical conclusions guide his biological work.[29] But that is not to say that he brings to that work specific views as to what, say, the lung's function is, or the female's role in generation is. He may bring such views as hypotheses, sometimes, but they are always to be checked against the data (as in the case of the general expectation that the right side of an animal will be stronger than the left[30]). And in fact, his philosophical conception of the living organism is itself based on observation of and reflection on certain facts about the world (as I'll explain later in regard to his teleology). As is clear from the above, this conception of the living organism is in no way a substitute for the meticulous *collection* of data about 'animal differences

[27] *Resp.* 471b23-29.

[28] *PA* III.4 664b4-19. The view being ridiculed is probably Plato's in the *Timaeus*: cf. *Tim.*78c and Cornford 1935, 303-315, esp. 303-4 and 312-13.

[29] I develop this point further in chapter 12 above, p. 288.

[30] *IA* 4 705b13-706a26.

and in general features'. Nor does it eliminate the need for the careful *organization* of this data, so let us move on to the second stage of Aristotelian science.

(2) The organization of the data

How, according to Aristotle are we to organize the data collected? Well, what are we trying to discover—what are we looking *for*? As before, our aim is ultimately to understand *what* in detail each animal is (i.e., the full range of its characteristics) and *why* it is as it is. Or, as we could say, *how* the animal is and *why*. This two-part aim calls for a two-part organization.

In order to grasp *how the animals are*, we need to grasp each of the features animals possess and their variations, and to grasp them *precisely*. For example, consider the mode of production of young at birth, which is an important feature by which animals differ, connected as it is with numerous other features. Some offspring come out in eggs, some 'live'. And there are other possibilities too—e.g., larvae—and variations in each of these, e.g., animals born 'live' come out some more ready to go, some less (e.g., kittens are born blind). But a close study of animal reproduction turns up the fact that of the ones that come out live, some are live within the womb, while others are in the egg stage there, and come out after hatching inside. And this difference too correlates with important other differences. If we are to grasp what the property of 'bearing live young' is, we must grasp the difference between these two types of it: that is what I meant by saying we must grasp animal features 'precisely'.

So, as mentioned before, the *History of Animals* is from the beginning organized by *differences*—parts, activities, modes of life, and characters (or 'dispositions'—in Greek: *moria, praxeis, bioi, ethē*), and within these the organization is careful. For example, parts are divided first into those of blooded animals and those of bloodless (our 'red-blooded' and 'non-red-blooded'), and each of these into external and internal parts; modes of reproduction are divided into sexual and spontaneous, then for the sexual, Aristotle reviews the types and frequency of copulation, the breeding seasons, then the different sorts of formation of egg or of fetus, the different conditions at birth, and so on.

In this way we get a precise grasp of what the differing animal features *are* (and thereby, as I put it, *how the animals, which have those features, themselves are*). But there is a second aspect to Aristotle's organization of his data in *HA*: we want to know not only 'how each animal is' but *why* it is as it is—what are the *causes* of the animal's being as it is. Here Aristotle's key idea is to ask *which differences go together (and which don't)*. This is a very important dimension of his method and of the organization of his treatises. But how would a grasp of which differences go together help us to find the causes of the differences being as they are? Well, consider: why do some animals have four stomachs, others only one? Well, the cause will be something that the four-stomached ones have that the one-stomached ones don't—or they'd be four-stomached too! If we can find out which features all four-stomached animals have and only four-stomached animals have, the cause will be among them. The search for which differences go with four stomachs will point us towards the cause.

(The actual explanation Aristotle gives is very clever. He observes that the four-stomached animals are all large, live-bearing, four-legged, have only one row of teeth in their front jaws, and, with one exception, have horns. His explanation: he knows from his theory of generation and his matter theory that there is a limited amount of earthy material available at the top end of the embryo after the skull is formed. Since defense is a crucial need, and these animals use *horns* for defense, the excess earthy material must go to make horns. There isn't enough, therefore, to go to the teeth, and so they can only get one row of front teeth. With only one row of front teeth to break up their food, they need—predictably—four stomachs. The exception is the camel, which is not horned, but needs the extra stomachs to break up the particularly tough food ['thorny and woody'] it must eat.)

So, having this methodology in mind, Aristotle gives a great deal of attention, in his layout of the data in *HA* and elsewhere, to the correlation of differences, to which features go with which, and which do not. The basic organization of *HA*, as I've indicated, is by feature, but within the discussion of each feature, Aristotle aims to indicate what other features it correlates with.

Analyzing and classifying the features, and then correlating them, positively and negatively, with each other, is what I call the 'organization' stage of Aristotelian science. Aristotle's two-part organization facilitates his two-part aim: organizing first by differences, and then by which differences go with which, facilitates grasping first exactly what the differences of animals *are*, and then, what the causes of these differences are. Getting the correlations does not itself yet explain anything—knowing that four-stomachs go with horns and one row of teeth doesn't tell you what causes what—but it facilitates it, by pointing you towards the features where the cause will be found. The actual finding of the causes—the actual discovery of the explanations as to why the animals that possess them do possess them (and the organization of these explanations into an integrated body of scientific understanding)—is the third and final stage of Aristotelian science, which I'm calling The Explanation of the Data.[31]

[31] Readers familiar with the distinction with which Aristotle begins the second book of his *Posterior Analytics*, between establishing regarding some putative fact being investigated *that* (*hoti*) it is so and *why* (*dihoti*) it is so, may already have associated my Organization stage with *hoti*-investigations and my Explanation stage with *dioti*-investigations, and they would be essentially right to do so. For the match between the Organization stage as I have described it and *hoti*-investigations, properly understood, see above, chapter 12, pp. 270–73 and chapter 14, pp. 312ff., 323 n. 40, and especially Lennox 1991 (= 2001b, ch. 2). My Collection stage is the lead-up that positions one to begin asking the questions involved in the *hoti*-investigations.

(3) The explanation of the data

This is a very complex enterprise, and a vast area of Aristotelian scholarship. I can only touch on the main points, each of which really deserves a lecture in itself.[32] In particular, I'd like to mention and briefly discuss five themes. (Again, I refer to *Philosophical Issues in Aristotle's Biology* for detailed studies of each of these themes.) They are:

1. The explanations are mostly *teleological.*
2. The explanations have an (implicit) *axiomatic* structure.
3. They are to be given at the *proper level of generality.*
4. They result in scientific *definitions.*
5. The explanations and corresponding definitions develop in *stages.*

Let's look briefly at each, with most attention to the first.

1. The explanations are mostly teleological[33] That is, just as we explain some human action, like taking a walk after dinner, by reference to the *aim* or *goal* of, say, facilitating a good digestion, so, Aristotle thinks, the development, structure, and functioning of living things is to be explained by reference to the end-result which that development, structure, or functioning is *aimed at* and has as its *goal.* The stages of development are occurring *because* they are needed for the production of the living organism which is to result, and the parts of that organism are present *because* they are needed for the life of that organism. This is so, he thinks, even though there's no mind or consciousness *directing* living processes, as there is directing human action.

How he thinks that's possible I'll come back to in a moment, but it's important to see first what he's denying. He's denying that we can understand why a part is present in an animal, or why a process takes place within it, merely by describing the physical and chemical interactions that lead to that result. Physics and chemistry, so to speak, are not enough. We need to understand why the relevant materials are there in the first place, and why they take the particular directions they do, *because* they are needed if a complex organism is to result at the end.[34] As a result, his treatise *Parts of Animals* explains the presence of any part (e.g., the four stomachs) in the animals that have that part by showing that the part is necessary (or best), if such animals are to survive in the environments in which they live, that they have just that part. And he explains any stage in the development of an organism,

[32] I will indicate for each theme where a fuller discussion can be found in the present volume.

[33] See above, Part I, chapters 1-6 and Part II, chapters 7 and 8.

[34] Strictly this is true for what Mariska Leunissen and I call 'primary teleology'. Teleological processes sometimes have material by-products that are not present for the sake of something, though the organism's nature *makes use of* such material for the sake of some further end, a process we call *secondary* teleology. See chapter 5 above, esp. pp. 130-32, with n. 35.

including when and where some part begins to be formed, by reference to those same needs of the mature organism, together with the natures of the available materials. (Thus, the presence of the earthy material at the head of the embryo doesn't guarantee that horns will develop; that will happen only if it's needed, but the presence of the material is still part of the explanation—a necessary but not a sufficient condition.) Explanations in *Generation of Animals* often have just that form, starting from the needs of the form of animal to be produced, and showing for any stage in development, how the available materials are formed in just such a way as to result in an animal of that form.

Where, according to Aristotle, does this directiveness come from? How can development be *aimed* at a result when there's no mind or consciousness to do the aiming?[35] I just want to say enough now to suggest some similarities with contemporary biology. For, contemporary biology agrees with Aristotle on two things. First of all, organic development is not haphazard or the result of a linear sequence of chemical interactions, like the collapse of a row of dominoes, in which once a domino falls, it is out of the process, over and done with. Rather (to put a complex set of interactions rather simply), some material—the DNA—is so structured as to be able to control—to regulate—the process so that it results in a complex whole of a certain kind. All the information, so to speak, is present in advance. And second, contemporary biology explains the presence of the particular DNA by reference to the survival value of the parts it is programmed to produce. It is because certain bodily parts have contributed to the survival of the parents that the DNA program for those parts is available to produce those parts. In short, the parts are present in organisms of that type *because* they contribute to the survival of organisms of that type. So, we get *directiveness* from the DNA program, and *adaptiveness* from natural selection.

Now, neither of these mechanisms was available to Aristotle. His chemistry, as you may know, is not a structural chemistry, in which subatomic particles combine into atoms, atoms into molecules, etc.; it is rather a chemistry of four homogeneous elements—earth, water, air, fire—that combine not by forming more complex structures but by *blending*. And the capacities for interaction are quite simple and limited. Water and earth, together with some fire, mix to become mud, or clay, and at the right consistency, metals. Substantial amounts of fire tend rather to tear things apart, burning or melting them. And so on. There was no basis in Aristotle's chemistry for the formation, by physical and chemical laws alone, of complex, heterogeneous, articulated structures (like fingers, hands, or the exact layout of the internal organs, say) able to function in complex ways. The movement to higher structure, such as is found in a living thing, required not only a very special blend of the ordinary material elements, but also a very special sort of heat, a 'vital heat' as it

[35] The subject is addressed in Part I of this volume.

came to be called, which was supplied by the parents, with a special contribution from the male.[36] This heat, which Aristotle said was not the heat of fire, but a much 'finer' type, had the capacity to interact with the special, cooked up material provided by the mother, the bloody material which when unused became menstrual fluid (Aristotle had no concept of a female ovum), in such a way that the embryo itself gradually took shape and got articulated in just the way required if the resultant offspring was to live and function. This heat was like our DNA in the sense of being the bearer of a capacity or potential to produce a complex living form at the end, but differed in that most of us believe that DNA can be fully analyzed chemically, and that it is complex feedback loops, and so forth, which provide the regulation, and the direction towards an end. Aristotle treated that capacity as basic, irreducible to capacities of the four elements. So for Aristotle the direction towards an end—the aiming at the form—was inherent in that irreducible capacity found in the heat, and needed no consciousness. And while contemporary biology explains the astonishing adaptiveness of animals to their environments through evolution by natural selection, Aristotle treats it as a basic fact about the world that organisms come well-adapted. Given that the parents are well-adapted, their capacity to produce an offspring like themselves ensures that the offspring too will be well-adapted.

Note, before we go any further, the *empirical*, or *scientific*, character of Aristotle's commitment to the need for biological explanations to be mostly teleological, as I have described that commitment. Examining the development and functions of living organisms to determine what basic natures and potentials are at work, Aristotle finds that, given what he already understands about the natures and potentials of their constituent elements, it is just not reasonable to expect the complexities of the living world to be explainable in terms of its constituent matter. (In one place, he even says it was not 'sober' for his predecessors to think they could explain it that way.[37]) That is an investigator of nature talking. It is on that basis that Aristotle postulates for each sort of living organism a basic nature or potential directed irreducibly to the organism's *form*, a potential irreducibly to develop into and maintain in being just such an organism. Likewise, although I won't argue it here, Aristotle's account of the basic natures and potentials of the material elements in the world is the result of a fact-based analysis of the source of the observable behavior of these elements.

These are the conclusions of a theoretical scientist, a student of nature. Were Aristotle to reappear tomorrow and be presented with the results of the last several

[36] Cf. above, Part I *passim*, esp. chapter 4.

[37] See the wonderful remark at *Metaph.* A. 984b10-18, in which Aristotle praises Anaxagoras for his 'sobriety' compared with his predecessors (*hoion nēsōn ephanē par' eikēi legontas tou proteron*, b17-18), who thought the order and beauty in the world could be due either to the material elements alone, or to spontaneity or chance.

hundred years of biochemistry and evolutionary theory, including the evidence which grounds them, my view is that (after recovering from the most awe-filled experience of his life) he would surely retreat from his insistence that the directiveness in nature is a *primitive* directiveness and the adaptiveness in nature a *basic* fact. Facts they would remain, but they would, for him, as for us, no longer be *basic*. The determination of what the *basic* level of explanation is, is for Aristotle ultimately an empirical or scientific matter.[38]

For the actual Aristotle, goal-directedness is a basic fact of living nature, and that fact requires that explanations of the features and processes of animals be mostly teleological, explaining their presence by reference to the contribution they make to the life of the animals that have them. Such is the first of my five themes of the Explanation stage of Aristotelian science.

2. *The explanations have an implicit axiomatic structure*[39] Like all scientific explanations, biological explanations, Aristotle thought, must proceed deductively from first principles. The model was geometry, and you remember how your geometry text began with axioms and then derived theorems from them. Aristotle did not seem to think that the working natural scientist has actually to lay out his treatise in the way the mathematicians did, but his explanations have to have the same logical structure: even if they aren't *put* in that form, they have to be *puttable* in that form. In my own work on this,[40] I've identified three sorts of principles that serve as *first* principles in Aristotle's biology, and you can see them at work in the four-stomachs example:

(1) *Principles about the nature and potentials of the elements.* It's only because earth when mixed with water and a little fire, and then gently heated becomes very hard, that horns can be made at the places they need to be made, and it's because the food

[38] I have discussed this matter in some detail above, in chapters 2 (pp. 60-61) and 3 (pp. 80, 83, 87); and see my first statement of the 'empirical' character of Aristotle's teleology in chapter 1 (pp. 9-10 n. 16, 13, 28-29). I should stress here that I do not think Aristotle's commitment to biological teleology rests on the detailed investigations reflected in his biological treatises. *Physics* II makes clear that Aristotle thought that the teleological character of living nature is inferable from facts about living things available to any thoughtful adult. In my view, treatises like *PA* II-IV and *GA* refine and confirm his commitment to teleology; they aren't the source of it. (See above, p. 251 n. 26). Note too that the case for the use of teleological explanation in the study of animals presented in *PA* I.1 is part of a description of what the *non-specialist* needs to understand before he can evaluate results he hears from a specialist. (*PA* I.1 *init.*)

This point regarding the empirical character of Aristotle's teleological approach to living nature (and his theory of elemental natures and potentials) carries over to the first two components of theme 2. My unargued claim about the basis of Aristotle's postulation of the ultimate natures and potentials of the elements would have to be argued (as I've already suggested) separately, from portions of *De Caelo* and elsewhere.

[39] See above, Part II, chapters 7-9 and Part III, chapter 10.

[40] Above, chapters 7 and 8.

has the consistency it has and teeth have just the hardness they have but gums without teeth don't, that more than one row of teeth is needed to break down the food fully for digestion in just one stomach.

But as our discussion of Aristotle's teleology has indicated, element natures and potentials by themselves are not sufficient to explain living things. There must be a separate potential for form—a potential to produce an organism of a certain form—supplied by the parent. It's a basic fact about the world that certain forms of animal exist, that these forms are self-maintaining and self-reproducing, and that they are well-adapted, or 'best-organized'. That gives our second set of first principles, which I call:

(2) *Principles asserting the* ***existence*** *of the various animal natures, their status as* ***ends*** *(which includes self-maintenance and reproduction), and their status as* ***best-organized ends***. Thus, e.g., it is for Aristotle a first principle of nature (i) that camels exist; (ii) that they are ends, that is that the development, structure, and functioning of camels is aimed at the production, maintenance, and reproduction of the camel; and (iii) that they are best-organized, which is to say that where a variation in a feature is possible, and makes the life of the camel go better—e.g., having two kidneys rather than just one—the camel will have it. Thus, to explain why, say, oxen have four stomachs, we will make reference to the ox's status as end and its organization for the best. But notice that we won't be able to explain why they have four stomachs in particular until we know what their basic nature is—which in this case is, say, *large live-bearing four-footed blooded animal*—because it is only from those features that we can derive their limited earthy material at the top end of the embryo, their need of horns, and thus their one row of teeth *and their four stomachs*, and that suggests a third set of first principles, which I call:

(3) *Definitions of these animal forms or natures.* These will state the *basic* essence or nature of each animal form or kind, just as in geometry we use the definition of the triangle as a closed figure with three straight sides to derive other facts about the triangle, such as its having an angle sum of 180 degrees. Certain features of animals are basic and provide the starting point for the explanation of other features.

As I said, these first principles are not all stated at the beginning of Aristotle's treatises, and are not always clearly labeled as first principles, although sometimes they are. But whether labeled or not, they are there, and give the biological explanation its implicit axiomatic structure.

The third basic theme of Aristotelian scientific explanation is that:

3. *The explanations are to be given at the proper level of generality*[41] That is, if we are to explain the possession of four stomachs, we must not ask separately why, say, oxen have four stomachs, and deer do, and antelopes, and so on, but must move to the

[41] See above, Part II, especially chapters 8 and 9.

widest class of animals which has them, whether or not that class is actually a single kind. Aristotle's favorite example here is that we will not understand why, say, isosceles triangles have an angle sum of 180 degrees, unless we understand why *all* triangles do, since their being isosceles triangles is not relevant to their angle sum. As we saw in talking about the aims and structure of Aristotle's treatise, the *History of Animals*, this move to the widest class of possessors of a feature is the key to discovering the causes: if we want to know the cause of a feature such as having four stomachs, we need to discover what other features all four-stomached animals have in common, because the cause will be among those other features. This method of looking for causes, by the way, was passed down by Renaissance Aristotelians into the scientific tradition, and was incorporated, along with other principles originating explicitly or implicitly in Aristotle, first by Francis Bacon and then by John Stuart Mill, into what came to be called 'Mill's methods', which are often taught in elementary courses on logic and scientific method. How this theme integrates with the previous theme, that the explanations should have an axiomatic structure, is an interesting and difficult question that I have discussed in chapter 8 above.

There is one other consequence of this theme worth mentioning: its implications for the *revisability* of Aristotle's scientific theorizing. I will return to this in connection with our fifth theme.

Although the fourth and fifth themes of the Explanation stage of Aristotelian science are distinct, they are related, and it will be useful to consider them together:

4. *The explanations result in scientific definitions*[42]

5. *The explanations and corresponding definitions develop in stages*[43] Here Aristotle's favorite example is from meteorology, the nature of thunder. Thunder is a certain noise that occurs in the clouds, Aristotle says. That's where we start. But the scientific study of thunder must investigate its *cause*, and set out that cause in an *explanation*. For the sake of illustration, Aristotle projects that the cause of thunder is the extinguishing of a flash-fire in the atmosphere that we see as lightning. Just as the noise a fire makes, when splashed with water, is sometimes called a 'sizzle', so on this theory (which Aristotle doesn't actually accept), thunder is an atmospheric sizzling. The explanation, written out in deductive form, goes something like this:

Noise is produced by the extinguishing of fire.
Extinguishing of fire is present in the clouds.
So, noise is present in the clouds.

[42] See above, Part II, chapter 7.
[43] Ibid.

The cause, extinguishing of fire, is given as the 'middle term', connecting noise with clouds. This explains the existence of thunder, but in so doing, Aristotle says, it also exhibits *what thunder is*. Thus, we can read off a scientific definition of thunder from this explanation: Thunder is noise in the clouds due to the extinguishing of fire. You'll recall my saying back at the beginning that the aim of Aristotelian science is to discover *how things are* and *why they are as they are*. Science starts with a preliminary account or definition of what things are so far as we can tell from observation and experience (e.g., thunder is noise in the clouds); it then explains why they are that way (e.g., because fire is being extinguished in the clouds); but in discovering why they are that way, science gets a much deeper understanding of what they are (e.g., thunder is noise in the clouds due to the extinguishing of fire).[44]

As Aristotle says:

(T14) What is thunder? Extinguishing of fire in cloud. Why does it thunder? Because the fire in the cloud is being extinguished. Cloud *C*, thunder *A*, extinguishing of fire *B*. *B* holds of *C*, the cloud (the fire is extinguished in it); and *A*, noise, holds of *B*—and *B* is indeed an account of *A*, the first extreme. (If there is another middle term [D] for this [*sc.* B], it will come from the remaining [i.e., further] accounts [of *A*].)[45]

But now look at the last line just quoted:

If there is another middle term [D] for this [*sc.* B], it will come from the remaining [i.e., further] accounts [of *A*].

That is, having discovered that thunder is, say, noise in the clouds due to the extinguishing of fire, one may through further discovery and analysis come to see that the extinguishing of fire itself is due to some deeper cause, say the overpowering of the elementary qualities that constitute fire, its heat and dryness, by the wetness and coldness of the moisture in the atmosphere. This 'further middle term' [D] gives us a further and deeper explanation of why there is thunder; but that deeper explanation also results in a deeper scientific definition of thunder, say: noise in the clouds due to the overpowering of hot and dry fiery stuff in the atmosphere by atmospheric moisture present in the clouds. Here we see that Aristotle recognizes that natural science *advances*, in stages; that a cause discovered at one stage may itself be explained by a deeper cause, and so on.[46] This process must of course be finite, and in some areas Aristotle clearly thought science had

[44] On the general relation between definition and explanation, see, e.g.: 'So, as we say, to know what something is is the same as to know why it is—either why it is *simpliciter* and not one of the items that hold of it, or why it is one of the items which hold of it (e.g. that it has two right angles . . .)' (*APo*. II. 90a31-33, tr. Barnes 1993.)

[45] *APo*. II.8 93b7-14, tr. Barnes 1993.

[46] Cf. Bolton 1987, 142-46, esp. 144-45. We might (as a first approximation) schematize it this way (I use the modern ordering of terms in a syllogism rather than Aristotle's own ordering,

reached the fundamental cause, or had come close to it; in others, as the quoted remark about the reproduction of bees (above, **T10**) makes clear, more work is yet to be done.[47]

There are more radical developments that sometimes occur in Aristotle's scientific work, which speak to his willingness to *revise* his theories in various ways beyond what we have just discussed. I will mention two sorts of cases. The first is a consequence of our third theme, that explanation should be at the proper level of generality. In *APo.* I.5, Aristotle notes that certain theorems regarding proportion were first established separately for numbers and magnitudes, but then it was discovered that the same cause was at work in the two cases, in virtue of their both being quantities. The theorem, and the general presentation of the theory of proportion, was then rewritten to apply to all quantities as such. There is a case that this is what Aristotle is doing at the opening of his *De Motu Animalium*, having realized that there are principles that apply to all motion as such, both animal and celestial, and that those studies proper to both realms must now build on or integrate with the principles that hold true of all motion as such.[48]

and avoid the controversy concerning the type or range of clouds Aristotle intends C ['the clouds'] to represent):

WHY IT THUNDERS:

A B
P_1 Noise (= thunder) is produced by the extinguishing of fire.
B C
P_2 Extinguishing of fire is present in the clouds.
A C
C Noise (= thunder) is present in the clouds.

DEFINITION OF WHAT THUNDER IS:
Noise in the clouds due to the extinguishing of fire.

DEEPER EXPLANATION OF WHY IT THUNDERS:

A D
P_3 Noise (= thunder) is produced by [a deeper cause].
D B
P_4 [Deeper cause] is involved in the extinguishing of fire.
A B
P_1 Noise (= thunder) is produced by the extinguishing of fire.

DEEPER DEFINITION OF WHAT THUNDER IS:
Noise in the clouds due to [the deeper cause].

47 Another case where it is clear that Aristotle must think the scientific work is incomplete may be found in chapter 7 above, p. 163 n. 34; the matter has to do with why certain animals have certain means of defense and not others.

48 One can imagine yet one more case of interest. In *PA* II.10, Aristotle makes a case for studying animals separately from plants, on the grounds that animals have much more variation than plants do. Let us assume that this is, on Aristotelian grounds, methodologically proper.

The second sort of case would be one in which the causal structure of a discipline is revised, in view of a set of discoveries that reveal that structure to be more complex than had been thought. The only example I have of that is a very speculative one. We have distinguished the organizational and explanatory stages of a science for Aristotle. In the study of animals we have *HA* at the organizational stage and (among others) *PA* and *GA* at the explanatory stage. But David Balme has shown that our *HA* was actually written *later* than *PA* and *GA*, and he has pointed out that all the data—all the properly organized data—on which *PA* and *GA* draw are already included in those treatises.[49] So, if *HA* was prepared as a *historia* aimed at facilitating the discovery of causes, then Aristotle may well have been planning a new body of causal theorizing.[50] In any case, there is ample room in both Aristotle's theory and practice of explanatory science for significant revision, where expanding knowledge calls for it. This feature of Aristotelian science, though it often does not get much attention, is worth stressing. For it is yet one more reason why the view with which we began, of Aristotle as 'armchair theorist', disregarding new evidence and holding back the course of science for millennia, is to be rejected.

However, at that time, Theophrastus had not yet done his extensive work on plants, including *On the Causes of Plants*. Let us now imagine Aristotle taking all that work on board and noticing that, in fact, there is quite a lot to say about what plants and animals have in common, in virtue of the fact that they are both living. Surely what he would do is to separate off a general study of living things. One could imagine him doing so on the model of a general study he does recognize, of 'universal mathematics', in which the object of study is those facts true of all quantities as such, whether arithmetical or geometrical. He doesn't have a term like 'biology', but one could imagine, if he did come to think there is a significant body of universal truths about all living things, him coining such a term (perhaps *zōōlogikē*), and establishing such a discipline. This discipline would theoretically precede what would still be distinct studies of animals as such and plants as such. See now Falcon forthcoming.

[49] Balme 1987a, 13. For a striking example of a text in *PA* where the (properly organized) data are presented first and then the causes given, see chapter 9 above. In support of Balme's claim to the relative lateness of *HA*, see Lennox 1996a.

[50] My own view, though it is at this point very speculative, is that Aristotle had come to recognize that proper causal accounts of the possession by animals of the features they have are significantly more complex than he had thought previously, involving a greater interplay of different sorts of causal differentiae than his initial work on *PA* had suggested. There is some hint of this complexity in passages like the ones one finds *passim* in the opening two chapters of *HA* VII, where he says, e.g., that animals' 'lives and activities differ according to their characters and nutrition', and that there are correlations between the sorts of food they need to eat, the habitats they live in, and their 'bodily blend'. This suggests, as I say, a more complex causal account, and just possibly a more complex structure to the new explanatory treatises that Aristotle might have been thinking *HA* might point to. This, if it was indeed contemplated by Aristotle, would be a case of a fairly extensive revision to the structure of a science. I am indebted to Karen Zwier and Andrea Falcon for raising the question of the extent to which revisability is a feature of Aristotelian science, both in its theory and in its practice.

Now, this third stage of Aristotelian science—what I have called 'the explanation of the data (and the organization of the explanations)'—is a vast subject, and there is much more that could be said about it, as there is about the two prior stages, the collection of the data and the organization of the data. But I hope I have said enough already to justify my verdict regarding Aristotle as scientist. For, by any reasonable criteria of what counts as a scientific enterprise—the aim of understanding *how* things are and *why* they are as they are, of achieving that understanding through an observation-based discovery of causes, of seeking those causes through a careful and systematic examination of the full range of relevant data, of basing one's conclusions only on the rational assessment of the evidence, and of organizing it all into a systematic whole, a set of theories with broad explanatory power going back to first principles, which are themselves established empirically, a set of theories that are significantly revisable if expanding knowledge should call for it—it seems to me that Aristotle was a *great* scientist, and among the greatest. For if Aristotle's practice in his biological works is not great science, then I don't know what could be.

References

Agassiz, J. L. 1856. 'The habits of the *Glanis* of Aristotle', *Proceedings of the American Academy of Arts and Sciences* 3, 325-34.

al-Imârati, J. ed. 2002. *Ibn Bâjja: Kitâb al-hawayân*. Beirut-Casablanca.

Amigues, S. 1988-96. *Théophraste: Recherches sur les plantes*, 5 vols. Assoc. Guil. Budé. Paris: Les Belles Lettres.

[Anon.]. 1864, 'Aristotelian Zoology'—Review C. J. Sundervall, *Die Thierarten des Aristoteles, von den Klassen, der Säugethiere, Vögel, Reptilien, und Insecten* (Stockholm 1863), *Natural History Review: A Quarterly Journal of Biological Science* 4, 493-96.

Arnott, W. G. 2007. *Birds in the Ancient World from A to Z*. London and New York.

Asúa, M. de. 1999. 'Medicine and Philosophy in Peter of Spain's *de animalibus*', in Steel et al. 1999, 189-211.

Asúa, M. de and French, R. K. 2005. *A New World of Animals: Early Modern Europeans on the Creatures of Iberian America*. Aldershot.

Ayala, F. 1972. 'The Autonomy of Biology as a Natural Science', in *Biology, History and Natural Philosophy*, ed. A. D. Breck and W. Yourgrau (New York, 1974), 1-16; orig. 'Biology as an Autonomous Science', *American Scientist* 56 (1968) 207-21.

Balme, D. M. 1961. 'Aristotle's Use of Differentiae in Zoology', in *Aristotle et les problèmes de méthode*, ed. S. Mansion. Louvain, 195-212; repr. with additions as Balme 1975 and Balme 1987b.

——1962a. '*ΓΕΝΟΣ* and *ΕΙΔΟΣ* in Aristotle's Biology', *Classical Quarterly* 12, 81-98.

——1962b. 'Development of Biology in Aristotle and Theophrastus: Theory of Spontaneous Generation', *Phronesis* 7, 91-104.

——1965. 'Aristotle's Use of the Teleological Explanation', Inaugural Lecture, London, Queen Mary College, University of London.

——1970. 'Aristotle: "PA" I, 2-3: Argument and Text', *Proceedings of the Cambridge Philological Society* n.s. 16, 12-21.

——1972. *Aristotle's* De Partibus Animalium *I and* De Generatione Animalium *I (with passages from II.3)*. Clarendon Aristotle Series. Oxford. (Rev. ed. Balme 1992.)

——1974. Review of Lanza and Vegetti 1971. *Isis* 65, 267-68.

——1975. 'Aristotle's Use of Differentiae in Zoology', in Barnes, Schofield, and Sorabji 1975, 183-93; repr. of Balme 1961 with additions; itself repr. with additions as Balme 1987b.

——1976. 'Form and Species in Aristotle's Biology'. Read to Princeton University Classical Philosophy Colloquium (unpublished).

——1977. 'The Place of Biology in Aristotle's Philosophy'. Read at Institute For Advanced Study. Princeton N.J. (unpublished; this is a different paper from Balme 1987a of the same name).

——1980. 'Aristotle's Biology was not Essentialist', *Archiv für Geschichte der Philosophie* 62, 1-20; repr. with additions as Balme 1987d.

——1984. 'The Snub', *Ancient Philosophy* 4, 1-8; repr. as appendix to Balme 1987d.

——1985. 'Historia Animalium Book Ten', in Wiesner 1985, 191-206.

Balme, D. M. 1987a. 'The Place of Biology in Aristotle's Philosophy', in Gotthelf and Lennox 1987, 9-20.

——1987b. 'Aristotle's Use of Division and Differentiae', in Gotthelf and Lennox 1987, 69-89; repr., with additions, of Balme 1961 and Balme 1975.

——1987c. 'Teleology and Necessity', in Gotthelf and Lennox 1987, 275-86.

——1987d. 'Aristotle's Biology Was Not Essentialist', in Gotthelf and Lennox 1987, 291-312 (orig. in *Archiv für Geschichte der Philosophie* 62 [1980], 1-20; two appendices added to 1987d: 'Note on the *aporia* in *Metaphysics* Z' [prev. unpubl.] and 'The Snub' [orig. in *Ancient Philosophy* 4 (1984) 1-8].)

——1990. 'Matter in the Definition: A Reply to G. E. R. Lloyd', in Devereux and Pellegrin 1990, 49-54.

——1991. *Aristotle's Historia Animalium, Books VII–X.* Prepared for publication by A. Gotthelf. Loeb Classical Library. London and Cambridge, Mass.

——1992. *Aristotle's* De Partibus Animalium *I and* De Generatione Animalium *I (with passages from II.3)*, rev. edn. with a report on recent work and additional bibliography by A. Gotthelf. Clarendon Aristotle Series. Oxford (orig. 1972).

——2002. *Aristotle: Historia Animalium*, Vol. 1, Books I–X: Text. Prepared for publication by Allan Gotthelf. Cambridge.

——forthcoming. *Aristotle: Historia Animalium*, Vol. 2, Books I–VII: Commentary. Prepared for publication by Allan Gotthelf. Cambridge.

Barnes, J. 1975a. *Aristotle's* Posterior Analytics. Clarendon Aristotle Series. Oxford.

——1975b. 'Aristotle's Theory of Demonstration', in Barnes, Schofield, and Sorabji 1975, 65-87.

——1981. 'Proof and the Syllogism', in *Aristotle on Science: The Posterior Analytics*. ed. E. Berti. Proceedings of the 8th Symposium Aristotelicum. Padova 1981, 17-59.

——1982. *Aristotle*. Oxford.

——1984. *The Complete Works of Aristotle: The Revised Oxford Translation*. Princeton.

——1993. *Aristotle's* Posterior Analytics, 2nd edn. Clarendon Aristotle Series. Oxford.

Barnes, J., Schofield, M, and Sorabji, R. 1975. *Articles on Aristotle: 1. Science*. Duckworth.

Barrett, P. H. et al., eds. 1988. *Charles Darwin's Notebooks, 1836-1844: Geology, Transmutation of Species, Metaphysical Enquires*. Ithaca.

Barrett, P. H. and Freeman, R. B., eds. 1988. *Variation of Animals and Plants under Domestication*. The Works of Charles Darwin, vols. 19-20. New York.

Beagon, M. 1992. *Roman Nature: The Thought of Pliny the Elder*. Oxford.

Beatty, J. 1990. 'Teleology and the Relationship between Biology and the Physical Sciences in the Nineteenth and Twentieth Centuries', in *Some Truer Method: Reflections on the Heritage of Newton*, ed. F. Durham and R. D. Purrington. New York.

Beavis, I. C. 1988. *Insects and Other Invertebrates in Classical Antiquity*. Exeter.

Berger, F. 2005. *Die Textgeschichte der Historia animalium des Aristoteles*. Serta Graeca 21. Wiesbaden.

Berryman, S. 2003. 'Ancient Automata and Mechanical Explanation', *Phronesis* 48, 345-69.

——2007. 'Teleology Without Tears: Aristotle and the Role of Mechanistic Conceptions of Organisms', *Canadian Journal of Philosophy* 37, 351-70.

——2009. *The Mechanical Hypothesis in Ancient Greek Natural Philosophy*. Cambridge.

Beullens, P. and Bossier, F. 2000–. *Aristoteles Latinus XVII 2.I.1-2: De Historia Animalium, Translatio Guillelmi de Morbeka*. Leiden.

Beullens, P. and Gotthelf, A. 2007. 'Theodore Gaza's Translation of Aristotle's *De Animalibus*: Content, Influence, and Date', *Greek, Roman, and Byzantine Studies* 47, 469-513.

Binswanger, H. 1990. *The Biological Basis of Teleological Concepts*. Marina del Rey, CA.

——1992. 'Life-Based Teleology and the Foundation of Ethics', *The Monist* 75, 84-103.

Bodson, L. 2004. *Index verborum in Aristotelis* Historiara animalium, 2 vols., Hildesheim: Georg Olms.

Bolton, R. 1976. 'Essentialism and Semantic Theory in Aristotle: *Posterior Analytics* II.7-10', *Philosophical Review* 85, 514-44.

——1987. 'Definition and Scientific Method in *Aristotle's Posterior Analytics* and *Generation of Animals*', in Gotthelf and Lennox 1987, 120-66.

——1993. 'Division, définition et essence dans la science aristotélicienne', *Revue philosophique*, 197-222.

Bourgey, L. 1955. *Observation et expérience chez Aristote*. Paris.

Bowen, A. C., ed. 1991. *Science and Philosophy in Classical Greece*. New York.

Bradie, M. and Miller, F. D., Jr. 1984. 'Teleology and Natural Necessity in Aristotle', *History of Philosophy Quarterly* 1, 133-45.

Brown, L. 2010. 'Definition and Division in Plato's *Sophist*', in Charles 2010a, 151-71.

Brugman, J. and Drossaart Lulofs, H. J., eds. 1971. *Aristotle: Generation of Animals: The Arabic Translation Commonly Ascribed to Yaḥyā ibn al-Biṭrīq*. Leiden.

Buffon, G.-L. Leclerc, Comte de. 1749-89. *Histoire naturelle, générale et particulière, avec la description du Cabinet du Roy*, 36 vols. Paris.

Burckhardt, R. 1904. 'Das koische Tiersystem, eine Vorstufe der zoologischen Systematik des Aristoteles', *Verhandlungen der Naturforschenden Gesellschaft in Basel* 15, 377-413.

——1905. 'Das erste Buch der aristotelischen Tiergeschichte', *Zoologische Annalen*. Würzburg, 1-28.

Burkhardt, F. et al. 1990. *The Correspondence of Charles Darwin*, vol. 6: *1856-1857*. Cambridge.

Burkhardt, F. et al., eds. 1994. *A Calendar of the Correspondence of Charles Darwin, 1821-1882, with supplement*. Cambridge.

Burnyeat, M. F. 2008. '*Kinēsis* vs. *Energeia*: A Much-Read Passage in (but not of) Aristotle's *Metaphysics*', *Oxford Studies in Ancient Philosophy* 34, 219-92.

Byl, S. 1973. 'Le jugement de Darwin sur Aristote', *L'Antiquité Classique* 42, 519-21; modified version, 'Note sur le jugement de Darwin', in Byl, 1980, xxx-xxxii.

——1980. *Recherches sur les grands traités biologiques d'Aristote: sources écrites et préjugés*. Brussels.

Caley, E. R. and Richards, J. F. C. 1956. *Theophrastus on Stones*. Columbus.

Camus, A.-G. 1783. *Histoire des Animaux d'Aristote*, 2 vols. [1. *Histoire des Animaux d'Aristote, avec la traduction françoise*. 2. *Notes sur l'Histoire des Animaux d'Aristote*]. Paris.

Carroll, P. T., ed. 1976. *An Annotated Calendar of the Letters of Charles Darwin in the Library of the American Philosophical Society*. Wilmington.

Cerami, C. 2012. *Nature et Sagesse*, Les rapports entre physique et métaphysique dans la tradition aristotélicienne (Festschrift Pierre Pellegrin). Louvain.

Charles, D. 1988. 'Aristotle on Hypothetical Necessity and Irreducibility', *Pacific Philosophical Quarterly* 69, 1-53.

Charles, D. 1990. 'Aristotle on Meaning, Natural Kinds and Natural History', in Devereux and Pellegrin 1990, 145-67.

——1991. 'Teleological Causation in the *Physics*', in Judson 1991, 101-28.

Charles, D. 1997. 'Aristotle on the Unity and Essence of Biological Kinds', in Kullmann and Föllinger 1997, 27-42.

——, 2000. *Aristotle on Meaning and Essence*. Oxford Aristotle Studies. Oxford.

——, ed. 2010a. *Definition in Greek Philosophy*, Oxford.

Charles, D. 2010b. 'Definition and Explanation in the *Posterior Analytics* (and beyond)', in Charles 2010a, 286-328.

——2010c. '*Metaphysics Θ*.7 and 8: Some Issues Concerning Actuality and Potentiality', in Lennox and Bolton 2010, 168-97.

Charlton, W. 1970. *Aristotle's* Physics *Book I and II*. Clarendon Aristotle Series. Oxford.

——1985. 'Aristotle and the Harmonia Theory', in Gotthelf 1985a, 131-50.

——1992. *Aristotle's* Physics *Book I and II, with a report on recent work*. Clarendon Aristotle Series. Oxford. (Rev. edn. of Charlton 1970).

Cherniss, H. F. 1935. *Aristotle's Criticism of Presocratic Philosophy*. Baltimore.

——1944. *Aristotle's Criticism of Plato and the Academy*. Baltimore.

Cleary, J. J. and Shartin, D. C. 1992. *Proceedings of the Boston Area Colloquium in Ancient Philosophy, v. 6 (1990)*. Lanham MD.

Code, A. 1987. 'Soul as Efficient Cause in Aristotle's Embryology', *Philosophical Topics* 15, 51-59.

Cohen, S. M. 1989. 'Aristotle on Hot, Cold, and Teleological Explanation', *Ancient Philosophy* 9, 255-70.

Cole, F. J. 1975. *A History of Comparative Anatomy from Aristotle to the Eighteenth Century*. New York (repr.; orig. edn. London 1949).

Coles, A. 1995. 'Biomedical Models of Reproduction in the Fifth Century BC and Aristotle's '*Generation of Animals*', *Phronesis* 40, 48-88.

Collingwood, R. G. 1945. *The Idea of Nature*. Oxford.

Cooper, J. M. 1975. *Reason and the Human Good in Aristotle*. Cambridge, Mass. (repr. with a new preface to the paperback edition, Indianapolis: Hackett, 1986).

——1982. 'Aristotle on Natural Teleology', in *Language and Logos: Studies in Ancient Greek Philosophy Presented to G. E. L. Owen*, ed. M. Schofield and M. C. Nussbaum. Cambridge, 197-222.

——1987. 'Hypothetical Necessity and Natural Teleology', in Gotthelf and Lennox 1987, 243-74.

——1990. 'Metaphysics in Aristotle's Embryology', in Devereux and Pellegrin 1990, 55-84 (orig. in *Proceedings of the Cambridge Philological Society* no. 214 [1988] 14-41); repr. in J. M. Cooper, *Knowledge, Nature, and the Good: Essays on Ancient Philosophy*. Princeton 2004, 174-203.

——1997. *Plato Complete Works*. Indianapolis.

Cornford, F. M. 1935. *Plato's Cosmology: The 'Timaeus' of Plato Translated with a Running Commentary*. London.

Corpus dei papiri filosofici greci e latini, I.1 (Florence, 1989).

Cosans, C. E. 1998. 'Aristotle's Anatomical Philosophy of Nature', *Biology and Philosophy* 13, 311-39.

Coutant, V. 1971. *Theophrastus, De Igne: A Post-Aristotelian View of the Nature of Fire*. Assen.

Cuvier, G. 1841. *Histoire des sciences naturelles, depuis leur origine jusqu'à nos jours, chez tous les peuples connus*, 5 vols. Paris.

Darwin, C. 1859. *On the Origin of the Species by Means of Natural Selection*. London ('Historical sketch' with note on Aristotle, p. xiii, added 3rd edn., 1861).

Darwin, F. 1887. *The Life and Letters of Charles Darwin*, 3 vols. London.

Davies, M. and Kathirithamby, J. 1986. *Greek Insects*. New York.

De Groot, J. 2008. '*Dunamis* and the Science of Mechanics: Aristotle on Animal Motion', *Journal of the History of Philosophy* 46, 43-68.

De Leemans, P., ed. 2011. *Aristoteles Latinus XVII 1 & XVII 2.II-III, De progressu animalium. Translatio Guillelmi de Morbeka / De motu animalium. Fragmenta translationis anonymae et Translatio Guillelmi de Morbeka*. Turnhout.

Dehmer, V. C. 2007. *Aristoteles Hispanus: Eine altspanische Übersetzung seiner Zoologie aus dem Arabischen und dem Lateinischen*. Tübingen.

Delbrück, M. 1971. 'Aristotle-totle-totle', in *Microbes and Life*, ed. J. Monod and E. Borek. New York, 50-55.

Depew, D. 1997. 'Etiological Approaches to Biological Aptness in Aristotle and Darwin', in Kullmann and Föllinger 1997, 209-30.

Desmond, R., ed., 1977. *Dictionary of British and Irish Botanists and Horticulturalists*. 3rd. edn. London.

Detel, W. 1997. 'Why all Animals have a Stomach: Demonstration and Axiomatization in Aristotle's *Parts of Animals*', in Kullmann and Föllinger 1997, 63-84.

Devereux, D. and Pellegrin, P. 1990. *Biologie, Logique et Métaphysique chez Aristote*. Paris.

Di Gregorio, M. A., ed., 1990. *Charles Darwin's Marginalia*, vol. 1. New York.

Drossaart Lulofs, H. J., ed., 1965. *Aristotle, De Generatione Animalium*. Oxford Classical Texts. Oxford.

——, ed., 1966. *Aristoteles Latinus XVII 2.V: De generatione animalium. Translatio Guillelmi de Moerbeka*. Bruges and Paris.

Dupree, A. H., ed., 1963. *Asa Gray: Darwiniana*. Cambridge.

Durant, W. 1926. *The Story of Philosophy: The Lives and Opinions of the Greater Philosophers*. New York.

Düring, I. 1943. *Aristotle's De Partibus Animalium: Critical and Literary Commentaries*. Göteborg.

Eichholz, D. E. 1965. *Theophrastus: De Lapidibus*. Oxford.

Einarson, B. and Link, G. K. K. 1976-90. *Theophrastus: De Causis Plantarum*, 3 vols. Loeb Classical Library. Cambridge, Mass.

Falcon, A. forthcoming. 'Aristotle and the Study of Animals and Plants', in Ferraces Rodríguez, Fischer, and Holmes forthcoming.

Ferraces Rodríguez, A., Fischer, K.-D., and Holmes, B. forthcoming. *Festschrift for Heinrich von Staden*. La Coruña.

Filius, L. 2007. 'The Arabic Transmission of the *Historia Animalium* of Aristotle', in Vrolijk and Hogendijk 2007, 25-33.

——2008. 'The *Book of Animals* by Aristotle', in *Islamic Thought in the Middle Ages: Studies in Text, Transmission and Translation, in Honour of Hans Daiber*, ed. A. Akasoy and W. Raven (Leiden, 2008), 267-73.

Filius, L., den Heijer, H. and Mattock, J. (in preparation). *Aristotle: History of Animals: The Arabic Translation Commonly Ascribed to Yaḥyā ibn al-Biṭrīq*. Leiden.

Fortenbaugh, W. W. et al., eds. 1992. *Theophrastus of Eresus: Sources for his Life, Writings, Thought and Influence*, 2 vols. Leiden.

Fortenbaugh, W. W. and Sharples, R. W., eds. 1988. *Theophrastean Studies: On Natural Science, Physics and Metaphysics, Ethics, Religion and Rhetoric.* Rutgers Studies in the Classical Humanities III. New Brunswick.

Frede, M. 1980. 'The Original Notion of Cause', in *Doubt and Dogmatism: Studies in Hellenistic Epistemology*, ed. J. Barnes et al. Oxford, 217-49; repr. in M. Frede, *Essays in Ancient Philosophy*, Oxford, 1987, 125-50.

Freeland, C. A. 1987. 'Aristotle on Bodies, Matter, and Potentiality', in Gotthelf and Lennox 1987, 392-407.

——1990. 'Scientific Explanation and Empirical Data in Aristotle's *Meteorology*', *Oxford Studies in Ancient Philosophy* 8, 67-102; also in Devereux and Pellegrin 1990, 287-320.

Freeman, R. B. 1978. *Charles Darwin: A Companion* Folkstone.

Freudenthal, G. 1995. *Aristotle's Theory of Material Substance: Heat and Pneuma, Form and Soul.* Oxford.

Frey, C. unpublished. 'From Blood to Flesh: Homonymy, Unity, and Ways of Being in Aristotle'.

Furley, D. 1983. 'The Mechanics of *Meteorologia* IV: A Prolegomenon to Biology', in Moraux and Wiesner 1983, 73-93.

——'The Rainfall Example in *Physics* II.8', in Gotthelf 1985, 177-82.

Furth, M., tr. 1985. *Aristotle Metaphysics: Books VII–X: Zeta, Eta, Theta, Iota.* Indianapolis.

Furth, M. 1987. 'Aristotle's Biological Universe: An Overview', in Gotthelf and Lennox 1987, 21-52.

——1988. *Substance, Form and Psyche: An Aristotelean Metaphysics.* Cambridge.

Galilei, G. 1890-1909. *Le Opere di Galileo Galilei, Edizione Nazionale*, ed. A. Favaro, 20 vols. Florence.

Gesner, C., tr. 1541. *Michaelis Ephesii Scholia Aristotelis libros aliquot.* Basel (cf. *Conrad Gesner Opera Omnia on Microfiche*, ed. H. H. Wellisch, Leiden. Microfiche H-2535/1).

Gesner, C. 1551-87. *Historiae Animalium* (vols. I–IV: 1551-1558, vol. V: 1587). Zürich.

Ghiselin, M. T. 1994. 'Darwin's Language May Seem Teleological, but his Thinking is Another Matter', *Biology and Philosophy* 9, 489-492.

Gigon, O., ed. 1987. *Aristotelis Opera*, vol. 3: *Librorum deperditorum Fragmenta.* Berlin and New York.

Gill, M. L. 1997. 'Material Necessity and *Meteorology* IV.12', in Kullmann and Föllinger 1997, 145-61.

Gill, T. 1905. 'Parental Care among Fresh Water Fishes', *Smithsonian Annual Report*, 403-531; repr. in T. Gill, *Contributions to the Life History of Fishes*, vol. I (1904-1907). Smithsonian Institution Reprints no. 1858 (1909). Washington, D.C.

——1906. 'The Remarkable Story of a Greek Fish, the Glanis', *The George Washington University Bulletin* 5/4, 5-13.

Gotthelf, A. 1975. *Aristotle's Conception of Final Causality.* Columbia University Dissertation.

——1976/77. 'Aristotle's Conception of Final Causality', *Review of Metaphysics* 30, 226-54 (= the original version of ch. 1, this volume).

——1980. Review of Nussbaum 1978, *Journal of Philosophy* 77, 365-78.

——1981/82. Review of Nussbaum 1978, *Review of Metaphysics* 35, 619-23.

——ed. 1985a. *Aristotle on Nature and Living Things: Philosophical and Historical Essays in Honor of David M. Balme's Seventieth Birthday*, Pittsburgh and Bristol (UK).

——1985b. 'Notes towards a Study of Substance and Essence in Aristotle's *Parts of Animals* II–IV', in Gotthelf 1985a, pp. 27-54 (= Ch. 10, this volume).

——1987a. 'First Principles in Aristotle's Parts of Animals', in Gotthelf and Lennox 1987, 167-98 (= ch. 7, this volume).

——1987b. Aristotle's Conception of Final Causality', in Gotthelf and Lennox 1987, 204-42 (= ch. 1, this volume).

——1988. '*Historiae* I: *Plantarum* et *Animalium*', in Fortenbaugh and Sharples 1988, 100-35. (= ch. 14, this volume).

——1992. 'Report on recent work', in Balme 1992.

——1997a. 'The Elephant's Nose: Further Reflections on the Axiomatic Structure of Biological Explanation in Aristotle', in Kullmann and Föllinger 1997, 85-96 (= ch. 8, this volume).

——1997b. 'Understanding Aristotle's Teleology', in Hassing 1997, 71-82 (= the original version of ch. 3, this volume).

——2010. 'Comments on Mariska Leunissen's "Aristotle's Syllogistic Model of Knowledge and the Biological Sciences: Demonstrating Natural Processes"', in Lesher 2010, 61-73.

——, unpublished. '"One Long Argument?"—The Unity of Aristotle's *Generation of Animals*'. Paper presented at the Fifth Pittsburgh-London (Ont.) Workshop on Aristotle's *Generation of Animals*, May 2010.

Gotthelf, A. and Lennox, J. G., eds. 1987. *Philosophical Issues in Aristotle's Biology*. Cambridge.

Gotthelf, A. and Salmieri G., eds. forthcoming. *Ayn Rand: A Companion to Her Works and Thought*. Oxford.

Grene, M. 1963. *A Portrait of Aristotle*. Chicago.

——1974. 'Is Genus to Species as Matter to Form?—Aristotle and Taxonomy', ch. 6 of M. Grene, *The Understanding of Nature: Essays in the Philosophy of Biology, Synthese* 28.1, 56-69.

——1985. 'On the Division of the Sciences', in Gotthelf 1985a, 9-13.

Günther, H.-C. and Rengakos, A., eds. 1997. *Beiträge zur antiken Philosophie: Festschrift für Wolfgang Kullmann*. Stuttgart.

Guthrie, W. K. C., 1965. *History of Greek Philosophy*, Vol. II: *The Presocratic Tradition from Parmenides to Democritus*. Cambridge.

Guyer, P. 2005. 'What is Living and What is Dead in Kant's Teleology', in P. Guyer, *Kant on Nature and Freedom: Selected Essays* (Oxford), 343-72.

Hamlyn, D. W. 1993. *Aristotle's De Anima Books II and III (with passages from Book I)*, with a report on recent work and a revised bibliography by C. Shields. Clarendon Aristotle Series. Oxford.

Hantz, H. 1939. *The Biological Motivation of Aristotle*. New York.

Hassing, R. F., ed. 1997. *Final Causality in Nature and Human Affairs*. Washington, D.C.

Hayduck, M., ed. 1903 *[Ps.-]Philoponus [=Michael of Ephesus] In Aristotelis De Generatione Animalium*. Commentaria in Aristotelicum Graeca 14.3. Berlin.

——, ed. 1904. *Michaelis Ephesii in libros De partibus animalium; De animalium motione; De animalium incessu commentaria*. Commentaria in Aristotelem Graeca 22.2. Berlin.

Healey, J., ed. 1991. *Pliny the Elder: Natural History: A Selection*. London.

Henry, D. 2005. 'Embryological Models in Ancient Philosophy', *Phronesis* 51, 1-42.
——2006a. 'Understanding Aristotle's Reproductive Hylomorphism', *Apeiron* 39, 257-87.
——2006b. 'Aristotle on the Mechanisms of Inheritance', *Journal of the History of Biology* 39, 425-55.
——2007a. 'How Sexist is Aristotle's Developmental Biology?', *Phronesis* 52, 251-69.
——2007b. Review of M. R. Johnson, *Aristotle on Teleology, Ancient Philosophy* 27, 191-200.
——2008. 'Organismal Natures', in *Aristotle on Life*, ed. J. Mouracade. Kelowna, BC, 57-74.
——2009. 'Generation of Animals', in *A Companion to Aristotle*, ed. G. Anagnostopoulos. Oxford 2009, 368-83.
Hon, G. and Goldstein, B. R. 2010. *From Summetria to Symmetry: The Making of a Revolutionary Scientific Concept*. New York.
Hort, A. F. 1916. *Theophrastus: Enquiry into Plants*, 3 vols. Loeb Classical Library. London and New York.
Huby, P. M. 2007. *Theophrastus of Eresus: Sources for his Life, Writings, Thought and Influence: Commentary Volume 2: Logic*. Leiden.
Hull, D. 1965. 'The Effects of Essentialism on Taxonomy: Two Thousand Years of Stasis', *The British Journal for the Philosophy of Science* 15, 314-26 and 16, 1-18; reprinted in *Concepts of Species*, ed. C. N. Slobodchikoff, Berkeley 1976; and in *The Units of Selection*, ed. Marc Ereshefsky, Cambridge, Mass. 1992.
Irwin, T. H. 1988. *Aristotle's First Principles*. Oxford.
Johnson, M. R. 2005. *Aristotle on Teleology*. Oxford Aristotle Studies. Oxford.
Judson, L. ed. 1991. *Aristotle's Physics: A Collection of Essays*. Oxford.
Kádár, Z. 1978. *Survivals of Greek Zoological Illuminations in Byzantine Manuscripts*, Budapest.
Kahn, C. H. 1985. 'The Place of the Prime Mover in Aristotle's Teleology', in Gotthelf 1985, 183-205.
Kant, I. 2000. *Critique of the Power of Judgment*, ed. P. Guyer, tr. P. Guyer, E. Matthews. Cambridge Edition of the Works of Immanuel Kant in Translation. Cambridge.
Kaufmann, W. 1968. *The Portable Nietzsche*. New York.
Keaney, J. J. 1963. 'Two Notes on the Tradition of Aristotle's Writings', *American Journal of Philology* 84, 52-63.
——1968. 'The Early Traditions of Theophrastus' *Historia Plantarum*', *Hermes* 96, 293-98.
Kitchell, K. and Resnick, I. M. 1999. *Albertus Magnus, On Animals: A Medieval Summa Zoologicav*. Baltimore and London.
Kniseley, R. M., ed. 1964. *Dynamic Clinical Studies with Radioisotopes: Proceedings of a symposium held at the Oak Ridge Institute of Nuclear Studies, October 21-25, 1963*. Washington, D.C.
Kosman, L. A. 1969. 'Aristotle's Definition of Motion', *Phronesis* 14, 40-62.
——1984. 'Substance, Being and *energeia*', *Oxford Studies in Ancient Philosophy* 2, 121-49.
——1987. 'Animals and Other Beings in Aristotle', in Gotthelf and Lennox 1987, 360-91.
Kripke, S. 1980. *Naming and Necessity*. Cambridge, Mass.
Kruk, R. 1979. *The Arabic Version of Aristotle's Parts of Animals: Book XI–XIV of the Kitāb al-ḥayawān*. North-Holland.
——1990. 'A Frothy Bubble: Spontaneous Generation in the Medieval Arab Tradition', *Journal of Semitic Studies* 35, 265-82.

——1996. 'Ibn Bâjja's Commentary on Aristotle's *De Animalibus*', in *The Ancient Tradition in Christian and Islamic Hellenism*, ed. G. Endress and R. Kruk, Leiden, 165-80.

——2003. 'Aristoteles, Les écrits zoologiques dans la tradition orientale', *Dictionnaire des philosophes antiques*, ed. R. Goulet. Supplément préparé par Richard Goulet et al. Paris, 329-34.

Kullmann, W. 1974. *Wissenschaft und Methode: Interpretationen zur aristotelischen Theorie der Naturwissenschaft*. Berlin.

——1985. 'Different Concepts of the Final Cause in Aristotle', in Gotthelf 1985a, 169-75.

Kullmann, W. and Föllinger, S., eds., 1997. *Aristotelische Biologie: Intention, Methoden, Ergebnisse*. Stuttgart.

Lanza, D. and Vegetti, M., eds. 1971. *Opera biologische di Aristotele*. Turin.

Le Blond, J.-M. 1939. *Logique et méthode chez Aristote: étude sur la recherche des principes dans la physique aristotélicienne*. Paris.

——, ed. and tr. 1945. *Aristote, philosophe de la vie. Le livre premier du Traité sur les parties des animaux*. Paris.

Lear, J. 1980. *Aristotle and Logical Theory*. Cambridge.

Lee, H. D. P. 1948. 'Place-Names and the Date of Aristotle's Biological Works', *Classical Quarterly* 42, 61-67.

——1985. 'The Fishes of Lesbos Again', in Gotthelf 1985a, 3-8.

Lennox, J. G. 1980. 'Aristotle on Genera, Species, and "the more and the less"', *Journal of the History of Biology* 13, 321-46; extensively revised as Lennox 1987b.

——1982. 'Teleology, Chance, and Aristotle's Theory of Spontaneous Generation', *Journal of the History of Philosophy* 20, 219-38 (repr. in Lennox 2001b, 229-49).

——1985a. 'Are Aristotelian Species Eternal?', in Gotthelf 1985a, 67-94 (repr. in Lennox 2001b, 131-59).

——1985b. 'Theophrastus on the Limits of Teleology', in W. W. Fortenbaugh, P. Huby, and A. A. Long, eds., *Theophrastus of Eresus: His Life and Work*, Rutgers University Studies in Classical Humanities, vol. ii (New Brunswick 1985), 143-63 (repr. in Lennox 2001b, 259-79).

——1985c. 'Demarcating Ancient Science: A Discussion of G. E. R. Lloyd's *Science, Folklore and Ideology: The Life Sciences in Ancient Greece*', *Oxford Studies in Ancient Philosophy* 3, 307-24.

——1985d. 'Plato's Unnatural Teleology', in *Platonic Investigations*, ed. D. J. O'Meara, Washington, D.C., 195-218 (repr. in Lennox 2001b, 280–302).

——1987a. 'Divide and Explain: the *Posterior Analytics* in Practice', in Gotthelf and Lennox 1987, 90-119 (repr. in Lennox 2001b, 7-38).

——1987b. 'Kinds, Forms of Kinds, and the More and the Less', in Gotthelf and Lennox 1987, 339-59 (update of Lennox 1980; repr. of 1987b in Lennox 2001b, 160-81).

——1990. 'Notes on David Charles on *HA*', in Devereux and Pellegrin 1990, 169-83.

——1991. 'Between Data and Demonstration: the *Analytics* and the *Historia Animalium*', in Bowen 1991, 269-94 (repr. in Lennox 2001b, 39-71).

——1993. 'Darwin *was* a Teleologist', *Biology and Philosophy* 8, 409-21.

——1994. 'Teleology by Another Name: A Reply to Ghiselin', *Biology and Philosophy* 9, 493-95.

——1995. 'The Disappearance of Aristotle's Biology: A Hellenistic Mystery', in *The Sciences in Greco-Roman Society*, ed. T. D. Barnes. Edmonton 1995, 7-24 (repr. in Lennox 2001b, 110-25).

Lennox, J. G. 1996a. 'Aristotle's Biological Development: The Balme Hypothesis', in *Aristotle's Philosophical Development: Problems and Prospects*, ed. W. Wians. Lanham, MD, 229-48.

——1996b. 'Putting Philosophy of Science to the Test: The case of Aristotle's Biology', in *PSA 1994: Proceedings of the Biennial Meeting of the Philosophy of Science Association*, Vol. 2, 239-47 (repr. in Lennox 2001b, 98-109)

——1997a 'Material and Formal Natures in Aristotle's *De Partibus Animalium*', in Kullmann and Föllinger 1997, 163-81; orig. in J. J. Cleary and W. Wians, eds., *Proceedings of the Boston Area Colloquium of Ancient Philosophy, Volume XIV (1995)*, Lanham, MD,1996, 217-40 (repr. in Lennox 2001b, 182-204.).

——1997b. 'Nature Does Nothing in Vain . . . ' in Günther and Rengakos 1997, 199-214 (repr. in Lennox 2001b, 205-23).

——2001a. *Aristotle on the Parts of Animals I–IV, Translated with an Introduction and Commentary*. Oxford.

——2001b. *Aristotle's Philosophy of Biology: Studies in the Origins of Life Science*. Cambridge.

——2004. 'Getting a Science Going: Aristotle on Entry Level Kinds', in *Homo Sapiens und Homo Faber* (Festschrift Mittelstrass), ed. G. Wolters, Berlin: Walter De Gruyter, 2004, 87-100.

——2010a. '*Bios* and Explanatory Unity in Aristotle's Biology', in Charles 2010a, 329-55.

——2010b. 'The Unity and Purpose of *On the Parts of Animals* I', in Lennox and Bolton 2010, 56-77.

Lennox, J. G. and Bolton, R. 2010. *Being, Nature, and Life in Aristotle: Essays in Honor of Allan Gotthelf*, Cambridge.

Lenoir, T. 1989. *The Strategy of Life: Teleology and Mechanics in Nineteenth-Century German Biology*. Chicago (orig. edn. 1982, Dordrecht).

Lesher, J., ed. 2010. *From Inquiry to Demonstrative Knowledge: New Essays on Aristotle's* Posterior Analytics (*Apeiron* 42.2 & 3). Kelowna, BC (Canada).

Leunissen, M. 2007. 'The Structure of Teleological Explanations in Aristotle: Theory and Practice', *Oxford Studies in Ancient Philosophy* 33, 145-78.

——2010a. 'Aristotle's Syllogistic Model of Knowledge and the Biological Sciences: Demonstrating Natural Processes', in Lesher 2010, 31-60.

——2010b. *Explanation and Teleology in Aristotle's Science of Nature*. Cambridge.

——2010c. 'Nature as a Good Housekeeper. Secondary Teleology and Material Necessity in Aristotle's Biology', *Apeiron* 42, 117-42.

——unpublished. 'Order and Method in Aristotle's *Generation of Animals* II'. Paper presented at the Fifth Pittsburgh–London (Ont.) Workshop on Aristotle's *Generation of Animals*, May 2010.

Leunissen, M. and Gotthelf, A. 2010. 'What's Teleology Got to Do with It?'—A Reinterpretation of Aristotle's *Generation of Animals* V', *Phronesis* 55 (2010), 325-56.

Lewes, G. H. 1864. *Aristotle: A Chapter from the History of Science, including Aristotle's Scientific Writings*. London.

Liatsi, M. 2000. *Aristoteles, De Generatione Animalium, Buch V, Einleitung und Kommentar*. Trier.

Lloyd, G. E. R. 1961. 'The Development of Aristotle's Theory of the Classification of Animals', *Phronesis* 6, 59-81 (repr. with new introductory comment in Lloyd 1991, 1-26).

——1975. 'Alcmaeon and the Early History of Dissection', *Sudhoffs Archiv* 59, 113-147; repr. in Lloyd 1991, 164-93.

——1983. *Science, Folklore, and Ideology: Studies in the Life Sciences in Ancient Greece*. Cambridge.

——1987a. 'Empirical Research in Aristotle's Biology', in Gotthelf and Lennox 1987, 53-63.

——1987b. *The Revolutions of Wisdom*. Berkeley.

——1990. 'Aristotle's Zoology and Metaphysics: The *status quaestionis*', in Devereux and Pellegrin 1990, 32-47; repr. in Lloyd 1991, 373-97.

——1991. *Methods and Problems in Greek Science*. Cambridge.

——1996a. *Aristotelian Explorations*. Cambridge.

——1996b. 'Theories and Practices of Demonstration', in Lloyd 1996b, 7-37 (orig. 'Theories and Practices of Aristotelian Demonstration', in Cleary and Shartin 1992, 371-401).

Lones, T. E. 1912. *Aristotle's Researches in Natural Science*. London.

Long, H. S., ed. 1964. *Diogenis Laertii Vitae Philosophorum*, 2 vols. Oxford Classical Texts, Oxford.

Longrigg, J. 1977. Review of Balme 1972, *Classical Review* 27, 38-39.

Louis, P. 1955. 'Remarques sur la classification des animaux chez Aristote', in *Autour d'Aristote: Recueil d'études de philosophie ancienne et médiévale offert à Msgr. A. Mansion*. Louvain.

——1956. *Aristote: Les parties des animaux*. Paris.

——1961. *Aristote: De la génération des animaux*, Paris.

——1975. *La Découverte de la vie, Aristote*. Paris.

Madkour, I. 1970. *Ibn Sīnā, Al-Shifā'. La Physique. VIIIe-Les Animaux (Fī Ṭabā'i' al-Ḥayawān)*. Texte établi et édité par 'Abd el-Halim Montasir, Sa'id Zayed, 'Abdallah Isma 'il. Revu et précedé d'une Introduction par Ibrahim Madkour. Le Caire.

Manier, E. 1978. *The Young Darwin and his Cultural Circle*. Dordrecht.

Manquat, M. 1932. *Aristote naturaliste*. Paris.

Matthen, M. 1989. 'The Four Causes in Aristotle's Embryology', in Penner and Kraut 1989, 159-80.

Mayhew, R. 2004. *The Female in Aristotle's Biology: Reason or Rationalization*. Chicago.

Mayr, E. 1988. *Towards a New Philosophy of Biology*. Cambridge, Mass.

McKirahan, R. 1992. *Principles and Proofs: Aristotle's Theory of Demonstrative Science*. Princeton.

McLaughlin, P. 1990. *Kant's Critique of Teleology in Biological Explanation: Antimony and Teleology*. Studies in the History of Philosophy. Lewiston, NY and Lampeter.

Mendell, H. 1984. 'Two Geometrical Examples from Aristotle's *Metaphysics*', *Classical Quarterly* 34, 359-72.

Meyer, J. 1855. *Aristoteles Thierkunde: Ein Beitrag zur Geschichte der Zoologie, Physiologie und alten Philosophie*. Berlin.

Meyer, S. S. 1992. 'Aristotle, Teleology, and Reduction', *Philosophical Review* 101, 791-825.

Miller, F. D., Jr. 1995. *Nature, Justice and Rights in Aristotle's, Jr.* Politics. Oxford.

Moraux, P. 1951. *Les listes anciennes des ouvrages d'Aristote*. Louvain.

Moraux, P. and Wiesner, J., eds. 1983. *Zweifelhaftes in Corpus Aristotelicum: Studien zu Eingein Dubia: Akten des 9. Symposium Aristotelicum*. Berlin.

Morsink, J. 1982. *Aristotle on the Generation of Animals: A Philosophical Study*. Washington D.C.

Muir, E. 2007. *The Culture Wars of the Late Renaissance: Skeptics, Libertines and Opera* (The Bernard Berenson Lectures on the Italian Renaissance). Cambridge, Mass.

Müller, H. 1852. 'Über das Männschen von Argonauta argo und die Hectocotylen', *Zeitschrift für wissenschaftliche Zoologie* 4, 1-35.

Müller, J. 1842. *Über den glatten Hai des Aristoteles und über die Verschiedenheiten unter den Haifischen und Rochen in der Entwickelung des Eies*. Berlin (orig. in *Abhandlungen der Akademie der Wissenschaften*, Berlin 1840, 187-258).

Nagel, E. 1961. *The Structure of Science*. New York.

Navarro Sánchez, F. 2009. *Petri Hispani: Questiones supra libro 'De Animalibus' Aristotelis: Estudio, edición y traducción del ms. 1877 de la B. N. de Madrid*, 2 vols. Tesis doctoral, UAB. Barcelona.

Nissen, L. 1984. 'Woodfield's Analysis of Teleology', *Philosophy of Science* 51, 488-94.

Nussbaum, M. C. 1976. 'The text of Aristotle's *De Motu Animalium*', *Harvard Studies in Classical Philology* 53, 111-59.

——1978. *Aristotle's* De Motu Animalium: *Text with Translation, Commentary, and Interpretive Essays*. Princeton.

——1980. 'Review of *Substance, Body, and Soul: Aristotelian Investigations*, by E. Hartman', *Journal of Philosophy* 77, 355-65.

——1982. 'Aristotle', in *Ancient Writers: Greece and Rome*, ed. T. J. Luce. New York, 377-416.

——1983. 'The "Common Explanation" of Animal Motion', in Moraux and Wiesner 1983, 116-56.

——1984. 'Aristotelian Dualism: Reply to Howard Robinson', *Oxford Studies in Ancient Philosophy*. 2, 197-207.

Ogilvie, B. W. 2006. *The Science of Describing: Natural History in Renaissance Europe*. Chicago.

Ogle, W. 1882. *Aristotle on the Parts of Animals*. London.

——tr. 1912. *Aristote: De Partibus Animalium*, in Ross 1908-52, vol. 5.

Owens, J. 1951. *The Doctrine of Being in the Aristotelian* Metaphysics: *A Study in the Greek Background of Medieval Thought*. Toronto.

Parker, R. 1984. 'Sex, Women, and Ambiguous Animals', *Phronesis* 29, 174-87.

Peck, A. L. 1961. *Aristotle Generation of Animals*. Loeb Classical Library. Cambridge, Mass.

——1965. *Aristotle Historia Animalium Books i-iii*. Loeb Classical Library. Cambridge, Mass.

——1970. *Aristotle Historia Animalium Books iv-vi*. Loeb Classical Library. Cambridge, Mass.

Peck, A. L. and Forster, E. S. 1961. *Aristotle Parts of Animals* (Peck) and *Movement of Animals, Progression of Animals* (Forster). Loeb Classical Library. Cambridge Mass.

Pellegrin, P. 1985. 'Aristotle: A Zoology without Species', in Gotthelf 1985a, 95-115.

——1986. *Aristotle's Classification of Animals: Biology and the Conceptual Unity of the Aristotelian Corpus*, tr. A. Preus. Berkeley. (Orig. edn.: *La Classification des animaux chez Aristote: statut de la biologie et unité de l'aristotélisme*. Paris 1982.)

——1987. 'Logical Difference and Biological Difference: The Unity of Aristotle's Thought', in Gotthelf and Lennox 1987, 313-38.

Penner, T. and Kraut, R., eds. 1989. *Nature, Knowledge and Virtue: Essays in Memory of Joan Kung, Apeiron* Special Issue 22.4.

Perfetti, S. 2000. *Aristotle's Zoology and its Renaissance Commentators 1521-1601*. Leuven.

Petit, G. and Théodoridès, J. 1962. *Histoire de la Zoologie. Des origines à Linné*. Paris.

Popper, K. R. 1945. *The Open Society and its Enemies*. London.

——1963. *Conjectures and Refutations: The Growth of Scientific Knowledge*. London.

——1998. *The World of Parmenides: Essays on the Presocratic Enlightenment*, ed. A. F. Petersen (with J. Meyer). London.

Preus, A. 1979. '*Eidos* as Norm in Aristotle's Biology', *Nature and System* 1, 79-101; repr. in *Essays in Greek Philosophy, Volume Two*, ed. J. P. Anton and A. Preus, Albany 1984, 340-63.

——1981. *Aristotle and Michael of Ephesus on the Motion and the Progression of Animals*. Hildesheim.

Putnam, R. 1989. *The Natural History of Deer*. Ithaca.

Rand, A. 1961. *For the New Intellectual: The Philosophy of Ayn Rand*. New York; repr. New York (Centennial Edition), 2005.

——1963. 'Review of J. H. Randall, Jr., *Aristotle*', *The Objectivist Newsletter*, May 1963, 18; repr. in A. Rand, *The Voice of Reason: Essays in Objectivist Thought*, New York, 6-12.

——1964. *The Virtue of Selfishness: A New Concept of Egoism*. New York; repr. New York (Centennial Edition), 2005.

Randall, J. H., Jr. 1960. *Aristotle*. New York.

Renn, J., ed. 2001. *Galileo in Context*. Cambridge.

Resnick, I. M. and Kitchell, K. 2008. *Albert the Great, Questions Concerning Aristotle's On Animals*. Washington, D.C.

Rist. J. M. 1965. 'Some Aspects of Aristotelian Teleology', *Transactions and Proceedings of the American Philological Association* 96, 337-49.

Robin, L. 1944. *Aristote*. Paris.

Robinson, H. 1983. 'Aristotelian Dualism', *Oxford Studies in Ancient Philosophy* 1, 124-44.

Roger, J. 1997. *Buffon: A Life in Natural History*, tr. S. L. Bonefoi. Ithaca (orig. French edn. 1989).

Ross, W. D., ed. 1908-1952. *The Works of Aristotle Translated into English*. 12 vols. Oxford (vols. 4, 5, and 8 under editorship of J. A. Smith and W. D. Ross).

Ross, W. D. 1924. *Aristotle's Metaphysics: A Revised Text with Introduction and Commentary*. 2 vols. Oxford.

——1936. *Aristotle's Physics: A Revised Text and Commentary*. Oxford.

——1949. *Aristotle's Prior and Posterior Analytics: A Revised Text with Introduction and Commentary*. Oxford.

——1955. *Aristotle: Parva Naturalia. A Revised Text with Introduction and Commentary*. Oxford.

Ross, W. D. and Fobes, F. H. 1929. *Theophrastus Metaphysics*. Oxford.

Rossi, P., ed. forthcoming. *Aristoteles Latinus XVII 1-2.IV: De partibus animalium. Translatio anonyma et Translatio Guillelmi de Morbeka*. Turnhout.

Russell, B. 1946. *A History of Western Philosophy*. London.

——1950. 'An Outline of Intellectual Rubbish', *Unpopular Essays*. London.

Russell, E. S. 1982. *Form and Function: A Contribution to the History of Animal Morphology*. Chicago (orig. edn. 1916, London).

Saint-Hilaire, J. B. 1883-87. *Histoire des animaux d'Aristote: Traduite en français et accompagnée de notes perpétuelles*, 3 vols. Paris.

——1885. *Traités des parties des animaux et de la marche des animaux d'Aristote. Traduits en français pour la première fois et accompagnés de notes perpétuelles*, 2 vols. Paris.

Saint-Hilaire, J. B. 1887. *Traité de la génération des animaux d'Aristote: Traduit en français pour la première fois et accompagné de notes perpétuelles*, 2 vols. Paris.

Salmieri, G. 2008. *Aristotle and the Problem of Concepts*. University of Pittsburgh Dissertation.

Scaliger. J. C. 1619. *Aristotelis Historia de Animalibus*, ed. P. J. Mausaccus. Toulouse.

Scaltsas, T. C. 1989. 'Commentary on Gotthelf's The Place of the Good in Aristotle's Natural Teleology, in *Proceedings of the Boston Area Colloquium in Ancient Philosophy Volume IV 1988*, ed. J. J. Cleary and D. C. Shartin (Lanham, MD, 1989), 140-47.

Schaffner, K. F. 1974. 'The Peripherality of Reductionism in the Development of Molecular Biology', *Journal of the History of Biology* 7, 111-18.

Schmitt. C. B. 1980. *Cesare Cremonini: Un aristotelico al tempo di Galilei*. Venice.

——1983. *Aristotle and the Renaissance*. Cambridge, Mass.

Sedley, D. 1991. 'Is Aristotle's Teleology Anthropocentric?' *Phronesis* 36, 179-96.

——2000. '*Metaphysics Λ*.10', in *Aristotle's* Metaphysics *Lambda: Symposium Aristotelicum*, ed. M. Frede and D. Charles, Oxford, 327-50.

——2010. 'Teleology, Aristotelian and Platonic', in Lennox and Bolton 2010, 5-29.

Sharples, R. W. 1994. *Theophrastus of Eresus: Sources for his Life, Writings, Thought and Influences: Commentary Volume 5: Sources on Biology (Human Physiology, Living Creatures, Botany: Texts 328-435)*. Leiden.

Sloan, P. R. ed. 1992. *Richard Owen The Hunterian Lectures in Comparative Anatomy, May and June 1837*. Chicago.

Smith, R. 1989. *Aristotle: Prior Analytics*. Trans., with introduction, notes, and commentary. Cambridge and Indianapolis.

Sollenberger, M. G. 1988. 'Identification of Titles of Botanical Works of Theophrastus', in Fortenbaugh and Sharples 1988, 14-24.

Sorabji, R. 1980. *Necessity, Cause, and Blame: Perspectives on Aristotle's Theory*. Ithaca.

Stadler, A. 1916-20. *Albertus Magnus, De animalibus libri XXVI, nach der Cölner Urschrift*. 2 vols. Münster i. W.

Stavrianeas, S. 2008. 'Spontaneous Generation in Aristotle's Biology', *Rhizai* 6, 303-38.

Steel, C., et al., eds. 1999. *Aristotle's Animals in the Middle Ages and Renaissance*. Leuven.

Steigerwald, J. 2006. *Kantian Teleology and the Biological Sciences*. Special issue of *Studies in the History and Philosophy of Science, Part C: Studies in History and Philosophy of Biological and Biomedical Science* 37.

Tarán, L. 1981. *Speusippus of Athens: A Critical Study with a Collection of the Related Texts and Commentary*. Leiden.

Taylor, C. 1964. *The Explanation of Behaviour*. London and New York.

Thomas Aquinas. 1963. *Commentary on Aristotle's Physics*, tr. R. J. Blackwell, R. J. Spath, and E. Thirlkel. New Haven.

Thompson, D. W., tr. 1910. *Aristotle: Historia Animalium*, in Ross 1908-1952, vol. 4.

——1936. *A Glossary of Greek Birds*, 2nd edn. Oxford.

——1942. *On Growth and Form*, rev. edn. Cambridge (orig. edn. 1917; 2nd edn. repr. New York 1992).

——1947. *A Glossary of Greek Fishes*. Oxford.

Toulmin, S. and Goodfield, J. 1966. *The Architecture of Matter*. New York.

Vallance, J. T. 1988. 'Theophrastus and the Study of the Intractable: Scientific Method in *De Lapidibus* and *De Igne*', in Fortenbaugh and Sharples 1988, 25-40.

Van der Eijk, P. 2000-2001. *Diocles of Carystus: A Collection of the Fragments with Translation and Commentary*, 2 vols. Studies in Ancient Medicine. Leiden.

Van der Eijk, P. 2005. *Medicine and Philosophy in Classical Antiquity: Doctors and Philosophers on Nature, Soul, Health and Disease*. Cambridge.

Van Oppenraaij, A. M. I. 1992-. *Aristotle De Animalibus. Michael Scot's Arabic–Latin Translation*, 4 vols. Leiden. (*Volume 1.1 Books I–VI: History of Animals Books I–VI* (forthcoming); *Volume 1.2 Books VII–X: History of Animals Books VII–X* (in preparation); *Volume 2 Books XI–XIV: Parts of Animals*, 1998; *Volume 3 Books XV–XIX: Generation of Animals*. 1992.)

Van Oppenraay, A. M. I., ed. (in preparation). *Avicenna De animalibus, Translatio Scoti* (with the collaboration of J. Janssens).

Vlastos, G. 1969. 'Reasons and Causes in the *Phaedo*', *Philosophical Review* 78, 291-325.

Volz, W. V. 1852. *Beiträge zur Kulturgeschichte. Der einfluss des Menschen auf die Verbreitung der Hausthiere und der Kulturpflanzen*. Leipzig.

Von Staden, H. 1997. 'Teleology and Mechanism: Aristotelian Biology and Early Hellenistic Medicine', in Kullmann and Föllinger 1997, 183-208.

Vrolijk, A. and Hogendijk, J. P., eds. 2007. *O Ye Gentlemen: Arabic Studies on Science and Literary Culture, in Honor of Remke Kruk*. Leiden.

Waterlow [Broadie], S. 1982. *Nature, Change, and Agency in Aristotle's Physics*. Oxford.

Wendland, P., ed. 1903. *Michael of Ephesus in Aristotelis Parva Naturalia*. Berlin.

Whewell, W. 1837. *History of the Inductive Sciences, from the Earliest to the Present Times*. 1st edn., 3 vols. London.

——1858–59. *History of the Inductive Sciences, from the Earliest to the Present Time*, 3rd edn., with additions, 2 vols. New York.

Whitehead, A. N. 1925. *Science and the Modern World*. Cambridge.

Wians, W. 1992. 'Comment on G. E. R. Lloyd, "Theories and Practices of Aristotelian Demonstration"', in Cleary and Shartin 1992, 402-12.

Wieland, W. 1970. *Die aristotelische Physik*, 2nd edn. Göttingen.

——1975. 'The Problem of Teleology', in Barnes, Schofield, and Sorabji 1975, 141-60 (tr. of Wieland 1970, §16 'Zum Teleologieproblem', 254-78.

Wiesner, J., ed. 1985. *Aristoteles: Werk und Wirkung. Paul Moraux gewidmet*, 2 vols., Berlin

Wilson, M. 2009. 'A Somewhat Disorderly Nature: Unity in Aristotle's *Meteorologica* I–III. *Apeiron* 42, 63-88.

Witt, C. 1985. 'Form, Reproduction, and Inherited Characteristics in Aristotle's *Generation of Animals*', *Phronesis* 30.1, 46-57.

Woodfield, A. 1976. *Teleology*. Cambridge.

Wright, L. 1973. 'Functions', *Philosophical Review* 82, 139-68.

——. 1976. *Teleological Explanations*. Berkeley and Los Angeles.

——1978. 'The Ins and Outs of Teleology: A Critical Examination of Woodfield', *Inquiry*, 21, 223-37 (Review of Woodfield 1976).

Zeller, E. 1897. *Aristotle and the Earlier Peripatetics*, tr. B. F. C. Costelloe and J. H. Muirhead, 2 vols. London.

Zonta, M. 1999. 'The Zoological Writings in the Hebrew Tradition: The Hebrew Approach to Aristotle's Zoological Writings and to their Ancient and Medieval Commentators in the Middle Ages', in Steel et al. 1999, 44-65.

Zwier, K. R. (unpublished). 'Aristotle on Spontaneous Generation'.

Index Locorum

Index of Names

Numbers in parentheses identify publications by year as given in the list of *References*.

Index of Subjects